Tanning Chemistry
The Science of Leather
2nd edition

Tanning Chemistry
The Science of Leather
2nd edition

By

Anthony D. Covington
The University of Northampton, UK
Email: tony.covington@northampton.ac.uk

and

William R. Wise
The University of Northampton, UK
Email: will.wise@northampton.ac.uk

ROYAL SOCIETY
OF **CHEMISTRY**

Print ISBN: 978-1-78801-204-1
EPUB ISBN: 978-1-78801-907-1

A catalogue record for this book is available from the British Library

The Royal Society of Chemistry is a charity, registered in England and Wales, Number 207890, and a company incorporated in England by Royal Charter (Registered No. RC000524), registered office: Burlington House, Piccadilly, London W1J 0BA, UK, Telephone: +44 (0) 20 7437 8656.

Visit our website at www.rsc.org/books

Printed in the United Kingdom by CPI Group (UK) Ltd, Croydon CR0 4YY, UK

Preface to the First Edition

Even in the 21st Century, the manufacture of leather retains an air of the dark arts, still somewhat shrouded in the mysteries of a millennia old, craft based industry. Despite the best efforts of a few scientists over the last century or so, much of the understanding of the principles of tanning is still based on received wisdom and experience. It has been my mission to contribute to changing the thinking in the industry, to continue building a body of scientific understanding, aimed at enhancing the sustainability of an industry which produces a unique group of materials, derived from a natural source.

This book is the culmination of a 40 year career in chemistry and the leather industry. My love of chemistry was fuelled at Ackworth School, Pontefract, UK, by Phillips Harris, a genius of a teacher. I obtained an HNC in chemistry at Newcastle Polytechnic (now the University of Northumbria), followed by the Graduateship of the Royal Institute of Chemistry (now the Royal Society of Chemistry) at Teesside Polytechnic (now the University of Teesside). This provided me with a firm basis in general chemistry, with a strong emphasis on practical skills – mine was the last year the GRIC involved four days of practicals as part of finals! My chemistry experience was enhanced in my doctoral studies in physical organic chemistry at Stirling University under the invaluable supervision of Professor R. P. Bell FRS, followed by post doctoral research in physical chemistry with Professor Arthur Covington DSc (coincidence, no relation) at the University of Newcastle upon Tyne – the city of my birth.

Tanning Chemistry: The Science of Leather 2nd edition
By Anthony D. Covington and William R. Wise
© Anthony D. Covington and William R. Wise 2020
Published by the Royal Society of Chemistry, www.rsc.org

In 1976 I joined the British Leather Manufacturers' Association (later to become BLC The Leather Technology Centre of Northampton, UK), where I spent 18 years engaged in research and development, industrial consultancy and problem solving, mostly under Director Dr Robert Sykes OBE. The value of this period was in my exposure to a wide range of aspects of the global leather industry, particularly the opportunity to work in different types of tanneries around the world. At this time, I was initiated into doctoral supervision, in collaboration with Dr Richard Hancock of the Royal Holloway campus of the University of London: the student was Ioannis Ioannidis, later to become a leading light in major European projects.

In 1995 I was appointed by Nene College, later to be The University of Northampton, to teach leather science in the British School of Leather Technology. It is this latter period that proved to be the most fruitful in the development of my understanding of the subject, stimulated in the main by the questioning of my students. It has been said that to understand a subject fundamentally, one should try teaching it: I can vouch for the truth of that observation. Only when I was confronted with classes of (mostly overseas) students, to whom I was trying to convey a logical approach to thinking about the subject, did I become aware of the inconsistencies and inaccuracies of received wisdom, but with which I had worked without critical review for two decades. Since few scientists in the last quarter of a century have been working on the principles of leather making, these problems in the body of knowledge had to be addressed by my research group.

Contrary to the belief expressed in some factions in higher education, leather technology and leather science are real subjects: they are full of technical and scientific rigour and difficulty, typically at the overlapping regions between conventional disciplines, because they are interdisciplinary subjects. Leather technology is fully deserving of recognition and respect in the field of the applied sciences. Those who beg to differ are challenged to examine the courses and examinations my students must face; dealing as they do with chemistry, biochemistry, biology, physics and materials science, not to mention management and legislation, and where additionally the topics are embedded with new concepts from emerging research.

In science and technology, we all stand on the shoulders of giants. I have been privileged to meet, befriend and work with the best. There are too many to mention them all, but here are a few, in no particular order of precedence: Eckhart Heidemann, Betty Haines, Gunther Reich, Bi Shi, David Bailey, Steve Feairheller, Richard Daniels, Jaume

Cot, Roy Thomson, Alberto Sofia, Sjef Langerwerf, Gabi Renner and Jakov Buljan. I must also include my old friend Shri Dr T. Ramasami, former Director of the Central Leather Research Institute, Chennai, India, to whom I am related – not by blood, but by academic ties – his PhD supervisor was a PhD student of my PhD supervisor.

For any academic, his or her relationship with their postgraduate students creates an unusual bond, quite unlike any other relationship. I have been fortunate in having had many highly gifted students, whose contribution to my thinking and understanding has been immense. Nor do I forget the contributions of my colleagues, both at BLC and in the British School of Leather Technology.

As I reflect on a lifetime in science, my only regret is that the future of leather science as a subject is uncertain. There are very few of us leather scientists around. Moreover, such is the current disinclination towards financing science, especially applied science, more especially fundamental applied science, whether in industry or in higher education, it seems unlikely that leather science will command the attention of many future practitioners in the field. Nevertheless, there will always be a need in the global industry for highly qualified practitioner leather technologists and it is to them that I direct this text.

My undergraduate students have often distinguished a clear demarcation between leather technology and leather science, on the basis that leather technology is the easier option, because it is mostly concerned with the practicalities of making leather; but leather science is perceived as more difficult, perhaps just because it is called science. I have always shared Pasteur's view that there is no real boundary between science and technology and especially leather technology and leather science; the former is merely the application of the latter, they are but two sides of the same coin. Indeed, my personal view is that leather science is easier, because it is entirely logical. But then I would say that, wouldn't I?

Tony Covington

Preface to the Second Edition

Sir William Thompson, aka Baron Kelvin of Largs, recognised and respected in his day for his contributions to physics at a global level, was once famously quoted in 1900 as saying 'There is nothing new to be discovered in physics now. All that remains is more and more precise measurement'. This bold statement has been recognised as laughably inaccurate, with countless physicists, including Einstein, proving that there will always be questions that need to be answered, fundamentals that need to be understood and improvements that need to be made beyond simply improving precision of measurement.

Because we have been making leather for over 5000 years, the outside world is similar in that it perceives the leather industry – and the science that underpins it – to be stagnant, with nothing new to discover. However, those within the industry know that nothing could be further from the truth!

As was stated in the First Edition of this book, much of what we do today remains more of an 'art' than a science. I have every intention of following in the footsteps of my predecessors and mentors in ensuring that we continue to gain a better scientific understanding of what has been practiced over centuries. I am firmly of the belief that from an increase in the understanding of the science we will progress the industry towards an increasingly sustainable future – from subtle changes that make gains in process efficiency through to exploring the paradigm shifts that move us entirely away from 'the norms' that we know today.

Tanning Chemistry: The Science of Leather 2nd edition
By Anthony D. Covington and William R. Wise
© Anthony D. Covington and William R. Wise 2020
Published by the Royal Society of Chemistry, www.rsc.org

Despite the desire to follow in the footsteps of some great scientists who have moved the area of leather science on enormously in the last half century, I have not always been involved with the leather industry; my background and higher education are in chemistry. My interests in 'science' started at an early age – encouraged by both my mother and father – and quickly developed into a passion. I went to university to read a BSc in chemistry, swiftly followed by an MSc in green chemistry and a PhD in organic chemistry. In 2014, after three years conducting postdoctoral research on developing sustainable alternatives to traditional polyolefin-based polymers, I took up my current post as a Senior Lecturer within the Institute for Creative Leather Technologies (ICLT) at the University of Northampton.

Prior to my appointment in 2014 and as 'an outsider' to the leather industry, I knew very little about the industry; I certainly would not have considered the amount of applied interdisciplinary science that it relied upon to make a material I took for granted. People often snigger when I explain what I now do as a job, but much of this stems from a lack of understanding – only when the scope of the science is explained do they begin to comprehend quite how complex the area is. I am now desperate to encourage more people to pursue leather manufacture/research as a career, to take leather science seriously and to give it the recognition it deserves. For me, this book acts as the starting point ...

Will Wise

Dedication

I dedicate this Second Edition to my wife Rebecca and my three daughters Emily, Ella and Sophie, because without their support I would not have been able to achieve so much in such a short space of time. They do not question where I am or what I'm doing, they just ensure that I know I'm loved and supported in every endeavour. I should also dedicate this book to my mother and father; it was because of them that I originally became interested in science and have the unquenchable thirst to understand why and how things work that now drives my research. My brother also deserves a dedication – those with siblings will understand the competitive drive that develops as a result of 'sibling rivalry' and the subsequent need to achieve the best that you are capable of in everything that you do. Finally, I dedicate this book to the friends and family, of whom there are too many to single out, that have supported me throughout my life, helped me when I needed it and remained a constant source of encouragement.

My thanks to you all.

Will Wise

Tanning Chemistry: The Science of Leather 2nd edition
By Anthony D. Covington and William R. Wise
© Anthony D. Covington and William R. Wise 2020
Published by the Royal Society of Chemistry, www.rsc.org

Acknowledgements

Rachel Garwood MSc, inspirational Director of the Institute for Creative Leather Technologies, The University of Northampton.
Our late beloved colleague, Amanda Michel.
The support of Professor Nick Petford DSc, Vice Chancellor of the University of Northampton.
The support of the Masters and Court of the Worshipful Company of Leathersellers of the City of London.

Tanning Chemistry: The Science of Leather 2nd edition
By Anthony D. Covington and William R. Wise
© Anthony D. Covington and William R. Wise 2020
Published by the Royal Society of Chemistry, www.rsc.org

Contents

Tanning Chemistry: The Science of Leather 2nd edition
By Anthony D. Covington and William R. Wise
© Anthony D. Covington and William R. Wise 2020
Published by the Royal Society of Chemistry, www.rsc.org

Introduction

This treatise is a discussion of the leather making process from the point of view of the scientific principles involved. The approach adopted here is to build an understanding of the steps involved, particularly the chemistry of tanning, by starting at the beginning of the industrial process and going on to the end. This means starting with an understanding of the nature, structure and properties of the material with which we make leather, then considering what to do with it, how to do it and what the outcomes of the transformations may be.

Leather is made from (usually) the hides and skins of animals – large animals such as cattle have hides, small animals such as sheep have skins. The skin of any animal is largely composed of the protein collagen, so it is the chemistry of this fibrous protein and the properties it confers to the skin with which the tanner is most concerned. In addition, other components of the skin impact on processing, impact on the chemistry of the material and impact on the properties of the product, leather. Therefore, it is useful to understand the relationships between skin structure at the molecular and macro levels, the changes imposed by modifying the chemistry of the material and the eventual properties of the leather. Consequently, these aspects of the subject will be addressed in the following order: collagen chemistry, collagen structure, skin structure, processing to prepare for tanning, the tanning processes and processing after tanning.

Tanning Chemistry: The Science of Leather 2nd edition
By Anthony D. Covington and William R. Wise
© Anthony D. Covington and William R. Wise 2020
Published by the Royal Society of Chemistry, www.rsc.org

Leather making is a traditional industry, after all it has been in existence since time immemorial, certainly over 5000 years, because the industry was established at the time of the Hammurabi Code (1795–1750 BC), when Article 274 laid down the wages for tanners and curriers.[1] Indeed, the use of animal skins is one of man's older technologies, perhaps only predated by tool making. In the modern world, the global leather industry exists because meat eating exists. Hence, most leather made around the world comes from cattle, sheep, pigs and goats. One of the byproducts of the meat industry is the hides and skins: considering that the annual kill of cattle alone is of the order of 300 million, such a byproduct, amounting to 10–20 million tonnes in weight, would pose a significant environmental impact if it were not used by tanners. Leather makers around the world perform the ultimate example of adding value: an unwanted, discarded byproduct from one industry is turned into one of the most desired and widely applied materials in the world. Unfortunately, one tonne of hide is converted into only 200 kg of leather: the difference is waste, pollutants or byproducts, depending on your point of view. So, the problems with which tanners must contend become apparent.

Here, notably, there is a difference between the leather industry and the fur industry. In general, tanners process the skins of animals killed for the consumption of their carcases: in the fur trade, the skin is the important part of the animal and the carcase may be the byproduct, albeit perhaps turned into fertiliser. The processes for preparing and stabilising furskins have similarities with tanning, but the chemistries can be significantly different. However, the industries have traditionally been separate. Within the leather trade, some skins are processed from animals sustainably farmed in part for their skins, *e.g.* ostrich and alligator. It is important to emphasise that any animals specified by CITES (Convention on International Trade in Endangered Species) are not part of the ethical global leather industry.

It might appear anachronistic that such an old established technology should still be relevant today, particularly because some of those traditional technologies for making specific leathers survive in the modern industry. However, it is worth considering the properties and performance of modern leathers, in comparison to other modern materials. The suitability of leather for shoe manufacture is based upon the twin abilities of being able to exclude water, but allow air and water vapour to pass through the cross section of the upper: this is the basis of foot comfort when shod. The properties are so important that attempts have been made to mimic them in synthetic materials, the so-called poromerics. To date, none has been successful: witness the failure of Corfam in the 1960s, documented by Kanigel.[2] Similarly, the reconstituting of leather fibres into a polymer matrix creates materials that cannot exhibit all the properties of leather, it merely creates other versions of poromerics. Strictly, all such materials cannot carry names connoting leather, since the definition of leather is based on the requirement that it must be made from the intact hide or skin of an animal.[3]

The properties of leather do not limit its use only as shoe upper material. The natural structure of leather, founded in the properties of the protein collagen with chemical modification, confers the unique ability to function within the temperature range at least −100 to +130 °C, unmatched by any other material: synthetic materials either soften at high temperature or become brittle at low temperature. However, the continuing popularity of leather is based on the way it feels, the way it looks, the way it interacts with and moulds to the body, the way it wears in use, even the way it smells (for information – 'essence of leather' is plant polyphenol extract plus a little fish oil). The aesthetics of leather ensure it is likely to remain the preferred material of choice for many modern applications: shoes, clothing, upholstery, belts, handbags, *etc.* Importantly, the choice is supported by its performance. For example, the high hydrothermal stability of modern chromium(III) tanned leather allows it to be used in rapid mass production processes, in which shoe uppers are moulded directly onto melted polymeric soles; other leathers can be used to make foundry workers' safety gaiters because they can resist splashing from molten metal streams, even liquid steel, clothing leathers can be steam pressed, *etc.* The best material for motorcyclists is leather, because, in the event of the rider coming off the bike, the frictional heat of the skidding action over the road surface will not cause the material of the clothing to melt and consequently it is not fused into the rider's soft tissues.

The properties of leather depend on the origin of the raw materials, how the pelt is prepared for chemical modification, how that modification is conferred chemically, how the leather is lubricated and finally how the surfaces are prepared. Leather can be made as stiff and as tough as wood, as soft and flexible as cloth and anything in between. It is the traditional art and craft of the leather technologist to control the parameters and variables of processing to make leathers with defined, desired or required properties. It is from the creativity of the leather scientist that the range of leathers that can be made is continually widening.

As a traditional craft, leather making has its own jargon, built up over centuries, but much of which has no clear origin. Some of the terms are set out in Glossary below and more can be found in the *British Standard Glossary of Leather Terms.*[3]

Leather scientists and technologists do not restrict their attention to leather alone. Leather is merely one set of examples within a range of collagenic biomaterials; there are many more already used in industry and many more can be developed: that should become clear by reading this text. Any application of proteinaceous materials, particularly collagen, and especially where the protein is modified in some way, is exploiting a biomaterial. For example, fixing of organic specimens in histology for microscopy study is using tanning technology, as is the stabilisation of medical sutures, *e.g.* in chromic catgut, when the rate of dissolution in the body is controlled by the degree of modification of the tissue by chrome tanning, the techniques of increasing the 'weight' of silk use tanning chemistry and so on. Therefore, clearly, the transfer of scientific and technological principles from the leather industry into other sectors is practical, feasible, desirable and full of potential.

It will become apparent from this text that the leather field has borrowed from other areas of science, often just using the principles rather than following developments in these subjects. In the past, tanners exploited technology and biotechnology long before the science was completely understood: so no change there! Hence, it is clear that some aspects of the development of the technology of the leather industry and progress in that technology have been slow, indeed Thomson has argued that the industrial revolution in leather making took place one hundred years after it began to transform the rest of industry.[4] The slow rate of change was compounded by the inherent conservatism of tanners, aware that change in processes and procedures can be detrimental to leather quality and performance; hence avoidance of change was the lesser of the twin evils of lack of technical progress or risking loss of money.

The continuing importance of leather production in the world cannot be under estimated. As a labour intensive operation, it is especially applicable to the developing economies. Moreover, in those countries, the industry offers significant employment opportunities for women: it is recognised that giving women the power to control the economics of the family has a potent stabilising effect on society. It has been argued that that movement of the global industry to the east and to the south has taken place to exploit cheaper labour and less stringent environmental legislation. The former is certainly the case, although the cheaper labour is becoming noticeably less cheap. The latter is not true: consent limits for discharges to the environment are surprisingly consistent throughout the world;[5] nowhere is the legislation on pollution significantly lax, it is merely the enforcement that varies.

Table I.1 Progression towards zero environmental impact.

CURRENT → CATNAP	CLEANER → BATNEEC	CLEAN BAT
Currently applied technology, narrowly avoiding prosecution	Best available technology, not entailing excessive cost	Best available technology

The perception that the leather industry is a dirty polluter is largely out of date. However, this does not mean that it can be assumed to be generally clean – unfortunately, there is still a long way to go. It is an important function of leather science to provide the technologists with sufficient understanding of the principles of the chemistry and biochemistry of processing for them to be able to make improving changes in the operations. Table I.1 characterises this progression towards zeroenvironmental impact.

Developments in this direction should ideally be economically attractive at the same time, including using waste streams, by adding value to the byproducts. Since we are dealing here with an applied, industrial science, the economics of the operation should never be far from the analytical mind. Making money is a big incentive for change.

The text of this book is designed to be scientific, to define as much as possible the principles underpinning the complex technology of leather making. Hence, it is part of the function of this text to point the way forward, to contribute to the sustainability of the global industry, always based on a fundamental understanding of those principles.

There are not applied sciences... there are only applications of the science and this is a very different matter. The study of the applications of science is very easy to anyone who is the master of the theory of it.

Louis Pasteur.[6]

References

1. R. Reed, *Ancient Skins, Parchments and Leathers*, Seminar Press, 1972.
2. R. Kanigel, *Faux Real*, Joseph Henry Press, Washington DC, 2007.
3. *British Standard Glossary of Leather Terms*, British Standards Institution, BS 2780, 1983.
4. R. S. Thomson, *Trans., Newcomen Soc.*, 1981, **53**, 139.
5. J. Buljan and M. Bosnic, *World Leather*, 1994, **7**(6), 54.
6. E. Haslam, *J. Soc. Leather Technol. Chem.*, 1988, **72**(2), 45.

Introduction to the Second Edition

This Second Edition builds on the First Edition, bringing the science of leather making up to date. It was noticeable that the pace of change was slow in the earlier phases of leather production (apart from the shock of sulfide), but more rapid in tanning and post-tanning, if the references are anything to go by. The content of the book reflects how the subject has changed in the last decade – and the introduction of a co-author.

The text has been expanded to place more emphasis on the biochemistry of processing, to address the subject of reagent delivery and to include the science of finishing. As before, environmental impact and sustainability are at the heart of leather science, as they must be when we consider the future of processing as we see it today.

Anthony D. Covington
William R. Wise

Tanning Chemistry: The Science of Leather 2nd edition
By Anthony D. Covington and William R. Wise
© Anthony D. Covington and William R. Wise 2020
Published by the Royal Society of Chemistry, www.rsc.org

Glossary of Terms

Alum	General term for aluminium(III) complexes used in tanning.
Angle of weave	The discernible average angle of the fibres in the corium: a diagnostic of the swelling/depletion history of the pelt.
Aniline dyed	Leather dyed with transparent colouring, traditionally dyes prepared with aniline as the precursor: the importance is the lack of opaque colouring on the leather surface, which would obscure the natural grain pattern. Semi-aniline has some colour opacity, to hide minor blemishes.
Barkometer	Also Baumé, Twaddle: scales of specific gravity, used to indicate concentrations of solutes, *e.g.* sodium chloride, vegetable tanning agents.
Basecoat	The first coat in finishing (analogous to the undercoat in conventional painting), the key for the topcoat.
Basification	Used in the context of mineral tanning, when the pH of the system is gradually raised, to increase the ionisation of the carboxyl sidechains of the collagen and thereby drive reaction/interaction with the cationic metal complex ions.

Tanning Chemistry: The Science of Leather 2nd edition
By Anthony D. Covington and William R. Wise
© Anthony D. Covington and William R. Wise 2020
Published by the Royal Society of Chemistry, www.rsc.org

Bating	A generic term for the use of proteolytic enzymes, to degrade non-structural proteins in skin; sources may be pancreatic material or from bacterial fermentation. Less common terms are puering (dog dung), mastering (pigeon guano) and bran drenching, referring to other animal or vegetable sources of enzymes, applied to pelt as warm aqueous infusions.
Beamhouse	The early stages of processing, to prepare the pelt for tanning, up to the point of initiating tanning.
BLC	The British Leather Confederation, formerly the British Leather Manufacturers' Research Association (BLMRA), latterly BLC The Leather Technology Centre, Northampton, UK and more recently Eurofins BLC.
Bloom	Insoluble residues from when hydrolysable tannins (ester-based plant polyphenols) are hydrolysed during tanning.
Boil test	Measuring the resistance of leather to boiling water: the period of the test varies from factory to factory, 1–5 min, and reflects whether or not the conventionally measured shrinkage or denaturation temperature of the leather lies above 100 °C and, if it is higher, by how much.
Bovine	Applied to ox or cow. Ovine (sheep), caprine (goats), porcine (pigs) and cervine (deer) are terms also used, but less commonly.
Break	The observed rippling of the grain when leather is bent inwards, grain uppermost: the effect may be fine or coarse, when the former is preferred.
Buffing	Applying sandpaper/glasspaper to a leather surface: buffing the grain surface removes material and flattens it in preparation for correcting or for making nubuck suede; buffing the split or flesh surfaces creates a suede nap.
Butt (or britch)	The central portion of a hide or skin: the remainder, bellies and neck, are self-explanatory.
Chamois	Leather tanned with oil (typically unsaturated cod liver oil capable of undergoing polymerisation), usually applied to sheepskin flesh splits.

Chrome	Refers to chromium salts, usually $Cr(III)$, but may include $Cr(VI)$ depending on the context.
Chrome oxide	Analysis of the chromium content of leather or other materials is usually expressed in the form of the trivalent oxide, Cr_2O_3, regardless of whether or not the oxide has anything to do with the chromium species analysed.
Chrome retan	Chrome-tanned leather retanned with vegetable tannin(s).
Chrome tanning	Tanning with basic chromium(III) salts, usually the sulfate.
Clearing	Sequestration of iron, when it causes staining on vegetable-tanned leather.
CLRI	Central Leather Research Institute, Chennai, India.
Condensed tan	Vegetable tannin with flavonoid structure; also called catechol tannin.
Conditioning	Adjusting the moisture content of leather, often upwards to a notional 14%, usually by spraying with water and holding perhaps overnight or longer. A traditional method is burying in damp sawdust. The purpose is to prepare leather for softening (staking).
Corium	Corium minor is the grain layer, the outermost layer of the pelt; corium major is the fibrous inner layer of the pelt.
Corrected grain	The grain layer is partially removed by buffing and replaced with polymer coatings, often with a new grain pattern impressed into it. The new grain pattern may not necessarily be the same as the original.
Crock	Particulate matter removed from leather by rubbing; it may be leather fibre or loose dye, depending on the processing conditions.
Crust leather	Leather processed to a point when it can be dried; this is usually after post-tanning, but may apply to leather given a primary tanning, possibly including some lubricant to prevent fibre sticking during drying.
Curing	A generic term for preserving raw hides and skins, often referring to salting.
Currier	A term no longer in use: one who took tanned hides and completed the processing by fatliquoring and crusting them.

Deep eutectics	A type of ionic liquid, characterised by a mixture of two solid components, with a low-melting eutectic such that it is liquid at ambient temperature.
Deliming	The step after liming, when the pH is lowered in readiness for the subsequent step, bating at pH 8–9. The commonest reagent is the ammonium salt, chloride is preferred to sulfate, but carbon dioxide is increasingly used.
Denaturation	Damaging the natural structure of collagen: it may be achieved chemically, but usually means heat damage of wet collagen or leather, when the observed effect is shrinking.
Detanning	Reversing the effect of tanning, usually chemically, lowering the shrinkage temperature to that of raw pelt.
DMTA/DMA	Dynamic mechanical (thermal) analysis: a technique for gathering data about molecular structure by stressing materials in a cyclical fashion while varying the temperature and/or relative humidity. Commonly the 'T' is dropped from the abbreviation DMTA – but DMA and DMTA are the same analytical technique.
Double face	Sheepskin tanned for clothing with the wool still attached: the wooled surface and the suede surface are equally important to the quality of the leather.
Double hiding/ skinning	The loosening of structure, usually of sheepskins, because of removal of non-structural components, such as fat cells. May also be a consequence of internal gelatinisation during sun drying for preservation.
Drenching	Bating with fermenting bran (less common).
Dry salting	A curing method involving salting followed by shade drying.
DSC	Differential scanning calorimetry: a technique for measuring the thermodynamic parameters associated with a reaction or transition, by determining energy absorption or energy produced as the temperature rises at a specific rate.
Enamel	The grain enamel: the outermost surface of the skin after the hair or wool has been removed. The most valuable feature of grain leather.

EXAFS
: Extended X-ray absorption fine structure: a technique in X-ray spectroscopy used to quantify the structure of a material in terms of the nearest atomic neighbours to an atom of interest.

Fatliquoring
: Lubrication of the leather, often using partially sulfated or sulfonated oils, which may be animal, vegetable or mineral in origin. The partially sulf(on)ated or 'sulfo' fraction is the emulsifying agent to carry the unreacted, neutral oil into the leather: it is the neutral oil that performs the lubricating action.

Fellmonger(ing)
: Literally a dealer in fells (sheep) skins, typically applied to processing raw sheepskins to the dewooled, pickled condition. Also used as a verb, referring particularly to the process of removing the wool: a thixotropic lime suspension with sulfide in solution is painted onto the flesh surface of the skin, allowing the sulfide to penetrate to the base of the follicle and attack the keratin of the prekeratinised zone of the wool fibre.

Finishing
: Application of a basecoat and topcoat of (usually) polymers, but may be protein, to protect and enhance the grain, to allow aesthetic effects to be created. Finishing may also be applied to suede surfaces.

Fixation
: A general term referring to the chemical binding of a tanning agent to collagen.

Flesh
: The muscle and fat of a carcase adhering to the pelt after flaying.

Flesh layer
: The part of the corium structure of hide or skin closest to the carcase.

Flesh split
: The layer closest to the carcase, produced by splitting hide or skin through the cross-section; the other part is the grain split.

Float
: The water used as process medium: the quantities relative to the amount of pelt (the 'goods') may be characterised as 'short' or 'long'.

Goods
: The load of pelt or leather in the process vessel.

Grain (layer)	The outer layer of the hide or skin, the more valuable part comprising the corium minor and the enamel outer surface.
Grain enamel	The surface of the pelt/leather after the epidermis has been removed: the most valuable feature of the leather.
Grain pattern	The pattern of the follicles on the surface of the skin or leather: it characterises the leather and it is a prized quality feature.
Green	Refers to rawstock that is fresh, *i.e.* without preservation; green weight may be used as the basis for chemical offers. Sometimes referred to as 'tail' weight.
Ground substance	Non-collagenous components of unprocessed pelt, comprising non-structural proteins, glycosaminoglycans, fats, minerals.
Growth/ growthiness	The pronounced natural rippling of skin or hide at the neck/shoulder.
Gums	The high molecular weight fraction of vegetable tannins, >3000 Da: complexes of carbohydrates with polyphenols. They are undesirable components because of their reaction on the pelt surface due to the high molecular weight.
Hair burn	The method of removing the hair by dissolving it, usually using alkali and a reducing nucleophile: the conventional reagents are lime and sodium sulfide at pH 12.6.
Hair save	The method of unhairing by attacking the hair at the base of the follicle, to allow the hair to be removed intact.
Handle	A general term referring to the sum total of the perceived properties of the leather when handled, flexed, *etc.*
Hide	The pelt of a large animal, *e.g.* cattle.
HIT	Hydrothermal isometric tension: a technique for measuring the forces developed when a collagenic material is heated in a solvent through the temperature at which the shrinking transition occurs.
Hide substance	The collagen content of part or fully processed pelt, as measured by the hydroxyproline content.
Hydrolysable tan	Vegetable tannin based on saccharide esterified by polyphenol – also called pyrogallol tannin.

Hydrothermal stability	The resistance of wet collagenic materials to heat. The commonest aspect is shrinkage temperature, T_s, but other techniques are common, such as boil tests.
ICLT	Institute for Creative Leather Technologies, University of Northampton, UK – formerly the British School of Leather Technology.
Ionic liquid	A generic term referring to ionic components that constitute a low-melting crystal lattice. The term also covers deep eutectic liquids.
Isoelectric point (IEP)	The pH at which a protein or derivative has zero net electric charge. The value is a reflection of the relative contents of carboxylic acid groups and amino groups. At pH values below the IEP, the protein is positively charged; at pH values above the IEP, the protein is negatively charged. The further away the pH is from the IEP, the more highly charged is the protein.
JALCA	*Journal of the American Leather Chemists Association*.
Japanese leather	Thin but strong leather, made by removing as much as possible of the ground substance, to allow the collagen structure to collapse; it is actually raw but oil dressed with rapeseed oil.
JSLTC	*Journal of the Society of Leather Technologists and Chemists* (UK).
Leather	Hide or skin with its original fibrous structure more or less intact, tanned to be imputrescible. The definition *excludes* any materials made from waste tanned by-products, such as shavings, reconstituted in a polymer matrix: these materials are often designated 'leather board' or by other names connoting leather.
Leathering	Stabilisation of collagen by tanning-type reactions, but the product does not possess all of the expected attributes of leather, although it does appear leather-like.
Lime/liming	A generic term for alkaline treatment, to cause swelling and hydrolysis in pelt – usually, but not exclusively, achieved with slaked lime, $Ca(OH)_2$.

Lipocytes   Cells containing triglyceride grease, most apparent in wool sheep, situated in the junction of the grain and corium.

Masking   Altering the properties of a mineral tanning agent by changing the ligand field of the metal ion complex, usually involving carboxylates.

Mastering   Bating with a warm infusion of bird guano (less common).

Metal oxide   It is conventional to express the metal content of a material in the form of the oxide, regardless of the chemical nature of the metal, whether as the element or any of its compounds.

Metal retan   Leather made by mineral tanning followed by vegetable tanning.

Middle split   For thick hides, this may be in addition to the grain split and the flesh split. Typically weak, such splits are used for cheap leather goods, where mechanical strength is not important.

Mineral tanning The use of metal salts for tanning. Other inorganic reagents would come under this heading, but are typically ineffectual at tanning.

Neutralisation   The process of raising the pH after main tanning, prior to initiating post-tanning reactions, to adjust the charge on the leather: the higher the pH, the more negatively charged the leather is. Despite the term, the final pH may be significantly below conventional neutrality (pH 7).

Non-tans   The low molecular weight fraction of vegetable tannin extracts, <500 Da. They are unreactive as tanning agents, but useful in the tanning process for solubilising the tannins. They may have use as the basis of organic tannages themselves.

Nubuck   Very fine suede created by buffing the grain surface of chrome-tanned leather.

Offer   The amount of a chemical introduced into the process vessel. Offers in the early stages may be based on the raw (green or tail), salted or limed weight of the pelt. After tanning, offers are usually based on damp, shaved leather weight.

Organic tanning Tanning without metal salts, notably without chromium(III) salts.

Organoleptic	A term referring to the handle characteristics of leather.
Paste drying	Wet leathers are stuck to glass plates with paste and dried in warm air. The paste is designed to allow release of the dry leather from the plate. Used to produce leather with a very flat grain surface.
Pelt	A general term for raw or part-processed hide or skin, before tanning.
Pickling	Acidification of pelt in brine for preservation or preparation prior to tanning.
Post-tanning	Usually applies to chrome tanning, but may apply to any other main tanning step: comprises retanning, dyeing and fatliquoring.
Pretanning	Any tanning step that precedes the main tanning process: usually applied to facilitate the tanning reaction.
Puering	Bating with a warm infusion of dog dung: used to be the industry standard, but now uncommon.
Putrefaction	The damage done to rawstock by the enzymes produced by proliferating bacteria, both saprophytic and opportunistic.
Rawstock	Hides and skins, fresh or preserved.
REACH	Registration, Evaluation, Authorisation and Restriction of Chemicals – European certification system for chemicals used in industry.
Reds	The phlobaphene aggregates in solutions of condensed tannin extracts.
Retanning	Following the first and main tanning process, usually chrome tanning, the leather is tanned again, typically to modify the handle properties. Any tanning agent may be used for this purpose, depending on the desired leather properties.
Samm(y)ing	Squeezing leather through felt-covered rollers, to remove moisture.
Scud(ding)	Residual epidermis, hair debris and melanin, left after beamhouse operations. The verb refers to the removal of scud, which might be by mechanical means or by the use of biochemicals.

Seasoning	See 'Conditioning'.
Semi-aniline	As for aniline dyeing, but including some pigment to limit the transparency of colour: typically to hide slight surface defects.
Semi-metal	Tanning process involving vegetable tanning, usually a hydrolysable tannin, followed by retanning with a metal salt. Options include semi-chrome using Cr(III), semi-alum using Al(III), semi-titan using Ti(IV) and semi-zircon using Zr(IV).
Setting	Squeezing (samming) the pelt, but with a spreading action over the grain surface from the rollers – to flatten the leather and remove creases.
(Shade) drying	A method of preserving hides or skins by drying: typically applies to tropical climates, but actually refers to drying protected from full sun.
Shrinkage temperature (T_s)	The (putative) temperature at which collagen undergoes the transition from intact helix to random coil. The most common measurement of hydrothermal stability: an increase in value of T_s indicates a consequence of tanning. Sometimes this term is used interchangeably with 'denaturation temperature' and occasionally with 'melting temperature'.
Shaving	Usually follows splitting: mechanically cutting through the surface to produce an accurate and consistent thickness.
Skin	The pelt of a small animal, *e.g.* sheep, goat.
Soaking	The first stage in processing, primarily to rehydrate preserved pelt: usually involves a preliminary dirt soak, one or more main soaks and rinsing.
Splitting, split	The act of cutting a hide layerwise, laterally through the cross-section. Also used to describe the layers after splitting: grain split, flesh split, sometimes middle split(s).
Spue	Spue usually refers to 'fatty acid spue', which is the migration of low molecular weight fatty acids to the (grain) surface of the leather, giving a white waxy appearance. 'Salt spue' is the migration of neutral electrolyte to the (grain) surface, often observed after shoes have been wetted and dried.

Staking	Softening following moisture adjustment (conditioning or seasoning). Traditionally achieved by manually bending the leather through a sharp angle over a blunt blade (such as an upturned spade); now, mechanical methods are used.
Staling	Delaying between flaying and curing, allowing bacterial proliferation; the damaging effect depends on the extent of the delay.
Striking out	The use of a blunt blade on a fast-rotating roller to spread and flatten tanned skins; used to remove creases.
Stuffing	Filling the leather structure with (usually) fatty materials, such as wool grease.
Suede	Leather that has been rubbed with glasspaper or sandpaper to raise the nap on the surface. A nap on the flesh surface makes conventional suede, a nap on the grain surface makes nubuck. Note that the suede from the split surface of hide or shin is usually too coarse to be of value.
Syntan	Synthetic tanning agent: usually polymerised aromatic hydroxyl compounds, designed to perform a variety of functions, depending on the structure, including assisting chemical processes and acting as solo tanning agents, such as vegetable tannins. May be characterised as 'auxiliary', 'retan' or 'replacement', depending on their astringency.
Tanning	The conversion of putrescible organic material into a stable material capable of resisting biochemical attack.
Tans	Usually applied to vegetable tannins: the molecular weight fraction 500–3000 Da of vegetable tannin extracts, distinguished from the 'non-tans' and the 'gums'.
Toggle drying	Leathers are clipped to a perforated plate with 'toggles' and dried in ambient or warm air.
Unhairing	The removal of hair, usually by chemical means, but biochemical treatments are also known. In this context, the general term 'unhairing' can include removal of wool or bristles.

Vegetable tan(nin)	Polyphenolic mixtures of plant origin: may be in the form of ground plant material, but now more usually refers to dried extracts of plant material.
Wet blue	Chrome-tanned pelt: it is wet because it is not usually dried before progressing to post-tanning and it is blue because of the nature of the chrome complexes bound to the collagen.
Wet salting	The commonest form of rawstock preservation: the term indicates that the treatment is applied to pelt in its natural moist state and the reagent is sodium chloride.
Wet white	Non-chrome-tanned leather or otherwise chemically stabilised pelt. Originally defined in terms of capability of being split and shaved. Now regarded as a tanning system in its own right: typically consisting of a combination of syntan (various, undefined) and (usually) glutaraldehyde or other aldehydic reagents.
Wool/hair slip	The loosening or detaching of wool or hair from the base of the follicle; an indicator of putrefaction.

1 Collagen and Skin Structure

1.1 Introduction

At the heart of the leather-making process is the raw material, hides and skins. As the largest organ of the body of mammals, the skin is a complex structure, providing protection against the environment and affording temperature control, but it is also strong enough to retain, for example, the insides of a 1 tonne cow. Skin is primarily composed of the protein collagen and it is the inherent properties and potential for chemical modification of this protein that offer the tanner the opportunity to make a desirable product from an unappealing starting material. It is part of the tanner's job and skill to purify this starting material, allowing it to be converted into a product that is both desirable and useful in modern life.

Collagen is a generic name for a family of at least 28 distinct collagen types, each serving different functions in animals, importantly as connective tissues.[1–4] The major component of skin is type I collagen – hence, unless specified otherwise, here the term 'collagen' will always refer to type I collagen. Other collagens do feature in leather making, however, and their roles are defined later.

Collagens are proteins, that is, they are made up of amino acids. They can be separated into the α-amino acids and the β-amino acids, as shown in Figure 1.1. Each one features a terminal amino group and a terminal carboxyl group, which become involved in the peptide link (see later), and a sidechain attached to the methylene group in

Tanning Chemistry: The Science of Leather 2nd edition
By Anthony D. Covington and William R. Wise
© Anthony D. Covington and William R. Wise 2020
Published by the Royal Society of Chemistry, www.rsc.org

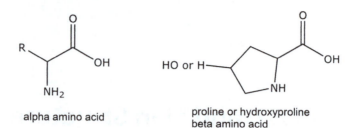

alpha amino acid

proline or hydroxyproline
beta amino acid

Figure 1.1 Amino acid structures, α and β.

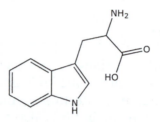

Figure 1.2 Tryptophan.

the centre of the molecule. When the amino acids are linked together to form proteins, they create an axis or 'backbone' to the polymer, from which the sidechains extend. It is the content and distribution of the sidechains that determine most of the properties of any protein. In the case of collagen, it is the sidechains that largely define its reactivity and its ability to be modified by the stabilising reactions of tanning. In addition, the chemistry of the backbone, defined by the peptide links, offers different reaction sites that can be exploited in some tanning processes.

All the common amino acids are found in skin or skin components. There are two notable aspects of the amino acid content of collagen. Hydroxyproline, represented in Figure 1.1, is almost uniquely present in collagen compared with other proteins, therefore offering the basis of measuring the collagen content in any skin or skin derivative. Tryptophan, shown in Figure 1.2, is absent, therefore making collagen deficient as a foodstuff.

In terms of leather making, some amino acids are more important than others, since they play defined roles, set out in Table 1.1: the roles of importance are either in creating the fibrous structure or involvement in the processing reactions for protein modification. Other amino acids, not included in the table, are important in defining the properties of the collagen, but play less defined roles in the leather-making processes.

Table 1.1 Amino acids of importance in leather making.

Name	Abbre-viation	Type	Sidechain: R =	Importance in leather making
Glycine	Gly	α, neutral	–H	Collagen structure
Alanine	Ala	α, neutral	–CH$_3$	Hydrophobic bonding
Valine	Val	α, neutral	–CH(CH$_3$)$_2$	Hydrophobic bonding
Leucine	Leu	α, neutral	–CH$_2$CH(CH$_3$)$_2$	Hydrophobic bonding
Isoleucine	Ileu	α, neutral	CH$_3$CH$_2$CH(CH$_3$)	Hydrophobic bonding
Phenylalanine	Phe	α, neutral	–CH$_2$C$_6$H$_5$	Hydrophobic bonding
Serine	Ser	α, neutral	–CH$_2$OH	Unhairing
Cysteine	CySH	α, neutral, S containing	–CH$_2$SH	Unhairing
Cystine	CyS–SCy	α, neutral, S containing	–CH$_2$SSCH$_2$–	Unhairing
Aspartic acid	Asp	α, acidic	–CH$_2$CO$_2$H	Isoelectric point (IEP),[a] mineral tanning
Asparagine	Asn	α, neutral	–CH$_2$CONH$_2$	IEP
Glutamic acid	Glu	α, acidic	–(CH$_2$)$_2$CO$_2$H	IEP, mineral tanning
Glutamine	Gln	α, neutral	–(CH$_2$)$_2$CONH$_2$	IEP
Arginine	Arg	α, basic	–(CH$_2$)$_3$NHC(NH)NH$_2$	IEP
Lysine	Lys	α, basic	–(CH$_2$)$_4$NH$_2$	IEP, aldehydic tanning, dyeing, lubrication
Histidine	His	α, basic		Aldehydic tanning, dyeing, lubrication
Proline	Pro	β, neutral	See Figure 1.1	Collagen structure
Hydroxyproline	Hypro	β, neutral	See Figure 1.1	Collagen structure, hydrogen bonding

[a]In biology, referred to as p*I*.

Amino acids create macromolecules, proteins such as collagen, by reacting *via* a condensation process as shown in the following equation, where the amide or peptide link is in bold:

$$H_2N – CHR – CO_2H + H_2N – CHR – CO_2H$$
$$\rightleftharpoons H_2N – CHR – \mathbf{CO \cdot HN} – CHR – CO_2H + H_2O$$

The condensation reaction, removing the elements of water, can be reversed by hydrolysis, by adding the elements of water. Clearly, hydrolysis, as set out in this equation, cannot be fast, nor does the equilibrium lie to the left, otherwise the protein would be unstable and useless as the basis of life. On the other hand, the hydrolysis reaction is catalysed by general acid and general base; importantly for leather making, it is catalysed by H$^+$ and OH$^-$. The impact on processing can be illustrated in the following way.

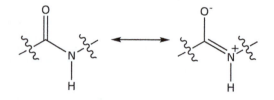

Figure 1.3   The electronic structure of the peptide link.

In the earliest stage of processing, hair is usually removed and at the same time the skin is given a prolonged alkali treatment, typically conducted over about 18 h for cattle hides, often in aqueous lime solution, $Ca(OH)_2$; longer treatment may result in detectable damage to the fibre structure. In saturated lime solution, $[OH^-] \approx 10^{-2}$ M. Conversely, pickling in brine solution with acid is routinely used as a preservation technique for more vulnerable sheepskins, enabling them to be transported across the world between Europe and Australia and New Zealand over a period of several months: here $[H^+] \approx 10^{-2}$ M. Hence, by this practical comparison, in which pickled pelt remains undamaged for months but limed pelt shows damage after a few hours, hydroxyl ion produces a much faster reaction than hydrogen ion – at least 100 times faster.

An important feature of the peptide link is that it is partially charged. The link can be expressed in two forms. The charged structure makes chemical sense, but Nature does not favour charge separation in this way. However, the electronegativity difference between the oxygen and the nitrogen means that the structure can be set out in a slightly different way, as shown in Figure 1.3.

The two parts of the peptide link each carry only a partial charge, but this still allows the peptide link to play significant roles in the interaction between the protein and water and in the fixation of reagents in the leather-making processes, most important in post-tanning, when the leather is dyed and lubricated (see Chapters 16 and 17).

1.2   The Hierarchy of Collagen Structure

It is a feature of all proteins that they have 'layers' of structure that provide a well-defined 'hierarchy' to form the final functional protein. However, simplistic definitions of the hierarchy of collagen do not always fit with the accepted definitions for the hierarchy of all other proteins:[5] the collagenic definitions often involve multiple collagen units (not sub-units) in the quaternary structure and arguably misdefine the tertiary structure also. The 'hierarchy' of collagen structure is discussed in the following and the difference between definitions is considered.

1.2.1 Primary Structure

The primary structure is the simplest level of protein hierarchy and refers to the sequence of amino acids (also called residues) in the polypeptide chain. It is held together solely by peptide bonds (as described in Section 1.1) that are made during the process of protein biosynthesis.

In the case of the collagens, the primary structure conforms to the universal protein definition and is characterised by a repeating triplet of amino acids: $-(Gly-X-Y)_n-$. Therefore, one-third of the amino acid residues in collagen are glycine. Furthermore, X is often proline and Y is often hydroxyproline: 12% of the triplets are –Gly–Pro–Hypro–, 44% are –Gly–Pro–Y– or Gly–X–Hypro– and 44% are –Gly–X–Y–, where X and Y are not defined. In this way, the helical shape of the molecule is determined (see later).

The amino acid compositions of bovine skin proteins are compared later in Table 1.3. In leather-making terms, the amino acid sequence plays only a minor role; indeed, the technologies for making leather are essentially the same for all animal skins: variations in technologies are much more dependent on the macro structure of the skin and its age than the details of the chemistry of the protein. The amino acid content determines the reactivity of the protein towards reagents such as tanning compounds and the sequence influences the formation of electrostatic links,[6] which is important for protein stability but which is also exploited in leather making: the amino acid content determines the isoelectric point (see later), which together with pH controls the charge on the protein. Therefore, these features of the protein influence its affinity for different tanning reagents, but there is little difference between animal species in terms of the outcomes of stabilising reactions. Here, stability or more commonly the hydrothermal stability (resistance to wet heat) is conventionally measured by the temperature at which the protein loses its natural structure; for collagen this is referred to as the 'shrinkage temperature', T_s, or sometimes as the melting or denaturation temperature.

The only discernible difference in the hydrothermal stabilities of collagens is observed in extreme variations in skin source, where there can be large differences in shrinkage temperature, depending on the natural environment of the animals concerned (see Table 1.4). When there are differences in shrinkage temperatures between collagen sources, the difference is also reflected in the shrinkage temperatures acquired from stabilising chemistries (tanning options).

1.2.2 Secondary Structure

In terms of a general definition, the secondary structure refers to the way in which the primary structure arranges into highly ordered local sub-structures. Linus Pauling and co-workers suggested that there are two main types of secondary structure, α-helices and β-sheets,[7] which are characterised by the arrangement of hydrogen bonds between the peptide groups in a single portion of the peptide backbone. In some cases, parts of the secondary structure are ordered, just not in a regular structure such as an α-helix or β-sheet. This lack of a regular structure should not be confused with a 'random coil', which is devoid of any stabilising interactions and forms an unfolded polypeptide chain deficient in any specific three-dimensional structure. In the leather industry, the term 'random coil' is often used to describe the structure of collagen after denaturation.

The secondary structure of collagen is dominated by α-helices, to the extent that the whole strand is often depicted as a single α-helix. It is perhaps for this reason that the traditional secondary structure for collagen refers to the whole primary structure rather than localised portions of the primary structure – as is the case with other protein structures, *e.g.* enzymes. This almost continuous α-helix is due to the high content of β-amino acids, which cause the chain to twist, due to the fixed tetrahedral angles, locking the twist in place, as shown in Figure 1.4. It should be noted that the twist is left-handed, *i.e.* anti-clockwise, because of the natural L conformation of the amino acids.

1.2.3 Tertiary Structure

The definition of a protein's tertiary structure more commonly refers to the complete configuration of a single (monomeric) polypeptide strand, where the α-helices and β-sheets are folded into a compact three-dimensional structure as defined by its atomic coordinates. The three-dimensional tertiary structure, in contrast to the secondary

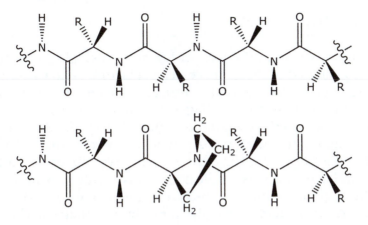

Figure 1.4 The peptide link: free rotation for α-amino acid chains, but locked conformation when β-amino acids are in the chain.

structure, also includes hydrophobic interactions and (where applicable) disulfide bonds.[5] This is especially the case for globular protein molecules such as enzymes.

It is at this point that the traditional definition for collagen deviates from the common norms for other proteins in that it considers collagen's tertiary structure in a multimeric form rather than the more conventional monomeric form as defined earlier. The deviation from the definition is likely due to the comparative simplicity of the secondary structure of collagen, the significant overlap between secondary and tertiary collagen structures (if adhering to these definitions) and the near identical and complementary nature of the three sub-units that create a triple helix.

1.2.4 Quaternary Structure

Quaternary structure is normally regarded as the three-dimensional arrangement of two or more combined sub-units (individual polypeptide chains) to form the fully functioning protein.[8] This is different from the traditional collagen definition, which describes the aggregation of multiple collagen molecules to form fibrils; however, it is not difficult to see how a new description of collagen hierarchy would conform to the given definitions. If it is accepted that the tertiary structure of collagen relates to the helical structure of a single sub-unit only, then the quaternary structure would simply relate to the interaction of the three sub-units to form a trimer and a subsequent fully functioning collagen molecule.

1.3   The Triple Helix

Irrespective of where in the level of hierarchy a single collagen molecule is considered to fit, its triple helix structure is now irrefutable. The notion of the triple helix structure of collagen was first proposed by Ramachandran.[9] In type I collagen, the monomeric molecule, protocollagen, contains three chains, designated αI(1) and αI(2): there are two αI(1) chains and one αI(2) chain, which differ only in the details of the amino acid sequence. These three chains twist about each other in a right-handed or clockwise triple helix – this is only possible because of the high glycine content, which has the smallest α-carbon sidechain, a hydrogen atom, so that glycine is always situated in the centre of the triple helix, as shown in Figure 1.5.

Each α-chain is about 1050 amino acids long, so the triple helix takes the form of a rod about 300 nm long, with a diameter of 1.5 nm – this is the monomeric unit from which the polymeric fibrous structure is

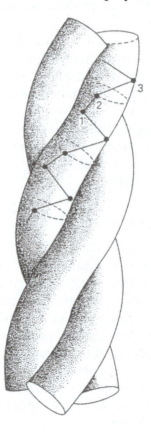

Figure 1.5   The triple helix. Reproduced from ref. 58 with permission from Edward Arnold, Copyright 1980.

created. In the production of extracellular matrix from the fibroblast cells of the skin, the triple helix is synthesised as the monomer pro-collagen and then the monomers self-assemble into a fibrous form. Before it becomes fibrous, soluble collagen can be isolated as triple helices and it is in this form that it can be extracted from immature skin under acidic conditions.

In the triple helix, there is an inside and an outside of the structure. The minimum size of the glycine sidechain and its occurrence every third residue in the α-helix allow it to fit into the inside part of the structure. If the sidechain was larger, the triple helix could not form.

At each end of the triple helices there are regions that are not helical, consisting of about 20 amino acids, called the telopeptide regions. If soluble collagen is isolated with the aid of pepsin, a proteolytic enzyme, triple helices are obtained without the associated telopeptide regions.[10] In the case of acid extraction of soluble collagen, individual triple helices are obtained with the telopeptide regions intact, as illustrated in Figure 1.6.[11]

In the context of leather making, the importance of the telopeptide regions lies in their role in bonding, to hold the collagen macromolecule together: the covalent bonds holding the triple helices together link the helical region of one triple helix to the non-helical region of another triple helix, as illustrated in Figure 1.6. Modelling studies have shown that, because of their flexibility, the telopeptide chains are capable of their own interactions between α-helices and triple helices.[12] In tanning terms, the telopeptide regions probably do not play a significant role in the preparative processes or in the stabilising reactions: in the latter case, the stability arises from the structure created between the triple helices, based on the high degree of structure already in place in the helical region, but which is not present in the more random telopeptide chains.

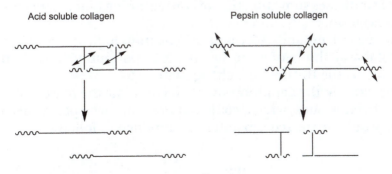

Figure 1.6 Extraction of soluble collagen.[11]

1.4   Isoelectric Point

Importantly, the triple helix is also held together by electrostatic bonds:

$$P-CO_2^- \cdots {}^+H_3N-P$$

These are the so-called 'salt links', formed by electrostatic reaction between acidic and basic sidechains of the protein, which together determine the isoelectric point (IEP) of collagen. The IEP is an important parameter, because it controls the charge on the protein at any given pH value: since the reactions in leather making are usually dependent on charge, they are in turn dependent on the IEP and the way in which the value of the IEP varies during processing. The IEP is defined in several ways, but the most fundamental definition is that it refers to the point on the pH scale at which the net charge on the protein is zero, illustrated as follows

$$P-CO_2H + H_3N^+ -P \rightleftharpoons P-CO_2^- \cdots H_3N^+ -P \rightleftharpoons P-CO_2^- + H_2N-P$$

acidic,	protein at IEP,	alkaline,
cationic	charge zero	anionic

Because the charge on the protein can be adjusted with acid or alkali, to make the protein positively or negatively charged, respectively, there must be a point on the pH scale where the net charge passes through zero. Therefore, the IEP depends on the relative availability of groups that can participate in reactions with acid and alkali, the carboxyl and amino groups. Hence IEP can be defined as follows:

$$IEP = \frac{f_1[NH_2]_T}{f_2[CO_2H]_T} = \frac{f_1\{[NH_2]+[NH_3^+]\}_T}{f_2\{[CO_2H]+[CO_2^-]\}_T}$$

where f indicates some function of concentration and T indicates total concentration.

The second, expanded form of the equation indicates that it is only the total availability of a species that is important; its form is unimportant, *i.e.* the IEP does *not* change with pH.

More strictly, the definition should incorporate the role of the pK_a or pK_b of these groups, which would require that each specific pH active group would be treated separately, exemplified as follows:

$$IEP = \frac{\sum_i f_i[NH_2]_i}{\sum_j f_j[CO_2H]_j}$$

In this way, all i and j species provide individual contributions to the IEP. Unfortunately, the nature of the functions f is not known, so calculation of the precise value of the IEP is not yet possible. However, this is clearly not an intractable problem for the future.

The equations are usefully presented in this form, with the amino function as numerator and the carboxyl function as denominator, because they then predict how varying the functions influences the IEP, in terms of the direction of change (Table 1.2).

At the IEP, the following phenomena occur, each of which might be used as the definition, as a whole or in part:

- net charge is zero
- content of intramolecular salt links is maximised
- swelling is at a minimum
- shrinkage temperature (and other aspects of hydrothermal stability) is maximum.

There are two points relating to the IEP and its relevance to leather making that are very important and worth emphasising.

1. IEP is a point on the pH scale, so it does not change with change in the pH of the system. The IEP of collagen is the same whether is in the alkaline, limed state or in the acidic, pickled state. The importance of this point is that the IEP can be changed only if there is a chemical change that alters the availability of active groups; this can occur in the beamhouse processes, in the tanning processes and in the post-tanning processes.
2. The charge on collagen is determined by the relative values of the IEP and the pH. If the pH is higher than the IEP, the collagen is negatively charged, and if the pH is lower than the IEP, the collagen is positively charged. Moreover, the further the pH is from the IEP, the greater is the charge, although it is limited by the availability of amino and carboxyl groups. The importance of this concept relates to the application of charged reagents, particularly post-tanning reagents and their interaction with the charged leather substrate.

Table 1.2 Effect on IEP of changing the content of active groups.

Change	Amino function	Carboxyl function
Increase content	Higher	Lower
Decrease content	Lower	Higher

1.5   Collagen and Water

An important part of the structure of collagen is the role of water, which is an integral part of the structure of collagen and hence of its chemically modified derivatives, as shown in Table 1.3.[13]

Gustavson[14] observed that the shrinkage temperature of raw skin depends on the pyrrolidine content, *i.e.* proline and hydroxyproline, as shown in Table 1.4; this observation was extended by Privalov,[15] who demonstrated that the relationship relies more on the hydroxyproline content than on the proline content. The entropy loss for the secondary amino acids in the denatured state is less than for other residues, because the ring component of their structures restricts steric conformations; therefore, this could contribute to the relationship between the higher stability and higher proline and hydroxyproline content of the protein. Clearly, the hydrothermal stability of the collagen is also a reflection of the environment in which the animal lives.

Berg and Prockop[16] extracted protocollagen, a non-hydroxylated version of collagen, and demonstrated that its structure is the same as that of collagen, based on the dimensions of the polypeptides and optical rotatory properties. It was found that the shrinkage

Table 1.3   Collagen and water.

Water content $(g\ g^{-1})$	Freezing point (K)	Desorption energy $(J\ mol^{-1})$	Absorption energy $(J\ mol^{-1})$	Structural function
0–0.07	not freezable	71	71	Bound to collagen by two H-bonds
0.07–0.25	<180	50	59	Bound to collagen by one H-bond
0.25–0.50	265	38	38	Bound to sidechains and peptide links
0.50–>2.0	265	Mechanically removable		Swelling water, interstitial within the fibre structure

Table 1.4   Dependence of shrinkage temperature on amino acid content of collagens.

Collagen source	Number of pro and hypro residues per 1000	Shrinkage temperature (°C)
Calf	232	65
Carp	197	54
Cod	155	40

temperature is 15 °C lower than that of hydroxylated collagen, indicating the importance of hydroxyproline to the stability of collagen.

Privalov believed that the hydrogen bonding by water at hydroxyproline is important in stabilising collagen,[17] but thought that the Ramachandran model[18] could not solely explain the high denaturation energy, but that the stabilisation probably included wider layers of water. He stated:[15]

> Having in mind the tendency of water molecules to cooperate with their neighbours, it does not seem improbable that the hydroxyprolyl can serve as an initiator to an extensive network of hydrogen bonds. This envelopes the collagen molecule and might be responsible for the exceptional thermodynamic properties of collagen.

In this way, he regarded the water bound to the triple helix as constituting a matrix structure. Model studies have confirmed that the triple helix is indeed surrounded by such a water structure, nucleated at hydroxyproline, as illustrated in Figure 1.7.[19] The crystal structure gives experimental proof of the existence of water bridges, whereby the triple helices are surrounded by a cylinder of hydration and the hydroxyproline residues appear to act as the 'keystones', connecting the water molecules to the polypeptide. It can be seen that the grooves in the triple helix are filled with solvent molecules directly bonded to the anchoring groups on the peptide or connected to those waters in the first hydrogen-bonded shell. The third shell of water molecules completes the cylinder of hydration, which is supramolecular solvation. Although it was realised that without the hydroxyproline residues there were only localised regions of water structure, Berman and co-workers later suggested[20] that the hydroxyproline acts as a nucleus for water bridges, extending the network far beyond what might be expected and that is critical for the lateral assembly and the supermolecular structure of collagen. Engel *et al.*[21] demonstrated that, in a comparison of peptides of the same length, those containing Hypro residues had higher thermal stability than those without Hypro.

The triple helices are linked through hydrogen bonding, as illustrated in Figure 1.8, in which water may form part of the hydrogen bonding system.[18] In this way, water is an integral part of collagen structure. The figure merely models the effect, emphasising that water is involved directly in the interactions between protein chains, linking

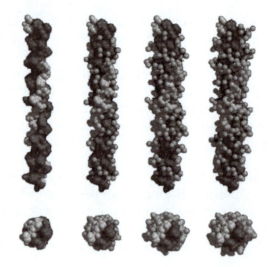

Figure 1.7 Supramolecular water around the triple helix: water molecules are bound to hydroxyproline, creating nuclei for additional water molecules to form a solvating sheath. Reproduced from ref. 19 with permission from Elsevier, Copyright 1995.

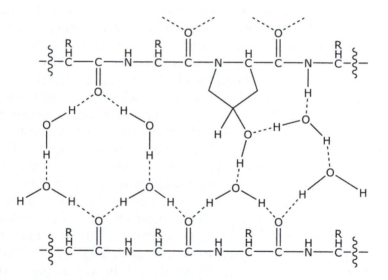

Figure 1.8 Model of hydrogen bonding in collagen, illustrating the way in which hydrogen bonding can link elements of the protein structure.

parts by hydrogen bonding, but the bound water is itself linked to water molecules. It is the build-up of layers of water structure, starting or nucleating at hydroxyproline, that creates the supramolecular solvent sheath. This is an important feature of the stabilisation of native collagen, but it is also the basis for chemical stabilisation, the tanning reactions, because the water illustrated in Figure 1.7 occupies the region in the collagen molecule where the tanning or stabilising reactions occur.

Privalov[15] calculated that the contribution of hydroxyproline to the entropy of water bound to the helix is 98 J K^{-1} mol^{-1}. In comparison with bulk water freezing, this represents the equivalent of freezing four molecules of water per hydroxyproline residue, illustrating the degree of association of water at the Hypro sidechain.

The involvement of water in structure is an important feature of collagen, because it influences the relationship between drying and subsequent leather properties. If the drying conditions are severe enough to remove water close to the triple helix, the fibre structure can approach close enough to allow the formation of additional chemical bonds[22] (see Chapter 18). This adversely affects the strength of leather by embrittling the fibre structure and the handle or feel of the leather is stiffened.

An alternative view of the role of solvent is based on inductive effects. It was thought by Cooper[23] that structural stabilisation might originate in solvent organisation around the triple helix, but he also suggested that the severe rotational restrictions placed on the polypeptide chain by the presence of pyrrolidone residues act to stabilise the triple helical structure by reducing the total conformational entropy change in the helix to non-helix transition. Raines and co-workers[24] agreed that the hydroxy group on proline helps to stabilise the collagen structure, but suggested that the important effect is due to induction. This group synthesised the polypeptide (Pro–Flp–Gly)$_{10}$, where Flp is a 4-(R)-fluoroproline residue; fluorine was used because it is the most electronegative element and because it is a poor hydrogen bond acceptor. Therefore, any stabilisation due to the fluorine must come from a source other than hydrogen bonding. By comparing circular dichroism spectra, they showed that the structure is similar to that of the comparable polypeptides in which the place of Flp is taken by Pro or Hypro. However, the difference in residues content did affect the unfolding transition temperature, as shown in Table 1.5, indicating the stabilising influence of an inductive effect.

It is quite possible that both views are correct and the net effect of the presence of the hydroxyproline residue is a combination of direct inductive and hydrogen bonding effects. Whichever is right, this view of the structure of the fundamental unit of structure also indicates the way in which the structure–reactivity relationships of

Table 1.5   Effect of polypeptide structure on the unfolding transition temperature.

Polymer	Unfolding transition (°C)
(Pro–Pro–Gly)$_{10}$	41 ± 1
(Pro–Hypro–Gly)$_{10}$	69 ± 1
(Pro–Flp–Gly)$_{10}$	91 ± 1

collagen can be addressed. In native collagen or in modified collagen prior to chemical stabilisation (tanning), the triple helices are held in an array in which the interstices are occupied by water, which interacts with the sidechains and backbone amide links of the protein. Chemical stabilisation of the protein can be regarded as the result of reactions occurring in this zone, acting at the surfaces of the triple helices. This model will be become clear as the reactions of leather making are addressed. Importantly, the model will allow the development of a general theory of tanning (see Chapter 23).

1.6   The Quarter Stagger Array

The pattern of the amino acid sequences is characterised by the clustering of acidic and basic amino acids, to create a pattern that is repeated four times within the triple helix, known as the *D*-period. Heidemann mapped the pattern of charged sidechains in the α_1 and α_2 chains[6] and demonstrated that the theoretical interactions between the sidechains are predominantly attractive, with few regions of charge repulsion. In order to allow the maximum formation of the salt links, there are two possible types of interaction, shown in Figure 1.9: the side-by-side interaction, called segment long spacing, and the alignment of the *D*-periods, resulting in staggering by a quarter of the length of the triple helices. It is possible to create the ribbon-like segment long spacing structure by precipitating collagen from weak acid solution, and this structure may play some part in fibrillogenesis,[25] but it is the quarter stagger array that is the basis of fibre formation.[26] Note the so-called gap region between the longitudinal arrangement of the triple helices. Not illustrated here is that there is a superhelical, left-handed twist to the array – at every level of structure there is a change to the direction of twist, a form of structure long exploited by humans for making twine or rope.

The triple helices are held together by salt links, hydrogen bonding, hydrophobic bonding and covalent bonding. The salt links, between

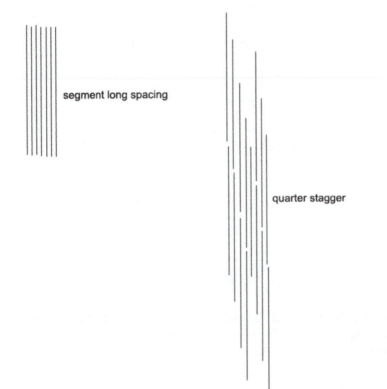

segment long spacing

quarter stagger

Figure 1.9 Model of possible interactions between triple helices.

cationic and anionic sidechains, occur along the α-helices, as do hydrogen bonds, involving water and charge–charge interactions. Similarly, less important hydrophobic interactions involve the electrically neutral sidechains, by the exclusion of water and induced charge attractions (Figure 1.10).

The covalent links are formed at the ends of the triple helices, between the telopeptide region of one triple helix and the helical region of another. In brief, the covalent bonding begins with the formation of a Schiff base, from the reaction between a lysine sidechain and an allysine sidechain, created by enzymatic oxidation of lysine by lysyl oxidase:[27]

$$P-\left(CH_2\right)_4-NH_2+OHC-\left(CH_2\right)_3-P \rightleftharpoons P-\left(CH_2\right)_4-N=HC-\left(CH_2\right)_3-P$$
$$\text{lysine} \qquad \text{allysine} \qquad\qquad\qquad \text{Schiff base}$$

Schiff base formation is the first step in the covalent bonding between triple helices and is characteristic of immature skin,

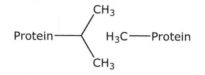

Figure 1.10   Illustration of hydrophobic bonding.

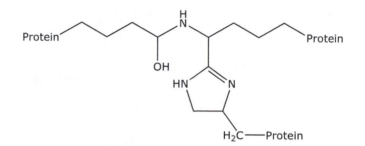

Figure 1.11   The maturation of the Schiff base bonding between α-helices.

particularly apparent in fetal skins. As the animal matures, the Schiff bases are reduced; the reaction is complex, but can be modelled as follows:

$$P-\left(CH_2\right)_4 - N = HC - \left(CH_2\right)_3 - P \rightleftharpoons P - \left(CH_2\right)_4 - NH - CH_2 - \left(CH_2\right)_3 - P$$

The actual reaction in Nature is complicated by the involvement of a histidine sidechain becoming attached to the allysine side of the new bond,[1] as shown in Figure 1.11.

This difference between immature and mature skins is important to the tanner, because they react differently during alkali treatment in the early stages of processing. Schiff bases are vulnerable to hydrolysis: if the bonds are broken then the triple helices can become detached from each other, *i.e.* the collagen is solubilised. Therefore, immature skins must be treated carefully when undergoing processing under alkaline hydrolytic conditions. On the other hand, the mature, reduced crosslinks are not hydrolysable, so can resist hydrolytic conditions. Immature skins can be artificially aged, by applying reducing agents, such as sodium borohydride, $NaBH_4$, or lithium aluminium hydride, $LiAlH_4$.[28] However, with those reagents the process is expensive and dangerous, owing to the evolution of hydrogen under the reaction conditions, so it is typically not used in industry. However, the principle has been proved and there could be economic benefits.

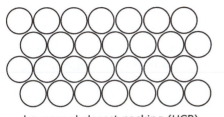

hexagonal closest packing (HCP)

pentafibril or microfibril

distorted hexagonal closest packing

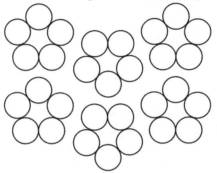

Figure 1.12 Packing of triple helices.

The quarter stagger model of collagen set out here so far has not defined the array of triple helices in the third dimension. Two models have been proposed, illustrated in Figure 1.12. Hexagonal closest packing (HCP) is the closest that circular sections can approach, common in crystal structures. (Try putting together seven coins of the same denomination – six will fit exactly around a central seventh.) The pentafibril model assumes that the triple helices are arranged in units of five, as a regular pentagon in section. However, it is likely that both views are correct, because the pentafibrils can form a sort of distorted HCP, if their regular structure is slightly flattened, as illustrated in Figure 1.12.[29]

The way in which the triple helices are packed together constitutes a contribution to the stability of the protein, referred to as a 'polymer-in-a-box' by Miles and Ghelashvili:[30] in their model, the confining of the triple helices within the fibre structure reduces the entropy of the thermally labile domain when it undergoes the transition to random coil and this stabilises the intact helix conformation.

The molecular structure at the triple helix level and interactions between these components of collagen have been investigated by molecular modelling.[31–36] Ultimately, this approach is likely to be the most fruitful in defining the details of structure, conformations and interactions with stabilising (tanning) species. Indeed, it may be the only reliable way to confirm the validity of theoretical views of tanning.

Cot[37] reviewed the current understanding of the structure of collagen. He included a discussion of the music that can be generated by assuming an amino acid scale of frequencies/notes; for example, the hydrophilic amino acids are assigned higher pitch, the hydrophobic amino acids are assigned lower pitch.[38,39] This use of the amino acid sequence generates music with recurring themes: true electronic music or music of the small spheres!

1.7 Fibrils

The triple helices are bound together in bundles called fibrils, shown in Figure 1.13. The repeating pattern of charges in the *D*-period banding becomes apparent in the macro structure of the fibrils; it can be visualised in raw collagen by reacting the charged sidechains with heavy metal salts, to create a corresponding pattern of high electron beam density. This is commonly done with, for example, anionic phosphotungstate to react with positive basic groups and counter staining with, for example, cationic uranyl ion to react with the carboxyl groups of acidic amino acids. Tanning with less heavy chromium(III) salts can suffice to show the pattern of charge distribution.

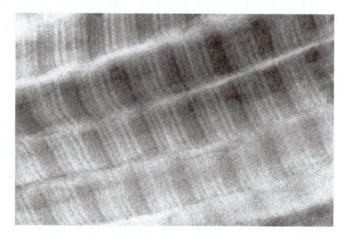

Figure 1.13   Fibrils shown by scanning electron microscopy, exhibiting the characteristic banding pattern of the *D*-period. Courtesy of *Journal of the Society of Leather Technologists and Chemists.*

Figure 1.14   Sub-structure in fibrils revealed by acid swelling.

Fibrils are the smallest units of collagen structure that are visible under the electron microscope, that is, unless the structure is disrupted chemically, for example by the effect of acid swelling, illustrated in Figure 1.14.[40] It can be seen that there is some sub-structure present in fibrils. However, the nature of the sub-units is not known; it might be thought that the structures are penta-fibrils, but the thickness of the units visible in the figure is much too large.

Moreover, in Figure 1.14 there appear to be layers of spiral structure, so they must represent further elements in the hierarchy of structure. It is possible that what is visible in the electron photomicrograph is unravelled surface. This would be in agreement with Koon's observations of apparent hollow structure in collagen fibrils recovered from bone,[41] shown in Figure 1.15. Here the demineralisation conditions are too mild to cause significant damage to protein, so the structure observed by transmission electron microscopy must be a reflection of the actual structure.

This clearly begs the question: what is the structure of fibrils? The question is complicated by the observation that sectioning through fibrils reveals no clear internal structure, as illustrated in Figure 1.16.[6,41]

Kronick *et al.* reported that the cores of fibrils can be melted at a lower temperature than denaturation of the surrounding sheaths[42] and that the inner material of the fibrils can be degraded by trypsin to leave hollow fibrils,[43] but this reaction is prevented if the fibrils are first stabilised with glutaraldehyde.[44] Other workers have suggested that the core material might be glycosaminoglycans[45] or glycoproteins.[46] Therefore, it is uncertain what lies within fibrils, if anything, but those latter observations are not inconsistent.

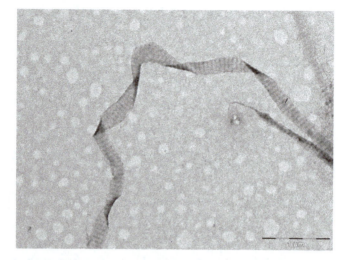

Figure 1.15　Fibrils recovered from bone by demineralisation with 0.6 M HCl. Courtesy of H.E.C. Koon.

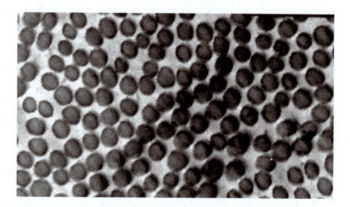

Figure 1.16　Cross-section through fibrils in chrome-tanned bovine hide, visualised by transmission electron microscopy.

1.8　Fibril Bundles

Fibrils are arranged in the form referred to here as fibril bundles, illustrated in Figure 1.17, in which the fibril bundles are a sub-structure or the constructing units of fibres.

This level of the hierarchy of collagen structure is important with respect to opening up of the fibre structure, in preparation for the tanning process. Splitting open the fibre structure at this level, at the junctions between fibril bundles, is important for creating

Figure 1.17 Scanning electron photomicrograph of fibril bundles in collagen fibres.

softness and strength in the final leather, particularly with regard to the deposition of lubricating agents, discussed further in Chapter 17.

1.9 Fibres

Fibril bundles come together to create fibres, as clearly seen under low-resolution light microscopy (see Chapter 2). The fibres characteristically divide and rejoin with other fibres throughout the corium structure. It is this variation in crosslinking or linking that provides the strength to the material. Some workers have used the term 'fibre bundle', but this merely reflects the observation that the fibres comprise a range of diameters, largely dependent on the position through the skin cross-section, illustrated in Chapter 2, Figure 2.3. Therefore, strictly, this does not constitute another level of structure in the hierarchy.

1.10 Other Collagens

1.10.1 Type III Collagen

Also called reticulin, type III collagen conforms to the typical characteristic of collagens: it is one-third glycine and has a high content of β-amino acids, hence it has the triple helical structure. It differs

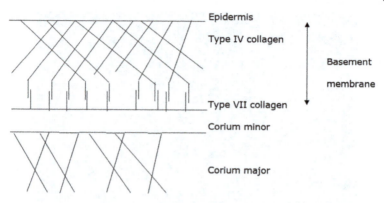

Epidermis

Type IV collagen

Basement

membrane

Type VII collagen

Corium minor

Corium major

Figure 1.18   Illustration of the structure of the basement membrane.

from type I in two respects: there are differences in the amino acid composition and the triple helix consists of three αI(III) chains. As a consequence, it forms finer fibres, although it can form mixed fibres with type I collagen,[27] about 40% in foetal skin, reducing to about 15% in mature skin, when the requirement for flexibility of the fibre structure during growth becomes less important.[47] It can be seen as fine fibres within the coarser fibre structure of the corium. It is understood to be a major component of the grain or corium minor enamel, which is the most valuable part of grain leather.

1.10.2   Type IV Collagen

Type IV collagen is found in the basement membrane, where it forms part of the structure by which the epidermis is fixed to the grain layer or corium minor. It is not a fibrous collagen, but instead has a network structure,[27] as illustrated in Figure 1.18. Type IV collagen can be selectively attacked by an enzyme mixture called dispase, which offers an option of unhairing mechanism, because the basement membrane extends down the hair follicles (see Chapter 3).

1.10.3   Type VII Collagen

Type VII collagen also is not a fibre-forming protein, but has a simple dimeric type of structure.[27] It acts as the link between the type IV collagen which is bound to the epidermis and the grain layer, as illustrated in Figure 1.18.

1.11 The Chemistry of Collagen

The titration curves for native and alkali-treated collagen are presented in Figure 1.19.[48] The regions of the curve are set out in Table 1.6, in which comparison is made between native collagen, in its natural state, and collagen that has received the typical hydrolytic alkali treatment at the early stage of processing, called liming.

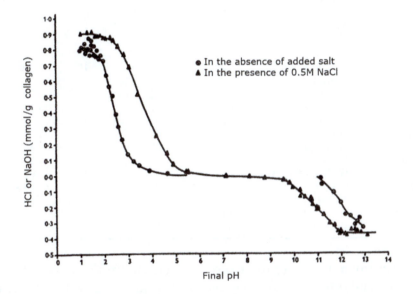

Figure 1.19 Titration curve for type I bovine dermal collagen. Courtesy of Eurofins BLC.

Table 1.6 Regions of interest in the titration curve for type I collagen, native or limed, *i.e.* prolonged treatment with saturated calcium hydroxide solution at pH 12.6.

pH range	Reactions	pK_a values	Amount of acid or alkali/mmol g^{-1} collagen	
			Native	Limed
1.5–5.5	Carboxyl sidechains	Asp 3.85 Glu 4.25	0.87	0.96
5.5–7.0	Uncharged		0.03	0.17
7.0–9.5	Histidine α-Amino terminal	His 6.5 8.5	0.04	0.10
8.5–12.5	Lysine amino	Lys 10.5	0.34	0.36
10.5–14.0	Arginine guanidino	Arg 12.5	0.49	0.47

Table 1.7 Amino acid contents of skin proteins: residues per 1000.

Type	Amino acid	Type I collagen	Type III collagen	Elastin	Keratin
Non-polar	Glycine	330	309	355	81
	Alanine	110	97	178	50
	Valine	22	27	175	51
	Leucine	26	36	59	69
	Isoleucine	12	18	19	28
Hydroxy	Serine	34	43	4	102
	Threonine	18	22	4	65
	Tyrosine	5	3	12	42
Carboxy	Aspartic acid + asparagine	47	53	2	60
	Glutamic acid + glutamine	73	76	12	121
Basic	Lysine	38	22	5	23
	Hydroxylysine	6	12	0	0
	Arginine	48	45	5	72
	Histidine	4	6	0	7
	Tryptophan	0	0	0	12
Imino	Proline	126	97	125	75
	Hydroxyproline	93	108	23	0
Sulfur	Cystine	0	0	1	112
	Methionine	4	8	0	5
Aromatic	Phenylalanine	14	18	22	25
	Tyrosine	5	3	12	42

It is useful to make comparisons between the amino acid contents of collagen and other proteins associated with skin, as shown in Table 1.7.

Compare the amino acid contents of types I and III collagen. There are differences, but not to the extent that significant differences in chemical properties might be expected: the carboxyamino acid contents are similar, so the reaction with chromium(III) in tanning will be the same. Therefore, technologically it is typically assumed that processing reactions will affect the collagen types equally and equivalently. For comparison, the amino acid compositions of elastin and keratin are included: these are proteins of importance in leather making, the former for the physical properties of skin and leather and the latter for hair and epidermis removal, which are contributors to leather quality. These proteins are discussed further in Chapter 2.

1.12 Hydrothermal Stability

One property that is routinely used to characterise collagen, whether native, structurally modified or chemically modified, is the hydrothermal stability, which is defined as the effect of wet heat on the integrity of the material, usually in terms of the denaturation transition. The value of this parameter, dependent on processing, preoccupies tanners and leather scientists, as will become apparent in this text. Typically, the discerned initiation of the transition is referred to as the shrinkage temperature, reflecting the observation that collagenic materials respond to wet heat by shrinking; in the case of native collagen, this is at 65–70 °C, depending on the conditions of the test. This is discussed further in Chapters 10 and 19.

Collagen has been described in terms of a block copolymer, incorporating hard and soft monomeric units,[49] in which there are crystalline regions, associated with the presence of hydroxyproline,[50] where the protein structure is supported by the supramolecular water. Where there is a deficiency in Hypro, the collagen is thermally labile, so that hydrothermal breakdown of the structure is initiated at those sites in the molecule.[51] Interestingly, type III collagen appears to be less stable than type I collagen, which may explain the suggestion by leather makers that the grain enamel is more vulnerable to hydrothermal damage than the remainder of the skin structures. Degradation of collagen by wet heat causes the transition

$$\text{helix} \rightleftharpoons \text{random coil}$$

where the product is well known in the form of gelatine. The gelling property of gelatine depends on the degree to which the helical structure is unravelled: the more this happens, the less strength the gel has, but if some helical structure is retained, the protein interacts in a state intermediate between solubilised and solid, because the helices can interact in a way analogous to self-assembly. Although the transition is expressed here as an equilibrium, it is an irreversible rate process, exhibiting reversibility only in the early stages of the reaction.[52–54] This is understandable by considering the progress of the reaction. In the early stages, when the triple helices have unravelled only slightly, the chains can re-register back into place. However, there comes a point when so much of the helical structure is lost that the randomised structure cannot spontaneously recreate the highly structured helical form. From a leather-making standpoint, the reaction is the same for raw collagen and leather; the hydrothermal reaction is seen as a shrinking of the fibre structure and is in practice an irreversible and highly damaging effect. The process can be summarised as follows:

$$\text{intact} \rightleftharpoons \text{shrinking} \rightarrow \text{shrunk} \rightarrow \text{gelatinisation} \rightarrow \text{solubilisation}$$

(a)   (b)

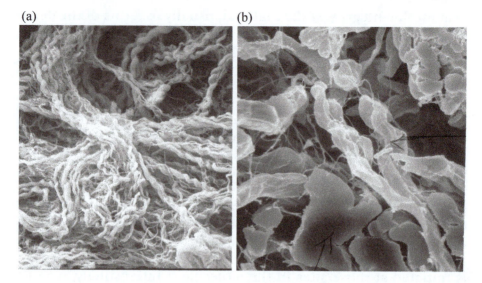

Figure 1.20   Crimp in corium fibres. (a) Caused by alkali + heat damage: obvious crimp; (b) caused by acid + heat damage: minor crimp and
gelatinisation.

Reversibility applies only in the first step. Subsequent steps constitute increasing heat damage to the collagen, which can occur at
any stage in the leather-making process. The potential for the reactions to go to completion depends on the state of the collagen. In
its raw state, collagen can go through all the stages, eventually to be
solubilised. Moreover, the conditions causing the shrinking result
in different intermediate states. Wet heat damage under alkaline
conditions causes rapid hydrolysis and accelerated solubilisation,
but before that happens the fibres exhibit a visible crimping effect,
shown in Figure 1.20.[40] Alternatively, under acidic conditions, crimping is less obvious because the hydrolysis reaction is slower, so the
damaged fibre structure is characterised by the presence of broken
collagen chains and collagen fragments: the outcome in dried pelt is
a cementing of the fibre structure causing the material to be embrit-
tled. When collagen is in a tanned state, depending on the chemistry
of the process, the reaction typically stops at the shrunk stage.

The phenomenon of partial reversibility of shrinking in wet heat is
reflected in the properties of some tanned collagen materials. When
the stabilising reaction involves the creation of an interpenetrating network within the collagen network, shrinking can be partially
reversed. This is called the Ewald effect: it is observed in oil-tanned
leather[55] (see Chapter 13) or if butadiene is polymerised within the
fibre structure. Here the re-registering of the chain interaction is

assisted by the scaffolding effect of the polymerised oil, shaped to the structure of the collagen.

The impact of hydrothermal damage reflects the nature of the starting material, whether raw or tanned, and, in turn, how the collagen has been tanned. Untanned collagen is eventually liquefied, which is the fate of collagen tanned by plant extracts, vegetable tannins, but collagen tanned with chromium(III) salts does not readily liquefy. The difference lies in the nature of the stabilising chemistry: in the former reaction, stabilisation comes from hydrogen bonding, whereas in the latter reaction, stabilisation comes from covalent complexation. The impact of tanning chemistry on outcome is developed further in Chapters 11–14.

It has been demonstrated that the shrinking transition is not the only hydrothermal transition, because there is another, higher temperature, transition observable by differential scanning calorimetry,[56,57] illustrated in Figures 1.21 and 1.22. In Figure 1.21, the native collagen of turkey tendon exhibits the expected low-temperature transition at about 60 °C, but has another transition at about 130 °C. The first transition is eliminated if the collagen is treated with lithium bromide, a well-known hydrogen bond breaker, but the higher temperature transition remains unaffected. Figure 1.22 shows the expected effect of the chrome tanning process; the shrinking transition is moved to higher temperature, but the same happens to the higher temperature transition.

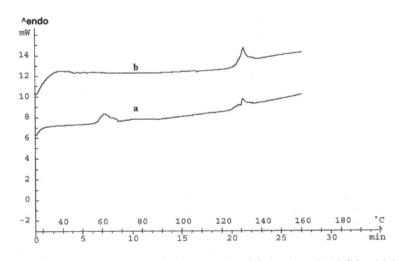

Figure 1.21 DSC thermograms of turkey tendon: (a) demineralised; (b) acid demineralised, denatured with LiBr.

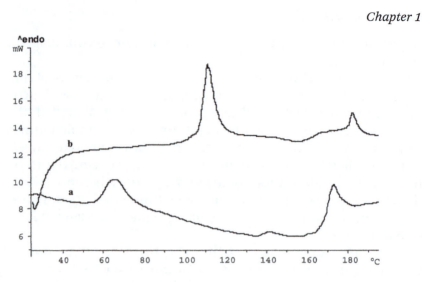

Figure 1.22 DSC thermograms of bovine hide collagen: (a) untanned hide powder; (b) chromium(III)-tanned hide powder.

It is not clear what reaction or reactions are responsible for the higher temperature transition: it may be a breakdown of other higher energy structures associated with the sidechains of the protein or decomposition of the amide links of the peptide chain.

References

1. A. J. Bailey and R. G. Paul, *J. Soc. Leather Technol. Chem.*, 1998, **82**(3), 104.
2. W. D. Comper, *Extracellular Matrix*, Harwood Academic Publishers, Melbourne, 1996, vol. 2.
3. C. M. Kiely *et al.*, *Connective Tissue and its Heritable Disorders. Molecular, Genetic and Medical Aspects*, Wiley-Liss, New York, 1993.
4. K. E. Kadler, *et al.*, *J. Cell Sci.*, 2007, **120**(12), 1955.
5. H. Lodish, A. Berk, S. L. Zipursky, P. Matsudaira, D. Baltimore and J. Darnell, *Molecular Cell Biology*, W. H. Freeman, New York, 4th edn, 2000.
6. E. Heidemann, *J. Soc. Leather Technol. Chem.*, 1982, **66**(2), 21.
7. L. Pauling, R. B. Corey and H. R. Branson, *Proc. Natl. Acad. Sci. U. S. A.*, 1951, **37**(4), 205.
8. K. Chou and Y. Cai, *Proteins: Struct., Funct., Genet.*, 2003, **53**(2), 282.
9. G. N. Ramachandran, *J. Am. Leather Chem. Assoc.*, 1968, **63**(3), 160.
10. N. D. Light, *Methods in Skin Research*, John Wiley and Sons Ltd, 1995.
11. D. Zeugolis, PhD thesis, University of Northampton, 2006.
12. E. M. Brown, *J. Am. Leather Chem. Assoc.*, 2004, **99**(9), 376.
13. K. J. Bienkiewicz, *J. Am. Leather Chem. Assoc.*, 1990, **85**(9), 305.
14. K. H. Gustavson, *The Chemistry and Reactivity of Collagen*, Academic Press, New York, 1956.
15. P. L. Privalov, *Advances in Protein Chemistry*, Academic Press, 1982, vol. 35.
16. R. A. Berg and D. J. Prockop, *Biochem. Biophys. Res. Commun.*, 1973, **52**(1), 115.

17. P. L. Privalov and E. I. Tiktopulo, *Biopolymers*, 1970, **9**, 127.
18. N. Ramachandran and C. Ramakrishnan, *Biochemistry of Collagen*, Plenum Press, 1976.
19. H. M. Berman, J. Bella and B. Brodsky, *Structure*, 1995, **3**(9), 893.
20. R. Z. Kramer, *et al.*, *J. Mol. Biol.*, 1998, **260**(4), 623.
21. J. Engel, H.-T. Chen and D. J. Prockop, *Biopolymers*, 1977, **16**, 601.
22. M. Komanowsky, *J. Am. Leather Chem. Assoc.*, 1991, **86**(5), 269.
23. A. Cooper, *J. Mol. Biol.*, 1971, **55**, 123.
24. S. K. Holmgren, *et al.*, *Nature*, 1998, **392**, 666.
25. A. G. Ward, *J. Soc. Leather Technol. Chem.*, 1978, **62**(1), 1.
26. D. J. S. Hulmes, A. Miller and D. A. D. Parry, *et al.*, *J. Mol. Biol.*, 1973, **79**, 137.
27. A. J. Bailey, *J. Soc. Leather Technol. Chem.*, 1992, **76**(4), 111.
28. BLC The Leather Technology Centre, UK, unpublished results.
29. J. P. R. O. Orgel, *et al.*, *Structure*, 2001, **9**, 1061.
30. C. A. Miles and M. Ghelashvili, *Biophys. J.*, 1999, **76**, 3243.
31. E. M. Brown, J. M. Chen and G. King, *Protein Eng.*, 1996, **9**(1), 43.
32. E. M. Brown and G. King, *J. Am. Leather Chem. Assoc.*, 1996, **91**(6), 16.
33. E. M. Brown and G. King, *J. Am. Leather Chem. Assoc.*, 1997, **92**(1), 1.
34. D. Buttar, R. Docherty and R. M. Swart, *J. Am. Leather Chem. Assoc.*, 1997, **92**(8), 185.
35. J. Fennen, *J. Am. Leather Chem. Assoc.*, 1998, **82**(1), 5.
36. L. Siggel and F. Molnar, *J. Am. Leather Chem. Assoc.*, 2006, **101**(5), 179.
37. J. Cot, *J. Am. Leather Chem. Assoc.*, 2004, **99**(8), 322.
38. http://www.whozoo.org/mac/music/collagen.htm.
39. http://algoart.com/forum.
40. K. T. W. Alexander, A. D. Covington, R. J. Garwood and A. M. Stanley, *Proceeding IULTCS Congress*, Porto Alegre, Brazil, 1993.
41. H. E. C. Koon, PhD thesis, University of York, 2006.
42. P. L. Kronick, B. Maleeff and R. Carroll, *Connect. Tissue Res.*, 1988, **18**, 123.
43. P. L. Kronick and B. Maleeff, *J. Am. Leather Chem. Assoc.*, 1990, **85**(4), 122.
44. P. L. Kronick, B. Maleeff and P. Cooke, *Ann. N. Y. Acad. Sci.*, 1990, **580**, 448.
45. N. Nakao and R. Bashey, *Exp. Mol. Pathol.*, 1972, **17**, 6.
46. S. Franc, *J. Submicrosc. Cytol. Pathol.*, 1993, **29**, 85.
47. E. M. Epstein and N. H. Munderloh, *J. Biol. Chem.*, 1978, **253**, 1336.
48. J. H. Bowes and R. H. Kenten, *Biochem. J.*, 1948, **43**(3), 358.
49. A. C. T. North, P. M. Cowan and J. T. Randall, *Nature*, 1954, **174**, 1142.
50. T. J. Wess, A. P. Hammersley, L. Wess and A. Miller, *J. Mol. Biol.*, 1998, **275**, 255.
51. C. E. Weir, *J. Am. Leather Chem. Assoc.*, 1949, **44**(3), 108.
52. C. A. Miles, *Int. J. Biol. Macromol.*, 1993, **15**, 265.
53. C. A. Miles, T. V. Burjanadze and A. J. Bailey, *J. Mol. Biol.*, 1995, **245**, 437.
54. A. D. Covington, R. A. Hancock and I. A. Ioannidis, *J. Soc. Leather Technol. Chem.*, 1989, **73**(1), 1.
55. J. H. Sharphouse, *J. Soc. Leather Technol. Chem.*, 1985, **69**(2), 29.
56. T. Green, PhD thesis, University of Northampton, 2004.
57. T. Green, A. D. Covington, J. Ding and M. J. Collins, *Tecnologie Conciare*, 2003, **93**.
58. J. Woodhead-Galloway, *Collagen Anatomy of a Protein*, Edward Arnold, 1980.

2 Skin and Its Components

2.1 Introduction

It is important to understand the nature of hide and skin, in order to rationalise the structure–function, structure–reactivity and structure–property relationships as they apply to pelt and to leather.

- What roles are played by the various structural features observed in skin?
- What roles are played by any other components naturally occurring in skin?
- What can be understood and rationalised by considering the structure of skin, in terms of the impact that the structure has on determining the way in which skin is processed into leather?
- How does the structure of skin influence the properties and performance of leather?
- How do modifications to the natural structure influence the properties and performance of leather?

The structure of skin is illustrated in Figure 2.1 and shown diagrammatically in Figure 2.2:[1] these images represent the generic structure of most types of mammalian skin – human, cattle, pig, goat and sheep skin, although there are minor differences between those main materials used in the global leather industry. The main difference is in the firmness or tightness of the fibre structure of the corium, in the

Tanning Chemistry: The Science of Leather 2nd edition
By Anthony D. Covington and William R. Wise
© Anthony D. Covington and William R. Wise 2020
Published by the Royal Society of Chemistry, www.rsc.org

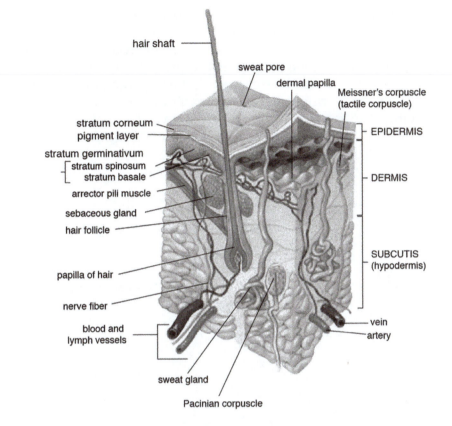

Figure 2.1   Representation of the structure of the upper part of the skin structure: epidermis, grain and junction.

following order of increasing firmness of tanning rawstock: wool sheep, hair sheep, cattle, pig, goat. Note that it is not unknown throughout history for human skin to be tanned – no change in technology is required. Pigskin is notable for the difference in follicle structure: unlike other mammals, where the follicles penetrate the skin structure only through the grain or corium minor, the coarse bristles of the pig penetrate to the base of the corium, to the underlying muscle. An actual photomicrograph of bovine hide is presented in Figure 2.3.

2.1.1 Epidermis

This is the outermost layer of the raw skin, the barrier between the animal and its environment: it is composed of so-called 'soft keratin', characterised by a relatively low content of cystine compared with cysteine, *i.e.* there are fewer crosslinking disulfide groups originating from the oxidation of cysteine (see Chapter 5).[2] The internal structure is illustrated in Figure 2.1. In the early stages of processing, when hair or wool is removed from the skin, particularly by chemical dissolution techniques, the epidermis is also removed, when the whole structure is lost by the single process step. Efficiency in this process is paramount for leather quality: the chemical differences in structure between keratin and collagen (Table 2.1) mean that they react differently in chemical fixation

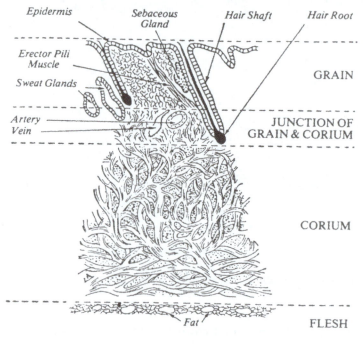

Cross-section of skin

Figure 2.2   Simplified illustration of the structure of skin. Courtesy of Eurofins BLC.

Table 2.1   Amino acid contents of collagen and keratin (residues per 1000).

Amino acid type	Collagen	Keratin
Non-polar	500	279
Acidic	120	181
Basic	95	114
Hydroxy	57	209
Cystine	0	112
Proline + hydroxyproline	219	75

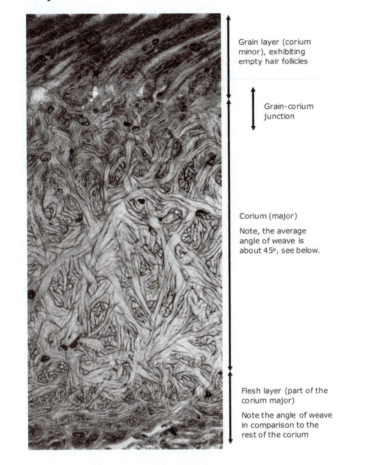

Grain layer (corium minor), exhibiting empty hair follicles

Grain-corium junction

Corium (major)

Note, the average angle of weave is about 45°, see below.

Flesh layer (part of the corium major)

Note the angle of weave in comparison to the rest of the corium

Figure 2.3 Photomicrograph of the cross-section of cattle hide.

reactions, such as dyeing. The consequence of residual epidermis is twofold: paler areas of dye resist, which are duller owing to light scattering from the fibrous surface, in comparison with the reflective surface of the grain enamel.

Even though the contents of basic sidechains are the same, the overall chemical reactivities are different (see Chapter 3), hence the presence of residual epidermis may lead to areas of dye resist and a dull appearance. Its presence is easily diagnosed by inspection under a microscope, seen as a rippled effect on the relatively flat grain surface, as shown in Figure 2.4.

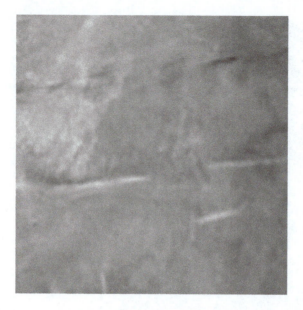

Figure 2.4   Epidermis on skin under the microscope. Courtesy of Leather Wise, UK.

grain

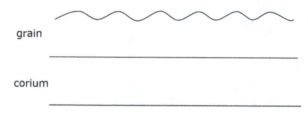

corium

Figure 2.5   Relationship between the grain and the corium.

2.1.2   Grain

The uppermost layer in unhaired or dewooled skin, the corium minor, is also referred to in the jargon as the grain layer. The structure is fibrous, but the fibres are so fine that the appearance is more like a solid. The lack of fibre interaction, in comparison with lower layers of the skin, makes the grain weak.[3,4] The macro structure is a convoluted sheet, because the grain layer is larger in area than the lower layers, so it has to be folded. This conformation is held in place by the presence of elastin, illustrated diagrammatically in Figure 2.5; see also Section 2.2.6.

There is a layer of collagen on the grain surface, the uppermost layer of the corium minor after the epidermis is removed, known in

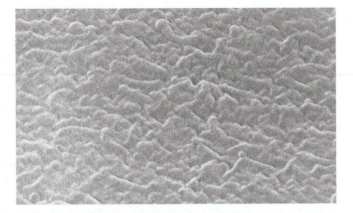

Figure 2.6 The surface of chrome-tanned bovine hide: high-resolution scanning electron photomicrograph (×30 000).

the jargon as the grain enamel: the structure may be based on type III collagen since it is known to concentrate in the grain.[5] The surface is shown in the photomicrograph in Figure 2.6 as an interwoven fibrous structure: the resolution is high, but the surface visible in the photomicrograph is not the convolutions, which are at much higher resolution. The enamel is the most valuable part of the skin, because it provides the desired appearance of the grain; in particular, the enamel confers the appearance to the naked eye of a continuous, reflective surface.

Damage to the grain enamel reveals the underlying fine fibres of the grain, which scatter light more than the enamel, therefore altering the perceived colour and appearing dull. Breaks in the enamel may create paler or darker areas, because either coloured molecules penetrate more easily through the surface or coloured particles lodge in the exposed fibre structure: water-soluble, hydrophilic dyes penetrate more easily to confer a paler appearance, whereas the less water-soluble or insoluble dyes cause colour build-up in the faults, to appear darker.

Chemically, the grain is the same as the more obviously fibrous corium, so the impact of chemical modification is the same. However, because the grain has a more solid structure, filling it with stabilising chemicals can embrittle it, allowing it to crack when stressed, especially if it is not adequately lubricated. An important performance characteristic is the force required to cause the grain to crack.

2.1.3 Junction

The grain–corium junction is the transition zone between the very fine fibres of the grain and the much larger fibres of the corium. It is an open structure, consisting of relatively small fibres and carrying other structural components of the skin; these include the veinous system and, in the case of sheepskins, lipocytes, which are the cells that contain triglyceride fat, illustrated in Figure 2.7.[6]

The junction is vulnerable to breaking by flexing through mechanical action in the process vessel or, for example, in staking/softening. The consequent defect is called looseness, in which the detachment of the layer becomes visible when the pelt or leather is flexed. The effect can occur in any skin or leather, but is facilitated by the presence and particularly the removal of the fat cells from sheepskin, causing large voids in the fibre structure. The fault affects the 'break' of the leather: this is the rippling in the grain surface observed when the leather is bent with the grain uppermost. Large ripples, coarse break, is a sign of inferior quality; fine break is desirable and is a feature of high-quality leather. The illustration in Figure 2.8 is an extreme case, where the junction has been damaged, causing separation or delamination between the grain and the corium; this is known in the jargon as 'double hiding' or 'double skinning'.

2.1.4 Corium

The main part of the skin is the obviously fibrous structure called the corium or the corium major. The fibre structure varies through the cross-section of hide or skin; the fibres vary in size, reaching

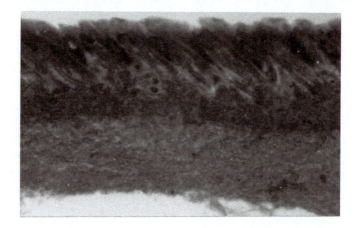

Figure 2.7 The fat layer of lipocytes within the junction in sheepskin. Courtesy of Eurofins BLC.

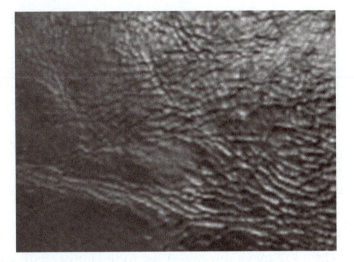

Figure 2.8   Looseness in leather. Courtesy of Leather Wise, UK.

maximum fibre diameter in the centre of the corium, decreasing as they approach the grain and flesh layers.

The network of fibres, often referred to as the weave, consists of fibres dividing and recombining with other fibres.[7] This makes the corium strong, able to resist stresses placed on it: the distribution of a stress imposed on the fibre structure from the point of stress over the surrounding area is reflected in resistance to the stress, observed as strength. An example of the consequence of the effect occurs when sewing leather. The use of a conventional needle, which acts by piercing the material, is likely to result in the needle breaking, because of the high resistance of the fibre structure to being torn apart as the needle passes through the cross-section. Therefore, a specialist needle is required for leather, where the sharp end takes the form of a blade, to cut the fibres, instead of puncturing through them.

An important feature of the corium structure is the angle of weave: experienced observers of corium structure are able to estimate the average angle and the magnitude of the angle can provide useful information regarding the process history of the pelt. The average angle of weave in raw skin is about 45°; a lower value indicates greater depletion or relaxation of the corium and a higher value indicates a degree of swelling (see Figure 2.3). This is addressed in Chapters 4–9.

2.1.5   The Flesh Layer

The so-called flesh layer is the layer of the skin closest to the flesh of the animal; although it has a distinct fibre structure, it is still part of the corium. Its structure is characterised by the low angle of weave, always lower than the corium angle of weave. Consequences of the lower angle and finer nature of the fibres are as follows.

1. The flesh structure is less able to relax, to spread out to create a greater area than in the corium, hence the flesh layer has a controlling effect on the area of the skin or leather.
2. The flesh layer is stronger than the main part of the corium, because the low angle of weave makes it harder to pull the structure apart.
3. The smaller fibres can be abraded to a fine nap for making suede leather, unlike the coarse fibres of the centre of the corium.

2.1.6 Flesh

Hides and skins are inevitably presented to the tanner with adhering flesh (muscle) and fat. These materials must be removed at the earliest stage possible in the processing programme, because they create a barrier to the uniform penetration of chemicals, which would cause non-uniformity of leather properties.

The traditional method of fleshing was to place the hide or skin over a wooden beam, inclined at about 45°, then scrape off the flesh and fat manually with a sharp, double-handled knife. This is the derivation of the term 'beamhouse processes', referring to the process steps conducted on the beam, leading to the tanning step, illustrated in Chapter 4. In modern tanneries, this process is done by machine, using a blade fixed to a rotating cylinder.

2.2 Skin Features and Components

2.2.1 Hair or Wool

The hair or wool is situated in follicles. It is common in the industry to remove the hair or wool, which can be achieved in a variety of ways (see Chapter 5). In some specialist applications, the hair or wool may be left *in situ*, for example, wool sheepskins for winter coats (referred to as 'double face', since both sides of the pelt are equally important for quality) or for rugs.

2.2.2 Follicles

The hair follicles provide the grain pattern that is so much a feature of the value of the leather. The follicle angle reflects the angle of weave of the corium: it is typically 45° in raw pelt, but rises if the corium angle increases and decreases if the corium angle decreases, so the angle of weave is a reflection of change in area. It is important that the follicle angle should be small in the final leather, because that determines the fineness of the grain pattern. The greater the angle of the follicle, the more the mouth of the follicle appears open, referred to as 'coarse, open or grinning grain', which is undesirable. This defect is likely to be enhanced by the dyeing step and in finishing rather than disguised.

In most animals, the grain pattern is uniform, in terms of both regularity and distribution over the pelt. In some animals, the pattern/appearance is distinctive, such as the arc pattern of pig bristles; in some it is the major feature of importance, such as the quill markings of the ostrich.

2.2.3 Erector Pili Muscle

Also known as the arrector pili muscle (Figure 2.9), this is the mechanism by which animals resist the cold, to trap a layer of air within the hairs: by contracting the muscle, the follicle is pulled across the upper part of the skin and the angle of the follicle is raised. In humans, this creates the phenomenon called 'goose flesh' or 'goose bumps'. It has been suggested[8] that starting processing under cold conditions in the preliminary soaking step may result in coarse grain by contraction of the erector pili muscles, which opens the follicle mouths by raising the follicle angle (see Chapter 4).

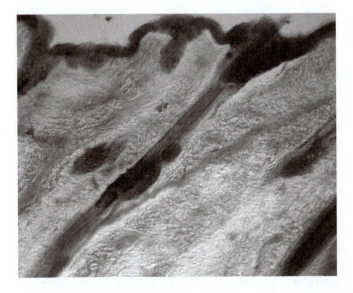

Figure 2.9 The erector (arrector) pili muscle. Courtesy of Leather Wise, UK.

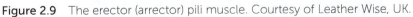

2.2.4   Sweat Glands

The sweat glands do not play any role in leather processing: they are structures that are degraded in the early alkali-based process and their removal does not influence leather properties or quality.

Dog skin is noteworthy, because it does not contain any sweat glands (dogs lose heat by panting). Leather from this source of raw material is now uncommon, although it was once the traditionally preferred material for riding gloves.

2.2.5   Veins and Arteries

The arterial system does not impact on leather making, because it is below the skin. However, the veinous system of the skin is located in the grain–corium junction, where the main vessels run parallel to the skin surface, fed by smaller vessels running vertically down into the body of the animal and into the arterial system. Structurally, veins have an outer layer, called the *tunica externa* or *adventitia*, which is a mixture of collagen and elastin, a middle layer called the *tunica media*, which is composed of smooth muscle, and an inner layer, the *tunica intima*, a layer of endothelial cells forming the barrier between the blood and the vessel wall.

At slaughter, efficient bleeding is required, to empty the blood vessels so as to allow them to collapse: if this is achieved, they will typically not affect the quality of the leather. However, four additional scenarios can occur that must be taken into account.

1. If complete bleeding is not achieved, the veinous system will not be collapsed, causing the structure to remain pumped up.
2. The blood in the veinous system will be degraded in the early stages of beamhouse processing, but the veins may not collapse completely.
3. The blood vessels will be only partially degraded during typical, conventional processing, because of the stability of the elastin, leaving the veinous system largely intact.
4. The veins may be degraded during processing if elastolytic bacterial enzymes are used in the opening up processes, leaving voids as a reflection of the intact system.

In consequence, if any of these circumstances apply, a defect called 'veininess' may be observed (Figure 2.10): the unmistakable appearance is a reflection of the blood vessel network, usually on the split (corium) surface, but occasionally in severe cases the pattern may be observed on the grain surface.

Figure 2.10   The pattern of veins on the grain (left) and suede (right) sides of leather, shown as a fault. Courtesy of Leather Wise, UK.

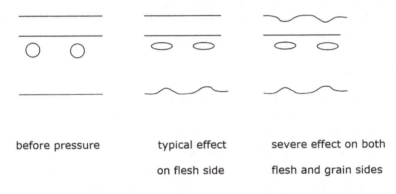

before pressure   typical effect   severe effect on both

on flesh side   flesh and grain sides

Figure 2.11   Representation of the effect of veins on leather appearance.

The appearance of a pattern of blood vessels may arise from two different mechanisms. Inadequate bleeding in the abattoir leaves the vessels expanded because they still contain blood. Although the blood is degraded and removed during liming, the veinous vessels do not collapse completely, creating voids in the hide or skin. Alternatively, because the veinous system is largely composed of elastin, due the requirements of flexing as the heart pumps blood around it, the use of elastolytic enzymes in processing will dissolve the blood vessels themselves, also leaving voids in the pelt cross-section.

The fault is developed during process steps that employ pressure, because the voids are compressed. Since the corium is more compressible than the grain, the effect is most commonly seen on the flesh side of the leather. Because the grain is less compressible, it is only in severe cases that the effect is seen on the grain surface, as illustrated diagrammatically in Figure 2.11.

The only guaranteed solutions are either to avoid or minimise pressure on the leather or to make the leather less compressible; the latter solution can be achieved by using filling retannages, such as vegetable tannins or polymeric resins (see Chapters 13 and 14). It has been assumed that the defect cannot be treated by filling the voids, since that cannot be accomplished without overfilling the remainder of the fibre structure and, anyway, the holes are too large on the molecular level to fill with tanning agent, even polymers. The TFL Company has now demonstrated that the problem can be rectified by treatment with microspheres, hollow 'blown' thermoplastic particles of size 5–100 μm, which collect selectively in the veinous voids during the application in the pickle, aided by an acrylic polymer dispersant;[9] the mechanism by which this is accomplished seems to depend on collapse of the vein structure, as illustrated in Figure 2.12, leaving voids around the veins, which than become filled by the microspheres penetrating through the pelt cross-section.

2.2.6 Elastin

The second most important protein in skin after collagen is elastin (Table 2.2). It contributes greatly to the physical properties of skin and leather, because it controls the elasticity of the grain layer. The material of the grain is weak, so it cannot stretch to accommodate stresses in the skin when, for example, a joint is flexed, hence it adopts a convoluted, rippled form, which can flatten as the corium stretches. The mechanism by which it returns to its convoluted state

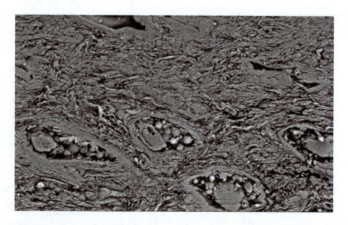

Figure 2.12   Hollow particles accumulating in the veinous structure of pickled hide. Courtesy of the International Union of Leather Technologists and Chemists Societies.

when a stress is removed is through the action of elastin fibres, which extend when the skin is stretched, then contract to the resting position when the stress is removed. The elastin fibres are centred on the follicles, with coarse fibres running parallel to the skin surface and finer fibres running at right-angles to the skin surface, as shown in Figure 2.13.[10]

Table 2.2 Comparison of elastin and collagen (residues per 1000).

Amino acid type	Elastin	Collagen
Glycine	355	330
Apolar	431	170
Acidic	14	120
Basic	10	96
Hydroxy	20	57
Proline	125	126
Hydroxyproline	23	93

(a)   (b)

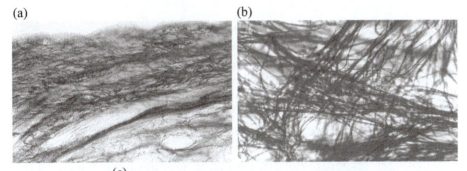

(c)

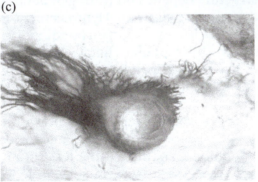

Figure 2.13 Light photomicrographs of the structure of elastin in the grain: (a) intact elastin in the cross-section of the grain of wool sheepskin; (b) elastin fibres in a lateral section through the grain of wool sheepskin; (c) degraded elastin fibres around a hair follicle. Courtesy of M. Aldemir.

Observations that can be made from the values in Table 2.2 are as follows.

1. The proteins have similar glycine content – by comparison with collagen, this indicates that the elastin structure could be helical, which it is.
2. Elastin has more apolar amino acids – therefore, the protein is more hydrophobic than collagen.
3. There are more acidic and basic amino acids in collagen than in elastin – making collagen relatively hydrophilic.
4. The proteins have similar proline contents, which supports the suggestion of helical structure in elastin.
5. There is less hydroxyproline in elastin – so the structure is less reliant on hydrogen bonding than collagen.
6. The lack of basic residues in elastin means there is little lysine and very little histidine: covalent crosslinking of the type found in collagen is not possible (see Chapter 1).

The following deductions can be made.

1. Although it is known that collagen structure depends on covalent bonding, to hold the triple helix units together, there is not enough information within the amino acid composition alone to be able to comment on the presence of covalent bonding in elastin.
2. Collagen structure depends on electrostatic, salt links from the charged sidechains, but they are not important in elastin.
3. Collagen structure depends on hydrogen bonding, based on the high Hypro content – since the Hypro content is low in elastin, its structure relies less on hydrogen bonding. This, too, supports the notion of the hydrophobic character of elastin.
4. Bonding in elastin is dependent on hydrophobic interactions, owing to the high content of apolar sidechains. Such bonding is relatively unimportant in collagen.
5. The differences in amino acid sidechains affect the chemical reactivity of the proteins: reactions dependent on the presence of charged sidechains will be less important in elastin than in collagen. Conversely, reactions dependent only on the backbone chemistry of the proteins will work the same in elastin and collagen. This difference can be exploited in the context of area yield (see later). Also of importance is the high proportion of apolar amino acids, because that indicates an

affinity for tanning agents, which function at least in part through hydrophobic interaction, such as vegetable tannins.

These deductions, based only on the amino acid composition, can be tabulated in terms of the relative importance of the bonding types, as shown in Table 2.3.

Elastin is an unusually stable protein, with a chain structure in the form of a single protein strand, twisted into a double helix: it is capable of resisting boiling water, so this offers a way of isolating it from collagen, because the less stable protein collagen is denatured, degraded and solubilised by boiling water. The macro structure of elastin is a crosslinked network, which adopts a random, disordered conformation, creating regions of hydrophobically bonded groups, by excluding water, whether in its native state or in leather, which still contains some water.[11] When elastin is extended, it has a more ordered structure, when the hydrophobic protein interacts with water, as illustrated in Figure 2.14. The protein acts just like more familiar elastomers, such as rubber, because the driving force to contract after stretching is entropic: the ordered nature of the fibres is of high energy and energy is released when the random structure is regained. In addition, the high-energy interaction of hydrophobic regions with water, both in Nature and in leather, will require the network to re-establish the energetically favoured hydrophobic bonding: this is an enthalpic driver, operating in the same direction as the entropic effect.[12]

The network is crosslinked by desmosine and isodesmosine groups.[13] The structures are presented in Figure 2.15; in each case, the aromatic crosslink is created from the condensation of three allysine sidechains and one lysine sidechain, apparent from the numbers of carbon atoms in the chains and in the ring. Isolation of these crosslinks provides the basis for assaying the elastin content in a material.

Table 2.3 Relative importance of bonding types in elastin and collagen: more asterisks means a greater role in protein structure.

Type of bonding	Elastin	Collagen
Covalent	**	***
Electrostatic	*	***
Hydrogen bonding	*	***
Hydrophobic bonding	***	*

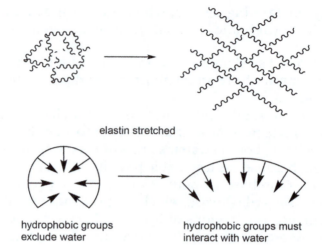

elastin stretched

hydrophobic groups
exclude water

hydrophobic groups must
interact with water

Figure 2.14   Mechanism of elastin elasticity.

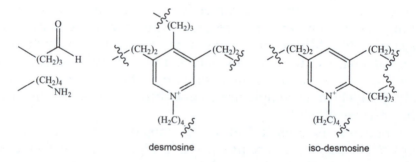

desmosine

iso-desmosine

Figure 2.15   Structures of linking groups in elastin, desmosine and isodesmosine.

In conventional processing, pancreatic enzymes may be employed in the so-called 'bating' step (see later), intended to break down the non-structural proteins of skin: under these conditions elastin is not degraded, nor does this occur under the alkaline hydrolytic conditions of liming (see later). However, elastin is readily degraded and dissolved by elastolytic enzymes, often found in bacterial fermentation formulations also used for bating. These enzymes can be used to break down the elastin structure, thereby allowing the grain to relax and flatten, so increasing the area of the skin (see later).

2.3 Non-structural Components of Skin

2.3.1 Glycosaminoglycans (GAGs)

Glycosaminoglycan is a generic term for a class of minor but significant components of skin, important in the leather-making process. They can take the form of free polysaccharides, called glycosaminoglycans, but which may also have a protein component, when they are called proteoglycans. There are five GAGs that should be considered during processing.[14-16]

2.3.1.1 *Hyaluronic Acid*

In terms of leather production, hyaluronic acid (HA) is probably the most important of this class of compounds in skin. The structure is a chain of alternating D-glucuronic acid and D-*N*-acetylglucosamine, joined by β-1,3-glycosidic links, with a molecular weight of the order of 10^6, corresponding to 3000–19 000 disaccharide units: the chemical structure is illustrated in Figure 2.16.

With regard to technological impact, HA can be considered to be a polymeric chain of aliphatic carboxylate groups with $pK_a \approx 4$, all ionised at physiological pH 7.4. It has been assumed that HA should be removed prior to the alkali treatment conducted with lime, $Ca(OH)_2$, because calcium ions will react with the carboxyl groups, precipitating it within the skin structure, thus preventing its removal; thereafter it would be reactivated as the free carboxylate in subsequent acid processing, to cause problems to the tanner. Since the presence of carboxylate groups in HA will function as sites for reaction with chromium(III) tanning salts or other mineral tanning agents, fixing HA within the fibre structure during tanning is likely to affect the physical

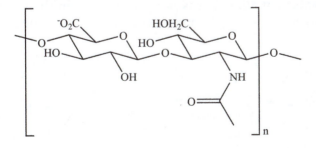

Figure 2.16 Structure of hyaluronic acid.

properties adversely. Little work has been done to establish the out-come of variation in HA removal from pelt. However, one study indi-cated that incomplete removal of HA in the soak, for example from fresh hides, is reflected in less effective opening up in the liming step; moreover, this affects the reaction in the chrome tanning step.[17]

The function of HA in skin is to provide resiliency, caused by the repulsive forces from the negative charges, which force the mole-cule to adopt an extended conformation, thereby creating a large exclusion volume in the form of a gel.[16] It is this effect that makes it difficult to rinse it out of native hide or skin, although it can be removed efficiently, albeit with difficulty, requiring prolonged washing because it is not bound to the fibre structure, merely entangled in it. However, removal is not a problem if the rawstock has been salted, because the ionic nature of the electrolyte causes the gel structure to collapse. This is illustrated in Figure 2.17, in which the attractions of ions for opposite charge effectively 'smear out' or weaken the magnitude of the full negative charge on the carboxyl groups. The effect is to allow the charged carboxyl groups to approach much closer, reducing their exclusion volume, and also reducing the ionic interaction with the protein of the pelt, so it is easily rinsed from the fibre structure. Alexander[15] showed that HA removal was 93% from salted stock in 24 h, compared with 37% from fresh hide; Siddique *et al.* claimed that removal from silicate-cured hide was 82%.[17]

The role of salt in the mechanism of removing HA is a useful spin-off from the commonest method of preserving hides; the use of sodium chloride is called 'salting' (see Chapter 3). In modern process-ing, it is becoming less common for rawstock to be preserved in this way: because of the undesirability of neutral electrolyte in effluent, the industry is turning to other methods of preservation, including no preservation at all. If salt is not present within the rawstock at the start of wet processing, the removal of HA is more difficult and the

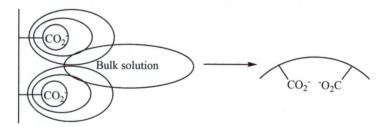

Figure 2.17   Smearing of charge by electrolyte.

evidence is that it must be addressed. Note that preservation using osmolytes that do not rely on an ionic effect will not assist the removal of HA (see Chapter 3).

Options are to use chondroitinase enzymes to degrade the molecule into smaller pieces for ease of rinsing out, but this is not established economic biotechnology, or to include some electrolyte in the soaking bath, which is a perverse contradiction to the reason for processing fresh rawstock in the first place.

2.3.1.2 *Dermatan Sulfate (DS)*

Dermatan sulfate proteoglycan or chondroitin sulfate B has a protein core and 2–3 sidechains of 35–90 repeating units, containing one molecule of L-iduronic acid and one molecule of N-acetyl-D-galactosamine-4-sulfate, as shown in Figure 2.18. The molecular weight is about 10^5, of which 60% is protein and 40% is polysaccharide.

DS is bound to the surface of the fibrils in the hierarchy of collagen structure by protein links and electrostatic interactions, shown diagrammatically in Figure 2.19. The role of DS in skin *in vivo* is concerned with hydrating the fibrils, but its role and importance in leather processing are not clear. However, it is unlikely to play a major

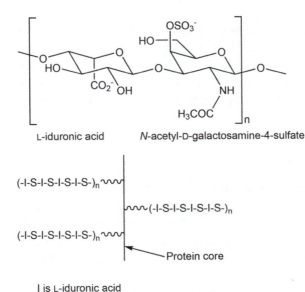

L-iduronic acid   N-acetyl-D-galactosamine-4-sulfate

I is L-iduronic acid
S is N-acetyl-D-galactosamine-4-sulfate

Figure 2.18 Structure of dermatan sulfate.

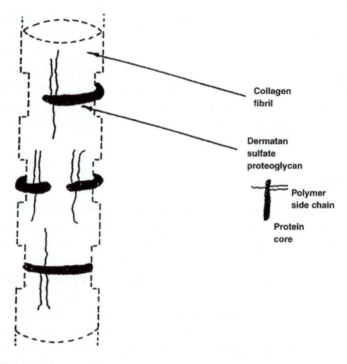

Figure 2.19 Relationship between dermatan sulfate and the fibril structure. Courtesy of the *Journal of American Leather Chemists Association.*[15]

role, since the opening-up reactions of the beamhouse apply primarily at the level of the fibril bundles. Nevertheless, since DS is bound to the collagen *via* peptide links, it is hydrolysed from collagen during liming, hence its removal has been used to measure the progress of hydrolytic reactions and thereby provide a marker for the extent of hydrolytic opening up.

2.3.1.3 Decorin

A minor component of skin is decorin, which is a low molecular weight (90–140 kDa) type of DS, also bound to type I collagen fibrils:[16,18] it has a protein core of 329 leucine-rich amino acid residues and a single glycosaminoglycan sidechain consisting of chondroitin sulfate and DS. Its fate in processing is the same as that of the more prevalent dermatan sulfate.

The effect of decorin on skin and leather properties is less well defined. Workers at the US Department of Agriculture[18] have claimed that decorin removal results in more stretchy, tougher leather. They have shown that it is more effectively removed in an oxidative unhairing process, using sodium perborate compared with conventional sulfide reaction, or by reliming with alkaline-stable protease and pickling in the presence of pepsin.

2.3.1.4   Chondroitin Sulfate A

Chondroitin sulfate A is a repeating polymer of D-glucuronic acid and N-acetyl-D-galactosamine-4-sulfate, shown in Figure 2.20. The molecular weight is of the order of 10^6, comprising a globular protein with 60–100 GAG chains, each with 10–100 disaccharide units (molecular weight 5000–50 000).

This is a minor component, a detached proteoglycan not bound to the fibre structure. As a sulfate derivative of hyaluronic acid, its properties will be the same, except for greater hydrophilicity and solubility, so its fate in processing is likely to be the same as for hyaluronic acid itself.

2.3.1.5   Chondroitin Sulfate C

Chondroitin sulfate C is a repeating polymer of D-glucuronic acid and N-acetyl-D-glucuronic acid-6-sulfate. The structure is shown in Figure 2.21. Like chondroitin A, this is a minor component of skin, which has no significant impact on the leather-making process.

2.3.1.6   Overview of Glycosaminoglycans

The fate of GAGs depends on the liming conditions, but it appears that there is removal to a greater or lesser extent. However, typically little attention is paid to the overall removal and certainly not to the specific GAG species removal, with the exception of dermatan sulfate

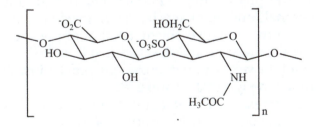

Figure 2.20   Structure of chondroitin A.

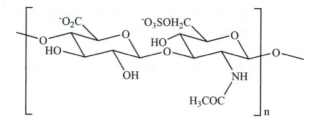

Figure 2.21 Structure of chondroitin C.

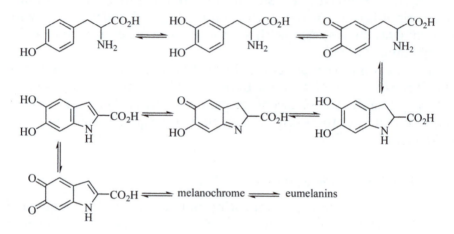

Figure 2.22 Synthesis of eumelanins.

as a marker of hydrolysis and the studies of decorin removal. Since it known that GAGs do play roles in the performance of skin *in vivo*, it is not unreasonable that those roles might be exploited in processing by removing or conserving those chemical species.

2.3.2 Melanins

Melanins are the pigmenting materials that provide much of the colour to mammals and other species. They are chemically inert pigments, based on the oxidative polymerisation of phenols and naphthols. Mammalian melanins may be classified into the brown to black eumelanins and the yellow to red–brown pheomelanins. The chemistries are outlined in Figures 2.22 and 2.23.

Figure 2.22 represents the Raper–Mason scheme of eumelanogenesis,[19–23] involving the hydroxylation of tyrosine to DOPA and subsequent oxidation to dopaquinone. Cyclisation creates leucodopachrome, leading to indole derivatives and polymerisation to the black pigment. Figure 2.23 shows the scheme for the synthesis of the

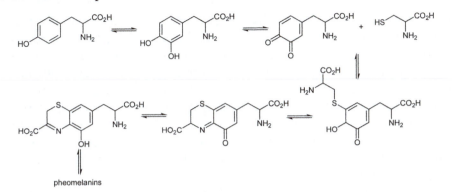

pheomelanins

Figure 2.23 Synthesis of pheomelanins.

(a) (b)

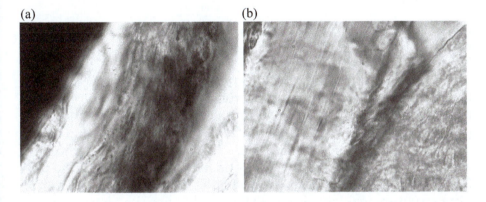

Figure 2.24 The presence of melanin in hair and skin at the follicle mouth and the melanin associated with the hair follicles. (a) Black haired hide; (b) White haired hide.

pheomelanins, starting with cysteine reacting with dopaquinone, leading to cyclisation and the formation of 1,4-benzothiazines.[24–27] Like lignin, discussed later, the melanins have complex structures; therefore, although the pathways to their synthesis have been elucidated and the precursors are known, the actual structures are not defined.

From a tanning viewpoint, the pigments occur in the epidermis, in the cortex of the hair and around the hair follicles in the form of granular melanocyte cells: in Figure 2.24, granular melanocytes are visible in the black-haired hide but not in the white-haired hide.[28] Much of the skin melanin is lost during liming, as the epidermis is solubilised, but it can be trapped within the grain when the pH of the pelt is lowered from the alkali-swollen state; this is discussed in Chapter 6, which is devoted to deliming. The presence of residual melanin is a quality issue, since any discoloration of the pelt is likely to be apparent after dyeing. The inertness of the pigments means that they cannot be chemically degraded without causing damage to the leather. Rao and co-workers at CLRI have demonstrated that melanin removal can be accomplished to completion in 6 h by treatment with 0.5% xylanase alone or less with 1% α-amylase after unhairing, possibly involving attack at an adhering mechanism involving proteoglycans.[29] This is an enzymatic reaction known to degrade hemicellulose, used to bleach wood products, but it has not yet been developed industrially for the leather sector. The biochemistry of xylan is discussed in detail later.

The role of melanin in leather making is limited to its influence on hair degradation (see Chapter 4) and in the subsequent process of deliming.

2.4   The Skin

As a natural material, required to perform different functions for the live animal, to deal with different stresses over its area, skin is anisotropic: its structure and properties vary over the area. The parts of a hide or skin can be defined in terms of the 'butt', the 'belly' and the 'neck'. The butt is defined by the region up to half way from the backbone to the belly edge and two-thirds of the way from the root of the tail to the neck edge: within this region, the fibre structure is relatively consistent and hence the physical properties of the skin or leather are relatively consistent. The remaining regions to the side of the butts are called the bellies and the remaining region beyond the butt towards the head is the neck. This is illustrated in Figure 2.25.

The regions can be characterised as follows. The butt has a tight fibre structure, making the skin relatively firm and stiff: it is thick compared with the belly, but thin compared with the neck. (In pigskin, the butt contains the shell area, which is particularly hard. Interestingly, of the commonly available collagens, pig collagen is the closest to human collagen and hence is often used to make wound dressings.) The bellies of all skins are the thinnest parts, with an open structure, making them relatively weak. The neck is the thickest part, also with a relatively open structure. It is an important aspect of the technology of leather making to try to make the non-uniform skin structure as uniform as possible in the final leather product.

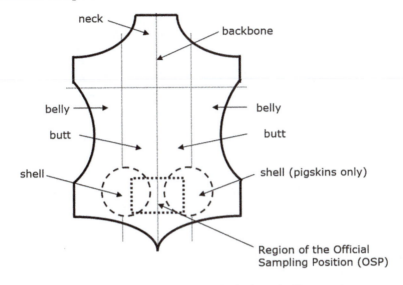

neck

backbone

belly

belly

butt

butt

shell

shell (pigskins only)

Region of the Official
Sampling Position (OSP)

Figure 2.25   The defined regions within the hide: butt, bellies, neck.

The butt region contains the official sampling position (OSP): because of the anisotropic nature of skin, to accommodate its physiological functions it is necessary to adopt the idea of an OSP, in order to make the optimal comparisons of leather properties. It is also important to note how the anisotropy varies over the whole skin, illustrated in Figure 2.26.[30] For this reason, sampling for physical testing is routinely done in two ways, parallel and perpendicular to the backbone. The region of the backbone extends from tail to neck and a few centimetres either side of that line. It is thicker than the butt and must be shaved to the thickness of the butt region.

The direction of the fibre weave, indicated in Figure 2.26, demonstrates the influence of the direction of physical testing. Because the skin has to stretch to accommodate stomach filling, the skin and its stabilised derivatives exhibit greater strength perpendicular to the backbone than parallel to the backbone. Another notable feature is the symmetry of features in the skin about the backbone: it can be assumed that the two sides of a hide or skin are symmetrically the same. This is useful in experiments or testing, when samples are matched from the two sides of the same hide: control samples might be taken from the left side and the experimental samples would then be taken from the same positions on the right side, or *vice versa*. In contrast, comparability between any two skins cannot be assumed,

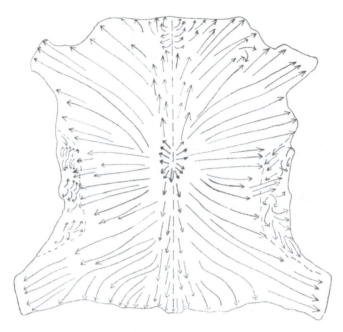

Figure 2.26   Anisotropy of fibre structure over the hide or skin. Courtesy of Eurofins BLC.

even if the same area of the skin is sampled, because animals vary for many reasons: breed, sex, age, diet, husbandry, curing, storage history, *etc.* Therefore, measuring the effect of a treatment must be performed statistically, with a large enough number of representative skins.

The appearance of the grain layer is generally consistent over the skin, with the exception of the neck region, where 'growthiness' may be encountered. The incidence and degree of this effect depend strongly on species and breed, being most apparent on bull cattle hides and merino sheepskins. It takes the form of ripples in the skin, across the neck region, held in place by the elastin content. The effect can largely be removed by degrading the elastin (see later). In the case of merino sheep, the effect can extend over the whole skin and the structure of these extreme growth ripples is illustrated in Figure 2.27; this natural feature was thought in the past to detract from the value of the pelt, hence these skins were used only for making chamois (oil-tanned) leather, where the grain is discarded prior to tanning. Nowadays, the feature of growthiness is regarded as a selling point for finished articles, such as ethnic jackets.

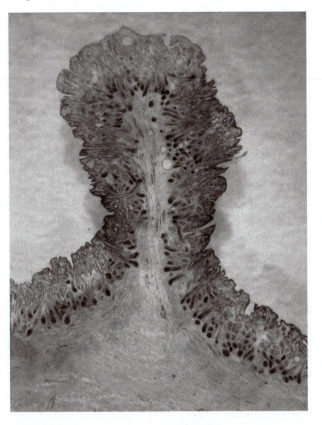

Figure 2.27 The characteristic feature of growthiness in a merino pelt. Courtesy of Leather Wise, UK.

2.4.1 Area of Skin or Leather

Chrome-tanned leathers from cattle, sheep, goat or pig are sold by area, so the area yield from processing is a critical feature. (Note that vegetable-tanned sole leather and other so-called 'heavy' leathers are sold by weight. Exotics may be sold by other characteristics of importance in the species.) The area of a piece of hide, skin or leather depends primarily on the angle of weave of the corium: a high angle of weave means a low area and a low angle of weave creates a larger area, as shown diagrammatically in Figure 2.28 – the same effect can be observed in trellis fencing.

The angle of weave is fixed within the structure by two controlling elements of the structure, the flesh layer and the grain. The area of the flesh layer can decrease, because the angle of weave can increase. On the other hand, its ability to reduce its angle of

Figure 2.28   Representation of the influence of fibre angle of weave on area.

weave is limited by its already low angle. Therefore, it has a limiting effect on allowing an area increase if it is present, as it usually is in leather made from skins, which are relatively thin. In cattle hides, this is usually not the case for most applications, since the thickness of the natural hide necessitates splitting through the cross-section to produce a usable thickness; this might be done in the limed state or the (chrome) tanned state. The products are the grain split, wherein lies the primary value of the hide, and the flesh split, typically used for making suede leather. When the flesh layer is separated from the grain–corium split, its limiting effect on area is removed and the corium may be able to relax its angle of weave, to give a greater area yield. This is particularly apparent when pelt is split at an early stage in the process, such as in the limed condition.

Features in the structure of skin, such as the angle of weave, are maintained or 'tanned in' by the tanning process, so it is useful to compare the effects of splitting after liming and after chrome tanning. In the former case, the open nature of the split surface allows easy access to process chemicals, resulting in more extensive opening up, yielding softer leather, suitable for upholstery. If the structure remains intact through tanning, opening up is less extensive and the leather is firmer, more suited for shoe uppers.

The area of the grain is determined by its convoluted nature, maintained by the presence of elastin. If the elastin is removed, then the grain can relax into a greater area, allowing the corium to follow, by lowering its angle of weave. This can be achieved by the application of elastolytic enzymes, which may be used in liming, in bating or at the wet blue (chrome-tanned) stage. The use of alkali-active enzymes in unhairing usually means that there is elastase activity associated with the bacterial proteases. However, the extended proteolytic effect typically results in looseness. This also happens when bating with bacterial enzyme formulations. In each case, the problem could be avoided by using pure elastase. However, it is rare for a microorganism to express elastase alone and the separation of enzymes is an expensive operation, making that option economically non-viable for the leather industry.

The economic option is to apply a protease–elastase mixture to wet blue:[31] this is commercially available as NovoCor AX. The argument goes as follows. Elastin can be degraded by elastase enzyme at any early stage in processing, but using purified enzyme is uneconomically expensive for an industry such as leather. However, elastase can be obtained cheaply in combination with protease: the unfortunate consequence of using the mixture is that while elastase is degrading the elastin, the protease is degrading the collagen. The result inevitably is loose, weakened leather with reduced value, despite the added value of increased area.

Following the beamhouse steps, chemical modification of protein alters its properties, including its ability to interact with the reaction sites of enzymes. In this way, chrome tanning protects the collagen from biochemical attack, which is the definition of tanning. Hence the protease part of the enzyme formulation has no damaging effect on wet blue leather and thus no adverse effect on the physical properties of the fibre structure. Conversely, because elastin contains fewer acidic amino acid residues, as already shown, the chrome tanning process causes little chemical change to elastin, so it remains vulnerable to degradation by elastase. Therefore, applying a mixture of proteolytic and elastolytic enzymes to wet blue will result in the degradation of elastin, but no change in the modified collagen. Under these circumstances, the presence of the protease is actually of positive benefit to the process, because it can accelerate the degradation of the elastin after it has been damaged by elastase. The removal of the elastin allows the grain to relax and spread, in turn allowing the corium angle of weave to be lowered. In this way, it is claimed that up to 9% extra area yield can be obtained, although 4–5% is typical; however, because the collagen is unaffected, the physical properties of the leather are unaffected, as is the quality of the leather. Even greater increases in yield may be obtained from pigskins, because of the naturally high content of elastin controlling the area. The argument applies only to chrome-tanned leather, because the tanning reaction applies to the carboxyl-bearing sidechains, with very little reaction with the protein backbone. Any alternative or additional tanning reactions that rely on reacting with the backbone of the protein or *via* hydrophobic interactions will render the elastin as resistant as collagen to enzymatic degradation and the biotechnology will not work.

German workers[32] have suggested that removal of elastin has no impact on either area or other leather properties. It is not clear how the observation arose, but the suggestion runs counter to the known functions of elastin *in vivo*. It is also known that the technology for the elastolytic effect to operate is subject to strict conditions that must be met, particularly addressing the vulnerability of elastin to degradation, otherwise the area effect is not obtained. Also, the success of NovoCor AX in the international industry raises a query about the validity of this study.

In this context, it is worth reiterating that tanning means resistance to microbiological attack. Bating enzymes have long been sought to modify the properties of wet blue, for example to create greater uniformity of properties between leathers obtained from different sources (common practice in large tanneries around the world) and claims have been made for opening up and softening of wet blue enzymatically.[33,34] However, such claims are doubtful. Studies of protease treatment for wet blue demonstrated that there is no observable effect unless the concentration of enzyme is abnormally high, the pH is close to the optimum, around 9, and the temperature of the reaction bath is close to the maximum that can be tolerated by the enzyme, about 50 °C. The last two conditions combine to produce incipient hydrolytic damage to the leather, reversal of tannage, after prolonged contact, for at least 1 h. Under these conditions, no effects are observed up to the point when there is sudden, catastrophic, fast and complete degradation of the leather.[35]

Area can also be controlled by physical processing. Applying stretching during drying increases the area yield. However, the gain is not all permanent and the physical properties are adversely affected: the leathers are stiffened, because the effect relies on fibre sticking to retain area gain.[36,37] See also Chapter 20.

2.5   Processing

Some process steps should be discussed with reference to the influence of the components of skin structure on the operations and their outcomes.

2.5.1   Splitting

Most cattle hides must be split at some stage of processing, because they are usually too thick for typical end-use applications. It should be noted that tanners in this field expect their economics to be determined by grain leather production – if any money can be made from the flesh split, it is a bonus.

It is usual for shoe upper leather to be split after chrome tanning: this is primarily because it is necessary to retain spring and solidity in the leather (a property known in the jargon as 'stand'), which might be lost if the opening-up processes were too effective. The hide is processed to this point as full-thickness rawstock, which means that all reagents must penetrate through the intact cross-section. This has attendant problems of uniformity of reaction, compounded by the variability of structure over the hide. The products of the operation are chrome-tanned grain split and chrome-tanned flesh split. Since the market for the suede leather is variable and the alternative is to make cheap coated (artificial grain) leather, the value of the flesh split is likely to be limited.

It is important to note that suede is leather also, even though it is often colloquially referred to as something separate from leather. The suede split will typically go through all the processing applied to the grain split, the opening up, tanning and post-tanning. At the end, the suede nap is prepared by 'buffing' the surface with glass/sand paper: as is the case with sanding a wood surface, the fineness of the fibre nap is controlled by the fineness of the grit on the paper. The nap is usually created on the flesh side of the leather, rather than the split surface, because the fibres in the middle of the corium produce a coarse nap, *i.e.* a rough suede compared with the fibres of the flesh layer. Alternatively, suede leather can be created from the grain split, by buffing the grain surface to remove the enamel and create a very fine nap: this product is called 'nubuck'.

The alternative to splitting after tanning is to split at the end of the liming step. The hide is thick and relatively stiff, because it is in the swollen state, so cutting is precise and easy. Less easy is handling the very slippery and heavy pelt. Nevertheless, lime splitting is routinely conducted for making upholstery leather. The benefit is the greater ease with which the reagent can penetrate through the cross-section, through the more open structure of the split corium surface, causing a greater degree of opening up. When this is combined with the loss of the flesh layer, area yield and softness are enhanced, although usually at the cost of looseness and often grain damage. Since the leather is required to be cloth-like, for its end use, and it is commonly pigment finished, these are not necessarily any drawbacks. The products of the operation are limed grain split and limed flesh split. In an untanned state, the limed split may be used in other sectors: an important application is to use the collagen for making sausage and salami casings by comminuting the skin and reconstituting it as a film.

The remaining option for splitting is in the raw state, either as uncured or as soaked salt-cured hide. This is uncommon, primarily because it is more difficult to split the thinner and softer material. However, another limiting factor is the presence of dung: if the hair side is not flat, the splitting knife will cut into the hide, causing area loss and hence a decrease in value. However, if the dung can be removed in the soaking step, then green splitting is an environmentally and economically sound operation (see later). The products of the operation are raw grain split and raw flesh split. The thin grain split is chemically processed more efficiently and effectively than if it is split later. The collagen of the flesh split is undegraded by highly alkaline treatment conditions and therefore has high value as a source of intact collagen. See also Section 2.5.4.

2.5.2 Grain-to-corium Thickness Ratio

The act of splitting thick hide, to create the valuable grain split, must take into account the nature of the starting material. Consider the production of shoe upper leather from a pack of tanned leather, in which the rawstock was non-uniform in weight. Since raw hide is traded by weight, hides are selected in weight ranges, but there is no guarantee that the weights are entirely consistent. The situation shown in Figure 2.29 could arise, in which hides of different starting

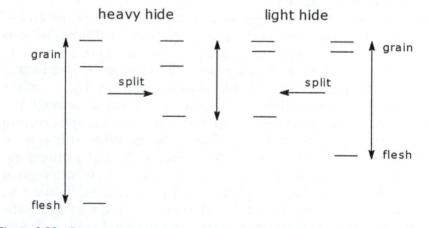

Figure 2.29 Representation of the effect of hide thickness and grain-to-corium ratio before and after splitting to the same thickness.

thickness are split to the same thickness for the same end use. It is generally the case that heavier hides are thicker, although they will often also be larger in area.

The outcome is splits of the same thickness, but with variations in the grain-to-corium thickness ratio. Because the strength of the leather depends only on the corium, the strength is determined by the grain-to-corium ratio: the lower the ratio, the stronger the leather will be.[3,4] The lesson to be learned from this analysis is that rawstock and resulting process packs should ideally be selected on the basis of area rather than weight and, strictly, by the grain-to-corium ratio in the raw material, as suggested by Betty Haines many years ago. In this way, the product would be more consistent in strength.

2.5.3 Fleshing

As stated above, the adhering flesh on raw hides and skins must be removed. The earlier this can be done, the more process steps will be conducted on clean pelt, so that penetration of chemicals and biochemical reagents will be uniform. There are two options open to the tanner.

Lime fleshing is the commonest practice, conducted after the hair has been removed and the hide has been subjected to many hours of treatment in a saturated lime solution at pH 12.5. The disadvantage is the limed state of the fleshings. If the fleshings are regarded as waste, there is an additional cost for disposing of the alkaline material to landfill. Alternatively, if the fleshings are treated to recover the tallow content, their value is diminished by liming, because alkali-catalysed hydrolysis reduces the tallow content and the content of free fatty acid is increased, which downgrades the quality of the tallow.

Green fleshing, after soaking, produces fleshings with high tallow content and low free fatty acid in the product; this improves the economics of the process. However, green fleshing can only be conducted if the hair side of the hide is free from dung. Developed technology allows dung removal in the dirt soak, making this process step practicable.

2.5.4 Dung

Dung is not a skin component, but it is commonly associated with skin (Table 2.4). It may appear to be an integral part of the structure, because it can form a composite material with the hair, which has

Table 2.4 Composition of UK Friesian cattle dung.

Component	Content/% on dry wt
Cellulose	30
Hemicellulose (xylan)	28
Lignin	21
Protein	6
Ether soluble	10
Cold water solubles	6
Hot water solubles	4

strength and water resistance. Indeed, composites of dung, clay and hair are the daub component of the traditional building material 'wattle and daub', still used in some developing economies. It is important to remove the dung as early as possible in the process, to allow the greatest flexibility in processing and opportunities for making financial savings in the operation. It can be removed in the lime–sulfide hair-burning step, when the dung is solubilised at the same time as the hair, but that adds to the environmental impact of the technology. A partial solution comes from American work on dried dung, using oxidative unhairing with 5% sodium perborate plus 2.5% sodium hydroxide in 200% float at 40 °C; the weakened hair releases the dung balls in 30 min.[38]

The relationship between dung and the skin is illustrated in Figure 2.30. What is clear from the image is that there is no direct contact between the dung and the surface of the skin. This is an important feature of the interaction, because any treatment of the live animal to remove dung, for example prior to slaughter, would not be an invasive procedure and hence should not cause distress to the animal. In the following discussion, it is shown that dung can be removed from hides in the tannery by the action of specific enzymes; however, in addition, the principle of safely treating live animals has also been demonstrated. It has been shown by rabbit eye testing that the enzyme mixture for dung removal described below has no irritating effect on mucous membranes, even at 10 times the required concentration;[39] this raises the prospect of spraying or misting cattle in the lairage prior to slaughter, depending on the national regulations for the state of animals presented for slaughter. Some countries allow wet animals to be slaughtered, *e.g.* in Australia, where misting keeps the temperature down in the lairage, but in the UK, for example, animals must be dry.

The analysis of typical dung reveals that almost 80% of the material is lignocellulosic: there is a little protein and some fatty components, indicated by the presence of ether solubles.[40] The composition reflects

Dung ball

region between dung and hide

hide cross section

Figure 2.30 Photomicrograph of dung on cattle hide. Courtesy of the *Journal of the Society of Leather Technologists and Chemists*.

Figure 2.31 Representation of lignocellulose: cellulose fibres are linked by xylan and wrapped around by lignin.

the diet of the animal, because the lignocellulosic content is similar to the composition of plant cell walls, with a slight excess of less easily digested lignin. The low levels of protein and fat explain the lack of efficacy of removal by proteases and lipases. The low solubility in water is shown by the levels of water solubles content. Consequently, it is clear that the removal of dung depends on degrading the main components, which must be the targets for specific enzymes: cellulase, xylanase and ligninase. The need for all three enzymes is illustrated in the representation of lignocellulose in Figure 2.31.[41] Lignin is degraded by ligninase to allow access to the crosslinking xylan, the degradation of which by xylanase releases the cellulose fibres, which are then solubilised by cellulose; the effect can be accomplished just by cellulose with xylanase, but the effectiveness is reduced.

Cellulose is the major polysaccharide of the cell walls of higher plants and is the most abundant polymer in the biosphere. It is a linear polymer made of glucose subunits linked by β-1,4 bonds (Figure 2.32). Each glucose residue is rotated by 180° relative to its neighbours, so that the basic repeating unit is in fact cellobiose.

The degree of polymerisation of cellulose is of the order of 10 000. Cellulose chains form numerous intra- and intermolecular hydrogen bonds, which account for the formation of rigid, insoluble microfibrils. The resistance of cellulose to hydrolysis is caused by both the nature of β-1,4 bonds and the crystalline state; however, β-1,4 linkages can be hydrolysed by cellulases, enzymes produced only by certain bacteria and fungi.[42] Cellulase is a multicomponent enzyme system, generally considered to be composed of three main components: endoglucanase (*endo*-1,4-β-D-glucan 4-glucanohydrolase; EC 3.2.1.4), exoglucanase (cellobiohydrolase; EC 3.2.1.91) and β-glucosidase (EC 3.2.1.21). Fungi tend to produce extracellular cellulase and those from *Trichoderma reesei*, *Trichoderma viride*, *Trichoderma koningii* and *Fusarium solanii* are the best characterised.[42,43] The endoglucanases attack cellulose

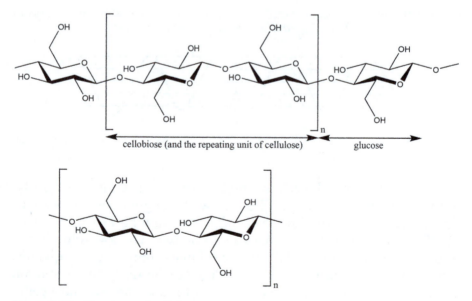

Figure 2.32 Structure of cellulose.

in a random fashion, resulting in a rapid decrease in the degree of polymerisation, and produce oligosaccharide. They have little apparent capacity to hydrolyse crystalline cellulose. The cellobiohydrolases (exoglucanases) are considered to degrade cellulose by removing cellobiose from the non-reducing end of the cellulose chain. β-Glucosidases hydrolyse cellobiose and some soluble cello-oligosaccharides to glucose.[41]

Figure 2.33 shows a generalised scheme for cellulolysis (modified from Eveleigh[44]). Cellulose is initially attacked in amorphous zones by endoglucanases, thus generating multiple sites for attack by cellobiohydrolase. The continued cooperative action between *exo-* and *endo*-splitting polysaccharidases continues, combined with the terminal action of cellobiase, to yield glucose. The cooperative action of cellobiohydrolase and endoglucanase is synergistic. Most individual enzymes do not promote effective hydrolysis, although certain cellobiohydrolysis reactions will completely degrade crystalline cellulose.

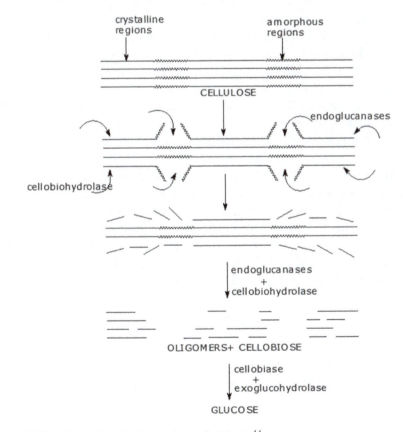

Figure 2.33 Generalised scheme for cellulolysis.[44]

Exo- and *endo*-polysaccharidases can each occur in multiple forms and can exhibit different specificities in relation to the degree of polymerisation of the substrate. Two forms of each endoglucanase and cellobiohydrolase are theoretically possible, due to the two stereoconfigurations of the cellobiosyl unit in the chain.[42]

Hemicellulose is a general term applied to a group of chemically heterogeneous carbohydrates found in the cell walls of plants.[42,45] They are closely associated with cellulose cementing the microfibrils together, and also covalently bound to lignin to give a lignocarbohydrate complex. The hemicelluloses, like cellulose, are polymers of anhydro sugar units linked by glycosidic bonds. Unlike cellulose molecules, hemicellulose molecules are much shorter and are branched and substituted, and as a result they are usually non-crystalline.

Hemicelluloses are classified according to their chemical structure and three predominant types are recognised: xylans, glucans and mannans. The xylans are the predominant type of hemicellulose found in plants. After cellulose, it is the most abundant polysaccharide in Nature. The basic structure of xylans is a main chain of β-1,4-linked D-xylose residues carrying short sidechains (Figure 2.34).

The enzymes capable of degrading xylans include β-1,4-xylanases (1,4-β-D-xylan xylohydrolases; EC 3.2.1.8) and β-xylosidases (1,4-β-D-xylan xylohydrolases; EC 3.2.1.37).[46] In general terms, the xylanases attack internal xylosidic linkages on the backbone and the β-xylosidases release xylosyl residues by endwise attack of xylooligosaccharides, as shown in Figure 2.35.[46,47]

Lignin is the second most abundant organic compound on Earth; it is closely associated with the cellulose fibres and chemically bonded to the hemicellulose.[48] From a chemical point of view, lignin is an amorphous, three-dimensional aromatic polymer, formed at the sites of lignification in plants by enzyme-mediated polymerisation of three substituted cinnamyl alcohols, *p*-coumaryl, coniferyl and sinapyl alcohols,[49] as shown in Figure 2.36.

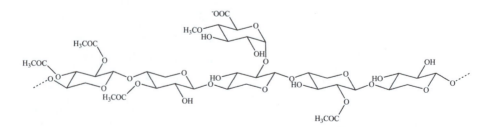

Figure 2.34   Structure of xylan.

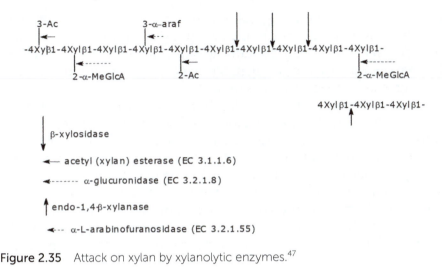

Figure 2.35 Attack on xylan by xylanolytic enzymes.[47]

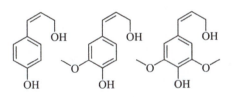

Figure 2.36 Structures of the main constituents of lignins: *p*-coumaryl alcohol (left), coniferyl alcohol (centre), sinapyl alcohol (right).

Certain fungi, mostly basidiomycetes, are the only organisms able to biodegrade lignin extensively; white-rot fungi can completely mineralise lignin, whereas brown-rot fungi merely modify lignin while removing the carbohydrates from the plant. The major enzymes isolated from ligninolytic fungi to affect lignin and lignin-model compounds are laccases (polyphenol oxidases), lignin peroxidases (LiPs) and manganese-dependent peroxidases (MnPs).[50] White-rot basidiomycetes such as *Coriolus versicolor*, *Phanerochaete chrysosporium* and *Phlebia radiata* have been found to secrete typical lignin-degrading enzymes.[51] Using β-1 and β-*O*-4 model compounds for studying lignin degradation, it has been established that both lignin peroxidase and laccase catalyse one-electron oxidations of either phenolic or non-phenolic compounds.[52,53] Lignin peroxidase converts both phenolic and non-phenolic moieties to their corresponding phenoxy radical and aryl radical cation intermediates, whereas laccase catalyses only phenoxy radical formation from phenolic substrates.[52] The aryl radical cations then undergo nucleophilic attack to generate aryl radical intermediates. Such reactive species, and also the phenoxy radicals, can then react with dioxygen radicals, leading to the cleavage of sidechains and aromatic ring opening of the compounds, or alternatively they can undergo free radical coupling reactions.[52,53] Thus two competing reactions can occur during lignin biodegradation: reaction with dioxygen radicals resulting in lignin degradation and free radical coupling leading to repolymerisation.[54]

A structural representation of lignin is given in Figure 2.37 (modified from Eveleigh[44]). The breakdown of lignin is of interest to the leather scientist because many attempts have been made to use the by-products of the Kraft paper process in creating useful syntan tanning agents (see Chapter 14). Although none have been successful so far, the availability of such large quantities of cheap materials that bear a resemblance to tanning agents is an incentive to keep trying.

Solid-state NMR studies[41] showed that treating the dung with either individual enzymes or mixtures of the three enzymes

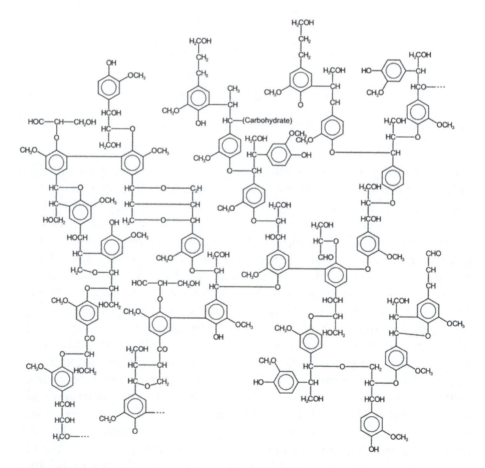

Figure 2.37   Schematic structural formula of lignin.[44]

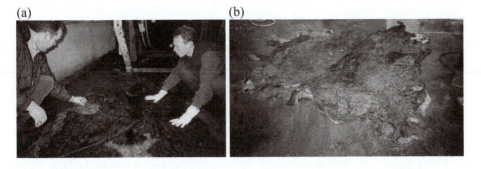

Figure 2.38 Dung-clad hide (a) before and (b) after enzymatic dirt soak.

degrades the dung composition as a whole, so that the effect is to solubilise the material rather than degrade selected components. In this way, the enzyme mixture effectively attacks the adhering mechanism, causing the dung to be removed from the hide intact. This is illustrated in Figure 2.38, in which a salted Irish hide is totally cleaned after only 1 h in the dirt soak treated with the three-enzyme mixture.

The presence of dung has important implications for the economics of leather production, because it affects the timing of the preliminary handling steps in processing. The impact of efficient dung removal is summarised in Table 2.5, in which some of the consequences of soaking technologies are presented.

To put green fleshing technology into context, some results are presented here from an independent cost analysis performed in 2002,[55] assuming complete dung removal in the dirt soak. The cost figures will change depending on inflation, the economic circumstances and the individual tannery, but the relative savings are worth noting and the principles still apply.

1. Throughput
   The presence of dung cladding adds to the hide weight, which must be included when determining the pack weight. It was estimated that the tannery could increase throughput by 15% as a consequence of removing its high level of dung cladding.

$$\text{Annual throughput} = 10000 \text{ hides per week}$$
$$= 5 \times 10^5 \text{ hides per year}$$
$$\text{Increase in throughput} = 7.5 \times 10^3 \text{ hides per year}$$
$$\text{Increase in area yield} = 3 \times 10^5 \text{ ft}^2 \text{ per year}$$

2. Chemical costs

The presence of 15% dung and the associated flesh on the hides is estimated to use up an additional 15% of chemical costs up to wet blue that would be saved.

3. Tallow recovery

The average production of fleshings can be assumed to be about 5 kg per hide, either green or limed. The recovery of tallow from limed fleshings is about 15% and it has reduced value, because of the high free fatty acid content. The recovery of tallow from green fleshings is higher at about 70% and the quality is high, because of the low free fatty acid content.

4. Fleshings disposal

In a total cost analysis of the effects of dung removal, the cost/ saving of disposing of limed fleshings should be included. The yield of fleshings, with its associated disposal cost, can be calculated as 5 tonnes for every 1000 hides processed per week.

Table 2.5 Comparison of soaking options for dung-clad hides.

Processing element	Conventional processing	Enzyme soak
Soaking	No dung removal	Complete dung removal
Fleshing	Not possible	Possible: • High-value by-product, yielding a large quantity of high-quality tallow • Savings in liming chemicals • Facilitates hair saving
Green splitting	Usually not possible	Options available: • Savings in liming chemicals • Green split as protein source
Unhairing	Possibly limited to hair burning: • Environmental impact of chemical oxygen demand (COD) • Contaminated hair, if hair saving	Options available: • Hair burn • Hair save – added value, clean by-product • Hair shaving
Liming	• Conventional, applied to full thickness • May be non-uniform, due to presence of flesh	• Uniform reaction • Better opening up • May be applied only to grain split, if desired
Fleshing	Low-value limed fleshings	Not needed
Splitting	Produces limed flesh split, limiting options of disposal	Conventional production

2.6   Variations in Skin Structure Due to Species

2.6.1   Hereford Cattle and Vertical Fibre

Many breeds of cattle are used in the global leather industry: in any pack of hides several breeds could be represented, but the distribution is not normally taken into account in the processing. There are subtle variations in hide and consequent leather properties that originate with breed or breed crosses.[56] One breed-dependent fault that must be taken into account is called 'vertical fibre', illustrated in Figure 2.39. This is a genetic fault, associated primarily with Hereford cattle and crosses.[57] The fault can be readily diagnosed by the appearance on the flesh side of the pelt of a pattern that looks like

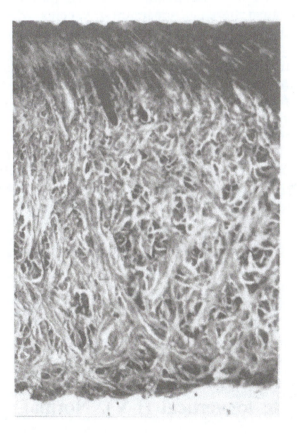

Figure 2.39   Vertical fibre in Hereford cattle. Courtesy of the *Journal of the American Leather Chemists Association*.

the surface of a cauliflower: this originates from the enhanced ease of reagents penetrating through the fibre structure and consequent increased opening up. Because the fibres are much less interwoven than in other breeds, the effect is to cause areas of weakness; no remedial action is possible.

2.6.2 Sheepskin

A feature of sheepskin is the open, loose nature of the fibre structure. Breed differences are clearest in the degree to which the fibre structure is loose and the presence of cutaneous grease in the grain–corium junction. The combination can lead to weakness at the junction, even double skinning separation of the layers. The properties of sheepskin can be compared with those of goatskin, because there is a spectrum of species, with wool sheep and goat at the extremes, and within the range of species there is hair sheep.

From Figure 2.40, it is clear that the preferred qualities in the skin relate inversely to the preferred qualities in the fibre: at one extreme there are merino sheep and crosses, which are primarily farmed for their fine wool, but the skins have limited use, and at the other end of the sheep scale is hair sheep, which have hair of no value, but skin that makes the finest gloving and clothing leather. In the same way, goats range from those with fine hair such as angora, where the fibre is the valued part of the animal, to those that have hair of no value but have very tough skin for

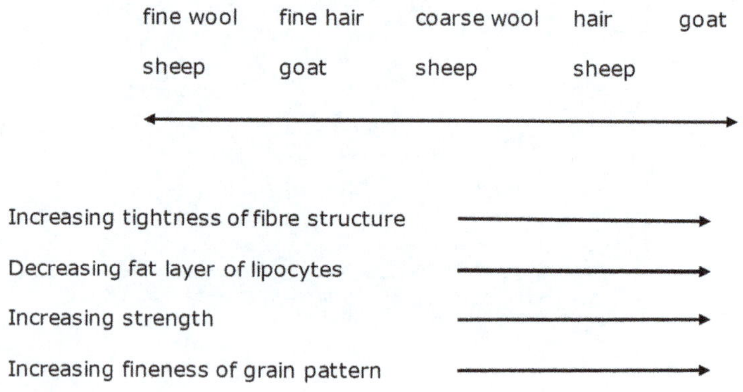

| fine wool | fine hair | coarse wool | hair | goat |
| sheep | goat | sheep | sheep | |

Increasing tightness of fibre structure

Decreasing fat layer of lipocytes

Increasing strength

Increasing fineness of grain pattern

Figure 2.40 Properties of sheep and goats.

making hard-wearing leather. Globally, the most important source of hair sheep is Ethiopia, providing the preferred raw material for gloving leather: the skin is characterised by the tight fibre structure, more like goatskin than sheepskin, but with the softness of sheepskin.

Note that some breeds of sheep and goats can look remarkably similar and the way to tell the difference between them is that sheep carry their tails down whereas goats hold their tails up.

2.6.3 Cutaneous Fat

Wool sheepskins, such as UK domestic, New Zealand or Australian, have a fat layer in the grain–corium junction, shown in the photomicrograph presented in Figure 2.41:[58] the structure is in the form of

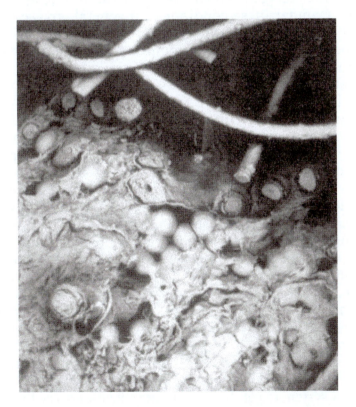

Figure 2.41 Scanning electron photomicrographs of sheepskin lipocytes. Courtesy of the *Journal of the Society of Leather Technologists and Chemists.*

large lipocytes, closely associated in a mass, which can act as a barrier to the penetration of aqueous reagents. The amount of grease is variable, breed dependent, and may range from 1 to 30% of the skin weight.

It is important that the grease should be removed from all hides and skin in which it occurs, not only to allow uniform reactions through the skin cross-section, but also to avoid odour problems if the fat turns rancid. For sheepskins, the traditional method is to treat pickled pelt with warm paraffin, then rinse with brine. This efficient method of degreasing has fallen into disrepute, owing to the environmental impact of using an organic solvent. An alternative is to use aqueous detergents. However, since the grease has a melting point of about 42 °C, processing at a temperature when the grease is molten is too dangerous, owing to the risk of heat damage to the collagen, which has a denaturation temperature of about 60 °C under these conditions. In general, grease dispersal below its melting point is inefficient, so aqueous degreasing of pickled pelt has not yet reached a universally satisfactory stage of development.

An alternative option is to use a lipolytic enzyme, lipase, to degrade and self-emulsify the grease by partially hydrolysing it.[6] Unfortunately, it is difficult for the enzyme to reach its substrate, because of the barrier formed by the lipocytes' cell walls. This can be overcome by using a formulation of lipase with protease, so that the protease can break down the protein component of the cell wall structure, to release the contents. The disadvantage of this biotechnology is that the prolonged treatment with protease causes too much damage, mostly to the corium, resulting in looseness, but also possibly in damage to the grain. A newer and more appropriate approach is to formulate the lipase with phospholipase, to target the other component of the cell wall structure, phospholipids.[58] The principle of the technology has been demonstrated, but the phospholipase enzymes are not available in industrial quantities and at a suitable price for economic application.

An option that has not been tested is the role of ionic liquids in this context. Here there are two possibilities: first, they can offer a different delivery system to get the appropriate enzymes to the lipocytes, or even into the lipocytes, and second, they may offer a new mechanism for releasing the grease and transporting it out of the skin.[59] See also Chapter 19.

References

1. J. H. Sharphouse, *Leather Technician's Handbook*, Leather Producers Association, Northampton, 1971.
2. K. Bienkievicz, *Physical Chemistry of Leather Making*, Krieger, Florida, 1983.
3. G. E. Attenburrow, A. D. Covington and A. J. Long, *J. Soc. Leather Technol. Chem.*, 1999, **83**(3), 167.
4. G. E. Attenburrow, A. D. Covington and A. J. Long, *J. Am. Leather Chem. Assoc.*, 1998, **93**(10), 316.
5. J. A. M. Ramshaw, *Connect. Tissue Res.*, 1986, **14**(4), 307.
6. V. L. Addy, A. D. Covington, D. A. Langridge and A. Watts, *J. Soc. Leather Technol. Chem.*, 2001, **85**(1), 6.
7. G. E. Attenburrow, *J. Soc. Leather Technol. Chem.*, 1993, **77**(4), 107.
8. G. D. McLaughlin, *et al.*, *J. Am. Leather Chem. Assoc.*, 1929, **24**(7), 339.
9. J. Fennen, *et al.*, *Proceedings IULTCS Congress*, Novo Hamburgo, Brazil, 2015.
10. M. Aldemir, MSc thesis, University of Northampton, 1999.
11. S. P. Robins, *Baillieres Clin. Rheumatol.*, 1988, **2**(1), 1.
12. J. M. Golsline and J. Rosenbloom, Elastin, in *Extracellular Matrix Biochemistry*, ed. K. A. Piz and A. H. Reddi, Elsevier, Amsterdam, 1988.
13. G. Francis, R. John and J. Thomas, *Biochem. J.*, 1973, **136**, 45.
14. P. D. Kemp and G. Stainsby, *J. Soc. Leather Technol. Chem.*, 1981, **65**(5), 85.
15. K. T. W. Alexander, *J. Am. Leather Chem. Assoc.*, 1988, **83**(9), 278.
16. P. L. Kronick and S. K. Iondola, *J. Am. Leather Chem. Assoc.*, 1998, **93**(5), 139.
17. M. A. R. Siddique, *et al.*, *J. Soc. Leather Technol. Chem.*, 2015, **99**(2), 58.
18. M. L. Aldema-Ramos and C. K. Liu, *J. Am. Leather Chem. Assoc.*, 2010, **105**(7), 222.
19. H. S. Raper, *Physiol. Rev.*, 1928, **8**, 245.
20. H. S. Mason, *J. Biol. Chem.*, 1948, **172**, 83.
21. S. Ito, Biochemistry and physiology of melanin, in *Pigmentation and Pigmentary Disorders*, ed. N. Levine, CRC Press, Boca Raton, 1993.
22. V. Hearing and M. Jimenez, *Int. J. Biochem.*, 1987, **19**, 1141.
23. A. Thompson, *et al.*, *Biochim. Biophys. Acta*, 1985, **843**, 49.
24. G. Prota, Structure and biogenesis of phaeomelanin, in *Pigmentation: Its Genesis and Control*, ed. V. Riley, Appleton-Century-Crofts, New York, 1972.
25. R. H. Thompson, *Angew. Chem., Int. Ed. Engl.*, 1974, **13**, 305.
26. G. Prota, *Melanins and Melanogenesis*, Academic Press, San Diego, 1992.
27. D. Sica, *et al.*, *J. Heterocycl. Chem.*, 1970, 7, 1143.
28. G. Nalbandian, BSc thesis, University of Northampton, 2001.
29. V. Punitha, *et al.*, *J. Am. Leather Chem. Assoc.*, 2008, **103**(7), 203.
30. G. O. Conavere, *et al.*, *Progress in Leather Science: 1920–45*, British Leather Manufacturers' Research Association, 1948, 34.
31. A. D. Covington, *UK Pat. Appl.*, GBA 0110695.4, May 2001.
32. M. Schropfer, *et al.*, *J. Am. Leather Chem. Assoc.*, 2014, **109**(9), 306.
33. M. Deselnicu, *et al.*, *J. Am. Leather Chem. Assoc.*, 1994, **89**(11), 352.
34. J. Mitchell and D. G. Ouellette, *J. Am. Leather Chem. Assoc.*, 1998, **93**(8), 255.
35. K. Filiz, MSc thesis, University of Northampton, 1999.
36. G. E. Attenburrow, *J. Soc. Leather Technol. Chem.*, 1993, **77**(4), 107.
37. G. E. Attenburrow and D. M. Wright, *J. Am. Leather Chem. Assoc.*, 1994, **89**(12), 391.
38. R. M. Marsic and C. K. Liu, *J. Am. Leather Chem. Assoc.*, 2017, **112**(3), 88.
39. A. D. Covington and C. S. Evans, Klenzyme Ltd., unpublished results.
40. N. Auer, *et al.*, *J. Soc. Leather Technol. Chem.*, 1999, **83**(4), 215.

41. A. D. Covington, C. S. Evans and M. Tozan, *J. Am. Leather Chem. Assoc.*, 2002, **97**(5), 178.
42. G. O. Aspinall, *Polysaccharides*, Pergamon, Oxford, 1970.
43. T. M. Wood, *Biochem. Soc. Trans.*, 1985, **13**, 407.
44. D. E. Eveleigh, *Philos. Trans. R. Soc. London*, 1987, **A321**, 425.
45. W. B. Betts, *et al.*, in *Biodegradation: Natural and Synthetic Materials*, ed. W. B. Betts, Springer-Verlag, London, 1991.
46. K. K. Y. Wong, L. U. L. Tan and J. N. Saddlar, *Microbiol. Rev.*, 1988, **52**, 305.
47. P. Bajpai, *Adv. Appl. Microbiol.*, 1997, **43**, 141.
48. R. L. Crawford, *Lignin Biodegradation and Transformation*, Wiley-Interscience, New York, 1981.
49. C. S. Evans, in *Biodegradation: Natural and Synthetic Materials*, ed. W. B. Betts, Springer-Verlag, London, 1991, ch. 139.
50. C. S. Evans, *et al.*, *FEMS Microbiol. Rev.*, 1994, **13**, 235.
51. U. Tuor, K. Winterhalten and A. Fiechter, *J. Biotechnol.*, 1985, **41**, 1.
52. T. Higuchi, *ACS Symp. Ser.*, 1989, **399**, 482.
53. T. Higuchi, *J. Biotechnol.*, 1993, **30**, 1.
54. E. Adler and T. M. Wood, *Sci. Technol.*, 1977, **11**, 169.
55. A. D. Covington, C. S. Evans and M. Tozan, *Proceedings IULTCS Congress*, Cancun, Mexico, 2003.
56. B. M. Haines, *J. Soc. Leather Technol. Chem.*, 1981, **65**(4), 70.
57. W. Pitchford, *et al.*, *J. Am. Leather Chem. Assoc.*, 2000, **95**(3), 85.
58. V. L. Addy, *et al.*, *J. Soc. Leather Technol. Chem.*, 2001, **85**(1), 6.
59. University of Leicester and University of Northampton, *Int. Pat.*, WO2015/159070 A1, 2015.

3 Curing and Preservation of Hides and Skins

3.1 Introduction

Preservation of rawstock has the objective of rendering the flayed pelt resistant to putrefaction, to allow transport and storage. Preservation is often referred to as 'curing', which implies some form of chemical treatment, although the terms are often used interchangeably. Preservation is accomplished either by destroying active bacteria, by preventing bacterial activity or by preventing bacterial contamination. Ideally, any treatment applied to a pelt should be reversible, without altering the properties of the pelt. This clearly eliminates any sort of tanning process as a curing option.

The mechanism of putrefaction is the production of proteolytic enzymes by bacteria. The source may be the natural saprophytic bacteria within the pelt, present in the living animal, where the purpose is to degrade and remove dead tissue. Following the death of the animal, these bacteria cause autolysis (self-hydrolysis) of the collagen. This breaks down the protein to amino acids, which further break down or mineralise, ultimately to produce carbon dioxide and ammonia. Alternatively, opportunistic bacteria from the environment may contribute to putrefaction. Other reactions include the hydrolysis of triglyceride fat to free fatty acids and glycerol by lipases and the breakdown of carbohydrates into sugars by carbohydrase enzymes.

Tanning Chemistry: The Science of Leather 2nd edition
By Anthony D. Covington and William R. Wise
© Anthony D. Covington and William R. Wise 2020
Published by the Royal Society of Chemistry, www.rsc.org

It is important to recognise the effect of 'staling', defined as the delay between flay and cure, when bacterial activity may proceed. Ideal practice in the abattoir would be to spread the newly flayed pelt on concrete, to remove the body heat as rapidly as possible, then to put it into cure or into processing as soon as possible. It has been shown that a delay of as little as 24 h will guarantee some grain damage.[1] This can be appreciated when it is known that the bacteria on hides can double in number in less than 4 h at 25 °C.[2] In reality, hides and skins are typically merely held in piles prior to curing or processing, so they are effectively kept warm – this is a reflection of the abattoirs' view that hides and skins are only a by-product of the meat industry, with little associated value to them compared with the value of the carcases (an argument often not recognised by opponents of the international leather industry). Consequently, because some incipient damage is often caused, whatever curing system is used, the process can only halt bacterial action and the damage caused after flaying cannot be reversed.

The following conditions favour putrefaction, so each may offer options for approaches to preservation.

- pH 6–10, preferably 7–9.
  Note that physiological pH is generally accepted to be 7.4, when enzymes in the body will function optimally. Altering the pH of the pelt might appear to be a sound option for preservation. However, acidification can result in acid swelling if there is no electrolyte present. In the case of wet salting (see later), acidic additives may be included. Alkali treatment is limited, because at pH 10 and above there is a risk of immunising the hair (see Chapter 4) and adversely affecting hair removal. Some alkaline additives for salt are known: it is not uncommon for soaking to be conducted at pH 9–10, partly to aid rehydration, but also in the expectation of reducing bacterial and enzyme activity. This cannot be relied upon.
- The presence of nutrients: protein, fat, sugar, blood, dung.
  The only way to reduce the availability of nutrients is to process immediately, *i.e.* to eliminate storage, by processing as soon as possible after flaying.
- Moisture content >25%.
  Fresh hide contains 60–70% moisture. The reduction of moisture is an important feature of preservation methods, so drying itself can be used for preservation, as discussed in Section 3.2. It also plays a role in salting, because of the osmotic effect of the sodium chloride dissolving in the water content of the pelt: the salt penetrates into the

pelt and the water moves out of the pelt as the mechanism to reach equilibrium (see Section 3.3). See also Chapter 9, in which the role of osmosis and the Donnan equilibrium condition play their parts in preventing swelling in pickling.

- Temperature >20 °C, particularly 30–40 °C.
  Controlling the temperature of the storage environment is an important feature of preservation methods. This can be done in several ways and to different extents (see Section 3.6).
- Calcium and magnesium ions in curing salt favour the growth of halophilic bacteria (see later).

It is useful to consider the features of the growth of bacteria, which apply both *in vitro* and to raw pelt *in vivo*, as illustrated in Figure 3.1. An exponential growth phase occurs when nutrients are plentiful, which is likely to be the usual scenario for hides and skins in storage prior to processing; the stationary phase refers to diminishing nutrients and this would only apply when the pelt is nearing complete degradation. Therefore, if the conditions of pH, temperature, moisture and lack of protecting chemicals are right for bacteria, they will soon reproduce at an exponential rate, thereby rapidly creating sufficient enzyme to cause noticeable damage.

Both fresh and salted hides have been shown to be contaminated with *Bacillus* sp., *Micrococcus* sp. and *Staphylococcus* sp., but the greatest danger of damage seems to be from *Bacillus* sp.[3] The

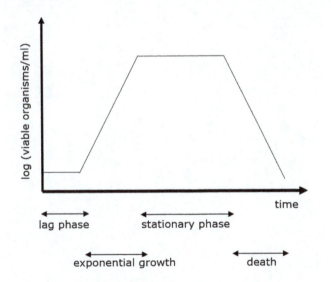

Figure 3.1 Bacterial growth phases: slow growth, followed by exponential rate of growth, a steady state of viable bacteria and then death.

observation of bacteria within the pelt is the clearest sign of pelt status and deteriorating quality: this can be done by fixing a sample of the pelt in formaldehyde or some other suitable agent, sectioning it for visualising the bacteria with Alcian Blue and observing under a light microscope, as illustrated in Figure 3.2. Debris in the pelt can also take up the stain and may be confused with bacteria; however, their identification is aided by the knowledge that bacterial growth occurs from the flesh side and tends often but not always to be seen as resembling rows of beads along the fibres. Bacteria only at the flesh surface are normal and not threatening, but bacteria within the corium are a danger sign: if they have reached the grain–corium junction, grain damage in the final leather is inevitable. It should be noted that the presence of bacteria is not a problem in itself, it is the proteolytic enzymes that the bacteria excrete; hence the danger zone associated with bacteria extends beyond the immediate vicinity of the cells themselves.

The presence of bacterial danger can be demonstrated by the gelatine film test.[4,5] This is based on obtaining a sample of liquid from the pelt (the pressure of a vice may be required to squeeze enough moisture out of the relatively dry pelt), then applying it to commercial monochrome photographic film (although this is likely not to continue to be a cheaply available testing material). After incubating

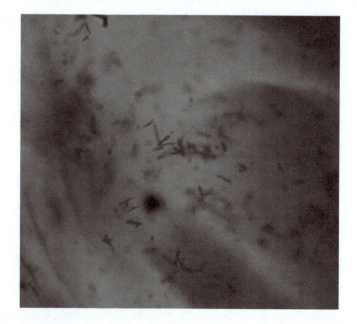

Figure 3.2   Bacteria in skin. Courtesy of Leather Wise, UK.

Table 3.1 Analytical results for salted hides.

Analysis	Immediately after salting	After typical storage
Volatile N per total N/%	0.4	0.75
Soluble N per total N/%	2.0	3.0

the treated film at about 37 °C for about 1 h, the film is washed with warm water to remove any degraded gelatine, which is easily seen. The test is not completely reliable, as it may yield false-negative results. However, it can be useful in the factory and may even offer insight into degrees of contamination, particularly if multilayered film is used.

The effects of inadequate preservation are as follows.

- Loss of structural fibre, shown by soluble nitrogen analysis.[6]
  As bacterial enzymes attack the skin components, particularly the protein, there is an increase in both volatile and soluble nitrogen; the former is essentially ammoniacal nitrogen and the latter is degraded protein. Table 3.1 provides some guideline values.

  If the measured levels are significantly higher than these values, there is evidence that the pelt has deteriorated, although the tests may not be able to discriminate between pre- or post-cure damage, so other clues need to be sought. The damage to the protein of the pelt is also apparent by the looseness of the fibre structure, particularly apparent in the looser flanks.

  Chapter 8 refers to a method for monitoring the progress of enzymes through the pelt, using the biuret reaction [basic copper(II) sulfate] to detect protein fragments from degradation; this colour test has been tried on stored and staled rawstock, but the method is too insensitive to indicate bacterial damage.[7]
- Damage to the grain occurs if the bacteria reach the grain–corium junction, because the enzymes can extend through the grain to the enamel–epidermis junction. Hence the signs of damage include lifting of the epidermis, loosening of the hair and loss of enamel, only visible after unhairing.
- Increase in free fatty acids, seen as yellowing of the fat. Fat breakdown has been proposed by Koppenhoeffer as a measurement of pelt quality. Within 30 days, the phospholipid is completely hydrolysed, in 30 days there is 15% loss of triglyceride and in 6 months there is 67% loss of triglyceride. The degree of hydrolysis of the triglyceride fat may give an indication of the history of the pelt, but does not yield information about the status of the protein. Oxidative rancidity does not become apparent until after 6 months of storage.[8]

- Prominent vein marking can result from loss of hide substance, which makes the intact veinous system more apparent at the surfaces.

It should be noted that the industry distinguishes between long- and short-term preservation. The former is exemplified by drying and salting, when the qualities of the rawstock can be retained over a period of many months, depending on the conditions of storage. Short-term preservation is usually thought of as a period of 1–4 weeks without deterioration, suitable for transporting the rawstock, for example from north to south Europe.

3.2   Drying

Drying hides and skins to preserve them was once common practice and the products were traded globally. However, the practice has become less common, abandoned in favour of salting, which is easier and therefore more reliable. The use of drying is limited to those countries where the climate is conducive to drying, *i.e.* the tropics, and in countries where salt is either scarce or expensive. A temperate climate is typically not warm enough to effect efficient and effective drying. There is a balance to be struck between the rate of drying and the likely rate of bacterial growth, which can be expressed as follows.

If the rate of evaporation of moisture from the surface is k_e and the rate of diffusion of moisture from the centre of the pelt to the surface is k_d, there are three possible situations and a fourth related relationship.

1. $k_e < k_d$
   Here the rate of evaporation is very slow, so that the equilibration between the moisture contents in the centre and at the surfaces is maintained. This means that the overall process is slow, so the likelihood of bacterial proliferation and consequent damage is high.
2. $k_e = k_d$
   In this case, the rates of evaporation and diffusion are the same, which means that the efficient removal of moisture is optimised. The hides or skins may be dried on the ground (flint drying) or suspended on wires, ropes or poles; alternatively, they may be

loosely fixed (nailed) to wooden frames. In all cases, they must be dried in open-sided sheds, for good ventilation, but protected from direct sunshine and rain. The reason for shade drying is that fast drying must be avoided.

3. $k_e > k_d$

   If the rate of evaporation is faster than the rate of diffusion, the outer layers may dry completely, causing fibre structure sticking, thereby preventing moisture from escaping at the surfaces, referred to as an example of 'case hardening'. In turn, this means that the inner layer remains wet and the conditions of moisture and warmth favour bacterial growth. The combination of bacterial degradation of the skin and the high ambient temperature may even cause gelatinisation in the centre of the hide or skin; this is usually not obvious during the preservation process, but becomes apparent when the pelt is rehydrated. When such hides are rehydrated in soaking, they can delaminate, separating into two outer layers: 'double hiding' or 'double skinning'. The damaged inner surfaces must be removed mechanically before tanning, because on drying the degraded collagen would 'glue' the surface, making it brittle and cracky.

4. $k_e \not< k_p$

   This is the requirement that the rate of evaporation is faster than the rate of bacterial proliferation, which in this case refers to the rate at which bacteria can grow from the flesh surface to the grain layer – bacterial growth the other way is hindered or prevented by the presence of the epidermis.

This apparently simple and 'low-tech' approach to preservation is actually difficult to control. The rate of evaporation of moisture from the surface depends on the following parameters: the surface characteristics, the air temperature, the surface temperature and the rate of air flow over the surface. The surface characteristics are created by the nature of the animal's skin and the impact of flaying practices. The air temperature is a weather feature and the surface temperature is moderated by the radiative conditions, hence the requirement for drying in the shade. The air flow is a weather feature, but may be controlled by the design of the facility and by the use of fans; the latter clearly has implications for capital and energy costs. In this way, process control can be exerted only indirectly and there is no simple way of checking successful progress (Table 3.2).

Table 3.2 Advantages and disadvantages of air drying as a preservation method.

Advantages	Disadvantages
Low capital cost – simple equipment and housing	Weather problems, *e.g.* during the monsoon Growth of bacteria, causing putrefaction, especially if drying is too fast at high temperature
Low running cost – free source of energy	Susceptibility to damage by insects, rats, *etc.* Difficult to rehydrate – must be done slowly
Lower freight charges – moisture content 10–14%	Difficult to assess quality of cure Storage in dry conditions required

Dried hides or skins must be folded while damp, to avoid cracking the dried pelt. Similarly, it is important to handle and process-dried hides or skins with care: the first part of the soak must be conducted under static conditions, to avoid creasing or cracking, then the remainder must be performed for a sufficiently long period to rehydrate to equilibrium. In both parts of the soak, the pelt must be protected by a bactericide. The horny appearance of dried hide does not at first sight indicate the easy reversibility of the preservation method. However, it is reversible because of the presence of the ground substance; this is a term used collectively for all the non-collagenous components of skin, the non-structural proteins, the glycosaminoglycans and the fats. These components protect the collapsed fibre structure from sticking, acting as a rehydratable barrier between the elements of the hierarchical structure. In contrast, if partially processed pelt is allowed to dry to this extent, the triple helices can approach close enough for new chemical bonds to form, making the structure resistant to rehydration.[9] If the drying process has been conducted well and the rawstock is adequately stored, particularly with regard to keeping it dry, it can be kept for many months. This means that the method can confer long-term preservation.[10]

There are variations in the dehydrating approach to preservation, based on partial dehydration.

1. Sawdust can act as a dehydrating agent, but the weak effect limits its consequence to preservation for only a few days, before bacterial activity causes damage, with or without added biocide. If the process is compounded by the presence of electrolyte, the preserving power is extended to 1 month. The Liricure process (developed by the Leather Industries Research Institute of South Africa, since closed) was a formulation of

35% sawdust, 40% sodium chloride and 25% tetrasodium eth-ylenediaminetetraacetate (EDTA), applied at the rate of 150 g per sheepskin.[11] Here, the sawdust acts as a drying agent, the salt acts as a drying and osmotic agent and the EDTA acts as a sequestrant for the metal component of metalloproteinases, enzymes commonly associated with putrefaction and which rely on the presence of a metal ion for their activity. If the metal ion is removed by a sequestering agent, the metallopro-teinases cannot operate in the same way, losing their catalytic function.

2. Silica gel can be used, as proposed by Indian workers.[12] This reagent, familiar as a laboratory desiccant, was used at up to 5%, together with 5% salt on pelt weight, with or without *p*-chloro-*m*-cresol. During the treatment, the moisture content of goatskins decreased from 65–70% to about 35%, conferring preservation for 2 weeks.

3. Superabsorbent acrylate polymer, the type of polymer used in babies' nappies/diapers, can be used in a similar way to silica gel. The reagent is effective at 3% on pelt weight, where the drying effect is similar to the effect of 45% sodium chlo-ride. This treatment is suitable for 1–2 weeks' storage, with easy rehydration at the end. Higher offers of polymer can cause enough dehydration for long-term preservation.[13] The cost per application is comparable to that with salt and the acrylic polymer is biodegradable, so the technology appears feasible, but has not been trialled at an industrial scale.

3.3 Salting

The use of salt for preserving foodstuffs is ancient and, globally, it is still the most common way of preserving hides and skins. In the context of clean technology, it is useful to calculate the amount of salt used in the global leather industry. If the global annual kill of cattle is 300 million and each hide requires an average of 10 kg of salt, the annual use of fresh salt is 3 million tonnes. If other rawstock is added to the calculation, at least another million tonnes is required (skins require an average of 1 kg of salt).

The status of the salt cure can be assessed easily by measuring the quantity of salt in the pelt, by extracting a sample of pelt into 0.2 M sodium acetate solution, then titrating the solution with 0.1 M

Table 3.3 Disposal of preserving salt.

Option	Pros	Cons
Shake off excess salt from hides	Saves ~10%	Salt is dirty, reuse is counter-productive disposal?
Solar pans for soak liquors	Separates salt from soak water by evaporation	Salt is even dirtier Cannot be reused Disposal to where?
Reverse osmosis	Allows recovery for reuse	Expensive – contamination of filtration materials
Landfill: solid or sludge	Simple, inexpensive	Water table contaminated
Disposal to sea	Simple, inexpensive	All marine life destroyed
Treat and reuse	Ecologically sound (apparently)	Expensive Use of energy, chemicals for sterilisation

silver nitrate solution, using sodium chromate as indicator. By determining the moisture content, the salt concentration in the pelt moisture can be calculated. The salt level should be not less than 90% saturation. If there is excess salt on the pelt, it should be removed as far as possible before conducting the analysis. Excess salt may indicate an adequate cure, but it is no guarantee that the curing was carried out properly or that the pelt was in good condition at the time of curing.

'Wet salting' is easy and convenient: the term refers to the application of dry sodium chloride to the freshly flayed hide or skin, which is wet with the natural moisture content. The only requirements for successful treatment are threefold: sufficient salt, the right kind of salt, distributed evenly over the pelt. Thereafter, storage is simple, merely requiring that the salt is not washed off with rain. However, all the advantages are outweighed by the disadvantage of the problem of disposal of the salt when the pelt is processed.

Consider the extent of the problem of neutral electrolyte and the options for its disposal, summarised in Table 3.3. The problem with salt stems from the consequences of allowing it to contaminate the land: the majority of plants cannot tolerate even brackish water, certainly not most food crops. Moreover, contaminated land becomes infertile, because the sodium ions will strip out the mineral micronutrients. A quick assessment of the simple options as set out in Table 3.3 reveals that there is no ecologically sound method of treating salt or disposing of waste salt. Couple that with the situation that salt is inexpensive, of the order of $100 per tonne, there is no economic incentive to recover salt for reuse. Current technologies include lagooning of

Figure 3.3 Perforated drum, for desalting preserved rawstock, demonstrated by Dr S. Rajamani (centre).

soak liquor, to allow solar evaporation of the water content, but that just leaves extremely dirty, unusable salt. Excess salt can be removed from pelts prior to soaking, using a perforated drum, as illustrated in Figure 3.3; the product is still contaminated salt, with limited use if it is untreated and no more ecologically disposable than clean salt. Dirty salt can be cleaned up, the chemistry is simple, but there is an associated cost, so nobody does it. Therefore, waste salt is a material that has truly negative value. Regardless of all the other perceived environmental issues, this is possibly the greatest problem facing the global leather industry.

Wet salting is typically achieved by distributing salt over static pelt. Spreading it by kicking a pile across the pelt on the ground is the common practice, but a more automated approach is to present the pelt by conveyor to a curtain of fluidised solid salt. In each case the

mechanism is essentially the same. The osmotic effect of a high concentration of salt on the outside of the pelt, with the surface acting as
a semipermeable membrane, causes the water in the pelt to migrate
to the outside and the salt to migrate to the inside. The removal of
water reduces the viability of bacterial activity and the high concentration of salt within the pelt creates an osmotic effect at the cell wall of
the bacteria, reducing its viability further. It is necessary to achieve at
least 90% saturation (*i.e.* >32% w/v in the moisture in the pelt, about
14% on hide weight) within the pelt, in order to ensure there is no
bacterial activity. Note that the effect of salt is not to kill the bacteria:
it does not act as a bactericide, but it does act as a bacteriostat. Collagenolytic activity is still rapid at 7% salt on hide weight, but low at
10%, although the bacteria still function.[14] This means that the bacteria are still alive, still viable and can be reactivated when the pelt is
rehydrated. Hence the problem of putrefaction can reappear during
soaking.

If the salting is done well, with regard to sufficient salt of the right
quality distributed over the whole surface area, and storage avoids
wetting the pelt, the rawstock can be kept for several months.

3.3.1 Stacking

This is the commonest method of wet salting. The freshly flayed hides
or skins are spread out on the ground, flesh side up, then salt is sprin-
kled or spread out over the whole of the surface. Alternatively, the salt
may be applied automatically, using a conveyor and automatic dosing.
In both cases, as a guide, hides are treated with about 25–30% salt,
typically about 10 kg; skins are treated with up to 40–50% of salt, typically about 1 kg. The salted pelts are stacked in piles 1.5–2.5 m high,
often folded into quarters, with additional salt in between each layer
and on the top of the pile. The piling helps to squeeze brine out of
the pelts, leaving 40–50% moisture. The distribution of salt is close
to equilibrium through the cross-section after 24 h and equilibrium is
effectively reached in 48 h.[14]

3.3.2 Drum Curing

The freshly flayed hides or skins are tumbled in a closed drum with
30–50% salt, which may contain an additive, so distribution is guaranteed. The rotational speed is kept low and the temperature is limited
to 25 °C. The process takes 6 h for hides and 3 h for skins. Compared
with static salting, it is less common to conduct salting in a drum:[15]

this produces more rapid saturation of pelt moisture than stacking, effectively complete in 4 h, compared with a required period of 1–2 days for the static process.

3.3.3 Types of Salt

Regardless of the type of salt used, there are necessary limits to the impurity content of the salt: calcium and magnesium 0.1% and iron 0.01%. Ca^{2+} and Mg^{2+} can react with phosphate and carbonate in the pelt, fusing the fibre structure and creating a fault known as 'hard spot', more usually found in calfskins.

3.3.3.1 Vacuum Salt

This is a high-purity salt (99.9% NaCl) and therefore a relatively expensive product, most commonly used as a reagent in leather making, *e.g.* for making brine solution. It is characterised by its small crystal size, <1 mm, and free-flowing nature. The crystal size makes vacuum salt unsuitable for preservation, because in the presence of moisture the crystals fuse and grow. This effect reduces the surface free energy and the result of this mechanism is that the salt can create a fused mass, where the area of contact between the salt and the pelt surface is actually reduced, therefore slowing the preservation process and consequently increasing the risk of bacterial growth.

3.3.3.2 Granular Salt

Granular salt is also of high purity, but the crystal size is larger, 2–4 mm, and it is cheaper than vacuum salt. The crystal size means the rate of crystal growth is slower and contact with the surface is maintained as it is dissolved into the pelt's moisture. The main use for this grade of salt is rawstock preservation.

3.3.3.3 Rock Salt

This grade of salt is mined and crushed to the required crystal dimensions. There may be appreciable concentrations of calcium and magnesium salts. The coarse crystals tend to have sharp edges, so indentation and cutting may occur during storage, making this grade of salt unsuitable for curing purposes.

3.3.3.4 Marine Salt

This grade of salt is prepared by evaporating sea water, a common source of salt for many purposes; consequently, it is highly contaminated with calcium, magnesium and other ions. Unlike sea salt prepared for the food industry, which is sterilised by the boiling and evaporation processes, the salt destined for industrial applications is evaporated by solar action and is likely to be contaminated with viable bacteria: these are halophilic (and halotolerant) bacteria, archaebacteria that evolved before mammals, examples of extremophiles, requiring extreme conditions for survival. Halophilic bacteria require saturated salt solution for growth: 5 M, 30% solution.[16] They are coloured pink, red or purple, having in their constitution rhodopsin, a molecule capable of switching between yellow and purple conformations (Figure 3.4). This molecule is also involved in vision, capable of changing colour in 10^{-15} s, the cause of 'red eye' in flash photography.

The colour gives rise to the effect known as 'red heat', although the effect has nothing to do with a rise in temperature. Proliferation of these bacteria in their aerobic phase causes a red coloration over the flesh side. This in itself does not necessarily create a problem, because it is not indicative of a rapidly damaging phenomenon, since the temperature of storage is usually sub-optimal. However, its appearance is a warning of a problem coming, especially damage to the grain.[17] If the bacteria within piled hides change to the anaerobic phase, the colour changes to purple, referred to as 'violet heat'. At this point, the damage occurs much faster, so that by the time it is observed, often in the centre of piles of hides or skins, severe damage has already been done, effectively rendering the rawstock useless and valueless. Halotolerant bacteria are colourless, and capable of growing in 20% salt solution, so they are more likely to thrive when the cure is incomplete for whatever reason. Therefore, marine salt is strictly unsuitable for preserving rawstock long term, but it is better than no curing at all and is useful for shorter than long-term preservation.

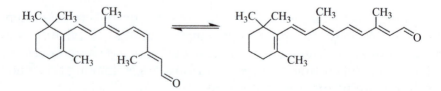

Figure 3.4 Isomers of rhodopsin.

3.3.3.5   Khari Salt

Strictly, this material cannot be considered as a grade of salt. It is the efflorescence found on the surface of soil after rain, especially after the monsoon. It is caused by ground/soil salts dissolving in rain water and migrating to the surface, where they are left after the water has evaporated. Khari salt is very impure, composed of variable amounts of sodium, magnesium, chloride, sulfate, bicarbonate and carbonate ions, typically a mixture of sodium and magnesium sulfates.[18] It is also contaminated with halophilic and halotolerant bacteria. The presence of alkaline impurities causes protein hydrolysis under warm conditions. These characteristics make this grade of salt generally unsuitable for rawstock preservation; however, it is still used in the Indian sub-continent, where it is probably better than no salt at all.

At the first sign of red heat, the problem can be effectively eliminated by starting the normal processing. By initiating the soaking step, the salt concentration around the rawstock is diluted below the critical concentration and the cells are killed. If required, the rawstock can be resalted safely with granular salt and preservation is restored.

3.3.3.6   Leaf Salt

Some plants can accumulate salt, so they become a source of salt, as an alternative to the options already set out. *Sesuvium portulacastrum* is such a halophyte, containing 18% sodium and 15% chloride on dry weight. Dried powder from the leaves, reconstituted as a paste with 20% water, was tested at 5–40% offer on pelt weight against 40% salt. At 10%, the moisture content of the pelt was lowered to 28% after 30 days, which is the same as the effect of salt alone after 14 days. Over the first 48 h, the bacterial content was reduced, an effect ascribed to the terpene content of the leaf.[19]

The technology has not been tested on a large scale; success is clearly dependent on the availability of the salty plant material. However, it is not a solution to neutral electrolyte in the effluent, but may have some usefulness for the developing economies.

3.3.4   Additives for Salt

In order to enhance or guarantee the action of salting or to discourage damage by insects, additives may be included in the salt for preservation. Some additives that have been used in the past include[17] sodium carbonate, sodium pentachlorophenate, sodium silicofluoride, boric acid, zinc sulfate and naphthalene (often formulated with the other salts in the list). The additives might be used to alter pH, to the alkaline or acidic side, or function as insecticides or bactericides, with different degrees of effectiveness. In modern processing, most of those traditional additives are no longer used owing to their toxicity or environmental impact. Modern additives include sodium metabisulfite and approved bactericides (see later).

In processes analogous to that outlined in Section 3.3.3, leaf materials have been used together with salt, effectively as plant-based additives, to exploit the bactericidal effect of the terpene content.

1. *Clerodendrum viscosum* as a 10% paste with 0–15% salt was applied to goatskins.[20] At 10% salt contribution, the biocidal effect of the mixture was better than 50% salt for curing up to 1 week after treatment; thereafter, up to 30 days, the bacterial counts were similar.
2. *Tamarindus indica* extract was prepared with methanol, then rotary evaporated.[21] A 15% extract with 10 or 15% salt was compared with 40% salt for preserving goatskins; lower offers of extract were not effective. The treatment with the leaf additive was effective for 3 weeks at 31 °C, where the bacteriostatic effect was attributed to the presence of the leaf product rather than dehydration.

3.3.5   Dry Salting

This is a combination of any type of wet salting and sun drying, used in hot countries where salt is plentiful and cheap. Here the moisture content is typically reduced to 20–25%. It has the advantages of reduced weight for transportation and the preserved rawstock is easier to rehydrate than rawstock that has only been dried. On the other hand, the common use of impure salt can produce red heat and drying may lead to rancid fat and alkaline hydrolysis. However, dry salting is not commonly practiced as a method of rawstock preservation.

3.3.6   Brining

An alternative to wet salting is to apply the salt in the form of a solution (Table 3.4). This is a method used almost exclusively in North America. Hides are tumbled in a paddle with saturated brine (36% w/v

Table 3.4 Effects of brining on the composition of hide.[24]

Component	Fresh/%	Brined/%
Hide substance	36–40	36–40
Water	60–64	34–38
Salt		10

at 15 °C) for 12 h; calculations of the changing diffusion coefficients during brining show the effects of the dynamic reduction of the salt concentration and requirement for readjustment during the process to achieve 85% saturation in the pelt. Note that 20–25% brine cannot reach the required concentration in the pelt; 30 and 35% brine require 440 and 280% float, respectively.[22,23] After draining, the hides may have additional salt applied to the flesh side, with or without additive: the latter is a strategy particularly used if there is to be prolonged storage or transportation.

Brined hides are wetter than wet salted hides, but the salt concentrations in the associated moisture are about the same. The preservation is comparable to wet salting and all the problems of halophilic bacterial contamination also apply to brined hides, particularly if the temperature reaches 37 °C.[17] It is not obvious where such species come from, but infection in premises handling salted stock is common and surprisingly difficult to eliminate. Birbir and Bailey[25] showed that halophilic bacteria in North American brining solutions can be controlled by bactericides, including a mixture of sodium bisulfite and acetic acid (see later).

3.4 Alternative Osmolytes

3.4.1 Potassium Chloride

In all the earlier discussion about the use of salt, it is understood that the salt is sodium chloride. However, any material that can exert an osmotic pressure might be a candidate for curing. The requirements are cheapness, availability and no damaging effects, *e.g.* lyotropic swelling. The obvious alternative options are other electrolytes.

The damaging part of salt is the sodium ions, so it is clear that any alternative electrolyte should ideally not be a sodium salt. On the other hand, plants require potassium ions as a nutrient, so the use of potassium chloride as a curing agent would not only solve the environmental impact problem, but also allow the waste stream to be disposed of to agricultural or horticultural land. Bailey's group

Table 3.5   Effect of temperature on the solubility of potassium chloride.

Temperature/°C	Solubility/m
0	3.6
10	4.0
20	4.4
30	4.8

has demonstrated the feasibility of using this technology.[26,27] The main drawback is the higher cost. From a North American perspective, an additional factor to take into account is the temperature dependence of the solubility in water, shown in Table 3.5. The aqueous solubility of NaCl does not display a marked temperature dependence, so brining operations are not affected by seasonal changes in the ambient temperature; however, the lower solubility of KCl at lower temperature would require adjusting the brining solution in cold weather, to avoid under-curing.

Potassium chloride is not immune from contamination with extremophiles. An opportunistic *Staphylococcus* sp. was found on preserved hides containing 4 m KCl;[25] this organism is proteolytic, so there is still potential danger of hide damage during storage.

3.4.2   Sugars

More speculative osmotic options in this application might include saccharides (sugars). These agents would not present an effluent problem, because they will easily break down in a tannery effluent treatment plant. However, the problem of mould growth would have to be addressed, but the addition of a fungicide should solve any problems. More important is the comparison with salt. Here the critical parameter is the water activity, defined as the ratio of the vapour pressure of a solution to that of the pure solvent; a comparison is given in Figure 3.5.[28] It can be seen that salt is more effective on a weight for weight basis, although both are used as dehydrating agents in food preservation: sugar is used in jam and preserve making and honey is an effective bacteriostat in wound management, known since the time of the ancient Egyptians.

The impact of water activity on preventing bacterial growth is illustrated in Table 3.6.[29]

3.4.3   Sodium Silicate

The use of sodium silicate was proposed by Munz:[30] sodium silicate (Wasserglass, water glass) solution is neutralised with sulfuric and formic acids, then dried. It is claimed to be a viable complete

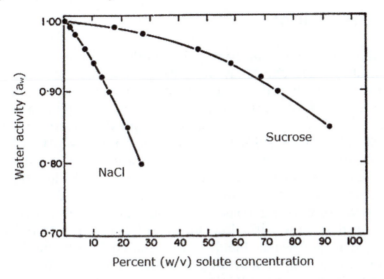

Figure 3.5 The water activities of salt (sodium chloride) and sugar (sucrose) solutions. Courtesy of the *Journal of the Society of Leather Technologists and Chemists.*

Table 3.6 Limiting water activities for growth of some bacteria.

Species	Limiting water activity below which no growth occurs
Bacillus mycoides	0.99
Escherichia coli	0.93
Bacillus subtilis	0.95
Micrococcus roseus	0.91
Pseudomonas aeruginosa	0.95
Staphylococcus aureus	0.86–0.88

substitute for salt and the soak solution from processing the preserved stock can be applied to growing plants without any adverse effects.

3.4.4 Poly(Ethylene Glycol)

Poly(ethylene glycol) (PEG) is a well-known reagent for preserving wet material by replacing water, for example in the case of the recovered Tudor wooden vessel *Mary Rose*. It has been tested for preserving rawstock over the molecular weight ranges 200–1000[31] and 200–8000;[32] in both cases, it was found that the lowest molecular weight was the most effective. The economics of the large scale application to rawstock are questionable.

3.5   pH Control

Treatment of rawstock for high-pH storage is risky, mainly owing to the rapid rate of base-catalysed hydrolysis of the collagen. Therefore, the only pH-based system in common usage is storage pickling, commonly used for woolskins in the UK, Europe, Australia and New Zealand (Table 3.7). The earliest storage pickling systems involved 12–14% salt and 1% sulfuric acid, corresponding to pH ~0.7. However, the high concentration of protons produced too rapid a rate of acid-catalysed hydrolysis. The modern version relies much less on acid and more on electrolyte. Pelts are processed conventionally to the point of pickling: soak, drain, samm (squeeze) or centrifuge, paint the flesh side with lime/sulfide, pile, pull the wool, relime, delime, bate and wash. Pickling is conducted with the following solution, processing in a drum or paddle: 12–15% salt on water volume at 15 °C, 0.7–0.8% sulfuric acid, corresponding to pH ~1.0, plus antimould. Note that the concentration of salt is relatively low compared with wet salting, but is more than enough to resist acid swelling (see Chapter 9).

Although storage pickling is conventionally used for sheepskins, it is not used for cattle hides. However, there is no good reason why this should not happen. It has been shown that a formulation of 100% float, 10–15% salt and 2% sulfuric acid will confer pH 2.5–2.7 to hides, which is suitable for effective preservation for 1 year.[33] It is probably only tradition that prevents this application of acid preservation to larger pelts. However, the opening up of bovine hide is probably too critical to be routinely left to others.

A salt-free pickling preserving system, using acetic acid and sodium sulfite, was proposed by Bailey and Hopkins and demonstrated to work.[34] Here the method relies on both acidity, pH < 4, to reduce bacterial activity, and the biocidal action of sulfur dioxide to sterilise the hide. Trials indicated that the method is suitable for short-term preservation and that it has the potential to provide long-term

Table 3.7   Advantages and disadvantages of storage pickle.

Advantages	Disadvantages
Savings on beamhouse processing: economic and environmental impact	Processing faults by fellmonger: Weak grain, scud/pigment, mottle, drawn grain, mould growth
Ease of sorting for quality immediately	Freight charges: ~45% moisture content
	Pelts sensitive to elevated temperature: store at <15 °C, rapid hydrolysis at 30 °C

preservation, but this approach has not been developed commercially. Valeika *et al.*[35] proposed a similar short-term preservation method, capable of working for 16 days at 22 °C: the formulation is 0.2–0.8% acetic acid with 0.1–1.0% benzoic acid in 20% water. It is clear that there are many potential short-term preservation systems that could be formulated using weak acids.

3.6 Temperature Control

Rawstock can be preserved by storing it at temperatures that slow bacterial activity. The Arrhenius equation is often used to estimate the effect on reactions of lowering the temperature:

$$k = Ae^{-E/RT} \text{ or } \ln k = \ln A - (E/RT)$$

where k is the rate, A is the pre-exponential factor (a constant), E is the activation energy, R is the gas constant and T is the temperature (K). Therefore, because the relationship is logarithmic, a rough guide is that for every 10 °C decrease in temperature, the rate is halved. The converse is true: for an increase of 10 °C, the rate is doubled.

The limit to holding untreated pelt is about 24 h at ambient temperature, if that is assumed to be about 20 °C. This period is relatively short, because it is affected by the way in which the rawstock is usually treated; the analysis also assumes that post-flay treatment is not ideal, *i.e.* the body heat is not removed immediately and efficiently from the pelt, compounded by holding the pelts in the abattoir in piles. The Arrhenius equation clearly indicates the limits to lowering temperature for preservation and therefore predicts that any method based on this change must be considered to be short-term preservation.

3.6.1 Chilling

Hides can be chilled to 3 °C within 30 min by blowing air at −1 °C at a rate of 100 m min^{-1}. There is a limit to how cold the air can be: for example, using air at −9 °C actually causes parts of the hides to freeze, which is inconvenient in this context,[36] discussed below. It is important to use moist air, otherwise the rawstock will dry, which could create problems when the hides are put into work. If the hides are stored at 2–3 °C, they will remain fresh for up to 3 weeks. This is sufficient time for transport within a continent, *e.g.* hides are routinely moved from north to south Europe in this state, using refrigerated lorries.

Table 3.8   Activity of bacteria at chilling temperatures.

Species	Growth at 2 °C	Growth at 4 °C	Gelatine liquefaction[a]	Collagenic hypro release at 4 °C[b]
Bacillus sphaericus	0 on day 21	Minimal on day 10	+	275
Bacillus pumilis	0 on day 21	Minimal on day 16	0	512
Bacillus subtilis	0 on day 21	Minimal on day 14	+	628
Macrococcus caseolyticus	Minimal on day 11	Minimal on day 10	++	842
Bacillus firmus	0 on day 21	Minimal on day 4	0	1024

[a]Scale of zero to +++.
[b]In $\mu g\ g^{-1}$ soluble collagen.

Although it has been assumed that chilling impacts equally on all bacterial organisms, that is not necessarily the case. Chandra Babu *et al.* analysed the effect of temperature on the growth of five species, set out in Table 3.8.[37] Some species remain active even at 2 °C, so biocide should be considered for inclusion in the treatment; this is especially indicated at higher temperatures or for storage periods longer than 9 days, when benzalkonium chloride or methylene bis(thiocyanate) (MBT) is effective.

There are alternative cooling systems in operation. In Australia, eutectic or chilling plates have been used to cool sheepskins rapidly after flay. Other coolants have been shown to work, *e.g.* carbon dioxide snow (−80 °C), which has been used to chill carcases; its viability depends on availability and cost. Liquid nitrogen (−100 °C) has been used in abattoirs for rapidly chilling carcases, but it is actually less effective on pelts than higher temperature coolants, because the rapid boiling of the liquid creates a gas layer between the coolant and the pelt surface, reducing heat transfer out of the pelt. The cooling methods generally have low environmental impact, but there is typically considerable capital cost involved for the cooling plant and subsequent cold storage.

3.6.2   Iceing

Ice can be applied to rawstock in the same way as salt is applied. There are similarities between the methods, especially with regard to the size of the ice particles. If the crystals are small, the ice will melt too rapidly, so the chilling period is limited; if they are too large, the contact with the pelt is limited. Iceing can achieve rapid cooling, but storage cannot prevent the ice melting. Therefore, this is a short-term method of preservation, unreliable for more than a few days – 5 days maximum at 15 °C. The best results have been achieved with woolskins, which provide an insulating effect for the centre of the pile. However, the edges of piles are soon unprotected. Ice is relatively cheap, but the costing must include not only the ice-generating plant, but also storage facilities.

An alternative form of ice is flo-ice, which is a slush of ice in a solution of salt, sugar or alcohol in water. Cooling is rapid, but so is melting.

It is worth noting that the wetting effect of cold water for cooling the pelt or the creation of melting ice may be counterproductive, because either condition can promote bacterial growth in the wetted pelt. In such cases, it may actually be better not to treat at all, that is, a drier pelt containing the natural level of moisture is a worse medium for bacterial growth than a wetted pelt.

3.6.3   Biocide Ice

To avoid the problems of rehydration by melting ice, the concept of using biocide ice was developed by BLC in the UK. The idea is to create ice from a biocide solution, so that melting releases the biocide, providing an additional element of protection. One difficulty is the availability of a suitable biocide. The criteria are as follows: the biocide must be effective as a bactericide and it must be soluble in water, to allow ice to be made. A biocide shown to be effective is 1,3-dihydroxy-2-bromo-2-nitropropane, also known as bronopol or Myacide AS (see Figure 3.6). Most bactericides currently used in the industry are hydrophobic chemicals, formulated to form emulsions with water: in these cases, freezing merely separates the biocide from the water.

The effective period of preservation still depends on the conditions of storage – the lower the ambient temperature, the longer is the preservation period. The application rate is similar to that in salting, with regard to the amount to be applied, and the costs have been calculated to be comparable. The major drawback of this technology is run-off from melting ice, which contains the biocide. This means that the piles of rawstock must be held in such a way that the run-off is contained, subsequently, the run-off must be treated as an effluent stream.

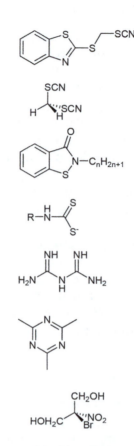

Figure 3.6 Structures of some biocides that may be applicable to rawstock preservation.

3.6.4 Freezing

Within the options for using temperature control, it might appear that the obvious extreme is to freeze the rawstock, by analogy with food preservation. This is certainly a practical option, since this technology is already commonplace in the food industry. However, there are drawbacks.

- Rawstock must be frozen quickly, to avoid the formation of large ice crystals, which can disrupt the fibre structure; this is not a simple procedure for large items such as hides. Also, the earlier the pelts are frozen, the less likely staling is to occur.
- The storage period is limited, because of the growth of ice crystals (as typically observed in domestic freezers).

- The pelts must kept at <0 °C, *i.e.* in a very large freezer, rather than cold storage.
- Thawing prior to processing is the defining problem. If a pallet of hides is frozen and stored at −30 °C, thawing at ambient temperature takes 48 h; if the hides are stored at −10 °C, thawing at 10 °C also takes 48 h.[38] The thawing operation is likely to allow time for bacterial growth and consequent damage. Thawing in the processing vessel is impractical. Loading the vessel with a stack of frozen hides is likely to cause damage to the vessel. Once loaded into the vessel, the absorption of the required latent heat is sufficient to freeze the float, even if it is in large excess. The mass of pelt and float then needs to be thawed – a non-trivial practical problem.

3.7 Biocides

It is possible to achieve short-term preservation by spraying hides on both sides with a solution of a bactericide. Agents that have been used include phenols, 2-(cyanomethylthio)benzothiazole (TCMTB), methylene bis(thiocyanate) (MBT), 1,2-benzisothiazolin-3-one (BIT), biguanide, sodium chlorite, boric acid[39] and 1,3-dihydroxy-2-bromo-2-nitropropane (bronopol) (see Figure 3.6 for structures).[40] Preservation periods may extend to about 2 weeks, depending on the agent and the concentration used, but 1 week is a safer period to assume, depending on the temperature of storage. In each case, there are problems associated with applying the solutions safely and disposal of the biocide after soaking.

An important factor to consider is the mode of action of the biocide. Some biocides work by reacting with the protein of the microorganisms, in effect tanning it; this is the case for aldehydic and triazine biocides. In the context of preservation of hides and skins, this can be counterproductive. If it is the intention to remove the hair or wool, particularly by hair-saving methods, tanning the pelt stabilises the proteins at the base of the follicle, preventing the detaching methods from working.

3.8 Radiation Curing

In a comprehensive review of modern preservation of rawstock, Bailey[24] included his groundbreaking work on the use of radiation. Gamma radiation provided complete sterilisation at a dose of just over 20 kGy, providing protection for 6 weeks. The radiation is capable of damaging the pelt, reducing the tear strength, but damage was not apparent below 30 kGy.[41] Electron beam radiation was also investigated: the procedure was to rinse hides with biocide, seal them in polythene bags and irradiate with 1.4 Mrad at 10 MeV, which provided protection for 3 weeks.[42] Both of these radiation techniques have been approved in the USA for eliminating *Escherichia coli* from hamburger beef. Issues arising from these technologies are the cost of the plant for treatment and keeping the hides sterile, *i.e.* avoiding recontamination.

3.9   Fresh Stock

The use of chilling notwithstanding, the alternative to salt curing is not to cure at all. This is an increasingly common approach to processing, particularly in those countries where there is a direct link between the abattoir and the tannery. It is non-trivial to point out that the difference is the presence or absence of salt: since salt has a profound influence on the fibre structure of the pelt by the dehydrating effect, this may have a lasting impact on the leather quality, compared with rawstock, which does not undergo such a treatment. Haines[43] made the comparison and found no difference between salted and fresh rawstock with regard to physical properties and area yield of the leather. She did, however, observe that leather made from fresh hide exhibited slightly coarser break. The only apparently related feature observed in the fibre structure was that the junction region of leather from fresh hides was more finely split, *i.e.* more opened up, making the region looser, which is a cause of coarse break, because it allows the grain to ripple (see Chapter 6). The reason for this difference in opening up is not clear; it appears that the salting somehow confers protection to the junction, perhaps related to the depleting effect.

References

1. R. G. Stosic, PhD thesis, University of London, 1995.
2. D. R. Cooper and A. C. Galloway, *J. Soc. Leather Technol. Chem.*, 1974, **58**(2), 25.
3. M. Birbir and A. Ilgaz, *J. Soc. Leather Technol. Chem.*, 1996, **80**(5), 147.
4. R. R. Schmitt and C. Deasy, *J. Am. Leather Chem. Assoc.*, 1963, **58**(10), 577.
5. R. R. Schmitt and C. Deasy, *J. Am. Leather Chem. Assoc.*, 1964, **59**(6), 361.
6. BLC, unpublished results, see also ref. 18.
7. A. D. Covington, unpublished results.

8. R. M. Koppenhoeffer, *J. Am. Leather Chem. Assoc.*, 1937, **32**(4), 152.
9. M. Komanowsky, *J. Am. Leather Chem. Assoc.*, 1989, **89**(12), 369.
10. P. J. Water, L. J. Stephens and C. Surridge, *Leather*, 1999, 38.
11. A. E. Russell, H. Tandt and R. Kohl, *J. Soc. Leather Technol. Chem.*, 1997, **81**(4), 137.
12. N. K. Chandra Babu, *J. Am. Leather Chem. Assoc.*, 2000, **95**(10), 368.
13. A. H. Quadery, MSc thesis, University College Northampton, 1999.
14. D. R. Cooper, *J. Soc. Leather Technol. Chem.*, 1973, **57**(1), 19.
15. G. W. Vivian and M. B. Rands, *J. Soc. Leather Technol. Chem.*, 1976, **60**(6), 149.
16. D. G. Bailey, M. Birbir, A. Ilgaz and W. Kallenberger, *J. Soc. Leather Technol. Chem.*, 1996, **80**(3), 87.
17. D. G. Bailey and M. Birbir, *J. Am. Leather Chem. Assoc.*, 1996, **91**(2), 47.
18. D. J. Lloyd, *Progress in Leather Science: 1920–45*, British Leather Manufacturers' Association, 1948, 130.
19. S. V. Kanth, *et al.*, *J. Am. Leather Chem. Assoc.*, 2009, **104**(1), 25.
20. M. A. Hashem, *et al.*, *J. Am. Leather Chem. Assoc.*, 2017, **112**(8), 270.
21. A. Tamil Selvi, *et al.*, *J. Soc. Leather Technol. Chem.*, 2015, **99**(3), 107.
22. E. Hernandez, *et al.*, *J. Am. Leather Chem. Assoc.*, 2008, **103**(5), 162.
23. M. Barinova, *et al.*, *J. Am. Leather Chem. Assoc.*, 2009, **104**(12), 397.
24. D. G. Bailey, *J. Am. Leather Chem. Assoc.*, 2003, **98**(8), 308.
25. M. Birbir and D. G. Bailey, *J. Soc. Leather Technol. Chem.*, 2000, **84**(5), 201.
26. D. G. Bailey, *J. Am. Leather Chem. Assoc.*, 1995, **90**(1), 13.
27. D. G. Bailey and J. A. Gosselin, *J. Am. Leather Chem. Assoc.*, 1996, **91**(12), 317.
28. A. E. Russell and A. C. Galloway, *J. Soc. Leather Technol. Chem.*, 1980, **64**(1), 1.
29. *Inhibition and Destruction of the Microbial Cell*, ed. W. B. Hugo, Academic Press, London, 1971.
30. K. H. Munz, *J. Am. Leather Chem. Assoc.*, 2007, **102**(1), 16.
31. K. C. Kannan, *et al.*, *J. Am. Leather Chem. Assoc.*, 2010, **105**(11), 360.
32. M. L. Aldema-Ramos, *et al.*, *J. Am. Leather Chem. Assoc.*, 2015, **110**(4), 109.
33. B. M. Haines and R. L. Sykes, *J. Soc. Leather Technol. Chem.*, 1973, **57**(6), 153.
34. D. G. Bailey and W. J. Hopkins, *J. Am. Leather Chem. Assoc.*, 1977, **72**(9), 334.
35. V. Valeika, *et al.*, *J. Soc. Leather Technol. Chem.*, 2013, **97**(3), 101.
36. M. A. Haffner and B. M. Haines, *J. Soc. Leather Technol. Chem.*, 1973, **59**(4), 114.
37. N. K. Chandra Babu, *et al.*, *J. Soc. Leather Technol. Chem.*, 2012, **96**(2), 71.
38. B. M. Haines, *J. Soc. Leather Technol. Chem.*, 1981, **65**(3), 41.
39. R. Hughes, *J. Soc. Leather Technol. Chem.*, 1974, **58**(4), 100.
40. *The Biocides Business*, ed. D. J. Knight and M. Cooke, Wiley-VCT, 2003.
41. D. G. Bailey, *J. Am. Leather Chem. Assoc.*, 1999, **94**(7), 259.
42. D. G. Bailey, *et al.*, *J. Am. Leather Chem. Assoc.*, 2001, **96**(10), 382.
43. B. M. Haines, *J. Soc. Leather Technol. Chem.*, 1981, **65**(3), 41.

4 Soaking

4.1 Introduction to Beamhouse Processing

Soaking is the first of the 'beamhouse' processes, so called because they traditionally included operations that were conducted over an angled wooden beam. Prior to the introduction of machinery, hides or skins would be draped over the beam, so that they could be hand fleshed, hand dehaired, with removal of residual epidermis and hair debris (scudded), and hand shaved to thickness, illustrated in Figure 4.1.

Nowadays, the term is used to encompass all of the processes conducted in the tannery leading to the tanning step. The purpose of the beamhouse is to prepare the pelt for tanning. Another way of putting that would be to say that the beamhouse is for purifying the pelt or opening up the pelt structure. Opening up is a generic term, which has two components.

1. The removal of non-collagenous skin components: hyaluronic acid and other glycosaminoglycans, non-structural proteins, fats. This is not done to completion (except in the special case of Japanese leather, where removal of non-collagenous components is conducted to extreme), so the processes in a tannery must be geared to the required degree of removing these, to produce the desired properties of the final leather.
2. Splitting the fibre structure at the level of the fibril bundles, to separate them.

Tanning Chemistry: The Science of Leather 2nd edition
By Anthony D. Covington and William R. Wise
© Anthony D. Covington and William R. Wise 2020
Published by the Royal Society of Chemistry, www.rsc.org

Figure 4.1 The traditional use of beams for hand fleshing and other tasks. Courtesy of the late J. Basford.

Table 4.1 Indicative process conditions for bovine hides.

Process step	Water/% on hide weight	pH	Time/h
Dirt soak	200	6–10	1–2
Main soak (often more than a single step, $n = 1$–4)	$n \times 200$	6–10	4–8
Unhair (lime/sulfide) with liming[a]	200	12–13	15–20
Rinse	200	12–13	1
Delime (ammonium salt)[a]	100	8–9	1–2
Bate (proteolytic enzyme)	100	8–9	1–2
Rinse	100	8–9	1
Pickle (sulfuric + formic acids)	50–100	50–100	1–2

[a]These are steps that are likely to change in the future, creating new norms (see Chapters 5–9).

These effects are the result of the complex reactions under the conditions in the beamhouse and the extent to which they happen depends on the precise conditions adopted throughout these preparative process steps.

The process steps in the beamhouse can be summarised as shown in Table 4.1: the indications of amounts of water on a raw weight basis and times merely represent typical industrial conditions, which can vary greatly.

It has been said that *leather is made or marred in the beamhouse.* What this means is that the fundamental properties and performance parameters of the final leather are conferred to the pelt during the beamhouse processing. Importantly, each step is important, insofar as subsequent process steps can never compensate for or overcome deficiencies in any given prior process step – at least, not without compromising leather quality. This is an example of the importance of the axiom 'get it right first time'.

4.2   The Soaking Process

Soaking is the first process applied to the rawstock. This and all subsequent chemical steps are most commonly conducted in drums, as shown in Figures 4.2 and 4.3, illustrating that the mechanics of chemical processing have not changed markedly in 100 years, except for the drive mechanisms. Alternatively, early steps may be conducted in open vessels, where the pelts are moved in the solution by a paddle.

There are several parameters that should be considered in the soaking process. In reviewing the variables, McLaughlin and Theis considered the following aspects of soaking:[1]

- history of the pelt and the type and degree of curing – at that time, salting was the primary curing process;
- water content of the rawstock;
- exchange of substances between the soak liquor and the pelt;
- nature of the protein solubilised in the soak;
- character/quality of the soak water;
- concentration of salt in the soak liquor and rate of diffusion from the pelt into the solution;
- pH of the soak;
- temperature;
- float-to-pelt ratio;
- time in soak;
- effect of changing the soak float;
- impact of biocide in the soak and the chemistry of the reagent.

Table 4.2 presents some selected data from their paper.

The effectiveness of the soak (here longer than typical processing) depends on the amount of water used: the efficiency of salt removal depends on the amount of water, since the salt is

Figure 4.2 Processing in drums in the first half of the twentieth century. Courtesy of the late J. Basford.

Figure 4.3 Modern drum processing – could be anywhere in the world.

effectively partitioned between the two phases and the solubilisation rate depends on the difference in concentration between the soak liquor and the rehydrated pelt. The salt removal figures can be used to estimate the effects of multiple soaking processes. If we assume that the total salt content is 15% and the efficiency of salt removal is defined by the values in Table 4.2, the accumulation of salt as a percentage of the total content in the pelt can be correlated with the number of separate soaking operations and the total amount of water used in those operations. The results are presented in Table 4.3.

Table 4.2 Effects of float-to-goods ratio and temperature on the outcome of soaking salted hide for 24 h.

Water-to-hide ratio	Salt concentration in soak/%	Salt in soak/% on hide weight	Swelling/%		
			4 °C	20 °C	37 °C
1:1	6.77	6.8			
1:2	4.75	9.5	31	32	28
1:3	3.50	10.5	28	33	28
1:4	2.63	10.5	40	33	27
1:5	2.34	11.7	36	47	42
1:6	1.95	11.7	48	48	38
1:7	1.79	12.5	48	42	32
1:8	1.64	13.1	48	51	30
1:9	1.36	12.2	41	41	34
1:10	1.35	13.5	31	31	26

Table 4.3 Calculated salt removal by multiple soaks.

Float-to-goods ratio	Accumulated salt removal/% and volume of water used/%									
	Soak 1		Soak 2		Soak 3		Soak 4		Soak 5	
1:1	45	100	70	200	83	300	91	400	95	500
1:2	63	200	86	400	95	600				
1:3	70	300	91	600	97	900				
1:4	70	400	91	800	97	1200				
1:5	78	500	95	1000						

Assuming that an acceptable level of salt removal in the soaking step is 95% of the total content, Table 4.3 shows that this can be achieved in only a few steps if the amount of float per soak is higher, but then the total amount of fresh water used is greater. These results indicate that the optimum float is 200%, but the actual value will depend on individual circumstances.

In each of the variables in soaking there are a number of contributors; in no category are the parameters given in order of importance here – the hierarchy must be judged in the context of the actual process and the specific requirements of the particular rawstock.

4.2.1 Rehydration

The function of rehydration is to fill up the fibre structure with water, ensuring that all the elements of the hierarchy of structure are wetted to equilibrium and beyond, with the purpose of facilitating movement of dissolved chemical reagents through the pelt cross-section. In the case of dried or dry salted hides, this is the dominant purpose of soaking, but the argument applies equally to wet salted stock. Even fresh hide requires filling with water, which is a faster reaction than any rehydrating process. Only when the pelt is completely wetted and filled with water can chemicals be transported throughout the structure, particularly down the hierarchy of the fibre structure. If this does not happen, chemical processing will be non-uniform and the effect will be reflected in the final leather, giving rise to non-uniform properties. See also Section 4.4.2.

4.2.2 Removal of Salt

All salt curing causes dehydration, so the fibre structure collapses, which consequently hinders the movement of solution within the pelt. The removal of salt occurs at the same time as rehydration happens. Therefore, the salt becomes increasingly diluted, reducing the osmotic effect, so the fibre structure is plumped up, eventually beyond the situation in fresh pelt. This process has been modelled by Blaha and Kolomazník[2] assuming a diffusion-controlled system of salt transport, neglecting adsorption of salt on the fibre structure. Further modelling by Kolomazník's group of the economics of salt removal in soaking demonstrated the potential for damage to the fibre structure by 'shock' desalination,[3] although it is not clear exactly what the consequences are from this alleged phenomenon. However, the majority of tanners around the world take no account of this, simply applying fresh water to their salted stock.

The removal of salt can require a lot of fresh water, depending on how the process is conducted. There are three options.

1. Running washing, in which water is continuously pumped into a rotating drum fitted with a lattice door. The rinse water is continuously discharged through the door to waste. The advantage is that the pelt is in contact with solution containing very little salt, so the rates of transfer of water into the pelt and salt out of the pelt are maximised. The disadvantage is that there is a very high use of fresh water.
2. Batch washing, in which the pelt is washed with several changes of fresh water. This is common when there is a short first or dirt soak, followed by one or more prolonged main washes, possibly with rinsing. The partitioning of salt

between the pelt and the solution depends on the ratio of water to pelt, the degree of mechanical action and the time allowed to move towards equilibrium and the overall effectiveness of the process also depends on the number of washes. The requirement for fresh water depends on the size and number of washes.

3. Countercurrent washing is an option for reducing the requirement for fresh water. Here, the only discharged solution is the dirt soak, the solution most contaminated with dirt and containing the highest concentration of salt. The principle is illustrated in Figure 4.4, showing a system of five washes. The first line of washes is for the first pack. The second pack uses the second soak liquor from the first pack as its dirt soak solution, its third soak is conducted with the previous second soak, and so on, until it is finally rinsed with fresh water. The third pack continues the pattern of reusing solutions and the system can continue indefinitely. It is called countercurrent because the water moves from fresh to increasingly contaminated and the hides move in the opposite direction, from heavily contaminated solution to fresh water. In this way, salt and dirt are accumulated in the small volume of water that is discharged to waste and the consumption of fresh water is minimised. Clearly, this does not change any problems associated with the environmental impact of the salt, but it

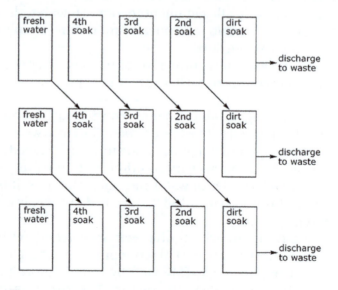

Figure 4.4 The countercurrent reuse of wash water.

can mitigate the required treatment, for example, by introducing reverse osmosis or membrane technology to isolate the salt.

The efficiency of salt removal still depends on the ratio of solvent to pelt, the degree of mechanical action, the time of contact and the number of treatments. As the number of packs of pelt treated increases, the concentration of salt in any given recycled soak solution will increase. However, the pattern is likely to reflect another unrelated example of recycling [chromium(III) from tannage] also when accumulated electrolyte is in equilibrium with pelt, in which it was found that a steady state was reached after five cycles.[4] The procedure may require more time than the conventional process, because it involves transferring solution from the process vessel into holding vessels, *via* a screen. Moreover, the rate of salt diffusion depends on the difference in the concentration between the bulk solution and the solution within the pelt: this difference will always be smaller than in a soaking procedure involving only fresh water.

4.2.3 Cleaning the Pelt

Cleaning means the removal of dirt, primarily from the surfaces, *i.e.* discolouring or abrading agents, such as mud or soil and grit. Cleaning includes removal of blood, urine and other body fluids from both surfaces and within the pelt.

4.2.4 Removal of Non-structural Proteins

This may occur,[2] indeed McLaughlin and Theis quantified the effect, determining that the extent depended on the number of changes made to the soak liquor.[1] However, unless the process is prolonged and if it is conducted without the assistance of chemical or enzymatic breakdown, it is unlikely to be a major contributor to opening up. In the absence of appropriate enzymes, it is assumed by tanners that the soaking step has little effect on the fibre structure and its components, other than rehydration.

4.2.5 Removal of Dung

Under conventional conditions of soaking, including the use of typical auxiliaries such as alkali and surfactants, dried-on dung is not removed. Only in the presence of lignocellulosic enzymes is dung removed quickly and efficiently, as discussed in Chapter 2.

4.2.6 Removal of Hyaluronic Acid

The structure of hyaluronic acid (HA) was discussed in Chapter 1; it is effectively a chain of carboxylate groups at physiological pH. The charge repulsion effect gives the molecule a large exclusion volume, making it difficult to remove it from the fibre structure, even though it is not bound to the collagen other than by electrostatic attraction. The presence of salt (neutral electrolyte) effectively smears out the charges, allowing the chain structure to contract in volume, causing the molecules to be washed out of the pelt.

For fresh hide, this mechanism is not available. The options are as follows.

1. Add salt to the soak solutions, which is ecologically counterproductive, but done in practice.
2. Use prolonged soaking, which has implications for damage by bacterial proliferation and the likely requirement for biocide addition. Only 10% of the HA is removed after 24 h of soaking, so for complete removal considerably more time is needed.[5]
3. Apply chondroitinase to break the chains into fragments for easier removal, but this has not been developed industrially.

The reference to options in processing fresh hide assumes the necessity for removing the HA. The argument is as follows: HA is removed very efficiently from salted hide and the outcome is leather of accepted and acceptable quality, hence HA removal is necessary. The argument is extended by the notion that HA can react with the calcium from the lime, carrying it through the processing into pickled pelt and contributing to filling the fibre structure. Furthermore, the HA can react with chromium(III), because of the high carboxyl content; this should result in polymeric crosslinking throughout the collagen structure, causing a decrease in mobility, which would be observed as reduced softness. Since the most common rawstock in recent years has been wet salted hide, the argument has not been tested. However, a less common raw material used to be dried hides and skins: because the pelt was soaked with little mechanical action, it is reasonable to suppose that the HA remained throughout the beamhouse processing and took part in the chrome tanning process. It is the experience of tanners of dried material that the product is always firmer than the same leather made from wet salted hide.[6] The difference was always assumed to originate in the collapse of the fibre structure during drying and the inability of rehydration to return the pelt completely to its natural state. However, it is also likely that the leather owes its handle, at least in part, to the extra element of structure, in a way analogous to the tanning of residual non-structural protein.

Reference was made in Chapter 2 to a study of the influence of HA and its removal on opening up and chrome tannage. The results were consistent with the predicted effects of HA remaining through processing to chrome tannage, although the small-scale study is not definitive.[7]

4.3   Conditions in Soaking

4.3.1   Long Float

Soaking float is typically 200–300% on salted pelt weight, depending on the raw material and its state of cleanliness. The float length tends to be consistent, even when multiple soaking steps are employed. The removal of salt is a diffusion- and osmotic pressure-controlled process, so the concentration of salt in solution determines the rate of salt removal, hence changes of float are necessary.

4.3.2   Change of Float

The options for changing float have already been addressed. One advantage of using only fresh water is that the requirement to include a bactericide may be reduced.

4.3.3   Temperature

Since the pelt is raw, with a denaturation/shrinkage temperature (T_s) of ~65 °C around neutral pH, the temperature of the soak must be limited to <30 °C; this applies to all pH conditions. It has been suggested that the erector pili muscle can still contract when the raw pelt is placed in cold water, as discussed in Chapter 2. Hence it may be better to use water at 20–30 °C rather than at 10–20 °C, to avoid increasing the angle of the follicle and thereby risk coarsening the grain. In any case, warm water is a better hydrating medium.

4.3.4 pH

Typically, soaking is conducted without pH change; from the combination of pelt at physiological pH ~8 and water at pH ~6 due to absorbed carbon dioxide, the bath is typically at around pH 7. However, some tanners add alkali, usually sodium carbonate, to the bath to raise the pH to about 10 in order to assist the wetting action. The efficacy of higher pH is dubious and it runs the risk of initiating hair immunisation (see Chapter 5).

4.3.5 Time

In conventional batch soaking, salted hides usually require 6 h or more to remove enough salt to ensure that the pelt is completely rehydrated in the centre of the cross-section and down the hierarchy of structure. Dried skins and hides require 24–48 h or more, depending on the thickness.

4.3.6 Mechanical Action

For salted stock, the degree of mechanical action is not critical for leather quality. However, violent mechanical action is not recommended in the earliest stages of processing, when the hides or skins are in a relatively stiff condition. As in any process involving the transfer of solution into and out of the pelt, efficiency and effectiveness increase with the degree of mechanical action. This is because the mechanisms of solution penetration and solute transfer rely on alternately compressing and relaxing the pelt. In the case of dried hides or skins, it is essential that the first stage of rehydration is conducted in a static bath, probably for at least 24 h, depending on the pelt thickness, when rehydration should occur only through diffusion. Only when the pelt has softened can the process be conducted with any mechanical action, to begin to accelerate the rate of rehydration.

The role of mechanical action with regard to the energy applied to the load is not easy to define: power input can be measured, but needs special electrical equipment. Mitton addressed the problem of drumming theoretically in 1953,[6] when he analysed the way in which hides tumble in a drum fitted with shelves or pegs. He concluded that in the absence of internal structure and under typical conditions, pelts are never carried higher than the drum axle. However, it was not until about 20 years later that the subject was addressed practically and reviewed by Longstaff.[8] One of the problems faced by tanners is how to compare the action in one drum with the action in another, particularly when the drums are not of the same size; this is the difficulty in achieving realistic results in small experimental drums. Lhuede[9] showed that there was dynamic similarity of action if the following relationship applied:

$$N_1D_1^{\frac{1}{2}} = N_2D_2^{\frac{1}{2}}$$

where N is the drum speed in revolutions per minute and D is the drum diameter in feet. This also assumes that the loading is similar, *i.e.* the drum is filled to the same degree and the float-to-goods ratios are the same. The optimum value for the term $ND^{\frac{1}{2}}$ was found to be 47–55, corresponding to the typical conditions for chrome tanning of 17 rpm for a drum 8 ft in diameter. As Lhuede commented:[9]

> *... the expenditure of equal quantities of energy by batches of falling skins of equal mass does not necessarily imply equally effective pounding.*

Therefore, the assumption that all the components of a pack of goods in the drum receive the same treatment may not necessarily be the case.

Hinsch[10] showed that the optimum speed for maximum mechanical action is about 66% of the centrifugal speed, *i.e.* when $ND^{\frac{1}{2}} \approx 28$; this refers to the speed at which the pelts are carried round the drum by centrifugal force to the highest point when they drop back down into the load. None of this takes into account the effects of any internal structure to the drum. All tannery drums help the tumbling action by having either horizontal shelves or inward-pointing pegs; if neither are present, the hides or skins merely roll around, causing them to tangle.[11] Longstaff experimented with peg design and demonstrated that longer pegs were more effective in causing mixing action.[8] Corning continued investigations of the role of mechanical action in processing on the small scale, but the work has not been openly published.[12] This is an area of research that needs to be developed.

It is worth noting that there is an alternative approach to conventional drum processing, namely the use of plastic beads as a tumbling medium, as addressed in Chapter 19. In a crossover from the laundry industry, the Xeros Company (now known as Qualus) showed that a substantial proportion of the float can be replaced with beads, resulting in significant savings of water.[13] The new process has not been trialled for soaking, but the modified mechanical action could prove useful in this context.

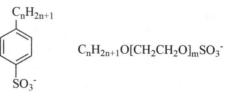

Figure 4.5   Anionic detergents.

4.4   Components of Soaking Solutions

4.4.1   Water

The quality of the water, in terms of its bacterial status, is an issue that needs to be addressed (see later).

4.4.2   Detergents

These agents accelerate rehydration by assisting the water to flow over or wet the fibre structure and thereby allow it to interact with the structure at the molecular level. Removing grease is a lesser effect, since the temperature of soaking is typically low and therefore the grease is not mobilised. Triglyceride grease is associated with the fibre structure, present in varying amounts dependent on the species of animal; it is effectively removed by non-ionic detergents of the alkyl poly(ethylene oxide) type:

$$C_nH_{2n+1}O[CH_2CH_2O]_mH$$

The melting point of triglyceride grease is 40–45 °C, so the actions of degreasing agents are greatly reduced at lower temperatures. The advantage of using this class of detergents is that they do not interact strongly with collagen; hence they are washed out and cannot interfere with subsequent processes such as dyeing. Note that detergents that are derivatives of nonylphenol are no longer used, since they have been shown to be endocrine disruptors.

Sebaceous grease is better removed using anionic detergents, which have the types of structure shown in Figure 4.5. Currently available anionic detergents can bind to collagen and survive into post-tanning, where they may adversely affect dyeing and water resistance. This ionic interaction is important in post-tanning (see Chapter 15).

Sebaceous grease compositions vary considerably between species, as indicated in Table 4.4.

Table 4.4 Compositions of examples of sebaceous grease.

Component	Sheep[14]	Bovine[a,15]	Human[16]
Squalene	0		15
Sterols	12		1
Sterol esters	46		2
Wax esters	10	11	25
Diesters	21		0
Glyceryl esters	0		0
Triacyl esters	0	53	41
Phospholipids		21	
Free acids	0	10	16
Free alcohols	11		0

[a]Fleshed hides fat content, *i.e.* sebaceous grease with some cutaneous lipids.

Sheep or wool grease is high in sterols and waxes, useful as lanolin in many applications. Bovine sebum is lower in sterols and higher in triglyceride lipids, therefore it is more vulnerable to degradation by lipases.[17]

4.4.3 Soaking Enzymes

Soaking is the process step in which enzymes are introduced. They are powerful reagents, capable of accelerating the rates of reaction by many orders of magnitude, as discussed in Chapter 18. Because skin is a complex material, requiring different reactions to prepare it for turning into leather, there is a range of enzymes that might be exploited for this purpose.

4.4.3.1 Proteases

Soaking proteases are essentially diluted bating enzymes. They are included in soaking not only to break down protein, particularly non-structural protein, to enhance the rehydration process, but also to contribute to the opening up of the fibre structure. Care must be taken not to loosen the pelt fibre structure by over-bating during the soaking steps.

4.4.3.2 Lipases

These enzymes hydrolyse triglyceride grease, thereby reducing the hydrophilicity of the fibre structure and aiding wetting. It has been claimed that attacking the sebaceous grease on the grain surface of cattle hides facilitates subsequent chemical processing, because the reagents are not hindered by the natural protection that the epidermis confers.[17]

4.4.3.3 Amylases/Carbohydrases

These enzymes are marketed for the soaking step. However, because enzymes are specific in their action on substrates, it is not immediately obvious how they can be effective. Amylases (named from amylum, an old name for starch) break down starch and starch is not an animal carbohydrate, it is a plant carbohydrate. However, there is increasing evidence that such enzymes can have a role to play in opening up, so they have a spectrum of activity for other carbohydrates.[18,19]

Fermentation of *Aspergillus terreus* produces a cocktail of carbohydrases that have been tested for their contribution to the soaking step of opening up.[20] The observations are summarised in Table 4.5.

The impact on opening up, as measured by component removal, is clear, although the trends are apparently inconsistent. The anomalous observations are attributed to glycation, *i.e.* a stabilising effect due to reaction between substrate and glucose. A marginal benefit to leather strength was measured.

4.4.3.4 Lignocellulosic Enzymes

Dung contains 80% lignocellulose, with little protein or fat. Therefore, it is broken down only by a mixture of enzymes that attack the primary composition, ligninase, cellulase and xylanase, addressed in Chapter 2.

Table 4.5 Removal of carbohydrate (sugar) and proteoglycan in the soak in comparison with a control of 10% lime in 150% float for 48 h.

		Sugar removal/%	Proteoglycan removal/%
Control		77	78
Enzyme offer	1%	80	77
	2%	83	87
	3%	86	85
1% enzyme offer	pH 6	84	78
	pH 8	96	91
	pH 10	93	87
	pH 12	74	76
1% enzyme offer, pH 8	0.5 h	95	90
	1.0 h	93	88
	1.5 h	89	86
	2.0 h	79	72

4.4.3.5   Chondroitinases

These enzymes might be used to break down the glycosaminoglycans hyaluronic acid and dermatan sulfate; in the former case, this would be a useful addition to the technology of processing fresh or chilled hides, when salt is not available to facilitate the removal of hyaluronic acid. The value of degrading dermatan sulfate is not certain, considering its function in skin (see Chapter 2).

4.4.3.6   Amidases

Whether or not liming/unhairing technology involves prolonged alkali treatment, amide sidechain hydrolysis can be achieved and enhanced by the inclusion of amidases to prepare the pelt for chrome tanning. The enzymes can specifically target asparagine or glutamine or both: since chrome tanning may preferentially occur at aspartic acid sidechains (see later), asparaginase might be preferred, but a cheaper option is a general amidase.

4.4.3.7   Phospholipase

The biotechnology of using phospholipase in degreasing[21] to target lipocyte cell walls has already been mentioned and it may be appropriate to consider starting the process in the soak.

4.4.3.8   Other Enzymes

There are many other enzymes that might be applied at this stage of processing, although the options are speculative, since they have not reached industrial acceptance. Examples are as follows.

1. Enzymes to target the basement membrane or the prekeratinised zone in follicles to start the unhairing process without loosening the pelt.
2. Enzymes to target the adhering mechanism of fibril bundles for opening up.

4.5   Biocides

At this stage of processing, the major risk is from bacterial activity; fungal damage is not a problem until the point of holding in the wet tanned state. Some tanners allow a degree of bacterial activity during soaking, as a contributor to opening up, by running a warm soak in which bacteria are allowed to proliferate. However, this is a high-risk strategy, because the conditions vary between packs and therefore the effect is inconsistent. Inconsistency in opening up at any stage of processing means that the product becomes inconsistent.

Bacterial activity can be controlled by the use of biocides. Note that there is usually no need for a fungicide at this stage of production, although some bactericides do possess fungicidal activity. The need for bactericide in soaking depends on the totality of all the prevailing conditions and circumstances – the more conditions that would contribute to danger, the great is the imperative for a bactericide to be used. Bactericide types and comments are set out in Table 4.6. The

Table 4.6 Biocides that might be used in soaking.

Biocide type	Comments
Phenolic	• Can be substantive to collagen and cause dyeing problems • Inhibitory to enzymes • Variable effectiveness, but typically not suitable for soaking
Oxidising, *e.g.* hypochlorite, chlorite	• May damage components of pelt • Require high concentration to be effective • Likely to be inhibitory to enzymes
Isothiazolone, *e.g.* octyl	• Very good fungicide • Less effective bactericide
2-(Thiocyanomethylthio) benzothiazole (TCMTB)	• Good substantivity to skin • Relatively low bactericidal activity • Currently the industry standard fungicide
Thiocyanate, *e.g.* bis (isothiocyanate) (BIT)	• Effective bactericide • Degraded at pH > 8 • Frequently found in formulations with TCMTB
Quaternary ammonium	• Relatively ineffectual
Biguanide	• Useful, but inhibitory to pancreatic types of enzymes
Thione/thiadiazine	• Good activity in the presence of organic matter • May adversely affect leather • Can have an unpleasant odour
Aldehyde	• Exhibits tanning effect, therefore unsuitable for applications when the hair or wool is to be removed • Not suitable for soaking
Triazine	• Very good biocidal activity in the presence of organic matter • Has a tanning action, *cf.* chemistry of reactive dyes (see Chapter 15)
Phosphonium, *e.g.* tetrakis (hydroxymethyl)phosphonium sulfate (THPS)	• Good biocidal activity • Has an aldehydic tanning action
Bronopol, *i.e.* 1,3-dihydroxy-2-bromo-2-nitropropane	• Effective bactericide • Good health and safety record

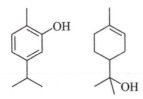

Figure 4.6   Terpene structures.

choice of biocide is determined by the requirements of the raw material and the product to be made (for example, those biocides that have a tanning mechanism are unsuitable for processes in which hair or wool is to be removed by hair save techniques) and compatibility with other agents present, *e.g.* enzymes.

Ozone has been investigated as a alternative approach to protecting the pelt in the soak.[22] It is claimed that the gas is more effective than sodium dimethyldithiocarbamide, depending on the dosing conditions: 0.4 mg L^{-1} allowed some bacterial growth, 0.65 mg L^{-1} completely controlled bacterial growth. The technology is claimed to be safe, despite the toxic nature of ozone gas, because the application is in aqueous solution, but it has not yet received acceptance in the industry.

Essential oils have been suggested as components of the soak.[23] Fennel oil was found to be inferior to oregano oil, the main constituents of which are carvacrol and terpineol, shown in Figure 4.6. Several essential oils have a similar mixture, at a ratio of 80:20, depending on the source. Essential oil terpenes have been proposed for preserving foodstuffs; the technology echoes the use of plant materials for hide and skin preservation, referred to in Chapter 3.

In deciding whether or not the use of a biocide in soaking is advisable, the prevailing conditions must be assessed. The following conditions might each indicate that a bactericide is necessary, but do not necessarily constitute a list of definitive threshold conditions that apply individually, nor are they listed according to priority.

4.5.1   pH

This is not strictly part of the problem, since pH can be adjusted to any value. Even soaking at physiological pH does not indicate the necessity for a bactericide, because other factors are overriding.

4.5.2   Temperature

The higher the temperature, the faster bacteria will grow and the faster their enzymes will react. Soaking at temperatures above 20 °C constitutes a risky condition.

4.5.3   Time

The longer the rawstock is held in the soaking solutions, the greater is the risk of damage. Longer than 6 h is regarded as likely to incur damage.

4.5.4   Fresh Hides or Skins

Salted hides produce a brine solution during soaking: in an analogy with salting for curing, the presence of salt in solution confers some bacteriostatic protection. In the case of fresh hides or skins there can be no bacteriostatic effect, so they are at greater risk of bacterial damage.

4.5.5   Dried Hides

Clearly, these need a bactericide during their long first soak, which is likely to last at least 24 h. Also, the conditions of drying are likely to have resulted in higher levels of bacteria than salt-cured stock.

4.5.6   Dirty Rawstock

The presence of contamination, particularly dung, provides additional bacterial loading and will heighten the risk of bacterial damage.

4.5.7   Putrefied Rawstock

If it is known or suspected that the rawstock has already been damaged, either by staling (delay between flay and cure) or by inadequate curing or preservation or by inadequate storing conditions, the use of a biocide is mandatory.

Note that the presence of red heat does not constitute a danger warning in this context. This problem can be dealt with by simply reducing the salt concentration in a conventional soaking process.

4.5.8 Nature of the Cure

Some salt curing processes include a biocide or biocidal additives, *e.g.* sodium metabisulfite. In these circumstances, additional bactericide is probably not needed, at least not in the dirt soak.

4.5.9 Water Quality

If the water is significantly loaded with bacteria, as may be the case with directly abstracted ground water, unlike processed town water, the likelihood of damage is increased.

4.5.10 Use of Enzymes

Proteolytic enzymes in the soak provide a controlled way to begin the opening-up process: the presence of additional, uncontrolled numbers of bacteria is likely to result in more enzyme action than is desirable and thereby cause excessive action leading to damage, particularly on the grain surface.

4.5.11 Soaking Procedure

The cleaner the water is at any stage of soaking, the less likely it is that bacterial damage can occur. This will depend in part on the number of changes of water during the soaking procedure. Recycling soak liquors would be high risk in this regard.

4.6 Role of the Erector Pili Muscle

In Chapter 2, the influence of the erector (arrector) pili muscle on fibre structure was mentioned as a possible effect of soaking conditions. Tables 4.7 and 4.8 demonstrate that there is indeed

Table 4.7 Effect of temperature and time in soaking on the angle of the erector pili muscle. Errors are standard deviations of 10 measurements.

	Angle/°				
Time/h	5 °C	10 °C	15 °C	20 °C	30 °C
1	8 ± 1	9 ± 1	10 ± 1	11 ± 1	12 ± 2
2	10 ± 1	14 ± 2	13 ± 2	15 ± 1	16 ± 1
3	12 ± 2	16 ± 2	17 ± 1	17 ± 2	19 ± 1
5	13 ± 2	17 ± 2	17 ± 1	18 ± 2	20 ± 2
8	14 ± 2	18 ± 1	18 ± 1	19 ± 1	20 ± 2

Table 4.8   Effect of calcium(ii) concentration in soaking on the angle of the erector pili muscle. Errors are standard deviations of 10 measurements.

Time/h	Angle/°				
	10^{-7} M	10^{-6} M	10^{-5} M	10^{-4} M	10^{-3} M
1	12 ± 2	16 ± 1	20 ± 2	16 ± 2	12 ± 2
2	14 ± 2	18 ± 1	23 ± 2	20 ± 2	16 ± 2
3	16 ± 2	20 ± 2	24 ± 2	22 ± 2	18 ± 2
5	18 ± 2	21 ± 2	24 ± 3	24 ± 2	20 ± 1
8	18 ± 2	21 ± 2	25 ± 3	24 ± 2	20 ± 2

Table 4.9   Effect of time in soaking at 20 °C on the angle of the hair follicle and the angle of weave in the corium. Errors are standard deviations of 10 measurements.

Time/h	Angle/°	
	Follicle	Corium weave
1	16 ± 2	19 ± 3
2	22 ± 2	28 ± 2
3	28 ± 2	39 ± 2
5	33 ± 3	57 ± 3
8	41 ± 2	57 ± 2

a measurable effect of temperature and calcium concentration on the muscle *post mortem*:[24] an increase in angle correlates with muscle contraction, elevating the angle of the follicle and thereby opening the mouth of the follicle, producing an undesirable coarsened grain pattern.

The relaxation of the erector pili muscle depends on both temperature and time: low temperatures produce residual contraction, although the effect is relatively small. Similarly, the presence of calcium ion, important in the muscle-contracting mechanism, indicates that the muscle can still react, exhibiting maximum relaxation at 10^{-4}–10^{-5} M concentration, corresponding to 0.4–4 ppm calcium ion; this effect is also small.

Table 4.9 shows the corresponding effects of soaking on the corium angle of weave and the follicle angle, indicating the effects of rehydration. The following inferences can be drawn.

1. The effects of hydration outweigh the effects of the action of the erector pili muscle.
2. The rate of rehydration of the corium is faster than the rate of rehydration of the grain layer. This is presumably a consequence of the difference in physical structure.

It may be concluded that the soaking process itself will affect the follicle angle, regardless of conditions; the only parameter that has a large effect is the time during which soaking is conducted. Therefore, since the erector pili muscle is degraded in liming, the continued rehydration and swelling during that process will dominate the grain quality, so the impact of the soaking step in that regard can be ignored. It is clear that the parameters controlling the follicle angle are the area of the corium as it changes during rehydration and, to a lesser extent, the area of the grain, which is constrained by the hydrophobic elastin structures within it.

References

1. G. D. McLaughlin and E. R. Theis, *J. Am. Leather Chem. Assoc.*, 1923, **18**(7), 324.
2. A. Blaha and K. Kolomazník, *J. Soc. Leather Technol. Chem.*, 1989, **73**(5), 136.
3. K. Kolomazník, Z. Prokopova, V. Vasek and D. G. Bailey, *J. Am. Leather Chem. Assoc.*, 2006, **101**(9), 309.
4. J. E. Burns, *et al.*, *J. Soc. Leather Technol. Chem.*, 1976, **60**(4), 106.
5. G. Stockman, *et al.*, *J. Am. Leather Chem. Assoc.*, 2008, **103**(2), 76.
6. R. G. Mitton, *J. Soc. Leather Trades' Chem.*, 1953, **37**(4), 109.
7. M. A. R. Siddique, *et al.*, *J. Soc. Leather Technol. Chem.*, 2015, **99**(2), 58.
8. E. Longstaff, *J. Soc. Leather Technol. Chem.*, 1974, **58**(2), 41.
9. E. P. Lhuede, *J. Am. Leather Chem. Assoc.*, 1969, **64**(4), 164.
10. H. Hinsch, *Leder Häutemarkt*, 1969, **21**, 278.
11. C. Pilard and D. Vial, *Technicuir*, 1969, **3**, 169.
12. D. R. Corning, BLC, restricted circulation.
13. S. Rostami, J. Steele and A. D. Covington, *Proceedings, Asian International Conference on Leather Science and Technology*, Okayama, Japan, 2014.
14. D. T. Downing, in *Mammalian Waxes*, ed. P. E. Kolluttukudy, Elsevier, Amsterdam, 1976.
15. J. Pelckmans, *et al.*, *World Leather*, 2008, **21**(5), 31.
16. J. J. Leyden, *J. Am. Acad. Dermatol.*, 1995, **32**, 815.
17. J. Christner, *J. Am. Leather Chem. Assoc.*, 1992, **87**(4), 128.
18. A. D. Covington, *et al.*, *J. Soc. Leather Technol. Chem.*, 1999, **83**(4), 215.
19. A. D. Covington, C. S. Evans and M. Tozan, *J. Am. Leather Chem. Assoc.*, 2002, **97**(5), 178.
20. J. Durga, *et al.*, *J. Am. Leather Chem. Assoc.*, 2015, **110**(1), 7.
21. V. L. Addy, A. D. Covington, D. Langridge and A. Watts, *J. Soc. Leather Technol. Chem.*, 2001, **85**(2), 52.
22. M. M. Mutlu, *et al.*, *J. Soc. Leather Technol. Chem.*, 2009, **93**(1), 18.
23. E. G. Bayramoglu, *J. Am. Leather Chem. Assoc.*, 2007, **102**(11), 347.
24. L. Ma, MSc thesis, University of Northampton, 2008.

5 Unhairing

5.1 Introduction

Unhairing and liming are often linked because the traditional processes of hair dissolution and alkaline hydrolysis combine the process steps into one. However, strictly they ought to be thought of as separate processes and, in modern processing, the steps are increasingly commonly conducted separately.

As the term implies, unhairing is the process of removing the hair from the pelt. It is one of the dirtier aspects of leather processing, from the point of view of the odour created (typically from the sulfide employed and the decomposed protein) and the polluting load generated.[1] The traditional method of removing the hair is to dissolve it, called 'hair burning'; this is an example of 'low-tech' processing, so called because it can be applied to closed drums and does not need any active process control as the reactions progress. It is technically more difficult to undertake the process of hair removal in alternative ways, 'hair saving', keeping the hair intact while removing it; each technology requires a different degree of process control. However, it is certain that in the future, tanners will be universally required to adopt some form of hair-saving technology. The two approaches can be compared as set out in Table 5.1.

Frendrup and Buljan[2] reviewed the technologies of hair saving and made the comparisons of pollutant production given in Table 5.2.

Tanning Chemistry: The Science of Leather 2nd edition
By Anthony D. Covington and William R. Wise
© Anthony D. Covington and William R. Wise 2020
Published by the Royal Society of Chemistry, www.rsc.org

Table 5.1 Comparison between hair burning and hair saving.

	Hair burning	Hair saving
Technology/ process control	Simple technology Minimum process control Relies on undesirable chemicals	More complicated technology More process control needed reagents may be safer
Hair removal	Incomplete – staple remains within the follicle	Complete – allows better dyeing of cleaner grain
Epidermis removal	Usually complete	Might need to be addressed separately, depending on the chemistry/biochemistry
Environmental impact	High, due to COD/BOD[a] from dissolved protein and process chemicals	Big reduction in BOD/COD
Environmental perception	Not clean	Cleaner/clean
Advantages	• Technically simple • Reliable • Capable of improvement, *e.g.* prior hair shaving • Combined with liming	• Cleaner • Better grain quality, *e.g.* for dyeing • Recovered hair has value
Disadvantages	• Dirty process • Residual hair can reduce leather quality	• Residual hair must be dissolved by hair burning technology • Liming is a separate process step

[a]COD, chemical oxygen demand; BOD, biochemical oxygen demand.

Table 5.2 Comparisons with hair burning of hair saving on pollutant production.

Pollutant	Discharge from hair saving/ kg tonne^{-1} raw hide	Reduction compared with hair burning/%
Total solids	60	30
COD	50	50
Total Kjeldahl nitrogen	2.5	55
Sulfide	0.6–1.2	50–60

5.2 Keratin and the Structure of Hair

Keratin differs from the other main proteins in skin in its elemental composition, as shown in Table 5.3.[3]

Keratins vary in structure and composition, depending on the application. The amino acid composition varies: keratins contain less glycine and proline than collagen and no hydroxylysine, whereas the contents of cystine and tyrosine are high. The feature of importance here is the ease with which cysteine can be converted to cystine by a crosslinking oxidation reaction (Figure 5.1); this takes place in the prekeratinised zone during the growth of hair, which is an equilibrium

Table 5.3 Elemental composition of skin proteins.

Protein	Total nitrogen/%	Total sulfur/%
Collagen	17.8	0.2
Elastin	16.8	0.3
Keratin	16.5	3.9

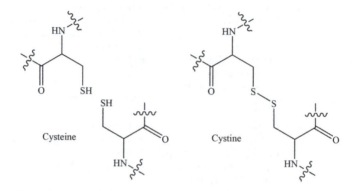

Figure 5.1 Interconversion of cysteine and cystine.

that can be exploited in the removal of hair from hides and skins (and indeed from humans by depilatory cosmetics). Keratin is usually encountered in the β-form, which is a stretched helix, in the way that collagen's fundamental structure is helical. It forms fibrils created from protein chains linked *via* disulfide bonds, which are the target of many of the unhairing techniques.

The structure of hair or wool is illustrated in Figure 5.2, using human hair as an example of mammalian hair, originally published in *Gray's Anatomy*; the figure also demonstrates the relationship between the hair, with its accompanying structures, and the epidermis.

- The bulb – containing the papilla, this is the structure from where the hair grows. Only in the case of young, papillary hair is the hair still attached to the base of the follicle; in the case of bovine hair, the mature or club hairs are shed two or three times per year.[1] The structure of the bulb is protein, but not keratin. It can be used as the point of attack in hair-saving technologies
- The prekeratinised zone – this is the region in the growing hair where the keratin is laid down and disulfide links are created; it is situated above the bulb and extends part way up the follicle. It is protein, with some of the characteristics of keratin. It also may be the point of attack in hair-saving technologies.

- The cuticle – this is the outer surface structure of the hair staple, illustrated in Figure 5.3. It is made of hard keratin, *i.e.* containing a high concentration of disulfide links, which is not fibrous in texture. The cuticle is composed of sheet-like cells that overlay each other, making it relatively chemically inert. In the time scale of hair burning, it is not degraded, merely broken off from the dissolved cortex, visible in solution as suspended solids.

The hair surface is shown in Figure 5.3, illustrating the scales that form the outer surface. The degree to which they lie flat determines the ability or otherwise to lock the shafts together, seen in the tangling of long human hair, usually a relatively mild effect, or in the creation of felt from fur or sheep fibre, an extreme effect. In the latter case, the cuticle scales can be chemically lifted for a great tangling effect, when the traditional material for hats is felt made from beaver or hare fur. The cuticle has an inner structure: layers are the epicuticle, the exocuticle and the endocuticle. The structure is illustrated in Figure 5.4 and later in Figure 5.10; in the former, alkaline degradation of the whole of the cuticle leaves the cortex exposed, whereas in the latter, sulfide and enzyme degradation leaves the cuticle layer, from which the scales can flake away.

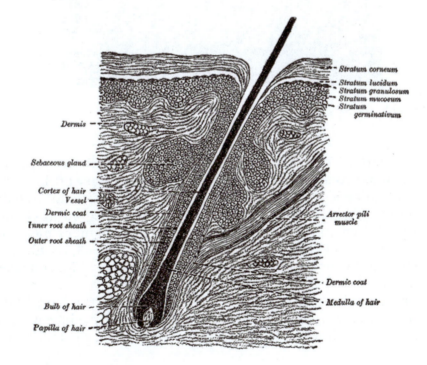

Figure 5.2 Structure of hair.

(a) (b)

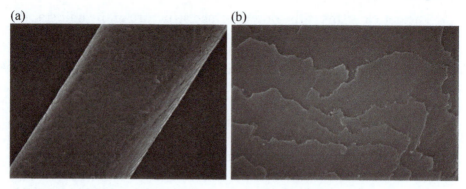

Figure 5.3 Cuticle of black bovine hair: magnification (a) ×600 and (b) ×3000. Courtesy of A. Onyuka.

Figure 5.4 Alkali-degraded cuticle of pig bristle (magnification ×200).

- The cortex – this is the inner structure of the hair, comprising the majority of the cross-section. Its structure is soft keratin, containing less sulfur than the cuticle, and it is fibrous. In hair burning, this is the main target of the keratin-degrading chemistry (often sulfide). This is where the hair pigment melanin is situated.[4] The cortex is illustrated in Figure 5.5; its fibrous structure is not immediately apparent because of its compact nature.
- The medulla – this is the central structure of the hair, illustrated in Figure 5.5. It is protein, but not keratin, since it does not contain any sulfur. In very fine hair or wool, *e.g.* merino or

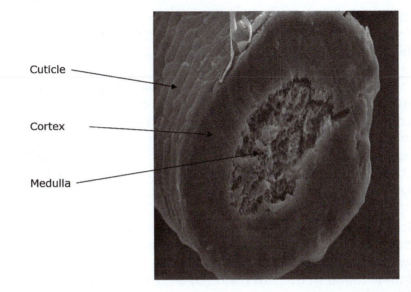

Cuticle

Cortex

Medulla

Figure 5.5 Cross-section of bovine hair (magnification ×900). Courtesy of A. Onyuka.

merino cross wool, the medulla may be missing, but it is typically present in other sheep breeds and in cattle hair. In hair burning, the medulla is not attacked, except for some alkaline hydrolysis, but it breaks up due to the mechanical action on the weak structure.

The structure of keratin does not impact greatly on leather making, for three reasons.

1. The hair may be left intact and chemically not altered, except when it might be dyed, *e.g.* in woolskins. In the case of wool dyeing, the substrate is intact cuticle; the physical and chemical structure of the cuticle means that it is resistant to the dyes most commonly used for colouring collagen, the acid and premetallised dyes. Therefore, the chemistry of the reactive dyes is required (see Chapter 16). Alternatively, keratin may be 'immunised' *i.e.* stabilised as part of a hair removal technology (discussed later), so the chemistry of keratin becomes important, but the physical structure is not exploited.
2. The hair is often dissolved, in which case the structure is lost and the only requirement is to understand the chemistry of its destruction.

3. The hair may be removed intact, when the reaction is usually focused on the structure within the follicle, rather than on the hair structure itself.

At the same time as hair is attacked in hair burning, the epidermis is also attacked. Since it is composed primarily of keratin, the structure is degraded under the conditions of hair burning, using a reducing nucleophile in alkali. It is worth considering that not all hair-saving processes include epidermis removal, but this must be undertaken at some time before tanning is initiated. The epidermis consists of five layers of structure, as illustrated in Figure 5.2.

- Stratum basale (stratum germinativum): the closest layer to the grain surface. It is a single layer of cylindrical cells that is the source of cells to create the other layers in the epidermis. Melanin is found in this layer, although it can also be found within the grain.
- Stratum spinosum: the layer above the stratum basale, it consists of several layers of polyhedral cells where keratin synthesis occurs.
- Stratum granulosum: the layer above the stratum spinosum. It is one to three layers of cells that participate in the keratinisation process.
- Stratum lucidum: the layer above the stratum granulosum. This is two to three layers of cells that have high gloss but no nuclei.
- Stratum corneum: this is the outermost layer. The cells are flattened and in the uppermost part of the layer they can flake off.

The keratinous nature of the epidermis and its physical structure mean that in the production of grain leather it must be removed (Figure 5.6). If epidermis is present, the difference in chemistry between keratin and collagen results in areas of dye resist. The flaking nature of the epidermal surface means that the surface is not reflective, unlike the grain enamel, so the scattering of light by the epidermis means that the surface appears dull.

The epidermis is attached to the grain by the basement membrane, consisting of type IV and VII collagens, as discussed in Chapter 1; this is also a potential site of attack for hair-saving unhairing. In hair burning, the combination of alkali and reducing nucleophile is usually sufficient to degrade and displace the epidermis efficiently.

Figure 5.6 Epidermis lifting from hide surface. Courtesy of LeatherWise, UK.

5.3 Hair Burning

Hair burning refers to the practice of dissolving the hair, usually with sulfide in the presence of lime. The mechanism can be summarised as follows.[5]

1. Hair burning occurs from the tip down to the pelt surface and beyond. This usually occurs faster than penetration of the reagents through the pelt from the flesh side. However, once the penetrating reagents encounter the hair at the base of the follicle, they begin to dissolve the prekeratinised zone. The net result is that a plug of hair remains in the follicle; although not attached, it is held in place by friction with the follicle wall. This is commonly observed by the thumbnail test: applying pressure across the grain surface squeezes the residual hair from the follicle. This can be done mechanically, to clean the grain surface, but as a separate operation 'scudding' is no longer common.

2. Hair burning chemistry acts on the softer keratins of the cortex and epidermis. The presence of melanin in the cortex gives rise to the observed hair colour, which is alleged to influence the rate of degradation;[6] there are indications that there may some truth in the suggestion that black hairs dissolve more slowly than white hairs.[7] However, the melanin is released into the solution and remains as a component of the 'scud', residual pigmented solids; scud can affect the grain quality by being driven into the surfaces of the pelt by mechanical action, causing uneven discoloration.

3. The cuticle is typically not dissolved under the normal sulfide concentration and time conditions of unhairing.[8] The scales flake off from the degraded cortex, to remain as 'scud', suspended in solution.
4. The medulla is not attacked by the unhairing chemicals (but it is attacked by proteolytic enzymes if they are present), so it merely breaks off, weakened by the hydrolytic effect of high pH.

The previously accepted mechanism for sulfide degradation of keratin at pH > 11, presented in Figure 5.7, can be summarised as follows.[3]

- The reducing nucleophilic sulfide ion attacks the disulfide bond of cystine.
- The disulfide link is broken, creating a cysteine moiety and an anionic cystine disulfide moiety.
- The disulfide moiety is attacked by sulfide and is reduced to cysteine.
- The sulfide is converted to polysulfide.

It is assumed that the polymerisation of sulfur will not progress much further, because the disulfide ion is a weaker nucleophile than sulfide and therefore is less effective in attacking the disulfide link. This can be seen in Table 5.4, which gives the oxidation potentials;[9,10] note that the reducing power of the reactions will depend on the precise chemical conditions, particularly pH.

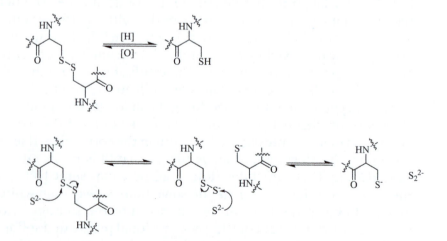

Figure 5.7   The previously accepted mechanism of sulfide hair burning.

Table 5.4 Standard oxidation potentials for sulfide species.

Species	Standard oxidation potential, $E°$/mV
$H_2S \rightleftharpoons S + 2H^+ + 2e^-$	-0.142
$HS^- + OH^- \rightleftharpoons S + H_2O + 2e^-$	$+0.478$
$S^{2-} \rightleftharpoons S + 2e^-$	$+0.476$
$S_2^{2-} \rightleftharpoons 2S + 2e^-$	$+0.428$

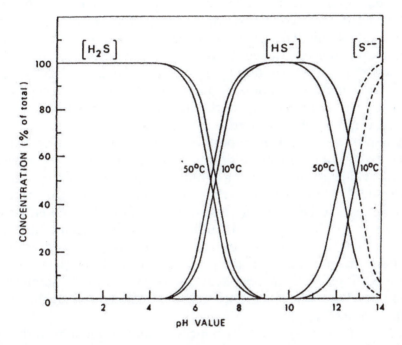

Figure 5.8 Relationship between sulfide species and pH, based on dissociation constant values of $pK_{a1} = 7.05$ and $pK_{a2} = 12.92$.[9]

The relative power of reduction can be judged because of the following equation, which relates the Gibbs free energy ($\Delta G°$) to the standard electrode potential ($E°$):

$$\Delta G° = -nFE°$$

where n is the number of faradays (the number of electrons in the reaction) and F is the Faraday constant. A positive electrode potential results in a negative free energy, which means a spontaneous or favoured reaction.

The mechanism presented in Figure 5.7 is based in part on the assumption of the relationships between the sulfide species, shown in Figure 5.8, in particular low pK_a of the second ionisation, which creates equilibria at pH values familiar in lime buffer[9] – hence the frequent assertion that the unhairing species is the sulfide dianion.

At this point, it must be noted that a problem has been raised regarding the chemistry of sulfide: the very existence of the S^{2-} species in aqueous solution has been questioned. The problem lies in the value of the pK_a of the dissociation of hydrosulfide to S^{2-}: the previously assumed value of 13 has been revised upwards to 19, which means that at pH 12.6 the concentration of sulfide ion in water can only be vanishingly small. The latest argument is based on high-resolution Raman spectroscopy, which could not confirm the presence of sulfide, whereas it was clear that hydrosulfide was detectable at very high pH.[11] The conclusion is not completely confirmed and accepted, but the evidence is compelling, so the new situation must be addressed and the implications reviewed.

Up to 2014, the literature reported the value for the second dissociation constant of hydrogen sulfide as 12.9. A more recent suggestion is that this value is not correct and a more accurate value for pK_{a2} should be about 17. There is controversy concerning the revised value and there appears to be some agreement that it should be 19; either way, this would in effect eliminate sulfide ion from consideration in an aqueous medium. There is supporting evidence for this proposal in the literature; it is not as yet accepted chemistry, although current and recent editions of the 'Rubber Book' quote pK_{a2} as 19.[10] Therefore, the possibility should be recognised and the implications for the leather industry assessed.

In 1948, Bowes assumed, without reference to pK values, that sulfide was hydrolysed to completion at pH 12:

$$Na_2S + H_2O \rightleftharpoons 2Na^+HS^- + OH^-$$

and, therefore, the unhairing agent is hydrosulfide.[12] It may be noted from Table 5.4 that the quoted redox potentials for sulfide and hydrosulfide are the same in older reference sources[9] and in the latest definitive ref. 10.

In 1956, Merrill reviewed the literature,[13] reporting that some sulfur is lost from degraded keratin, and Bieńkiewicz stated that sulfide-based hair dissolution is negligible below pH 11.[3] Sodium hydroxide is ineffective as a pulping agent below pH 13, so there is a clear pH effect on unhairing. The implication is that the rate can be expressed in the following generalised ways:

$$\text{Rate } (S^{2-}) = k_s[S^{2-}]^a$$
$$\text{Rate } (HS) = k_{HS}[HS^-]^a[OH^-]^b$$

However, the order of the reaction has not been defined. If the latter case applies, the mechanism of attack at the disulfide link and formation of polysulfide will have to be redefined; it may include an analogous attack at the disulfide links in keratin by the nucleophilic hydrosulfide, perhaps with activation by hydroxyl ion.

The theoretically required amount of sulfide can still be calculated as follows:

$$\text{Cystine content of hair} = 50 \text{ mmol per 100 g}$$

If the requirement for hair burn is 2 mol of sulfide per cystine:

$$= 100 \text{ mmol } S^{2-} \text{ or } HS^- \text{ per 100 g of hair}$$
$$= 7.9 \text{ g } Na_2S \text{ or } 5.6 \text{ g NaHS per 100 g of hair}$$

If the hair is 10% of the hide weight, then the sulfide requirement

$$= 0.8\% \text{ } Na_2S \text{ or } 0.6\% \text{ NaHS}$$
$$\equiv 1.1 \text{ or } 0.8\% \text{ sulfide flake}$$

where 'flake' is the designation for industrial sodium sulfide or sodium hydrosulfide, both quoted by the manufacturer as 70% Na_2S or NaHS, respectively.

The typical industrial offer is 2–3% 'sulfide' flake, *i.e.* at least double the theoretical requirement, but since this is considerably higher than the calculated amounts range, the doubt concerning the sulfide species does not affect the unhairing technology.

The rationale for the excess can be expressed as follows.

1. Loss from oxidation. Sulfide species are oxidised in the presence of atmospheric oxygen, although the rate is not fast enough to use up a significant proportion of the total offer within the timescale of typical unhairing.
2. Errors in calculation, particularly the assumption of weight of hair, which can vary seasonally.[14]
3. The rate of reaction depends on the sulfide concentration and the extent of the reaction depends on rate multiplied by time; the available time is fixed, so the extent of the hair dissolution reaction is controlled by the rate of reaction. Since the available time is controlled by production conditions, the outcome must be controlled by the chemical conditions.
4. The impact of a revision in the mechanism of sulfide hair burning in favour of hydrosulfide will not affect the argument significantly.

Although the new view of sulfide speciation does change the science and the mechanism of hair degradation, it does not change the observation made about the reaction, nor does it change the technology of the process step. It was always an option to combine lime with hydrosulfide, since the equilibrium pH is 12.6, so there need be no change there. It begs the question, what is the difference between reagents labelled NaSH and Na$_2$S? If the new findings are accepted, a reagent labelled sodium sulfide is merely sodium hydrosulfide plus some sodium hydroxide. The commercial consequence is: buy the cheaper version and carry on regardless!

Regarding the relationship between hydrosulfide and hydrogen sulfide, there is no suggestion in the literature that there is any error in pK_{a1}. The conditions that might lead to the production of hydrogen sulfide gas are critical for safety. Note that H$_2$S attacks the nervous system – it is more toxic than hydrogen cyanide (HCN) and therefore should not be underestimated. If the pH is allowed to decrease, the unhairing reaction will slow and ultimately stop;[12] see Section 5.4.

It is worth reflecting on the contention that using sulfide is technologically the safest of the traditional hair burning processes. Unhairing can be achieved with alkali alone (see later), but it was recognised long ago that the rate of hair dissolution could be increased by the addition of so-called 'sharpening' agents, of which sodium sulfide was one. In the Victorian leather industry, it was not uncommon to use other reagents, such as sodium cyanide and arsenic sulfides.[1]

It has been common practice in the industry to use a mixture of sodium hydrosulfide and sodium sulfide products as an alternative to sulfide alone. The technological reason was never clear, but the current development would suggest that the only difference would be in the pH profile of the process step. This is discussed in Chapter 6, where it is reported that there is no discernible impact on the outcome of processing.

5.3.1   The Role of Swelling

In hair burning, the pelt will be swollen by the alkaline conditions (see Chapter 6). This can contribute to a problem if the rate of swelling is faster than the rate of hair degradation. Consider the following sequence of events.

1. The pelt swells quickly. The outcome is that an additional portion of the hair shaft becomes hidden by the follicle in the swollen grain layer.

2. The degradation of the hair proceeds from the tip down to the surface. At the surface, the rate slows because of the reduced surface area, so the hair is degraded only a little way into the follicle. Also, by this time, the rate of keratin degradation has slowed as sulfide is used up. The net effect is that the amount of hair shaft degraded is less than it would be if the grain had not swelled as quickly.
3. In deliming, when the pH is lowered, the pelt is depleted. Once the depletion is complete, some of that hair trapped by the swelling now protrudes from the mouth of the follicle. If it is visible, as it would be if it is coloured, there is a loss of value, because the grain may have to be corrected (buffed to remove faults).
4. If the result of a mismatch of swelling and hair burning rates is that hair is left at or below the mouth of the follicle, then dyeing will be affected. The surface of the leather has effectively become a combination material, part collagen and part keratin. The dyeing characteristics of these two proteins are different: under the usual conditions of chrome leather dyeing, keratin is less reactive than collagen, so the colour struck is typically paler than it should be.

The problems associated with swelling in hair burning can be overcome by the so-called Herfeld drum painting method.[15] The process is as follows:

- soak
- drain
- 0.75–1.0% sodium hydrosulfide flake (70% NaHS)
- 0.25% surfactant
- run 15 min, stand 15 min, to give pH 11.0
- add 2.5% sodium sulfide flake
- run 150 min intermittently, to give pH 12.5
- check hair pulping
- add 250% water at 25–28 °C
- add 3% lime
- run conventionally for about 15 h.

The key to this technology is that the hair burning is conducted in the minimum amount of water, therefore the pelt cannot swell – swelling can occur only in the presence of water to plump up the structure, regardless of the chemical status of the fibre structure. Hair burning proceeds below the grain surface, part way inside the follicle. If the pelt is not swollen, then the residual hair cannot protrude from the mouth of the follicle, as it might if swelling occurred before hair pulping is completed. In this way, the grain quality is not compromised by visible hair. However, tumbling hides in a drum without water can cause grain abrasion, particularly under alkaline conditions, so this process is rarely used.

5.3.2 Chemical Variations

5.3.2.1 Lime-free Hair Burning

The calcium hydroxide in conventional hair burning can be replaced with sodium hydroxide, to provide the high pH required to maintain the reaction rate. However, as a strong base, the pH will tend to be high unless controlled by dosing, thereby running the risk of over-swelling the pelt, *i.e.* creating irreversible swelling (see Chapter 6). It does not have a buffering effect, like lime. Unhairing can be conducted with sodium hydroxide alone, so-called 'caustic soda unhairing'. This has been used commercially, but the requirement for pH > 13 means that the risk of over-swelling is high.

5.3.2.2 Sulfide-free Hair Burning

Other reducing agents have been proposed and a comparison of the reducing power of candidate reagents is presented in Table 5.5[16] using the standard oxidation potentials. In practice, the effects of the reagents will depend on the chemical conditions, particularly the pH.

Amines have been used successfully,[17] but then fell into disuse when it was found that they could be converted into nitrosamines, which are potent carcinogens.[18] Thioglycolate is a viable option;[19] although not established in the leather industry, it is

Table 5.5 Standard oxidation potentials of some unhairing agents.

Chemical	Formula	$E°/mV$
Trimethylamine	$N(CH_3)_3$	1.19
Thiourea	$(H_2N)_2C{=}S$	0.90
Dithionite	$S_2O_4{}^{2-}$	1.12
Hydrosulfide	HS^-	0.48
Cysteine	$H_2N{-}CH(SH){-}CO_2H$	0.05

commonly used in cosmetic depilatory formulations. Sodium aluminate ($NaAlO_2$) can be used as a hair burning reagent, together with sodium sulfide, at 2–35 and 1.3–1.5%, respectively, plus 0.5% sodium hydroxide, but the results were not as good as those for the control.[20]

5.4 Immunisation

Under conditions of moderate alkalinity, pH 11 and above, *e.g.* in lime solution, another reaction can take place with keratin. Unlike the degradation reaction, this is based on hydroxyl ion attack at the methylene group on the cystine sidechain. The result is the creation of cysteine and serine; the latter loses water to form dehydroalanine. The products can then react to form a new crosslink, methionine.[21,22] The overall effect is for the disulfide link to become a single sulfur or thioether link; since the thioether link is much less susceptible to attack and scission by reducing nucleophiles, the keratin becomes resistant to conventional hair burning and is said to be 'immunised' against degradation.

The generally accepted mechanism of immunisation is presented in Figure 5.9. There are, however, some questions arising from the mechanism that hitherto have not been addressed, but would be useful in understanding the chemistry and hence the technology.

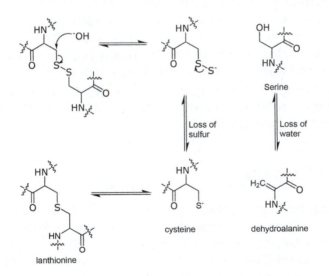

Figure 5.9 Mechanism of immunisation.

1. Why does hydroxyl ion attack at the methylene group rather than a sulfur in the disulfide bond?
2. What is the mechanism that drives the loss of sulfur from the broken disulfide bond?
3. Does hydroxyl ion function in a catalytic manner?

An accompanying reaction is between dehydroalanine and lysine, to create lysinoalanine:

$$R_2C=CH_2 + H_2N-(CH_2)_4 - R \rightleftharpoons R_2CH-NH-(CH_2)_4 - R$$

Mackay proposed that part of the immunising effect in the conventional lime/sulfide unhairing process 'gone wrong' was the part played by calcium ion:[23]

$$R-CH_2-SOH + Ca^{2+} + HS-CH_2-R$$
$$\rightleftharpoons R-CH_2-S-Ca-S-CH_2-R + H_2O$$

In this way, immunisation in the presence of a low concentration of calcium ion (0.01 M) can be very effective in stabilising the keratin structure, completely preventing the normal supercontraction reaction.

Valeika *et al.* reported that the application of 20 g L^{-1} of sodium metasilicate, $Na_2SiO_3·5H_2O$, for 3 h was effective for immunisation in the absence of divalent ions;[24] hair saving can be achieved with subsequent attack by 0.9% sulfide at the prekeratinised zone.[25]

The chemistry of immunisation is exploited in some better known hair-saving processes (see later). Tanners may experience the implications of immunisation if they make the mistake of adding the lime before the sulfide. The result is difficulty or inability to remove the hair by conventional hair burning; the use of high concentrations of unhairing agents is likely to damage the grain. The only option open is to shave the hair and burn the remainder conventionally (see later).

5.5 Hair Saving

As the term hair saving indicates, there is the possibility of removing the hair intact, *i.e.* saving the hair structure. All these processes target the structure of the hair within the follicle, either the bulb or the prekeratinised zone.[2] The effect is to ensure detachment of the hair from the base of the follicle; the hair may be partially degraded, reducing the frictional forces holding the fibres within the follicles. Some technologies rely on immunisation to protect the hair staple from degradation. Probably the best known method in the industry is the Sirolime process, developed at CSIRO in Australia.[26] The process can be summarised as follows, with comments in italics.

- Soak.
- Impregnate with sodium hydrosulfide for 2 h; the pH falls from 11.2 to 8.5. (*At low pH there is no keratin damage as the reagent penetrates the cross-section, to reach the base of the follicle.*)
- Drain. (*Now the only sulfide in the system is within the pelt.*)
- Add calcium hypochlorite, run 5 min. (*Any residual sulfide on the outside of the pelt is oxidised, to prevent reaction on the hair when the unhairing agent is activated.*)
- Add lime, raise the pH to 12.5. (*At this point the hair may be immunised; this is not strictly required, but it is not counterproductive in the process. Rapid penetration by hydroxide ion from the flesh side – the epidermis remains a barrier – causes high-pH attack at the prekeratinised zone to sever the link between the hair and the follicle. The hair is loosened after 1 h.*)
- Run to remove the hair and filter the float. (There is no mechanism by which the hair will come out of the follicle of its own accord – it is held in place by friction. Therefore, the rubbing action of the pack is needed to pull the hairs out. This is an effect that depends on the hair staple remaining intact; it does not apply to residual hair in a hair burning process.)
- Relime. (There are always short hairs remaining after hair-save processing – it is a measure of the success of a technology how much reliming, typically with sulfide, is required after hair saving to dissolve residual hair.)

The process was assumed to depend on the controlled conversion of sulfide species, but revising the thinking to a pH effect does not invalidate the technology in any way.

5.6 Variations in Unhairing Technologies

5.6.1 Heidemann's Darmstadt Process

This technology was developed as part of a rapid processing approach to tanning: the objective was to produce wet blue in 6 h from raw.[27] The method does work, although the quality of the product is not ideal, as may be judged by the absence of process steps that would allow the transport of unwanted components out of the pelt. The process can be summarised as follows.

- Soak/wash. (*In this technology, it is essential to remove the dung: here, this is a more important step than green fleshing, since dung can protect the hair from chemical degradation.*)
- Green flesh. (*Necessary to ensure sulfide penetration, as part of the opening-up process.*)
- Hang hides over bars.
- Submerge in 10% sodium sulfide solution or spray sulfide solution on the hair side. (*The time in sulfide must be limited, otherwise hair accumulates in the solution. Some swelling occurs: 4–6% on hide weight.*)
- Mechanically press to remove excess sulfide.
- Hang for 10 min.
- Scrape off epidermis and hair.
- Hang in 10% sodium peroxide solution or spray on the solution. (*Conventional times are heavy ox 60 min, calfskin 35 min, goatskin 25 min. There is an exothermic reaction that removes residual hair and epidermis, has a bleaching effect and opens up the fibre structure. Recycling is limited to about 50 times, owing to buildup of sodium sulfate.*)
- Mechanically press to remove excess sodium peroxide.
- Neutralise the surface to pH 5.0 over 5 min with a 10% solution of ammonium sulfate and hydrochloric acid.
- Clean and split.
- Complete neutralisation.
- Pickle conventionally for 1 h.
- Chrome tan in concentrated 50% basic chromium(III) sulfate.
- Hot age. (*Boil fast leather is achievable.*)

5.6.2 Oxidative Unhairing

This is a technology that has been used commercially, based on the product sodium chlorite, sold as Imprapell CO. It is conducted under acidic conditions, as follows. Note that this reaction involves the changing of an oxidation state to both higher and lower oxidation states, known as disproportionation.

$$4ClO_2^- + 2H^+ \rightleftharpoons 2ClO_2 + ClO_3^- + Cl^- + H_2O$$
$$\text{Cl oxidation state: +III +IV +V −I}$$

$$4R-S-S-R+10ClO_2 +4H_2O \rightleftharpoons 8R-SO_3^- +8H^+ +5Cl_2$$

The fire risk of solid sodium chlorite is exchanged for the toxicity of chlorine dioxide gas, which is the actual unhairing reagent, and then one of the products of the reaction is chlorine, used in the past as a chemical weapon. Because of the strong oxidising chemicals involved, the process cannot be conducted in wooden vessels, so stainless-steel or polypropylene vessels have to be employed.

Other oxidising agents that have been proposed are hydrogen, sodium and calcium peroxides and peracids, *e.g.* peracetic acid, perborate, hydrogen peroxide with potassium cyanate or urea.[28–30] In the last case, the oxidising agent is assisted by the action of hydrogen bond breakers. It has been reported that 5% sodium percarbonate with 7% sodium hydroxide in 200% float unhairs in 4 h.[31] Similarly, magnesium peroxide with potassium peroxymonosulfate will unhair, but requires 20% sodium hydroxide to achieve the required pH of 13.4.[32] Although alternative options frequently appear in the literature and claims of safe operation and technologically favourable outcomes have been made, none has achieved industrial acceptance; the formulations briefly set out above indicate why not.

5.6.3 Reductive Unhairing

This chemistry of reduction is known in principle from the use of reagents discussed earlier, but more powerful chemical reactions have not found commercial acceptance. The mechanism can be summarised as follows, using sodium borohydride as an example:

$$BH_4^- +3H_2O \rightleftharpoons 4H_2 +BO(OH)_2^-$$
$$R-S-S-R+H_2 \rightleftharpoons 2R-SH$$

The reaction may be undertaken with sodium borohydride or lithium aluminium hydride.[33] In either case, the reactants are pyrophoric, which means they will spontaneously combust in air, thereby posing a fire hazard. As with artificial ageing of pelt (see Chapter 2), the principle is proved, but the technology is undesirable.

5.6.4 Acid Unhairing

The technology of exploiting the lyotropic effect for unhairing was developed in Australia for fellmongering sheepskins.[34,35] Concentrated acetic acid and salt are applied on the flesh side of fresh skins (the process does not work on dried or salted pelts); overnight the wool is loosened and the epidermis separated from the grain by the combined actions of the acetic acid and the autolytic enzymes within the pelt. Schlösser and Heidemann reported that the same reaction could be achieved for cattle hides by the *in situ* fermentation of *Lactobacillus*, their so-called 'sauerkraut process': here the active agent is lactic acid, which loosens the hair and destroys the basement membrane, allowing the epidermis to be removed almost intact.[36]

5.7   Enzymes in Unhairing

5.7.1   Enzyme-assisted Chemical Unhairing

Conventional hair burning can be accelerated by the presence of so-called 'alkali-stable' proteolytic enzymes, obtained from bacterial fermentation. In fact, such enzymes are no more stable in alkali than any other, but their activity profile indicates they have significant activity at high pH, so they can be used in liming.[37] When protein has been chemically damaged, protease will accelerate the breakdown. This means that keratin breakdown is faster, as is the degradation of the non-structural proteins and also the collagen itself. Hence there is a danger of loosening the corium structure if the reactions go too far. This makes the process high risk. The outcomes of this process are as follows.

- cleaner grain, better for dyeing;
- more opening up, so the leather is softer;
- allows a reduction in sulfide offer or a reduction in process time;
- removes the need for a bating step.

A typical feature of these bacterial enzymes is that they exhibit elastolytic activity, *i.e.* they degrade elastin. This changes the characteristic of the leather and produces additional outcomes.

- 'Growth' (ripples in the skin, particularly in the neck) is removed, so the leather is more uniformly flat.
- The skin is thinner, with greater surface area, because the elastin is degraded.
- Strength per unit thickness is increased.
- Leather is less stretchy.

These changes are clearly suitable only for some leathers, such as upholstery or clothing, where softness and drape are useful properties.

Other leathers that require stiffness, stretch or resilience, referred to in the jargon as 'stand', such as shoe upper leather for structured shoes, would not benefit from this biotechnology.

5.7.2 Chemical-assisted Enzyme Unhairing

This a more speculative application of enzyme technology, developed by BLC,[33] a version of hair saving that has not been used in industry to date The argument is that the shortest way to the prekeratinised zone is through the grain. However, the epidermis, as in life, is a barrier to the penetration of chemicals and microorganisms and their products. Therefore, the approach here is to damage the epidermis sufficiently to allow a proteolytic enzyme to penetrate to the base of the follicles. The process can be summarised as follows.

- Conventional soak.
- Disrupt the epidermis with a conventional unhairing agent, *e.g.* sulfide. (*This is a short process, taking only a few minutes, exploiting the reactivity difference between the epidermis and the hair. If the time is short, the hair is relatively unaffected.*)
- Drain the solution for recycling.
- Apply alkali-stable enzyme.
- Run and filter loosed hair.
- Relime for short hairs.

This too is a high-risk process, which requires a high level of process control. It may also cause looseness in the leather because of the use of alkali-stable enzymes with their elastolytic action.

5.7.3 Enzyme Hair Saving

All enzyme options for hair saving involve proteolytic species, but their effectively uncontrolled impact on the pelt has ruled them out for routine production. One suggestion is to eliminate associated collagenolytic activity, for example from the enzymes expressed by *Bacillus licheniformis*: the bacterium produces a serine protease and a collagenolytic metalloproteinase. By including 10 mM ethylenediamine tetraacetate (EDTA) in the process to sequester the metal ion, 91% of the collagenolytic activity is lost; at the same time, 67% of the collagenolytic activity by the serine protease is lost, but 90% of its proteolytic activity is retained. In this way, the enzyme action is much more targeted.[38]

Aspergillus tamarii also expresses a serine protease, active at high pH, which can attack the hair from the hair side. When the reaction was assisted by sulfide there was grain damage, but when the reaction was assisted by sodium percarbonate there was no grain damage and removal of the fine hairs.[39]

An alternative approach to the use of enzymes is to attack another feature of the hair structure in the follicle, namely the epidermis. The argument is that by degrading the structure by which the epidermis is attached to the dermis (grain enamel), the basement membrane the hair will also be removed. One of the components in the basement membrane is type VII collagen, which can be degraded by dispase, a mixture of proteolytic enzymes. Trials showed that the hair could be loosened and removed without causing associated damage to the fibre structure. Dispase also has elastolytic activity, hence the elastin content of the pelt was also degraded.[40]

5.7.4   Keratinase

Since the hair is largely composed of keratin, it might seem obvious to attack it with keratinase enzymes. However, there has been a traditional reluctance to use these reagents in industry because of the potential danger to operatives, associated with possible degradation of the epidermis. The feasibility of this biotechnology for removing residual hair from wet blue leather has been demonstrated,[41] but the idea has not been pursued.

5.8   Painting

For the special case of woolskins, the technology of painting has been developed. The idea is to drive the unhairing process, more strictly called 'fellmongering', from the flesh side of the pelt. In this way, the chemicals do not make contact with the wool and devalue it. In these circumstances, although prices fluctuate, it is often the case that the wool is more valuable than the pelt. Paint compositions are summarised in Table 5.6.[42]

Table 5.6   Paint compositions.

Component	Typical/g L^{-1}	New Zealand Quikpul/g L^{-1}
Sodium sulfide flake (64% Na_2S)	100–150	185
Calcium hydroxide	400–600	200
Sodium hydroxide		30

The paint is mixed and, importantly, left overnight to thicken – this is a function of the lime content. (Note that if lime is mixed with a limited amount of water, it will thicken to a paste, which can be used as a cement for building – known as lime mortar, used for centuries before the advent of modern Portland cement, which is based on calcium sulfate hemihydrate). The following day, the paint is applied to the flesh side of the pelt; this may be done by hand or mechanically. The pelts are then stacked, flesh to flesh, for several hours, usually overnight. The following day, the pelts are 'pulled': this means the wool is removed, either by hand or by machine. If the reaction has been successful, the wool is wiped off the pelt with ease. Processing continues with conventional liming and beamhouse processing.

The principles underpinning this technology are as follows.

- The paint is thixotropic: it can be sprayed onto the pelt, but does not run off. A Newtonian liquid is one in which the resistance to a shear force (stirring) is proportional to the force, *i.e.* the faster one stirs, the harder the work becomes. Alternatively, a thixotropic liquid is non-Newtonian, *i.e.* the resistance to a shear force is inversely proportional to the force. This means that if the liquid is semi-solid, a gel, then stirring it makes it runny, but as soon as the stirring stops it becomes a gel again: this is the principle of non-drip paint.
- The concentration of sulfide ion is high enough to penetrate the pelt by diffusion alone in the time allowed, because there is no mechanical action to drive penetration. Sulfide reacts at the pre-keratinised zone of the wool shaft, to degrade the keratin and detach the wool in a hair-saving process.
- The pH of the system is maintained by the buffering action of the lime for maximum unhairing rate and alkali-catalysed hydrolytic reaction can occur within the pelt, for opening up.
- The concentration of water is low, hence swelling is minimised.

The composition of the paint can be varied as follows.

1. Replace the sulfide with another unhairing reagent, *e.g.* thioglycolate, $-S-CH_2-CO_2-$. Thioglycolate is marginally less effective than lime and sulfide; in addition, in trials the workers did not trust it because it had a different smell.[43]
2. Replace the lime with another thickener, *e.g.* synthetic polymer, together with another alkaline compound, *e.g.* sodium hydroxide.

The use of unhairing paint is not confined to woolskins – it can be used for any thin rawstock, which allows the reagents to penetrate through the cross-section in the available time. Hence it has been used commercially for immature skins such as calf, so that the liming reaction is controlled. Another useful application is for pigskins, in which the shell region of the skin is particularly tough. However, if the pelt were to be conventionally limed enough to soften this region, the rest of the pelt would become over-limed. The solution is to treat the flesh side of the shell areas with a lime–sulfide paint, leave overnight, then lime in the normal way. In this way, only the hard shell is doubly limed and hence selectively opened up and softened.

5.9   The Role of Shaving in Unhairing

The simple act of cutting the hair completely changes the technology of hair removal. El Baba *et al.*[5,8] showed that removing the majority of the hair from cattle hide with electric clippers (therefore not a particularly close shave) was sufficient to reduce the COD and suspended solids to nearly half the levels obtained by conventionally hair burning the full staple. The hair cutting allows a reduction in the sulfide used, which can be compounded by including alkali-stable enzyme, to accelerate the breakdown of degraded hair. Because the hair burning is initiated closer to the pelt surface, the reaction proceeds further down the hair shaft than in the conventional process, resulting in less hair close to the surface. Closer examination of the outcome of hair burning cut hair revealed a new mechanism of unhairing. Figure 5.10 illustrates the outcome of hair burning technology applied to cut hair: the reaction is concentrated in the cortex, effectively hollowing out the hair, accelerated by the presence of proteolytic enzyme, which targets not only the degraded keratin but also the intact medulla.

The intact cuticle is not strong enough to resist collapse, which reduces the friction with the follicle wall. Since the hydrolytic attack reaches the base of the follicle, there is no anchoring mechanism and the residue of the hair can be squeezed out by normal mechanical action. In this way, the outcome is grain quality normally associated with hair-saving technologies but obtained with simple hair burning technology.

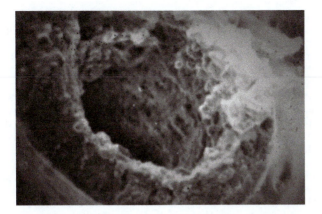

Figure 5.10   Residual hair, showing loss of cortex and medulla, but remaining cuticle. Courtesy of *Journal of the Society of Leather Technologists and Chemists*.

References

1. C. A. Money, *J. Soc. Leather Technol. Chem.*, 1996, **80**(6), 175.
2. W. Frendrup and J. Buljan, *Hair-Save Unhairing Methods in Leather Processing*, US/RAS/92/120, UNIDO, 2000.
3. K. J. Bieńkiewicz, *Physical Chemistry of Leather Making*, Krieger, Florida, 1983.
4. P. Hallegat, R. Peteranderl and C. Lechene, *J. Invest. Dermatol.*, 2004, **122**, 381.
5. H. A. M. el Baba, A. D. Covington and D. Davighi, *J. Soc. Leather Technol. Chem.*, 1999, **83**(4), 200.
6. F. O'Flaherty and W. T. Roddy, *J. Am. Leather Chem. Assoc.*, 1932, **27**(4), 126.
7. A. Onyuka, PhD thesis, University of Northampton, 2008.
8. H. A. M. el Baba, A. D. Covington and D. Davighi, *J. Soc. Leather Technol. Chem.*, 2000, **84**(1), 347.
9. *Handbook of Chemistry and Physics*, ed. R. C. Weast, CRC Press, 1987.
10. *Handbook of Chemistry and Physics*, ed. W. M. Haynes, 95th edn, CRC Press, 2014.
11. P. M. May, *et al.*, *Chem. Commun.*, 2018, **54**, 1980.
12. J. H. Bowes, *Progress in Leather Science: 1920-45*, BLMRA, London, 1948, p. 158.
13. H. B. Merril, *The Chemistry of Leather Manufacture*, ed. F. O'Flaherty, W. T. Roddy and R. M. Lollar, American Chemical Society, 1956, p. 257.
14. R. M. Hayman, in *Biology of the Skin and Hair Growth*, ed. A. G. Lyne and B. F. Short, Angus and Robertson, 1965.
15. H. Herfeld and B. Schubert, *Das Leder*, 1963, **14**, 77.
16. N. Weinberg and H. R. Weinberg, *Chem. Rev.*, 1968, **68**(4), 449.
17. I. C. Sommerville, *et al.*, *J. Am. Leather Chem. Assoc.*, 1963, **58**(5), 254.
18. D. G. Bailey, *et al.*, *J. Am. Leather Chem. Assoc.*, 1982, **77**(10), 476.
19. C. S. Cantera, *J. Soc. Leather Technol. Chem.*, 2001, **85**(2), 47.
20. S. H. Feairheller, *et al.*, *J. Am. Leather Chem. Assoc.*, 1972, **67**(3), 98.
21. S. H. Feairheller, *et al.*, *J. Am. Leather Chem. Assoc.*, 1976, **71**(8), 360.
22. J. Sirvaityte, *et al.*, *J. Am. Leather Chem. Assoc.*, 2016, **111**(11), 406.
23. R. Mackay, *J. Soc. Leather Trades' Chem.*, 1951, **39**, 328.
24. V. Valeika, *et al.*, *J. Soc. Leather Technol. Chem.*, 2015, **99**(5), 223.

25. J. Sirvaityte, *et al.*, *J. Soc. Leather Technol. Chem.*, 2015, **99**(5), 231.
26. R. W. Cranston, M. H. Davies and J. G. Scroggie, *J. Am. Leather Chem. Assoc.*, 1986, **81**(11), 347.
27. E. Heidemann, *Fundamentals of Leather Manufacturing*, Eduard Roether, Darmstadt, 1993.
28. S. Broneo, *et al.*, *J. Am. Leather Chem. Assoc.*, 2005, **100**(2), 45.
29. W. N. Marmer and R. L. Dudley, *J. Am. Leather Chem. Assoc.*, 2004, **99**(9), 386.
30. K. Uehara, *et al.*, *J. Am. Leather Chem. Assoc.*, 1986, **81**(10), 331.
31. W. N. Marmer and R. L. Dudley, *J. Am. Leather Chem. Assoc.*, 2005, **100**(11), 427.
32. A. G. Gehring, *et al.*, *J. Am. Leather Chem. Assoc.*, 2006, **101**(9), 324.
33. BLC The Leather Technology Centre, unpublished results.
34. J. R. Yates, *Aust. J. Biol. Sci.*, 1968, **21**, 1249.
35. C. A. Money and J. G. Scroggie, *J. Soc. Leather Trades' Chem.*, 1971, **55**(10), 333.
36. L. Schlösser and E. Heidemann, *Das Leder*, 1986, **37**, 33.
37. K. T. W. Alexander, *J. Am. Leather Chem. Assoc.*, 1988, **83**(9), 287.
38. J. Song, *et al.*, *J. Soc. Leather Technol. Chem.*, 2015, **99**(3), 115.
39. D. Anandan, *et al.*, *J. Am. Leather Chem. Assoc.*, 2008, **103**(10), 338.
40. R. G. Paul, I. Mohamed, D. Davighi, V. L. Addy and A. D. Covington, *J. Am. Leather Chem. Assoc.*, 2001, **96**(5), 180.
41. A. D. Covington, C. S. Evans and F. Foroughi, *UK Pat. Appl.*, GB 05191721, 2005.
42. M. Dempsey, *et al.*, *J. Soc. Leather Technol. Chem.*, 1978, **62**(5), 108.
43. A. D. Covington, unpublished results.

6 Liming

6.1 Introduction

As in Chapter 5, account must be taken of the likelihood of the non-existence of sulfide ion under normal conditions of processing, in favour of hydrosulfide. Here, the term 'sulfide' will be used to refer to the species present under the conditions as defined in the particular context.

In typical industrial processing, when the hair is removed by dissolving it with a solution of calcium hydroxide and sulfide, the liming step is not separated from unhairing. The unhairing step is dominated by the reaction of sulfide and could be accomplished in the absence of lime; typically, unhairing commences with sulfide alone at high pH to avoid the possibility of immunising the keratin, but lime is soon added and so the combination of steps merely extends the period of treatment with the two reagents. The hair is usually dissolved completely within a few hours, often 2–4 h, then the liming is continued, usually to a total of around 18 h. The value of using lime in this process step is that the pH is controlled only by the solubility of the sparingly soluble alkali; excess is routinely used, to allow for variations in conditions, so the solution and hence the process are self-regulating.

A feature of the conventional liming process is the accompanying swelling of the pelt, due to the high pH (see Figure 6.4 later). It should be recognised that the swelling typically achieved in industrial processing is less than the swelling that might be achieved if the pelt were to be allowed to swell to equilibrium; at the maximum extent that the conditions allow, there would be structural damage and the swelling would not be completely reversible. The extent of swelling in conventional liming is restricted by conducting the process under the conditions set out in the following.

6.1.1 Float

Typically the float length is 200% on raw weight, sometimes more, but rarely less. In this way, the pelts will tumble over and around each other, without creating the pressure/squeezing type of mechanical action characteristic of shorter float processing. Also, using a relatively long float minimises the possibility of abrasion damage to the grain. The minimum mechanical action is provided by the process vessel, either by processing in paddle-driven vats or by running conventional drums slowly and usually intermittently.

6.1.2 pH

Amide (peptide) hydrolysis is general base catalysed[1] and, in order to have a sufficiently fast reaction for industrial processing conditions, the pH must be 12–13. If the pH is lower, the unhairing chemistry does not work; as explained in Chapter 5, this is traditionally understood to arise from the equilibrium between the hydrosulfide ion and the sulfide ion being unfavourable. However, if the presence of sulfide ion cannot be invoked, the reaction mechanism is more dependent on hydroxyl ion concentration than previously thought.[2] Moreover, the liming rate would be slow because the hydroxyl ion concentration is reduced; a reduction of one pH unit means a 10-fold decrease in hydroxyl ion concentration. At higher pH, the rate is faster, but the danger of over-swelling is greater, since swelling is pH dependent.

   The theoretically required amount of alkali can be calculated as follows. To start, it is necessary to refer to the titration curve for collagen, given in Figure 1.16 in Chapter 1.[3]

   A 100 kg amount of hide contains about 40 kg of collagen and 60 kg of water. From Figure 1.16, collagen at neutrality requires about 0.4 mol NaOH for adjustment to pH 12.5. Therefore, if the hide contains 40 kg of collagen, it requires 0.64 kg of NaOH.

$$\text{At pH } 12.5, [OH^-] = 0.03 \text{ M} = 1.26 \text{ g NaOH L}^{-1}$$

Therefore, 60 kg of moisture requires 0.08 kg of NaOH.
For 100 kg of pelt, 200% process float requires 0.25 kg of NaOH.

However, if it is assumed that the contention about sulfide speciation set out in Chapter 5 is true and the role of sulfide ion can be ignored, the distribution of sulfide between sulfide ion and hydrosulfide ion can therefore be ignored, and there would be no contribution to the mass balance for alkali from the sulfide species.

Total amount of alkali = 0.64 + 0.08 + 0.25 kg NaOH = 0.97 kg NaOH
$\equiv$ 0.90% Ca(OH)$_2$ on hide weight

If the contention is ignored, the contribution of sulfide species would be equivalent to only 0.05% lime. The technological offer of lime is usually 2% on hide weight in 200% float, sometimes more; hence the offer is always in excess of the calculated amount. Therefore, the argument concerning the speciation of sulfide in solution makes only a small difference to the mass balance and no difference to the technology. The calculation makes other assumptions that may have an associated error; additional requirements, such as the presence of dung, are not included.

It is essential for consistent buffering by lime that the solution should be saturated, which means that there must be excess lime suspended in the solution; the amount in excess varies, but larger offers have no effect on the conditions in solution. For example, the Official Method (IUF 151)[4] for making standard chrome-tanned leather (wet blue) specifies the use of 10% lime in 500% float; from the analysis set out above, this represents an excess of about 90% of the offer. Any excess is a cost that creates no benefit but merely contributes to the suspended solids component of the effluent and can act as a damaging abrasive.

6.1.3 Temperature

The temperature of the liming solution does not influence swelling, except insofar as it affects the solubility of the lime and hence may alter the pH of the solution, but the temperature will control the kinetics of hydrolysis. Since the hydroxyl ion is a very effective catalyst, any acceleration of the rate would increase the sensitivity of the process to cause damage.

A feature of this high-pH process step is that the shrinkage temperature is lowered to 50 °C or less (see Figure 6.1).[5] Therefore, it is important to control the process temperature to <30 °C, so that it does not encroach within 25 °C of the denaturation temperature and cause hydrothermal damage.

Features to note from Figure 6.1 are as follows.

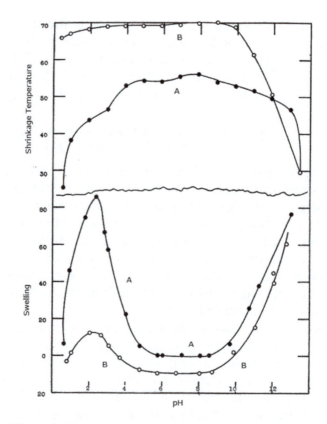

Figure 6.1 Effect of pH on hydrothermal stability: (a) without salt and (b) with 2 M sodium chloride. Courtesy of the *Journal of the American Leather Chemists Association*.

1. Studies of this kind always measure effects at equilibrium. Practical technology avoids reaching equilibrium, so the actual situation is slightly different from that presented in the graphs – the variations in industry depend on the specific conditions used.
2. The hydrothermal stability of collagen is very sensitive to extremes of pH.
3. The presence of salt influences both swelling and hydrothermal stability, discussed later and in Chapter 9.

The temperature during liming can affect the rate of hydrolysis in two ways. First, there is the influence of temperature on the solubility of lime, which therefore affects the pH of the buffer solution (Table 6.1).

Table 6.1 Effect of temperature on lime in solution.

Temperature (°C)	Solubility (g L^{-1})	pH	[OH] (M)	Relative molar concentration
0	1.85	12.7	0.050	1.2
20	1.6	12.6	0.043	1.0
100	0.79	12.3	0.021	0.5

By using these limited data, the impact of temperature in the liming working range can be interpolated in round terms. Between 20 and 10 °C, there would be an acceleration of 5% in the rate of reaction, based solely on the calculated hydroxide concentration. Between 20 and 30 °C, the effect would be a deceleration in the rate of 10%.

However, there is a general effect of temperature on the reaction rate, as defined by the Arrhenius equation, introduced in Chapter 3:

$$\ln k = \ln A - E/RT$$

Substituting the temperatures of interest (283, 293, 303 K) into the equation, using 20 °C/293 K as the reference condition, the effect of cooling by 10 °C would be to slow the reaction to 40% of the original rate; similarly, warming by 10 °C would speed up the reaction by 260%. These results are in line with the 'rule of thumb' that the rate changes by a factor of two for every 10 °C change in temperature. Therefore, the thermodynamic effect on the kinetic far outweighs the catalytic effect.

The two factors can be combined by multiplying them: the overall effect of cooling from 20 to 10 °C is to slow the reaction by 60%, but the overall effect of warming from 20 to 30 °C is to speed up the reaction 2.3-fold.

6.1.4 Time

The longer the process goes on, the greater is the degree of hydrolysis and hence the greater is the amount of damage to the protein components of the hide or skin. Therefore, limits are typically set to the time allowed for this step. For medium ox hide, the normal liming period is 18 h, with commensurately shorter periods for more vulnerable skins. Thicker or tougher rawstock may need longer, *e.g.* bull hides and goatskins.

The swelling reaction itself is a rate process, but typically rapidly reaches a kind of steady state, which is dependent on the conditions, particularly the degree of mechanical action. Therefore, this is strictly not a true equilibrium, because an increase in the level of mechanical action would change the degree of swelling. Therefore, the swelling can be assumed to be complete during the usual period of liming, restricted by controls imposed by the liming conditions.

6.2   Purposes of Liming

The purposes of liming are set out in the following, and together constitute controlled damage to the pelt and to the collagen in particular. This is illustrated in Table 6.2, showing the time-dependent changes in bovine pelt due to alkali treatment.[6]

From Table 6.2, there is relatively rapid amide hydrolysis during the first 6 h of liming. From isometric tension measurements,[7] illustrated in Figure 6.2, in which the specimen is heated while clamped under tension, the hydrolysis is accompanied by rapid destabilisation of the collagen, as suggested by the lowering of the shrinking onset temperature, T_o, which is related to the conventional shrinkage temperature. This destabilisation is reflected by the decrease in the temperature at which maximum force in generated during shrinking, T_m, associated with a decrease in the energy of the (hydrogen bond) crosslinks, and

Table 6.2   Analysis of bovine hide, treated with lime alone at ambient temperature.

Time in lime (h)	0	6	18	336
Non-collagenous protein, fat, salt (%)	14	5	6	5
Amide N:collagen (relative %)	100	64	68	49
Isometric tension: T_o (°C)	67	62	61	61
T_m (°C)	96	89	91	84
F_m (g mm^{-2})	3.5	2.1	1.8	0.6

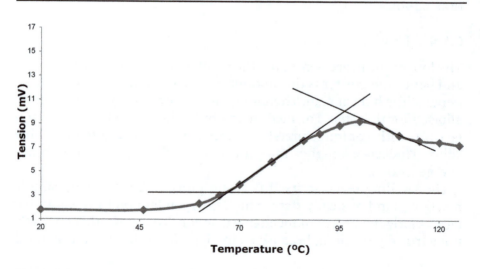

Figure 6.2   Hydrothermal isometric tension measurements on bovine hide.

by the maximum force generated during shrinking, F_m, related to the number of crosslinks. It is noteworthy that the majority of changes to pelt that relate to structure appear to be achieved in the first 6 h of liming.

In reviewing the totality of the beamhouse processes, it will be clear that the properties desired in the final product are largely conferred by the beamhouse operations in general and by the liming process in particular. The effects are as follows.

6.2.1 Removal of the Non-collagenous Components of the Skin

The hydrolytic action is aimed mainly at the non-structural proteins, the albumins and globulins within the fibre structure. The degrading effect of the liming step on the non-collagenous protein is so effective that it dominates the opening-up processes. However, it should be noted that the removal of the fragments barely takes place at all during liming. This is because the trend of water movement is from the bulk solution into the pelt. The usual rinsing mechanism of squeezing and relaxing the pelt is not available during liming, because that is also the mechanism of swelling. Furthermore, the swollen nature of the pelt makes it less likely that the degraded protein can diffuse through the thickened cross-section. Other affected components include keratin, in the form of residual hair and epidermis; any triglyceride grease present will be degraded in a saponification reaction (see later).

6.2.2 Splitting the Fibre Structure at the Level of the Fibril Bundles

A feature of the liming process is splitting of the fibres at the level of the fibril bundles. Under the microscope (Figure 6.3), the splitting can be characterised as coarse or fine, depending on the degree of the effect.[6] Fibre splitting allows better penetration of reagents, for more effective reaction. This is especially true in the case of lubrication (fatliquoring), when fibre splitting is a prerequisite for soft, strong leather.

6.2.3 Swelling the Pelt

It is received wisdom that a necessary contributor to the opening-up process is swelling, when (unspecified) features of the fibre structure are damaged and chemical bonds are broken. This view is not

(a)   (b)

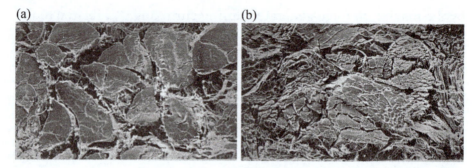

Figure 6.3   Scanning electron photomicrographs of the effects of liming on fibre splitting. (a) Not opened up; if the structure is not treated with alkali for splitting at the level of the fibril bundles, the fibres remain in a solid form. The effects on the leather are stiffness and weakness, with the probability of looseness in the corium. (b) Opened up; after the treatment with alkali, the fibres are split at the hierarchical level of the fibril bundles. This allows optimum tanning and post-tanning process steps, which should optimise the physical properties of the leather. Courtesy of the *Journal of the American Leather Chemists Association.*

unreasonable, since the effect of water is to push the structure apart, a fibre-splitting effect.

There are three mechanisms by which pelt can be swollen.

1. Charge effects, based on breaking salt links and creating charge in the protein structure.
2. Osmotic swelling, caused by the imbalance between the ionic concentration outside and inside the pelt.
3. Lyotropic swelling, caused by disruption to the structure by species that can insert into the hydrogen bonding.

In the case of alkali swelling, the mechanisms operating are charge effects and lyotropy. The effect of alkali on collagen is to break the natural salt links, to make the protein anionic:

$$-CO_2^- \cdots H_3N^+ - \xrightleftharpoons{OH^-} -CO_2^- + H_2N^-$$

The repelling effect of the anionic centres causes the structure to open up, allowing water in, which is observed as swelling. In breaking the salt links there is no electrolyte bound by the collagen (unlike the situation in pickling; see Chapter 9); this means that there is no imbalance in electrolyte concentration between the inside and outside of the pelt, hence the osmotic mechanism does not operate in the same way as it does under acidic conditions. Consequently, salt does not reverse the swelling effect under

Water uptake g per 100g collagen

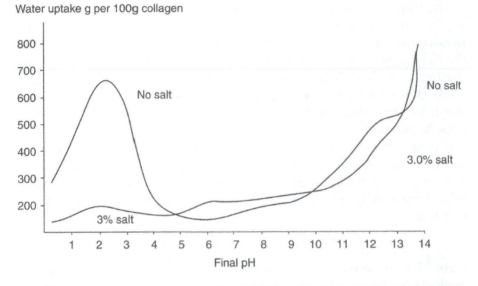

Figure 6.4 Dependence of swelling on pH and the effect of added neutral electrolyte, 2 M sodium chloride. Courtesy of Eurofins BLC.

alkaline conditions. However, from the swelling curve (Figure 6.4), it can be seen that salt does have a small effect on the degree of swelling. This is the same effect as salt has on hyaluronic acid removal: the ions effectively smear out and diffuse the charges, reducing their repelling interactions. In this way, some of the swelling is prevented, but the mechanism is not powerful enough to reverse it completely.

It is frequently assumed in the technology that large changes in pH carry a risk for the quality of the pelt and the subsequent leather; moreover, it is assumed that the risk and deleterious effect are exacerbated if the pH change is rapid. In the case of liming, there is a suspicion that the fast change from soaking at neutrality to liming at pH 12.6 cannot be a good thing for the pelt integrity. In order to test that view, trials were conducted in which the pH was elevated from soaking to liming in stages.

1. Soak at pH 6.
2. Add sodium carbonate, to continue soaking at pH 8.
3. Refloat with sodium hydrosulfide, raising the pH of the pelt to 10.
4. Add sodium sulfide, to initiate unhairing at pH 13, falling to pH 11.
5. Add calcium hydroxide, to buffer at pH 12.6.

The trials incorporated all the combinations of rate of pH change, so the rate varied from as slow as the technology allowed to the maximum rate of change of the conventional process. It was found that there was no statistical difference in the properties of the leathers.[8] Although there was no evidence from these small-scale trials that rapid pH changes made any difference to the outcome, the intuitive assumption of cause and effect is worth confirming on a larger scale.

It is asserted by some tanners that liming can only be conducted in the presence of lime. In effect, they are saying that calcium ion is a necessary component of a liming process. The effect can be modelled by considering the reaction between calcium ion and a hydrogen bond:

$$>C=O^{\delta-}\cdots^{\delta+}H-N<+Ca^{2+}\rightleftharpoons>C=O^{\delta-}\cdots Ca^{2+}\leftrightarrow^{\delta+}H-N<$$

Here, the attractive force between the slightly charged negative and positive centres, which forms the hydrogen bond, is turned into a repulsive force between the positively charged centres. This lyotropic reaction will contribute to the whole swelling effect, which results in the opening up of the collagen structure.

The lyotropic effect of cations can be modelled in the following way:

$$-C=O^{\delta-}\cdots^{\delta+}H-N<\rightleftharpoons-C=O^{\delta-}\cdots A^{+}\leftrightarrow^{\delta+}H-N<$$

Similarly, anions can be inserted into the structure:

$$-C=O^{\delta-}\cdots^{\delta+}H-N<\rightleftharpoons-C=O^{\delta-}\leftrightarrow B^{-}\cdots^{\delta+}H-N<$$

Lyotropy does not always depend on charge – any other species capable of hydrogen bonding can also act in this way, including electrolytes with significant covalent character, *e.g.* lithium bromide, but illustrated in Figure 6.5 using an aliphatic carboxylic acid. Figure 6.5 demonstrates that unionised carboxylic acids can pose a danger to collagen of causing disruption to the structure, ultimately capable of solubilising the protein.

The effectiveness of electrolytes in this reaction depends on the charge-to-radius ratio, the charge density or polarising power: the more intense the charge density, *i.e.* smaller ions or higher charge, the better the ion can interact with a charged centre in protein. The

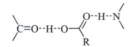

Figure 6.5   Model of the lyotropic effect from an electrically neutral hydrogen bond breaker.

outcome is a solubilising or 'peptising' effect on the protein. The effect is analogous to the phenomenon of 'salting out', where the solubility of a gas is less in an aqueous solution of an electrolyte than it is in pure water; this is due to the effect of polarising power on the structure of the solvent, when it may be preferentially organised around the electrolyte ions. The relationship is defined by the following equation:

$$\log S = \tfrac{1}{2} \log S_0 - KI$$

where the solubility S of a protein in a salt solution is related to the solubility in water, S_0, and the ionic strength, I. K is the salting-out constant.

The ionic strength is given by

$$I = \tfrac{1}{2}\Sigma c_i z_i^2 = \tfrac{1}{2}(c_1 z_1^2 + c_2 z_2^2 + c_3 z_3^2 + ...)$$

where Σ is the continued sum, c_i is the concentration of the ith ion and z_i is the charge on the ith ion, with

$$c_i = a_i \gamma_i$$

where a_i is the activity and γ_i is the activity coefficient of the ith ion.

The activity coefficients of dilute solutions of strong electrolyte ions are the same when the ionic strength of the solutions is the same. In practical leather science, the concentrations of solutions are generally high, usually in the range 10^{-3}–1 m, so that thermodynamic treatments or analysis are not possible or strictly valid.

The relative effects of polarising power are set out in the Hofmeister or lyotropic series. The measurement of the effects of ions can be variable in this context, so the following series represent examples of Hofmeister series, where the order is the increasing polarising power, dependent on atomic weight and ionic charge:

$$Cs^+ < Rb^+ < NH_4^+ < K^+ < Na^+ < Li^+ < Ba^{2+} < Sr^{2+} < Ca^{2+} < Mg^{2+}$$
$$\text{citrate} < \text{tartrate} < SO_4^{2-} = S_2O_3^{2-} < CH_3CO_2^- < Cl^- < NO_3^- < Br^-$$
$$= ClO_3^- < I^- < CNS^-$$

Combinations of ions can be viewed in the same way, *e.g.* $Na_2SO_4 < NaCl < CaCl_2$.

Hence ammonium chloride deliming should produce softer leather than ammonium sulfate, due to the effect of the chloride ion (see also Chapter 7).

Rabinovich[9,10] has long maintained the importance of the lyotropic influence of the Hofmeister series in causing lyotropic disruption to the collagen structure. Controlling the equilibria in the salt links can make a large difference to the reactivity of the collagen in tanning reactions. Also, the view can be widened to consider the effect of preferentially reacting sidechains in the protein in order to release other reactive groups 'buried' in the protein structure;

this has been exploited in reactions such as aldehydic reactions to facilitate the titration of carboxyl groups. This approach to protein science echoes the original application of the Hofmeister series to so-called 'salting in' and 'salting out' of colloidal proteins, which referred to swelling/destabilising and deswelling/stabilising, respectively.

From Figure 6.4, swelling is pH dependent, with a positive correlation in the alkaline region of the pH scale. It was shown earlier that the primary cause of swelling is the repulsive effect of the negative charges generated by the neutralisation of the basic groups. The pK_a of lysine sidechain amino groups in proteins is 9.4–10.6 and the pK_a of the sidechain guanidino groups of arginine in proteins is 11.6–12.6, hence the charge continues to increase up to the end of the pH scale, as indicated in the titration curve.

The effect of neutral electrolyte is to reduce slightly the swelling at moderately alkaline pH values by a charge diffusion effect. However, at higher pH values, the effect of neutral electrolyte is to enhance swelling by the lyotropic influence of the ions in solution. It is not uncommon for tanners to include salt in the lime bath, in the hope of controlling swelling. It can be seen that this practice may actually be counterproductive, producing the opposite effect to the desired outcome. Alternatively, any positive effect is likely to be small. Heidemann *et al.* asserted that swelling may not be a prerequisite of opening up.[11] Since the disruption of the natural structure of pelt is the result of different types of reaction, it may be a practical proposition, although it has not yet been tested by developing an alternative biochemical process.

6.2.4 Hydrolysis of Peptide Bonds

The liming step is dominated by the breaking of peptide links, *i.e.* liming does cause damage to protein, but in a controlled way. There is a fundamental difference in the effect of hydrolysis on collagen and on the non-structural proteins: it is the latter that must be degraded and removed (to the required extent) by the action of alkaline hydrolysis. These proteins are linear, although folded, so when a peptide link is broken, the chain is severed into two fragments – continued degradation produces more and more, smaller and smaller fragments, making their removal easier and easier. On the other hand, if a peptide link in collagen is broken, the effect is negligible, because the structure is hardly affected, owing to the hierarchy of structural elements. This means that many peptide links must be broken before significant damage is observed. In this way, collagen can resist the effect of hydrolysis, so the non-structural proteins are damaged relatively much more quickly.

6.2.5   Hydrolysis of Amide Sidechains

Amide sidechain hydrolysis occurs in the same way as hydrolysis of peptide links; indeed, peptide bonds are often called amide bonds, to emphasise the similarity to amide groups. The reaction is as follows:

$$P-\left(CH_2\right)_n CONH_2 + H_2O + OH^- \rightleftharpoons P-\left(CH_2\right)_n CO_2^- + NH_3$$

where $n = 1$ for asparagine, converted to aspartic acid, and $n = 2$ for glutamine, converted to glutamic acid.

A by-product of the reaction is ammonia, which is readily perceived at the end of the reaction by its smell. This reaction converts the neutral amide groups to carboxyl groups, which has a major implication for the isoelectric point (IEP). The IEP is defined by the relative numbers of pH-active basic and acidic groups. Therefore, the dependency does not include amide groups, because they are electrically neutral, but does include carboxyl groups, the product of hydrolysis. Therefore, the extent to which hydrolysis occurs will determine the impact on the IEP. In conventional processing, lasting about 18 h, around half of the available amide sidechains are hydrolysed. The deamidation reaction causes the relative proportion of the total carboxyl groups to increase with respect to amino groups; this shifts the IEP of collagen from physiological pH 7.4 to pH 5–6, depending on the degree of hydrolysis.

In proteins, the hydrolytic reaction of amide sidechains has two possible mechanisms: direct hydrolysis by hydroxyl catalysis or intramolecular catalysed hydrolysis.[12,13] Heidemann and Deselnicu[14] found that prolonged liming changes the conformation of L-amino acids into their D-isomers. In the case of asparaginyl and aspartyl residues in peptides and proteins, the mechanism for this racemisation is *via* a succinimide ring, formed from an asparaginyl sidechain and the adjacent peptide-bound nitrogen in the collagen backbone,[15] shown in Figure 6.6. After ring closure, the racemisation at the chiral carbon (indicated by an asterisk) occurs by general base catalysis, removing and replacing the proton. This mechanism, however, is highly dependent upon the rigidity or intactness of the collagen backbone: the more degraded the collagen is and hence more flexible, the easier it is for the reaction to occur.[13] Using molecular dynamics, it has been estimated that the rate of asparagine deamidation by intramolecular

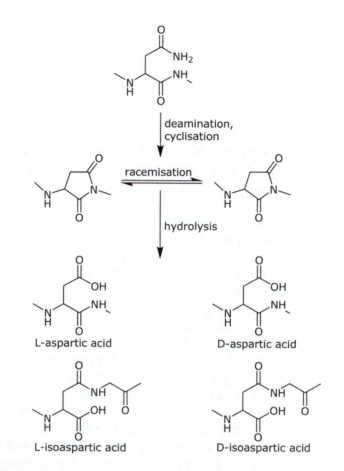

Figure 6.6 Intramolecular catalysed amide hydrolysis by succinimide ring formation.

catalysis at 37 °C would be 10^4 times slower in an extended α-chain than in a random coil.[16] The accompanying reaction, forming the racemised isoaspartate residues, is likely to lead to localised disruption of the triple helix.[13]

Total ammonia production results from hydrolysis of asparagine and glutamine amide groups and the guanidine groups of arginine: all these reactions can take place by direct catalysis with hydroxyl ion, at similar rates.[16] Production of racemised aspartic acid itself can take place by the same intramolecular mechanism, set out in Figure 6.6, but it is at least an order of magnitude slower than the reaction of asparagine hydrolysis, based on peptide studies.[15] Because the formation of a six-membered ring is less favoured than the formation of the five-membered succinimide ring in this reaction, the racemisation of aspartic acid is faster than the racemisation of glutamic acid.[17,18]

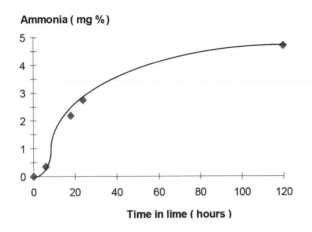

Ammonia (mg %)

Time in lime (hours)

Figure 6.7 Ammonia production during white liming. Courtesy of the *Journal of the Society of Leather Technologists and Chemists.*

In practice, liming typically takes up to 24 h; it is evident that pro-longed liming causes some changes in the collagen structure, but it is still not known to what extent conformational changes occur within collagen's three-dimensional structure during conventional liming periods.

From Figure 6.7,[12] ammonia production conforms to (pseudo) first-order kinetics during the reaction period of 120 h and the data fit a rate constant of 8.7×10^{-6} s^{-1} ($r^2 = 0.98$). The calculated infinity value of 4.7 mg% NH_3 compares well with the observed yield at 120 h and with the theoretical ammonia content of collagen, 4.5 mg% on dry weight, calculated from the known amino acid sequence of type I bovine collagen; the difference in those values for ammonia content is accounted for by the additional, non-collagenous protein content of the hide. The kinetics predict a half-life of 22 h; this confirms that, during a conventional liming process lasting 18 h, typically about half of the available amide groups are converted to carboxyl groups.

From Figure 6.8,[12] the percentage of D-aspartic acid in the limed samples initially increases and then decreases; this may be explained by all of the non-structural, non-collagenous aspartic acid changing its molecular conformation within a period of 18 h. The rate of race-misation is low during the early part of the liming process, increas-ing only after 24 h. Some racemisation by acid hydrolysis always occurs during the processing of the samples, and this explains why, in Figure 6.9 (see later), there is a slightly positive measurement for the untreated collagen, at time zero. From the literature, deam-ination of sidechain amide groups by the intramolecular mecha-nism also follows first-order kinetics: in synthetic polypeptide the

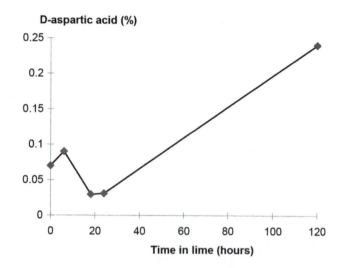

Figure 6.8   Rate of racemisation of L-aspartic acid in collagen during white liming. Courtesy of *Journal of the Society of Leather Technologists and
Chemists.*

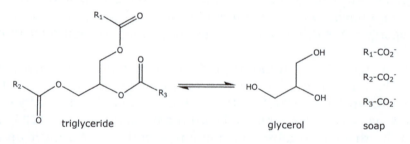

Figure 6.9   Hydrolysis of triglyceride.

reaction displays a rate constant of 1.0×10^{-6} s^{-1} at 25 °C, but the
observed rate in collagen decreases rapidly below the melting temperature,[19] calculated to be over 1000 times slower.[15] Therefore, the
observed rate of ammonia production from the liming reaction of
intact hide is much faster, so it probably originates from an intermolecular mechanism.

Conventional liming processes do not cause a marked increase
in racemisation of aspartic acid, so it is concluded that hydrolysis
does not cause major damage to the high order of collagen structure
during the first 24 h of liming. This means that the collagen fibres
remain intact, with no relaxation of the triple helix structure. If there
was significant conformational damage to the α-chains, then the formation of succinimide rings within the peptide chains would be easier and therefore faster. This kind of destabilisation of the collagen

structure is obtained only by prolonged liming. Therefore, the mechanism of amide hydrolysis during conventional liming is by a direct base-catalysed mechanism, but prolonged liming introduces the intramolecular catalysis mechanism. The inference is that the collagen helical structure remains effectively intact for up to 24 h in lime, but then starts to denature significantly. This observation has implications for enzyme-assisted unhairing and opening up. If enzymes are to be used in liming, their application should be restricted to the earlier part of the process. Intact collagen is resistant to attack by general proteases, because of its high degree of structural order, but chemically damaged collagen can be attacked by those same enzymes.[20-22] Therefore, if enzyme action is allowed to continue to the end of the normal liming period or beyond, there is a likelihood of enzymolysis damage, which would rapidly adversely affect leather integrity and strength.

It is interesting that there is a lack of an intramolecular contribution to the hydrolysis mechanism of this process step; this means that there is little discrimination between asparagine and glutamine in the reaction, that is, no preferential reaction at asparagine. Since there is evidence to show that the chrome tanning reaction may be favoured at aspartic acid carboxyl groups, compared with glutamic acid carboxyl groups,[23] the liming reaction cannot be expected to change the chrome tanning reaction. Although liming increases the efficiency of chrome uptake, by increasing the number of fixation sites, the effectiveness of the reaction, in terms of the rise in shrinkage temperature per unit bound chrome, is unlikely to be different when typical liming periods are used. Therefore, the consequence of abnormally prolonged liming will be a more efficient but not more effective tannage, with accompanying loss of leather performance due to reduced strength.

6.2.6 Hydrolysis of Guanidino Sidechains

The guanidino groups on the sidechains of arginine residues can also undergo hydrolysis, but in a stepwise manner, when the by-products are urea, ammonia and carbon dioxide:

$$P-\left(CH_2\right)_3 - NH - C(=NH) - NH_2 + H_2O \rightleftharpoons P-\left(CH_2\right)_3 - NH - C(=O) - NH_2$$

arginine   citrulline

$$\rightleftharpoons P-\left(CH_2\right)_3 - NH_2$$

ornithine

The arginine group can participate in the IEP determination, because it is capable of acting as a basic group:

$$P-(CH_2)_3-NH-C(=NH)NH_2+H^+ \rightleftharpoons P-(CH_2)_3-NH-C^+(NH_2)_2$$
$$\leftrightarrow P-(CH_2)_3-NH^+ = C(NH_2)_2$$

During liming, the guanidino group is hydrolysed to an amino group (see earlier). Therefore, the conversion of one type of basic group to another might, as a first-order effect, not markedly alter the IEP. However, the pK_b values are different (9.0 for free arginine and 8.7 for free ornithine), so there will be a small second-order effect on the IEP.

6.2.7   Removal of Dermatan Sulfate

In Chapter 2, dermatan sulfate proteoglycan and its structure were presented. Because the protein core of the compound is bound to the surface of the fibrils, it is hydrolysed and released during liming, accelerated by the presence of alkali-stable enzyme, shown in Table 6.3.

The compound can be easily quantitatively analysed and hence has been used a marker or measure of the extent and progress of liming. However, its removal must not be regarded as an important part of the opening reaction, since that operates at the level of the fibril bundles; there is no evidence that opening up or splitting at the level of the fibrils contributes significantly to the leather-making operation. The function of glycosaminoglycans in tissue is to control hydration by their hydrophilic anionic chains. Since the function of dermatan sulfate *in vivo* is therefore concerned with hydration of the fibrils, this raises the question of what its effect might be if it were to remain in the leather.

Table 6.3   Removal of dermatan sulfate in liming.

Liming conditions	Removal of dermatan sulfate (%)
6 h	22
6 h + enzyme	53
18 h (conventional)	50

6.2.8   Fat Hydrolysis

Triglyceride fat, both cutaneous and subcutaneous, is likely to be present in the pelt at the beginning of liming, particularly if the pelts have not previously been fleshed. The fat is hydrolysed (saponified, *i.e.* converted to soap and glycerol) during liming in the way shown in Figure 6.9. The odour of soap is invariably apparent during and at the end of liming.

6.3   Variations in Liming

The variations in liming that might be used depend to a great extent on the unhairing process. If there is a departure from conventional lime/sulfide hair burning, then other reactions may be used to achieve the limed outcome.

6.3.1   Chemical Variations

Since the reactions required for opening up are hydrolytic, they could be conducted using any other alkaline agent, capable of reaching the required pH range. The use of strong alkali, such as sodium hydroxide, is possible, because of the high solubility, but control would have to be imposed so that the pH does not reach too high values. Since the alkali may be used up in achieving equilibrium, it is necessary to keep adding it, to maintain the required pH, because there is no buffering effect with strong alkali chemistry. The danger of over-swelling is increased, even though mechanical action may be minimised. However, it is possible to monitor and control the pH using conventional pH-stat technology, although that presumes that the process vessel includes recycling into a so-called 'lab box', where chemical dosing can be conducted.

Indian workers[24] introduced the concept of ionic liquid into the opening-up process – however, not as a process medium, but as a reagent. Using an offer of 0.2% 1-butyl-3-methylimidazolium chloride in 30% float to follow unhairing/liming, the effect on opening up was assessed after chrome tanning. Variations in unhairing procedure were lime–sulfide, enzyme alone, sulfide–enzyme and lime–sulfide–enzyme; the differences in leather properties were small. However, better chrome uptake was ascribed to the opening influence of the ionic liquid. Since the ionic liquid is offered in dilute aqueous solution, it is not clear how much of the observed effect can be attributed to its ionic liquid status.

6.3.2 pH Variations

Higher pH can be achieved with buffering, using the limited solubility of lime, but influencing the solubility with a sequestrant:

$$\left[Ca(OH)_2\right]_{solid} \rightleftharpoons \left[Ca(OH)_2\right]_{solution} \rightleftharpoons Ca^{2+} + 2OH^-$$

The solubility product $K_p = [Ca^{2+}][OH^-]^2 = 4.7 \times 10^{-6}$.

If the equilibrium between dissolved lime and its ions is changed by sequestering calcium ions, more lime must dissolve in order to maintain the solubility product. In this way, the hydroxyl ion concentration can be increased, raising the pH. This can achieved in a number of ways, but the technologically accepted approach is to add sugar. Monosaccharide (glucose) or disaccharide (sucrose) can complex with calcium ions *via* their hydroxyl groups, causing the pH to rise, when the equilibrium pH depends on the quantity of sequestrant. The effect of raising the pH is to accelerate the hydrolysis reactions, but still have lime buffering, together with the lyotropic effect of calcium ions. Clearly, the potential danger of over-swelling arises if pH values are elevated above conventional conditions.

6.3.3 Biochemical Variations

Hydrolytic reactions in proteins can be achieved enzymatically, using proteases, hence it is theoretically possible to conduct liming biochemically. Three points must be considered in this regard.

1. Chemical liming is non-specific; all reaction sites are vulnerable to attack and may be attacked equally. Enzyme reactions are relatively much more specific, hence the biochemical approach presumes that the required reactions are known and targeted by an appropriate choice of enzymes. Since this is not the case, because the precise outcome of chemical liming is not known, the use of enzymes as the only opening-up agent is not currently achievable.
2. Proteolytic enzymes are used industrially in liming. They are so-called alkali-stable enzymes; actually, they are not stable in

alkali, because they are degraded as rapidly as any other enzyme, but they do have enough reactivity at high pH for them to operate sufficiently to contribute to the process reaction.[25] Typically, they are used to enhance the keratin degradation rate, but clearly can contribute to hydrolysis reactions, as evidenced by the faster rate of dermatan sulfate removal.

3. It might be thought possible for the degradation of non-structural proteins to be selected by using non-collagenolytic proteases, since intact collagen is damaged only by collagenase. It is true that the highly structured nature of collagen requires a specific enzyme to attack it; however, the damaging effect of liming does make collagen increasingly vulnerable to general protease attack.

6.3.4 Recycling

It is possible to reuse the liming float by recycling it – the technology was in use in South Africa at the end of the last century. Because they were operating a hair burning process, typical technology at that time, they filtered the float for recycling to remove suspended lime, but could not remove the hair debris. The consequences were twofold: some savings of water, lime and sulfide were made, but remaining scud was effectively driven into the grain of the next pack and made the leather dirty.

Now that processes for hair saving have been developed, it has become possible to recycle the lime liquor without risking making the leather dirty with scud. Such an approach was reported by Daniels *et al.*, as part of a systematic approach to closing the loop, to eliminate waste water discharge.[26]

6.4 Limeblast

An important reaction associated with limed pelt is the damaging fault called 'limeblast'. The reaction is as follows:

$$[Ca(OH)_2]_{solid} + [CO_2]_{gas} \rightleftharpoons [CaCO_3]_{solid} + H_2O$$

This heterogeneous reaction occurs between solid lime, which has been driven into the grain surface of the pelt (the flesh surface is not important in this context), and gaseous carbon dioxide in the atmosphere, resulting in the formation of solid crystals of calcium carbonate. The fault arises if the limed pelts are removed from the liming drum, *e.g.* for fleshing or splitting, when they are usually held in piles. Since the concentration of carbon dioxide in air is low, the reaction can proceed only if there is a constant supply of air, *i.e.* if the piles are uncovered, in a draught. Therefore, the fault can be easily prevented if the piles are covered, *e.g.* with polythene sheeting.

The damaging effect is caused by the distortion of the surface by the growing crystals and that effect remains on the grain surface, even if the crystals are subsequently removed by solubilisation in the deliming or pickling processes. The fault can be seen as dull areas on limed hide, but becomes more apparent when the resulting leather is dyed and dried. The change in surface characteristics is represented in Figure 6.10 and illustrated in Figures 6.11 and 6.12. The intact grain acts like a smooth surface, reflecting light uniformly, but the distorted grain causes light scattering from the surface, observed as dullness.

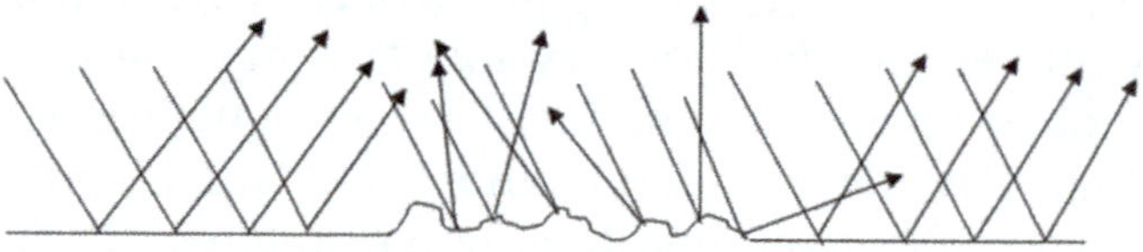

Figure 6.10   Effect of limeblast on the surface of pelt or leather.

Figure 6.11   Light photomicrograph of limeblast. Courtesy of Leather Wise, UK.

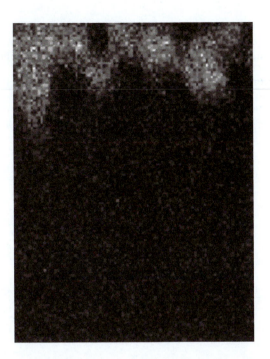

Figure 6.12 X-ray backscatter electron photomicrograph of a section through limeblasted pelt, showing the presence of calcium in false colour (the light area at the top). Courtesy of Leather Wise, UK.

References

1. W. P. Jencks, *Catalysis in Chemistry and Enzymology*, McGraw-Hill, 1969.
2. P. M. May, *et al.*, *ChemComm*, 2018, **54**, 1980.
3. J. H. Bowes and R. H. Kenten, *Biochem. J.*, 1948, **43**(3), 358.
4. *Official Methods of Analysis*, Society of Leather Technologists and Chemists, UK, 1996.
5. E. R. Theis and R. G. Steinhardt, *J. Am. Leather Chem. Assoc.*, 1942, **37**(9), 433.
6. A. D. Covington and K. T. W. Alexander, *J. Am. Leather Chem. Assoc.*, 1993, **88**(12), 241.
7. B. M. Haines and S. G. Shirley, *J. Soc. Leather Technol. Chem.*, 1988, **72**(5), 165.
8. A. D. Covington, unpublished results.
9. D. Rabinovich, *World Leather*, 2008, 26.
10. D. Rabinovich, *J. Am. Leather Chem. Assoc.*, 2011, **106**(8), 242.
11. E. Heidemann, J. Sagala and A. Hein, *Das Leder*, 1987, **38**, 48.
12. O. Menderes, A. D. Covington, E. R. Waite and M. J. Collins, *J. Soc. Leather Technol. Chem.*, 1999, **83**(2), 107.
13. M. J. Collins, E. R. Waite and A. C. T. van Duin, *Philos. Trans. R. Soc.*, 1999, **354**, 1379.
14. E. Heidemann and M. Deselnicu, *Das Leder*, 1972, **23**, 17.
15. T. Geiger and S. Clarke, *J. Biol. Chem.*, 1987, **262**(2), 785.
16. A. C. T. van Duin and M. J. Collins, *Org. Geochem.*, 1998, **29**(5), 1227.
17. H. T. Wright, *Crit. Rev. Biochem. Mol. Biol.*, 1991, **26**(1), 1.
18. A. B. Robinson and C. J. Rudd, *Curr. Top. Cell. Regul.*, 1974, **8**, 247.

19. F. M. Sinex, *J. Gerontol.*, 1960, **15**, 15.
20. J. W. Giffee, *et al.*, *J. Am. Leather Chem. Assoc.*, 1965, **60**(6), 264.
21. R. G. Donovan, *et al.*, *J. Am. Leather Chem. Assoc.*, 1967, **62**(8), 564.
22. A. L. Rubin, *et al.*, *Biochemistry*, 1965, **4**(2), 181.
23. A. D. Covington, *J. Am. Leather Chem. Assoc.*, 1998, **93**(6), 168.
24. M. Azhar, *et al.*, *J. Am. Leather Chem. Assoc.*, 2016, **111**(6), 206.
25. K. T. W. Alexander, *J. Am. Leather Chem. Assoc.*, 1988, **83**(9), 287.
26. R. P. Daniels, *et al.*, *J. Soc. Leather Technol. Chem.*, 2017, **101**(3), 105.

7 Deliming

7.1 Introduction

The process steps of deliming and bating are traditionally considered as inextricably intertwined, because the former is thought of as only preparing the pelt for the latter. Indeed, it is common for bating to be conducted in the deliming solution. However, from an analysis of the specific functions of the steps, it is useful to regard them as separate.

The functions of the deliming step can be summarised as follows.

1. Lowering the pH, in preparation for bating.
   In preparing the pelt for chrome tanning, the pH will be adjusted from 12.6 (in liming) to 2.5–3.0 (in pickling); between those stages, the pelt is adjusted to pH 8.5–9.0, suitable for the biochemical action of protease.
2. Removing the lime.
   It is an often quoted feature of deliming that one of its main purposes is to remove lime. It is clearly part of the reaction of lowering the pH to neutralise residual alkali from the liming process:

$$Ca(OH)_2 + 2H^+ \rightleftharpoons Ca^{2+} + 2H_2O$$

Tanning Chemistry: The Science of Leather 2nd edition
By Anthony D. Covington and William R. Wise
© Anthony D. Covington and William R. Wise 2020
Published by the Royal Society of Chemistry, www.rsc.org

It is also assumed that there might be a risk of creating another insoluble or sparingly soluble calcium salt, for example if ammonium sulfate is the deliming agent:

$$Ca^{2+} + SO_4{}^{2-} \rightleftharpoons CaSO_4$$

This would cause a problem analogous to limeblast (see Section 7.4 and also Section 7.2.5).

It is less clear what happens to the calcium ions that are associated with the ionised carboxyls or are engaged with the hydrogen bonding system as a lyotropic effect during liming. It is certainly the case that the bound calcium is usually largely eliminated by the time the pelt is pickled.

3. Depleting the pelt, partially reversing the swelling.

   From the swelling curve, reproduced in Figure 7.1, the change from pH 12.5 to 8.5 is accompanied by a significant degree of deswelling or depletion. Even though the swelling may not have been conducted to equilibrium, the change is marked.

There are two features of importance in the depleting.

1. The appearance of the pelt becomes less translucent, more opaque: as the water is lost, the fibre colour reappears.
2. There is a net transfer of water from the inside of the pelt to the outside. This may seem a trivial restatement of the effect of the process step, the reversing of swelling, but it is of profound importance to opening up, because this is a rare occasion in the leather-making process that such a net flow occurs.

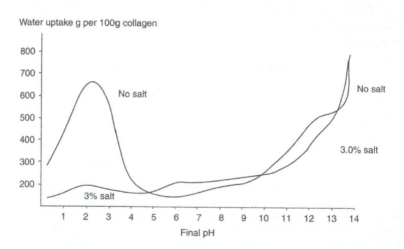

Figure 7.1   Effect of pH on collagen swelling. Courtesy of Eurofins BLC.

Consider any rinsing or washing operation: it relies on transferring water into the pelt, dissolving the component of interest, then squeezing the solution from the inside out into the bulk solution. This mechanism therefore includes transporting solutes not only out of the pelt, but also back in again. In the case of deliming, the transport is effectively one way, so the removal of the degraded protein and other products from the liming reactions is conducted efficiently. Therefore, deliming is not merely adjustment of the pH, it is one of the more important steps in opening up.

The process of depletion also occurs in pickling, but it can be seen from Figure 7.1 that the effect is much smaller from pH 8.5–9.0 to 2.5–3.0 than from pH 12.6 to 8.5–9.0. In addition, the impact of depletion will also depend on the rate of the process: the faster the water flows out of the pelt, the smaller is the reversing effect of the conventional washing mechanism on the removal of debris from within the pelt.

7.2 Deliming Agents

Deliming can be conducted using any chemical capable of reducing the pH to the required range. The options are as follows.

7.2.1 Water

Continued washing with water will lower the pH by two mechanisms. The first mechanism is merely dilution of the hydroxyl content and the second is neutralisation of the hydroxyl with the bicarbonate content in the water, as shown in Figure 7.2, illustrating the difference between the inside and the outside of the pelt, with the surface acting as a semipermeable membrane.

The dilution effect is a factor of 10^3, hence a lot of water is needed to reach equilibrium between the inside and the outside of the pelt, and the slow effect of diffusion will contribute to the need for even more

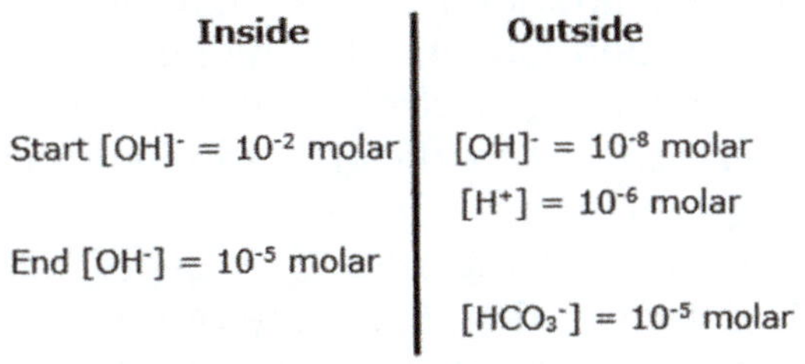

Inside	Outside
Start $[OH]^- = 10^{-2}$ molar	$[OH]^- = 10^{-8}$ molar
	$[H^+] = 10^{-6}$ molar
End $[OH^-] = 10^{-5}$ molar	
	$[HCO_3^-] = 10^{-5}$ molar

Figure 7.2 Deliming with water.

water. The concentrations of protons and bicarbonate are too low to provide a driving chemical potential for diffusion from the outside to the inside.

There are two other good reasons why deliming by washing is undesirable or risky.

1. Fresh water is a resource that should be conserved; this applies more in some countries than in others, but it is generally the case that fresh water is neither so cheap nor so abundant that it can be used profligately.
2. It must be remembered that deliming applies to limed hide that has not reached swelling equilibrium, hence it is capable of additional, damaging swelling. Mechanical action, particularly at the beginning of deliming in fresh water, runs the risk of over-swelling the grain. Since pelt is most vulnerable to abrasion damage in an alkali-swollen state, there is an extra risk of grain damage.

7.2.2 Strong Acids

Any strong acid might be a candidate for acting as a deliming agent, provided that the counterion does not react with calcium ion. The undesirable consequence would be the creation of an insoluble salt, which can disrupt the grain structure by precipitating within the grain layer, in the same way as the problem of limeblast is caused. This effectively rules out sulfuric and phosphoric acids. More importantly, the use of strong acids might lead to pH values in solution that are low enough to cause acid swelling; this type of damage could be avoided by the addition of neutral electrolyte, but that would be counterproductive in terms of economics and environmental impact. The pH achieved in solution depends on the concentration of acid added into the system: if the total offer were added in one aliquot, acid swelling would be highly likely. An offer of 1% sulfuric acid in 100% float gives a concentration of 10 g L^{-1}, equivalent to 0.1 M, so that the hydrogen ion concentration is 0.2 M, equivalent to pH 0.7. Even though this concentration falls into the region of reduced swelling (see Figure 7.1), owing to the electrolyte effect (discussed in Chapter 9) consumption of acid will drive the pH into and through the zone of maximum swelling at pH 2.

The alternative is to add the acid in aliquots small enough to maintain the pH not lower than 5, to stay within the zone of minimum swelling (see Figure 7.1). However, this strategy means that the concentration of hydrogen ions in solution is always low, hence the rate of deliming will be slow. It is this effect on the kinetics that usually rules out the routine use of strong acids as solo deliming agents, although they may be included in a mixture of agents.

7.2.3 Weak Acids

It might be assumed that weak acids would be safer than strong acids, because they would not cause the pH to decrease to a low enough value to cause acid swelling. However, that assumption cannot be relied upon, because even weak acids can cause acid swelling:

1% HCO_2H in 100% float is 10 g L^{-1} = 0.2 M; the pK_a of formic acid is 3.75, *i.e.*

$$K_a = \frac{[H^+][HCO_2^-]}{[HCO_2H]} = \frac{[H^+]^2}{[HCO_2H]} = 10^{-4} M$$

For a weak acid, assume that the concentration of unionised acid is the same as the total concentration. Then, $[H^+]^2 = 2 \times 10^{-5}$, *i.e.* $[H^+]$ = $4.5 \times 10^{-3} M \equiv pH$ 2.3. Therefore, the solution pH is close to the pH of maximum swelling.

In particular, unionised carboxylic acids can cause lyotropic swelling, as discussed earlier. In addition, carboxylate ions often react with calcium ion to produce an insoluble salt. An exception is lactic acid, $HOCH(CH_3)CO_2H$, which produces a soluble calcium salt; it has been used in industry as a deliming agent.

Boric acid, H_3BO_3, is an example of a weak acid, which has been tested in a mixture with sodium hexametaphosphate, $Na_4[Na_2(PO_3)_6]$; equilibrium at pH 8 was reached in 1 h, with the lowest pH of 7.1. The mixture produced less hydroxyproline in solution than ammonium sulfate. It was found that 3% sodium hexametaphosphate plus 1% boric acid removed the same amount of calcium as 3.5% ammonium sulfate:[1]

$$Na_4[Na_2(PO_3)_6] + Ca^{2+} \rightleftharpoons Na_4[Ca(PO_3)_6] + 2Na^+$$

The notion of using weak acids extends to mixtures, where the same argument applies, even when the mixture comes from an unusual source. It has been reported that an infusion of *Hibiscus sabdariffa* contains ascorbic, glycolic, citric and dihydroxybenzoic acids, which delimes to pH 4 in about 1 h.[2]

The application of aliquots of weak acids leads to the problem of too slow a rate of reaction, in the same way applies to strong acids.

7.2.4   Acidic Salts

There are many acidic salts that might be used for deliming. Sodium bicarbonate, $NaHCO_3$, has been proposed,[3] but it has not been applied industrially, probably because of cost and low solubility (see carbon dioxide deliming, Section 7.2.8). Sodium metabisulfite, $Na_2S_2O_5$, is often included in ammonium salt deliming, but its function there is primarily to scavenge residual sulfide after lime/sulfide unhairing. The reaction is complicated and will yield several sulfur species, for example elemental sulfur, thiosulfate ($S_2O_3^{2-}$) and other sulfur oxyanions.

At pH 9, during deliming, the dominant sulfide form is unequivocally HS^-.

$$S_2O_5^{2-} \rightleftharpoons SO_3^{2-} + SO_2$$

Hence, under these conditions, the dominant form would be HSO_3^-:

$$2HS^- + HSO_3^- \rightleftharpoons 3S + 3OH^-$$

7.2.5   Ammonium Salts

Ammonium salts are technologically preferred because they readily yield the conditions required for conventional pancreatic bating enzymes to operate ideally. Ammonium salts are weak, with $pK_a \approx 9$, so they buffer around pH 9:

$$pK_a = pH + \log \frac{\left[NH_4^+ \right]}{\left[NH_3 \right]}$$

A significant advantage of the buffering effect is that any error in the offer will have only a small effect on the pH. The effect of, for example, a 20% error can be calculated, assuming here the pK_a of NH_4^+ is 9.0.

Assume that half of the ammonium salt is used up in the reaction, *i.e.* 1% ammonium chloride reacts with the residual lime:

$$NH_4^+ + OH^- \rightleftharpoons NH_3 + H_2O$$

Then, normally, $[NH_4^+] = [NH_3] \approx 0.2$ M, giving pH 9.0.

If 20% too much is added, $[NH_3] = 0.20$ M and $[NH_4^+] = 0.28$ M, hence $[NH_4^+]/[NH_3] = 1.4$ and pH = 8.85.

If 20% too little is added, $[NH_3] = 0.20$ M and $[NH_4^+] = 0.12$ M, hence $[NH_4^+]/[NH_3] = 0.6$ and pH = 9.22.

The deliming reaction is very fast, because the whole offer of ammonium salt can be added at the beginning of the process step and the pH cannot fall below the value fixed by the equilibrium condition of the reaction products. Also, ammonium salts are very soluble in water, so the offer may even be added to the process vessel after draining the lime liquor or any wash liquor from that process, without adding any fresh float. This creates an additional contribution to the fast diffusion rate, driven by the osmotic effect of the concentrated solution. The water released from the alkali-swollen pelt by the pH change can function as the float for the deliming step.

In industry, it is common to use either ammonium sulfate or ammonium chloride. It is important to know that they are not equivalent reagents: for example, a 1% offer of ammonium chloride contains 0.34% NH_4^+, but the same offer of the sulfate contains 0.27% NH_4^+. However, if an operator ignored the difference, the error of about 25% would only give rise to a difference in pH as calculated above. Technologically, the main difference is the effect of the counterion on the residue of the liming step. In the presence of sulfate, calcium ion can form the sparingly soluble salt:

$$Ca(OH)_2 + SO_4^{2-} \rightleftharpoons CaSO_4 + 2OH^-$$

This may be a problem, because crystals can grow within the fibre structure and, more importantly, below the grain surface. The disrupting effect on the surface distorts the surface by stretching it, so even if the cause of the fault is removed by solubilising the crystals, the effect remains and can adversely affect uniformity of dyeing. A greater danger is the conversion of limeblast crystals of calcium carbonate into a more insoluble salt, which may survive pickling and be present in tanned leather:

$$CaCO_3 + SO_4^{2-} \rightleftharpoons CaSO_4 + CO_3^{2-}$$

The occurrence of this reaction depends on the conditions in the vessel with regard to concentration. There are three possibilities that determine the outcome, shown in Figure 7.3.

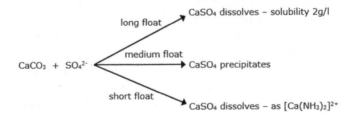

Figure 7.3 Formation of calcium sulfate.

In long float, the solubility of the sparingly soluble salt is exceeded; at short float, the high concentration of ammonia in solution allows the formation of the calcium complex, which keeps it in solution in the presence of sulfate ion. Only in the intermediate situation does calcium sulfate precipitate – which is the typical industrial condition for deliming with ammonium sulfate. In the case of ammonium chloride, this danger does not exist, because chloride ion does not form an insoluble salt with calcium ion. The lyotropic effect of chloride ion is greater than that of sulfate ion, so the outcome will be enhanced opening up, although the difference in effects may not be significant enough to be discernible after the relatively brief period of contact.

There are two environmental problems associated with using ammonium salts. The first is the release of ammonia gas into the working environment; although it does not create a serious toxicity hazard, it does create a nuisance, causing discomfort to workers. Second, the discharge of ammoniacal nitrogen in the effluent is commonly subject to stringent consent limits, because ammonia is poisonous to fish and other aquatic life. Therefore, there is an imperative to develop an alternative technology.

7.2.6 Alternative Buffers

One option, suggested by Koopman, is to use Epsom salt (magnesium sulfate), which can operate in the following way:[4,5]

$$MgSO_4 + 2OH^- \rightleftharpoons Mg(OH)_2 + SO_4^{2-} \rightleftharpoons Mg^{2+} + 2OH^-$$

The reaction product between the salt and hydroxyl ions is sparingly soluble magnesium hydroxide. As with lime (calcium hydroxide), the pH is defined by the solubility; in this case, the system is at equilibrium at about pH 10. Although this value is slightly high for many enzymes that might be used in bating, it still offers conditions that could be used as the basis for preparing for bating.

The technology was developed in The Netherlands and, at the instigation of the US Environmental Protection Agency (EPA), trials were conducted by the American Leather Chemists Association (ALCA).[4] The process was to use 50% float at 35 °C, 7% magnesium sulfate and 10% sulfuric acid added in 10% float, resulting in pH 6.4 in solution. The author, Constantin, reported a residual high-pH profile through the cross-section, based on the use of phenolphthalein indicator. The trials were not successful, because the author could not get the chrome tanning salt to penetrate the hide, so he reported that he was obliged to terminate the trials. Consequently, the originator of the technology, Koopman, responded that the trials were flawed, because the indicator should have been thymolphthalein, which would have shown that the centre of the pelt was at pH > 10.5 and therefore should have been more carefully acidified in pickling prior to chrome tanning, *i.e.* the trials could not succeed because of the way in which they were conducted.[5] See also Chapter 11 for the effect of pH on the chrome tanning reaction. This drew a tart response from Constantin and the upshot was that the technology has never been pursued. It is probable that the Epsom salt technology is feasible, if the underlying science is understood.

A combination technology has been suggested, using magnesium lactate as deliming agent;[6] the reagent was available from waste whey from the dairy industry. It is reported that it can be applied as an offer of 3% for 30 min, to give solution pH 9.7, when the bate can be added directly into the solution.

7.2.7 Hydroxyl 'Sinks'

Any chemical system that absorbs hydroxyl ions has the effect of lowering the pH; this has been exploited in the chemistry of ester hydrolysis for deliming. The reaction principle is a follows:

$$R^1CO_2R^2 + OH^- \rightleftharpoons R^1CO_2^- + R^2OH$$

Hydrolysis is the addition of the elements of water, to recreate the organic acid and alcohol, but under alkaline conditions the product is the acid anion. The R groups have to be selected to ensure solubility of the reactant and the products. The preferred reagents are methyl lactate and ethylene glycol mono- or diformate, when the claimed effect is 70% deliming in 30 min and a cut that is colourless to phenolphthalein after 2 h, giving solution pH 6.4.[7] This patented technology has not been taken up by industry, presumably because of issues of cost and rate of reaction.

The rate aspect could be addressed, since it is dependent on the details of the structure of the ester and the influences of the inductive effects of the acid and alcohol moieties. This is illustrated in Table 7.1, which contains a selection of data from the older literature.[8]

Table 7.1   Rates of hydrolysis of some esters at 25 °C relative to methyl acetate.

Ester	Formula	Relative rate
Methyl acetate	$CH_3CO_2CH_3$	1.00
Ethyl acetate	$CH_3CO_2CH_2CH_3$	0.60
Isopropyl acetate	$CH_3CO_2CH(CH_3)_2$	0.15
tert-Butyl acetate	$CH_3CO_2C(CH_3)_3$	0.008
Methyl formate	HCO_2CH_3	223
Ethyl propionate	$CH_3CH_2CO_2CH_2CH_3$	0.55
Methyl chloroacetate	$ClCH_2CO_2CH_3$	761
Methyl dichloroacetate	$Cl_2CHCO_2CH_3$	1.6×10^4

7.2.8   Carbon Dioxide

In the last quarter of the twentieth century, the use of carbon dioxide as a deliming agent was developed. It works as follows:

$$CO_2 + H_2O \rightleftharpoons HCO_3^- + H^+ \xrightleftharpoons{OH^-} CO_3^{2-} + H_2O$$

The gas is sparingly soluble in water, producing acid that can neutralise the alkalinity of the limed pelt. However, a limiting factor is the solubility, which means that the availability of acid is limited and hence the reaction rate is slow; typically, skins take 1 h to delime and 25 kg of ox hide can take several hours to delime completely. This applies even under the conditions of supplying excess gas as quickly as possible into the system.

Technologically, the application of CO_2 deliming is simple: the gas is stored under pressure in liquid form and is led by pipe through the hollow axle of the drum. It is not important to introduce complexity with regard to leading the gas into the float – merely supplying it to the headspace is sufficient.

However, in industrial processing, it is important to lead the gas from the storage vessel through a small heater, because the release of pressure causes the gas to cool. The consequence is that unheated gas will solidify and block the pipe, possibly causing it to burst under the pressure. (Note that this effect of gas cooling is observed when CO_2 fire extinguishers are discharged, when the gas stream is accompanied by some CO_2 'snow'.) The feed rate is dependent only on the rate at which the gas dissolves in the float. The concentration achieved is dependent on temperature and pressure, as shown in Table 7.2.

Although the use of carbon dioxide for deliming is increasingly common, the use of pressurised vessels is rare, so the solubilities of interest correspond to ambient pressure. Additionally, it is common

Table 7.2 Solubility of carbon dioxide in water.

| Temperature (°C) | Solubility (g L^{-1}) | |
	1 atm	5 atm
0	3	17
16	2.5	10
32	2	7
49	1	3

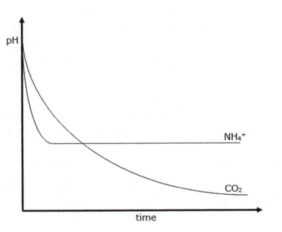

Figure 7.4 Relative effects on pH in deliming by ammonium salt and carbon dioxide.

to conduct deliming at 30–40 °C, in preparation for the bating step. Although there is a reduction in solubility because the water is warm, the rate of deliming through the cross-section is faster than processing with colder float.

Apart from the rate difference, the greatest difference between ammonium salt deliming and carbon dioxide deliming is the pH effect. There is a buffering effect with CO_2, but it is at a much lower pH than ammonium salts achieve; CO_2 deliming can lower the pH to 5.[9] This is illustrated in Figure 7.4.

This clearly has implications for bating, because most conventional bates have no activity at such a low pH. Therefore, the tanner faces two options to deal with this problem.

1. Change the bate.
   This is possible, because there are acid-acting bate formulations on the market. However, there is no doubt that the biochemistry

will be different, so the outcome may not be the same as that with the more usual pancreatic bating. That problem notwithstanding, there is the problem of adjusting the offer and conditions for the new process to match the broad effects of the conventional process. This can be done by matching the actions through assaying (see Chapter 8).

2. Control the acidification.

In essence, this means limiting the supply of gas, so the pH is not allowed to fall below pH 9 or making sure the reaction finishes at that pH. However, this is only achieved by reducing the availability if the gas in solution, limiting the concentration, with the consequence of slowing the rate of reaction, which is already slower than desirable.

The relative rates of deliming using ammonium chloride and carbon dioxide can be estimated from the solubility in solution, as follows.

Assume that 2% NH_4Cl is added to drained pelt, 10% residual float. $[NH_4Cl] = 200$ g $L^{-1} \approx 4$ M. Solubility of $CO_2 = 2.5$ g L^{-1}, which at ambient temperature and 1 atm pressure ≈ 0.06 M. The calculation indicates that the difference is almost two orders of magnitude.

Biochemistry offers a novel solution to the problem: carbonic anhydrase accelerates the solubilisation of carbon dioxide in water in the pH range 4–10, optimally at pH 7.5–8.0, therefore speeding up the deliming reaction.[10] (Carbonic anhydrase is the enzyme that controls the excretion of carbon dioxide in the breath.) The problem of the slow rate of deliming may be understood by considering the reactions involved: they can be defined as shown in Figure 7.5.

From reaction (7.1), the availability of acid is *via* hydrated carbon dioxide and reaction with or ionisation of the unstable (transition) carbonic acid species. Therefore, the available acidity depends on the solubility. Alternatively, from reaction (7.2), there may be direct reaction between the gas molecule and hydroxyl ion. However, this assumes that the gas can be transported to the required sites in the substrate or that the alkalinity diffuses out of the pelt; such

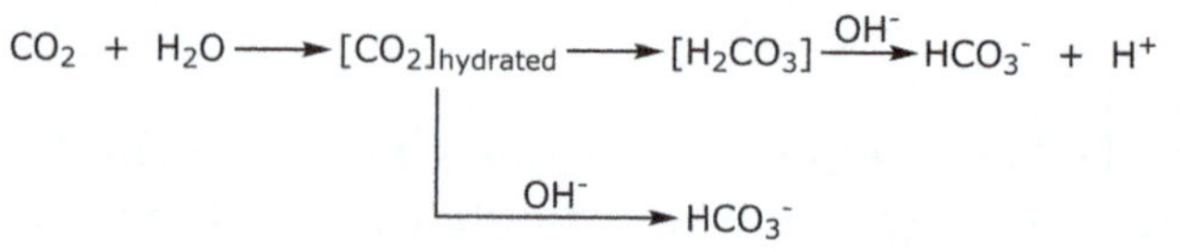

$$CO_2 + H_2O \longrightarrow [CO_2]_{hydrated} \longrightarrow [H_2CO_3] \xrightarrow{OH^-} HCO_3^- + H^+ \qquad \text{eqn (7.1)}$$

$$\xrightarrow{OH^-} HCO_3^- \qquad \text{eqn (7.2)}$$

Figure 7.5   Reaction of carbon dioxide with water.

equilibration is slow. Enlisting the aid of enzymology using carbonic anhydrase, the rate of CO_2 deliming can approach the rate of ammonium salt deliming.

There is a 'low-tech' solution to this problem, based on delivering all of the gas required for the reaction in a single offer, rather than feeding the gas into the system as the reaction is conducted. The technique is to sparge the process vessel of air, filling the headspace with carbon dioxide.[11] In this way, the rate of reaction is maximised and the process time can match the reaction with ammonium salt. However, the technology assumes modifications to the process vessel, to prevent loss of gas during sparging; this is not always feasible with most conventional processing drums.

A remaining problem with carbon dioxide deliming is the effect of the low solution pH on residual sulfide from liming: at the typical pH values of 5–6, any sulfide in solution is in the form of hydrogen sulfide. Although the amount of sulfide left in the pelt is relatively small, the toxicity of the gas constitutes a deadly risk to workers. However, sulfide in solution can be scavenged efficiently and effectively by oxidation. The application of an offer of hydrogen peroxide as low as 0.1% easily maintains the concentration of H_2S in the headspace at a safe level, below the level of detection by smell, *i.e.* below 1 ppm. Importantly, it should be recalled that swollen pelt is particularly vulnerable to damage by oxidising agents; therefore, the sequence of events should be as follows.

- Begin CO_2 deliming.
- Continue for 5 min – this is sufficient to delime only the surface, avoiding leaching out much of the sulfide held within the pelt.
- Add the hydrogen peroxide – the sulfide in solution is oxidised, but the grain is protected against oxidation damage because the pH is lowered by the initial deliming.
- Resume deliming – any sulfide leached into solution is scavenged by the peroxide:

$$S^{2-} + 4H_2O_2 \rightleftharpoons SO_4{}^{2-} + 4H_2O$$

7.3 Melanin

A feature of the CO_2 deliming process is the effect that it has on the melanin content of pelt. The skin pigment, melanin, is present in cells called melanocytes, mainly in the epidermis, but also within the grain, associated with the hair follicles, as discussed in Chapter 2. In conventional processing with ammonium salts, the pH is rapidly lowered to about 8.5–9.0, where there is still some swelling in the pelt, with consequent elevation of the follicle angle. The effect of this residual swelling is to allow the melanocytes to be removed from the pelt by the depleting action, resulting is relatively pigment-free grain. However, in carbon dioxide deliming, when the pH is lowered relatively slowly to 5, it has been observed that the melanin remains, clearly visible as coloration on the grain.[12] The residual melanin retains the pattern of pigmentation in black hides, particularly apparent in Friesian hides. From microscopic examination, it has been demonstrated that the effect is due to the presence of the melanin within the grain, caused by the depleting effect of the lower pH, not as an effect of rate of depletion. Furthermore, the critical pH for the effect to be manifested is 7: deliming to no lower than pH 7 results in clean grain, but deliming below pH 7 results in melanin remaining within the grain. This creates a dilemma for the tanner: in order to maintain pH >7 in solution, the availability of carbon dioxide should be restricted and there is a consequent slowing of the deliming rate, which is already too slow.

The solution for tanners of melanin-pigmented hides is to delime as rapidly as possible, with or without carbonic anhydrase, but then to degrade any residual melanin. It was indicated in Chapter 2 that chemical methods would be too violent, but that a biochemical option is available: Rao and co-workers reported that a solution of 0.5% xylanase can completely remove melanin from hide within a few hours.[13] Other biochemistries may be of assistance, such as polyphenol oxidases, capable of degrading the pigment.

7.4   Limeblast

It was described in Chapter 6 how the action of carbon dioxide in the air can cause the problem of limeblast, the growth of calcium carbonate crystals below the grain surface. However, in carbon dioxide deliming, limeblast is not observed. Therefore, this begs the question: can CO_2 deliming cause limeblast? The reactions of importance are as follows.

At high pH, insoluble calcium carbonate is formed, potentially creating limeblast:

$$Ca(OH)_2 + CO_2 \rightleftharpoons CaCO_3 + H_2O$$

At lower pH, with continued addition of carbon dioxide, soluble calcium bicarbonate is formed: this is the consequence of carbon dioxide deliming, lowering the solution pH to 5–7:

$$CaCO_3 + H_2O + CO_2 \rightleftharpoons Ca(HCO_3)_2$$

Therefore, carbon dioxide deliming is unlikely to cause limeblast.

This chemistry is the basis of the traditional school experiment in which students blow through a straw into lime water, a clear solution of calcium hydroxide. From the carbon dioxide in the breath bubbled through the solution, first a precipitate forms, seen as cloudiness in solution, then the precipitate redissolves and the solution becomes clear.

References

1. Y. Zeng, *et al.*, *J. Am. Leather Chem. Assoc.*, 2011, **106**(9), 257.
2. J. D. Putshak'a, *et al.*, *J. Am. Leather Chem. Assoc.*, 2013, **108**(1), 11.
3. E. Heidemann, A. Hein and P. Herrera, *Das Leder*, 1988, **39**, 141.
4. J. M. Constantin, *J. Am. Leather Chem. Assoc.*, 1981, **76**(1), 40.
5. R. C. Koopman, *J. Am. Leather Chem. Assoc.*, 1982, 77(7), 356.
6. K. Kolomaznik, *et al.*, *J. Am. Leather Chem. Assoc.*, 1996, **91**(1), 18.
7. E. Hahn, B. Magerkurth, D. Lach and L. Wuertele, *UK Pat. Appl.*, 2026538A, 1979.
8. C. K. Ingold, *Via Structure and Mechanism in Organic Chemistry*, G. Bell and Sons, London, 1953.
9. M. J. Klaase, *J. Am. Leather Chem. Assoc.*, 1990, **85**(11), 431.
10. A. D. Covington, K. B. Flowers and A. Onyuka, *UK Pat. Appl.*, 06063176, 2006.
11. K. B. Flowers and C. A. Jackson-Moss, *Proceedings IULTCS Congress*, Capetown, South Africa, 2001.
12. A. D. Covington, unpublished results.
13. V. Punitha, *et al.*, *J. Am. Leather Chem. Assoc.*, 2008, **103**(7), 203.

8 Bating

8.1 Introduction

Bating is a generic term that refers to the use of enzymes in an early stage of leather making. This chapter addresses the technology of applied enzymology, while Chapter 18 addresses the science of enzymology.

The purpose of bating is to break down specific skin components; usually the non-structural proteins are the target. These proteins can be degraded by general proteases, because they do not have highly defined structure, even though they may be folded specifically. Their lack of structure means that scission of the peptide chain creates smaller molecules, making them more easily removable by rinsing. Intact collagen is more resistant to such attack, because of its highly structured triple helix, which requires specific collagenases to break it down. However, it is important to recognise that alkali-damaged collagen is vulnerable to attack by general proteases, so limed collagen can be significantly degraded during conventional bating with non-specific proteases. The usual effect of this type of damage is grain enamel loss and/or corium loosening.

Bating has been carried out in different ways over the centuries. Traditional sources of enzymes have typically been animal faeces, where the required component was the pancreatic or digestive enzymes;

Tanning Chemistry: The Science of Leather 2nd edition
By Anthony D. Covington and William R. Wise
© Anthony D. Covington and William R. Wise 2020
Published by the Royal Society of Chemistry, www.rsc.org

using bird guano was called 'mastering' and using dog faeces was called 'puering'. There was also a botanical version, called 'bran drenching', where the origin of the bating material is apparent. Clearly, using warm infusions of such materials must have contributed greatly to the tanners' reputation of dealing with unpleasant products and making smells! However, shortly after the beginning of the twentieth century, the first enzymes isolated from animal sources were made available for leather-making purposes by Otto Röhm, although it was about 50 years before the use of puer was discontinued, at least in Europe. More recently, bating enzymes from bacterial fermentations have become available.

Alexander[1] reviewed the use of enzymes in the beamhouse, signalling the new era of biochemistry-based processing, which will characterise the leather industry of the twenty-first century. Enzymes are biochemical catalysts: they are the products of microorganisms, but they are not microorganisms themselves – they are not alive because they cannot reproduce. Their function is to catalyse organic reactions, which they can do very effectively: reactions may be accelerated up to 10^{20} times faster than the uncatalysed reaction. The definition of a catalyst is a reagent that can accelerate a reaction to equilibrium but is chemically unchanged in the process, although it may be physically changed. Note that a catalyst cannot change or displace the point of equilibrium, it can only affect the rates of a reaction, both forward and back.

8.2   Factors Affecting Enzyme Catalysis

Enzyme catalysis can be highly specific regarding the substrates with which an enzyme will or will not interact; moreover, the conditions under which enzymes function are typically also critical. This section deals with these parameters in a light-touch manner; more details about how these parameters affect the rate of reaction can be found in Chapter 18.

8.2.1   Temperature

The rate of the reaction depends on the origin of the organism that produced the enzyme. Since most bates are produced from bacteria, the optimum temperature relates to the environment of the host, *i.e.* body temperature. This is illustrated in Figure 8.1.

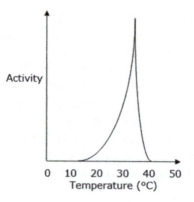

Figure 8.1 Rate of an enzyme-catalysed reaction as a function of temperature.

The shape of the profile in Figure 8.1 is asymmetric. On the low-temperature side of the maximum there is a rapid increase in the rate of reaction as the temperature rises to reach the optimum value, which is typically close to the natural environment of the organism that produced the enzyme. Therefore, the optimum temperature is usually 35–40 °C, although enzymes are known that operate at much lower or much higher temperatures, the so-called extremophiles. However, enzymes used in leather making are typically conventionally reactive.

On the high side of the temperature maximum, there is a steep fall in activity. This is because the reaction site of the enzyme is highly sensitive to temperature and, indeed, the protein may be denatured at a temperature slightly higher than the optimum temperature. In either case, the enzyme ceases to function. The steepness of the activity curve as temperature rises is indicative of the sensitivity of the enzyme to temperature and hence the action of the enzyme in bating. A small deviation of the temperature from the optimum value means a large decrease in activity and therefore a loss of value from the product.

8.2.2 pH

The pH–activity profile of an enzyme is illustrated in Figure 8.2.

It should be noted that real profiles may differ from the model in Figure 8.2 in the following respects:

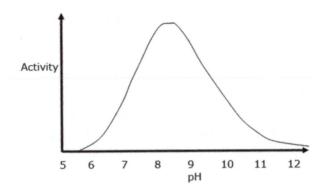

Figure 8.2 Profile of enzyme activity as a function of pH.

- the profile may or may not be symmetrical;
- the profile may be more narrow or more broad;
- the maximum activity may not be at pH 8–9.

Unlike the temperature dependence of activity, there is likely to be significant activity below and above the optimum pH, without the same catastrophic decrease in activity due to damage to the reactive site on the enzyme; in this case, the effect of pH is to change the charge on the protein, causing a change in the electrical field of the active site, with roughly similar changes either side of the pH optimum.

8.2.3 Concentration: Bate Formulations

The activity of enzymes is a function of concentration, because they are catalysts, as presented in the kinetic equations in Chapter 18. Therefore, offers in leather making are not made on the usual weight fraction basis on pelt weight, as with other reagents. Offers depend on the amount of water being used.

Commercial bate products are not pure enzymes. They are not pure in terms of being a single enzyme: proteolytic enzymes from any source are likely to be a mixture of enzymes, so the precise reactions that they catalyse may not be known. They are also not pure in terms of the formulation of the product. Since the activities are so high, the amount of pure, isolated enzyme that would be needed in a bating process would be very small. Therefore, there could be a large error in the weighing, hence there would be large variations in the consistency of the effect on the pelt. Consequently, bates are supplied in a diluent, which can have an effect on the process.

8.2.3.1 Ammonium Salt

If the deliming step is conducted with ammonium salt, the presence of more ammonium salt in the bate formulation will contribute to the buffering effect. Because the salt is very soluble in water, the enzyme is immediately available for reaction.

8.2.3.2 Wood Flour

An alternative diluent is wood flour, which is lignocellulose. It has the advantage of not contributing to the ammoniacal nitrogen in the effluent, but it may contribute to the suspended solids loading of the effluent. Because there is no great affinity between the diluent and the enzyme, the enzyme is immediately available for reaction.

8.2.3.3 Kaolin

The use of kaolin, a form of clay, as a bate diluent is less common, but it is useful to know that there are two consequences if it is part of the formulation. First, the diluent will contribute to the suspended solids in the liquid effluent and/or the sludge from mixed effluent streams. Second, there is often a strong affinity between protein and clay, so there will be a delay after adding the bate to the solution before the enzyme is desorbed from the diluent surface. Therefore, the reaction is not immediate and so allowance must be made for the delay.

8.2.4 Time

The extent of the bating effect depends on the time allocated to the process step. This is typically about 1 h. It is important to know that enzymes will usually penetrate through only about 1 mm of pelt in 1 h. For skins, this means that the bating effect will probably extend through the cross-section, causing damage to the non-collagenous proteins and to the collagen. Importantly, the available time is insufficient for the enzymes to penetrate through hide completely, even in the case of light ox or cow. Therefore, in most processes, the bating effect is confined to the grain layer and the outer part of the flesh side. This is usually deliberate: it avoids the risk of a loosening effect in the corium, but takes advantage of the potent enzymatic reactions to clean the grain of residual epidermis and damaged hair/keratin.

In reviewing industrial practices, it is not uncommon to find that bating is not conducted under optimum conditions for enzyme activity, pH and temperature. It is as though tanners recognise the inherent danger of the efficiency of enzyme catalysis and consequently opt for lesser effects than bates are capable of delivering. A cheaper option, of course, would be a lower offer at optimum conditions.

8.2.5 Origin

The origin of the bating enzymes will determine the actual reaction applied to the proteins in the pelt. These reactions are many and varied: some are set out in Table 8.1.[2] The other important aspect of origin is the difference between pancreatic and bacterial bates. The critical difference lies in their ability to degrade elastin. Pancreatic or digestive enzymes do not degrade elastin at all, at least not within the timescale of conventional liming and bating, but bacterial bates often contain elastolytic enzymes. Therefore, a

Table 8.1 Reaction mechanisms of some bating enzymes.

Enzyme	Site of attack
Trypsin	Pancreatic serine protease. High specificity
	Cleaves the peptide link on the carbonyl side of basic amino acids, lysine and arginine. Only applies to the telopeptide region, not the helical region
Chymotrypsin	Cleaves the peptide link on the carbonyl side of amino acids with aromatic sidechains, *i.e.* phenylalanine, tyrosine, tryptophan (note: there is no tryptophan in collagen)
	Optimum pH 7–9
Nagarse	Isolated from *Bacillus subtilis*
	Non-specific cleavage. Preferably active against small polypeptides
Subtilisin	As for nagarse
Papainase	Sulfhydryl protease from papaya fruit
	Active against long polypeptides containing Phe–X–Y
Leucine aminopeptidase	α-Aminoacyl exopeptidase, active towards peptides containing a free α-amino group
	Metal activated, 3–4 mM Mn(II), Mg(II)
Prolidase	Isolated from pig kidney
	Metal activated, cleaves peptide bonds containing proline and hydroxyproline
Pronase	Non-specific enzyme, isolated from *Streptomyces griseus*
	Very broad reactivity

comparison of these bates would reveal the impact on the leather character of degrading the elastin. The effect is important, because it results in very different leather, in terms of area, but more importantly with regard to the physical properties and performance, as discussed in Chapter 2.

Note that the appearance of bacterial bates to replace the more common pancreatic bates coincided with changes to the production of insulin for diabetics. The traditional source of insulin was extraction from bovine pancreas, so the by-product was pancreas tissue and the pancreatic enzymes. However, more recently, human insulin has been produced by gene engineering of bacteria, so bovine pancreatic tissue is no longer a cheap by-product of another industry. The obvious cheap source of proteolytic enzymes is by fermentation of bacteria, hence the trend of change in formulation of the product.

8.3   Monitoring Bating

There are two traditional, qualitative ways of monitoring the progress of the bating process.

1. The thumbprint method for hides or skins. This relies on the effect of bating to remove the resiliency of the pelt when the interfibrillary proteins are degraded. The pressure of a finger, usually the thumb, squeezes the bated, opened-up fibre structure, which does not spring back, leaving a thumbprint that recovers only slowly. This is a pass–fail test, with no degrees of bating easily discernible.
2. The air permeability method for skins. This test relies on the effect of opening up on the ability of the pelt to allow the passage of air through it. The tanner folds the pelt to create a bubble by trapping air. Holding the pelt tightly, the bubble is gradually made smaller, but the amount of air is maintained. This creates pressure within the bubble, forcing air from the inside to the outside, but only if transmission is allowed by the degree to which the fibre structure is opened up. This is a pass–fail test, with no degrees of bating easily discernible.

The progress of the bating reaction can be followed more accurately using the biuret test, which is a test for protein, using basic copper(II) sulfate.[3] The argument is that at the site of enzyme action there will be degraded protein, so there will be more reaction with the reagent than in unaffected regions. The reagent, which can be made *in situ*, is applied to a freshly cut edge: a drop of concentrated copper(II) sulfate solution is placed on the cut surface, followed by a drop of concentrated sodium hydroxide solution. Where there is enzyme activity, there is a purple colour reaction. Even though the colour developed in this way is against a blue background, it can be easily discerned.

References

1. K. T. W. Alexander, *J. Am. Leather Chem. Assoc.*, 1988, **83**(9), 287.
2. O. Menderes, PhD thesis, University of Leicester, 2002.
3. O. Glezer, BSc thesis, Nene-University College, 1997.

9 Pickling

9.1 Introduction

The pickling process is primarily conducted to adjust the collagen to the conditions required by the chrome (or any other) tanning reaction. The traditional, indicative recipe for pickling is as follows, based on limed weight of pelt:

- 100% float
- 10% salt
- 1% sulfuric acid.

The timing of the pickling step is variable, depending on circumstances, not least of which is the thickness of the pelt. Typically, it is conducted for about 1 h before adding the chrome tanning salt.

The order of addition of the components is as follows.

1. To drained pelt, add the salt offer and run to allow the salt to be absorbed into the pelt as concentrated solution – to avoid acid swelling.
2. Add the majority of the float, nominally 90%.
3. Add the acid, previously diluted 10:1 in water and cooled – to avoid localised high concentrations of (hot) acid, which can cause hydrolytic damage (see later).

Tanning Chemistry: The Science of Leather 2nd edition
By Anthony D. Covington and William R. Wise
© Anthony D. Covington and William R. Wise 2020
Published by the Royal Society of Chemistry, www.rsc.org

The offer of acid in pickling is determined by the pH required for the chrome tanning process; this varies, but for chrome tanning typically lies in the range pH 2.5–3.0, since a solution of 33% basic chrome tanning salt is at pH 2.7–2.8.

The function of the acid is to acidify the collagen, to protonate the carboxyl groups: in this way the reactivity is modified, because the chrome tanning reaction only involves ionised carboxyl groups (see Chapter 11). The effect of acidification can be calculated from the Henderson equation, applied to the ionisation of a weak acid:

$$HA \rightleftharpoons H^+ + A^-$$

The dissociation constant K_a is given by

$$K_a = \frac{[H^+][A^-]}{[HA]}$$

Taking logarithms and multiplying by −1 (which is the operator 'p'),

$$-\log K_a = -\log[H^+] - \log[A^-] + \log[HA]$$

$$pK_a = pH + \log\frac{[HA]}{[A^-]} \quad \text{(Henderson equation)}$$

Since the pK_a values for aspartic and glutamic acids are 3.8 and 4.2, respectively, the overall pK_a for collagenic carboxyl groups must lie between those values, assuming that reaction with chromium(III) applies to both types of sidechain carboxyl groups. Therefore, for simplistic calculation purposes, the pK_a for collagen carboxyls can be assumed to be 4.0, as shown in Table 9.1.

It should be noted that another aspect of the pickling reaction is the contribution to opening up. In the same way as hydrolysis of peptide bonds is accelerated by hydroxyl catalysis, the same reaction is catalysed by hydrogen ion:

$$-CO-NH- + H_2O \underset{OH^-}{\rightleftharpoons} -CO_2^- + H_2N-$$

$$-CO-NH- + H_2O \underset{H^+}{\rightleftharpoons} -CO_2H + H_2N^+ -$$

Table 9.1 Calculation of the degree of carboxyl ionisation in collagen.

pH	Log([HA]/[A⁻])	[HA]/[A⁻]	[A⁻]/%
1.5	2.5	316	0.3
2.0	2.0	100	1
2.5	1.5	31.6	3
3.0	1.0	10.0	9
3.5	0.5	3.2	24
4.0	0.0	1.0	50
4.5	−0.5	0.32	76

For most processes in conventional rapid pickling this reaction is unimportant because it is relatively slow. However, there are two situations when the reaction can be important.

1. Storage pickling, as a preservation technique.
2. As an alternative to alkali opening up for woolskins, when the wool is vulnerable to detachment if the process goes too far, too rapidly. Under conditions of alkalinity or proteolytic enzyme action, the wool on woolskins can become loosened or even detached, referred to as 'woolslip'. This is clearly undesirable if the pelt quality depends on the presence of wool, *e.g.* for rugs or for 'double face'. In this case, woolskins can be stored for several weeks in the conventionally pickled condition, to allow protein hydrolysis to take place, making the resulting leather softer. Under these conditions, the opening-up reactions are much slower than alkaline opening up and hence are more easily monitored/controlled for the delicate rawstock.

An additional function of the pickling process is to complete the depleting of the pelt, following alkaline swelling. From the swelling curve reproduced in Figure 9.1, the pelt goes through minimum swelling at the isoelectric point and the depletion is maintained by the effect of neutral electrolyte. In this way, damage to the pelt structure is avoided and the subsequently applied

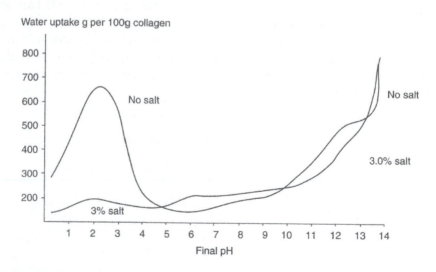

Figure 9.1   Relationship between pH and acid swelling and the role of neutral electrolyte.

reagents have the minimum thickness of pelt to penetrate (without collapsing the structure). The problem of cationic collapse is most commonly observed in the form of 'pickle creasing', caused by adhesions occurring as the pelt dries out, particularly along folds or creases during storage. Light adhesions can be reversed by tumbling the pelt in warm brine with non-ionic detergent; intractable creases may be removed by tumbling the drained pelt first in paraffin, then in warm brine and non-ionic detergent. Some creases may not be removable at all if new bonds are formed by the fibre structure collapse, as described by Komanowsky,[1] when the protein chains approach closely on the molecular scale, allowing reactions to occur.

9.2 Processing Conditions

9.2.1 Float

The overall float length is not critical to the efficiency and effectiveness of the pickling process. The traditional and typical amount of water reflects more the requirement of the subsequent process, in which the concentration of the chrome tanning reagent exerts some control over the efficiency of the reaction.

9.2.2 Salt

In order to avoid acid swelling, the concentration of sodium chloride in the whole of the available solution (added water + water carried over from the previous step + water within the pelt) should not be less than 1.0 M, corresponding to 6%. The relationship between salt and acid swelling is shown in Figure 9.1. The relationship assumes that the swelling is prevented by the presence of the salt, rather than the swelling being reversed by the electrolyte. There is a danger that acid swelling may cause disruptive damage that cannot be reversed by the addition of electrolyte, hence there is the requirement for a specified order of addition of pickling components.

The swelling curve is not symmetrical over the pH scale, so the variations in the shape of the curve must be rationalised in order to understand the role of conditions in leather making and their effects on collagen. We can begin by considering the changes that occur within collagen when the pH is changed, indicated in Figure 9.2, in which P represents protein.

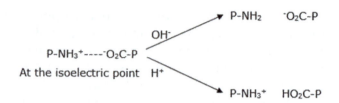

Figure 9.2   Effect of pH on the charge status of protein.

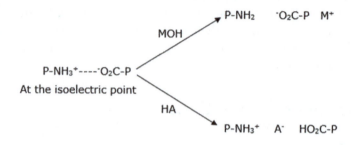

Figure 9.3   Effect of acidification and basification on protein.

The development of charge and its effect on swelling must be the same at both ends of the pH scale. However, acid swelling clearly has some different features compared with the alkaline conditions, because:

- the curve exhibits a turning point;
- the effect of salt is to reverse or prevent acid swelling – the extent of the reversal depends on the salt concentration: the higher the salt concentration the less swelling is observed.

From the Henderson equation, the ratio of acidic to basic form of the carboxyl groups can be calculated at any pH value, assuming that the pK_a of collagen is 4.0, given in Table 9.1. At pH 2 the titration is effectively complete, so there is maximum positive charge, corresponding to maximum swelling.

The observation of the effect of salt on acid swelling clearly indicates that there is another influence beyond merely smearing out the charge on the collagen. Consider again the effect on the charge of the protein of changing the pH, shown in Figure 9.3.

When collagen is acidified, the acid effectively binds electrolyte to the protein; therefore, in an acidification reaction in water, there is a net increase in electrolyte within the pelt compared with outside the pelt. If the surface of the pelt can be regarded as a semipermeable membrane, across which electrolyte and water can pass, an osmotic pressure will be developed. Therefore, water will pass into the pelt, to reduce the chemical potential within the pelt, because the electrolyte concentration is higher than the concentration in the external solution. This is observed as swelling.

The phenomenon of diffusible and non-diffusible salts separated by a semipermeable membrane was analysed by Donnan and co-workers.[2] In the case of pelt in acid, the electrolyte quantities inside and outside the pelt are separated by the pelt surface and the internal structure of the pelt constitutes the non-diffusable electrolyte. When salt alone is present, the situation at equilibrium can be summarised as follows:

$$
\begin{array}{ccccccc}
\text{outside} & & : & \text{inside} & & & \\
\text{Na}^+ & \text{Cl}^- & : & \text{Na}^+ & \text{R}^- & \text{Cl}^- \\
c_1 - x & c_1 - x & : & c_2 - x & c_2 & x
\end{array}
$$

The system must be electrically neutral on both sides of the surface and the chemical potentials must be equal:

$$\mu_{\text{NaCl}(1)} = \mu_{\text{NaCl}(2)}$$

The chemical potential of an electrolyte is the sum of the potentials of the ions:

$$\mu^\circ_{\text{Na}^+} + RT\ln a_{\text{Na}^+(1)} + \mu^\circ_{\text{Cl}^-} + RT\ln a_{\text{Cl}^-(1)} = \mu^\circ_{\text{Na}^+} + RT\ln a_{\text{Na}^+(2)} + \mu^\circ_{\text{Na}^+} + RT\ln a_{\text{Na}^+(2)}$$

where a is the activity. Therefore,

$$a_{\text{Na}^+(1)} \times a_{\text{Cl}^-(1)} = a_{\text{Na}^+(2)} \times a_{\text{Cl}^-(2)}$$

Under dilute conditions, the activities can be substituted for concentrations; strictly, leather processes are not conducted under dilute conditions, but in order to indicate this phenomenon, ignoring the activity coefficients is a simplification. Therefore,

$$(c_1 - x)(c_1 - x) = (c_2 - x)x$$

If we now consider the situation of a protein in acid, as is the case in pickling, the acid creates cations at the sites of the amino groups. Note that although the cationic nature strictly arises because of the breakdown of the salt links, the reaction can be regarded as fixation of acid on the protein. Referring to the protein as P, equilibrium can be represented as follows:

$$
\begin{array}{cccccccc}
\text{outside} & & : & \text{inside} & & & & \\
\text{H}^+ & \text{Cl}^- & : & \text{H}^+ & \text{Cl}^- & \text{PH}^+ & \text{P} \\
x & x & : & y & y+z & b & a
\end{array}
$$

If the protein has n basic groups, then $b = z/n$. Therefore, as before,

$$x^2 = y(y + z)$$

or

$$z = \frac{x^2 - y^2}{y}$$

where z is the excess acid bound to the protein.

In pickling, swelling must be avoided, so the Donnan equilibrium condition must be met, otherwise there would be a net osmotic diffusion of water into the pelt. The solution is to have an excess of electrolyte, which can equilibrate across the semipermeable membrane. This changes the situation as follows:

$$
\begin{array}{ccccccccc}
 & \text{outside} & & : & \text{inside} & & & & \\
Na^+ & H^+ & Cl^- & : & Na^+ & H^+ & Cl^- & PH^+ & P \\
A & x & x+A & : & A & y & A+y+z & b & a \\
\end{array}
$$

$$(A+x)(x+A) = (A+y)(A+y+z)$$

If A is large compared with x, y and z, then the equation is effectively balanced and the Donnan equilibrium condition is met. Typical pickling conditions include 6% offer of salt and the float is 100%.

Assume that the concentration of added salt, inside and outside, is 1.0 M, *i.e.* $[Na^+] = [Cl^-] = 1.0.M$. Then, the concentration of pickling acid $= 10^{-2}$ M $= [H^+] = [Cl^-]$, applying to both inside and outside.

If we assume that all amino groups on collagen are protonated, the counterion (Cl^-) concentration in solution balances the amino content of 0.6 mol kg^{-1} dry collagen.

Assuming that the float is 100% for collagen containing 75% moisture, the anion concentration is 0.15 M. Therefore, the relationship with regard to the Donnan situation can be calculated.

In the absence of added electrolyte:

$$\text{Outside: } 0.01 \times 0.01 = 1.0 \times 10^{-4} \text{M}$$
$$\text{Inside: } (0.01 + 0.15) \times 0.01 = 1.6 \times 10^{-3} \text{M}$$

which is a difference of 16-fold or 1600%.

In the presence of added electrolyte:

$$\text{Outside: } (1.0 + 0.01) \times (1.0 + 0.01) = 1.02 \text{M}$$
$$\text{Inside: } (1.0 + 0.01 + 0.15) \times (1.0 + 0.01) = 1.17 \text{M}$$

which is a difference of only 16%.

The Donnan equilibrium condition can be achieved using any electrolyte, although the side effects of lyotropic swelling must be borne in mind. For any 1:1 monovalent salt, the calculated amount is based on molar activities, which can be simplified to concentration. In the case of multivalent ions, the concentration can be corrected for activity coefficients from the calculation of ionic strength, I:

$$I = 0.5\Sigma c_i z^2_i = 0.5(c_1 z^2_1 + c_2 z^2_2 + c_3 z^2_3 + ...)$$

where c_i is the molar concentration of ion i and z is its charge.

Among the alternative neutral electrolytes that might be used in place of sodium chloride, the most likely candidate is sodium sulfate. There are associated environmental impact problems with its use, because sulfate ion can leach calcium ions from concrete and hence weaken it. The effect on sewerage systems means that consent limits for sulfate discharge in effluent are likely to become increasingly stringent. Sulfate can interact with chromium(III) species, affecting the hydrolysis rate and influencing the nature of the species in solution and hence the tanning effect (see Chapter 11).

The concentration of salt in pickling must be subject to control, because too much will constitute a problem. In the same way as salt causes dehydration in curing, if the concentration in pickle is higher than the minimum requirement to meet the Donnan equilibrium condition, the semipermeable pelt system will allow water to move from the pelt into solution, causing the fibre structure to collapse, as it does in wet salting curing. The lower moisture content of the pelt will restrict the free movement of solubilised reagents, possibly resulting in non-uniform chrome tanning through the pelt cross-section.

Strictly, the Donnan condition cannot be met at equilibrium in pickling, but the application of the electrolyte before the acid ensures that the equilibrium is approached from the side corresponding to depletion rather than the side corresponding to swelling. Equilibrium is approached sufficiently closely that swelling is avoided.

The remaining issue is a rationale for the turning point in the swelling curve: swelling reaches a maximum at about pH 2, but additional acid causes it to diminish. The pelt cannot swell any more at pH values below 2 because the negative charge is at a maximum. If acid continues to be added, to lower the pH, this is equivalent to adding electrolyte, even though one of the ions is H^+. At pH 1.0, the concentration of acid is molar, comparable to the situation calculated above, for the addition of the conventional amount of salt in pickling, so the swelling curves with and without salt converge at low pH.

9.3   Lyotropic Swelling

There is another contributor to swelling, namely lyotropic effect. Here, species are inserted into hydrogen bonds, thereby opening up the structure, as introduced in the discussion of liming, and indicated in Figure 9.4. The effect does not depend on pH, so the effects of ions and electrically neutral compounds apply equally under acidic and alkaline conditions.

In the case of the lyotropic ions, the insertion of the ion converts the attractive interaction into a repulsive interaction, indicated by the double-headed arrows. Alternatively, a neutral compound capable of involvement in hydrogen bonding can insert itself into the protein hydrogen bonding. In each case, the natural structure is disrupted and the protein may even be solubilised. Here, the effect of union-ised weak acids must be taken into account. For example, comparing formic and acetic acids under conditions of chrome tanning, starting at a pickle pH of 2.75 and ending at basified pH of 3.75, the relative amounts of neutral and charged species can be calculated, as shown in Table 9.2.

Both at the start and at the end of chrome tanning, acetic acid equilibrates with a higher proportion of unionised, lyotropically active form, thereby constituting a greater lyotropic swelling danger.

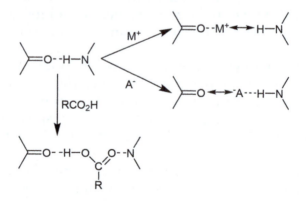

Figure 9.4   Lyotropic interactions.

Table 9.2   Ionisation of formic acid (pK_a = 3.75) and acetic acid (pK_a = 4.75).

	[HA]:[A$^-$]	
pH	Formic acid	Acetic acid
2.75	10	100
3.75	1	10

9.4   Sulfuric Acid

The commonest acid used in industry is sulfuric acid, because of its ready availability and low cost. A feature of the properties of concentrated sulfuric acid is that it releases a lot of energy when it is diluted – indeed, the solution can often boil. It is imperative that the acid is diluted and cooled prior to application in this step: failure to cool the solution will result in rapid, acid-catalysed hydrolytic damage, particularly to the grain. This type of damage is undetectable in wet pelt or leather; it is only manifested in crusted leather, when the fault appears as 'cracky' grain, a weakness in the grain manifested as cracking open if it is stressed. The damage is variable and may only be observed by a double fold: the appearance of cracky grain is almost certainly due to hot acid damage.

9.5   Hydrochloric Acid

Of all the available alternative strong acids, only hydrochloric acid has found industrial application. For all other commonly available acids, issues of cost, damage to the collagen or interference with the chrome tanning reaction rule them out.

As is the case for sulfuric acid, hydrochloric acid also must be diluted before adding it to the pickle float. There is no heat of dilution, because the acid is already in aqueous solution, but the damaging effect of localised high hydrogen ion concentration must be avoided, because of its catalytic effect on protein hydrolysis.

It is claimed that hydrochloric acid pickling makes flatter leather: this may be a reflection of the relative lyotropic effects of the chloride and sulfate ions. However, a review of ionic concentrations indicates that such an argument is weak, because the pickling period is short, then the addition of chrome tan powder introduces 50% of its weight as sodium sulfate, not to mention the contribution of the sulfate as counterion for the chromium(III) (see Table 9.3). The short period of pickling (usually 1 h) is followed by a longer period of low-reactivity chrome tanning, when the differences between the pickling systems are effectively eliminated. In each case, it does not make chemical sense to argue whether the conditions refer to hydrochloric acid + sodium sulfate or to sulfuric acid + sodium chloride, because the acids are strong electrolytes and so are the salts, hence they only exist in solution as separated ions. This argument can apply because the pickling process is rarely conducted to equilibrium prior to starting

Table 9.3 Ionic content of pickle formulations (molar), assuming the following: 100% float, 6% NaCl, 0.2 M [H⁺], chrome tan powder offer of 8%.[a]

Source of ions	HCl pickle					H$_2$SO$_4$ pickle				
	H$^+$	Na$^+$	Cl$^-$	SO$_4^-$	Cr(III)	H$^+$	Na$^+$	Cl$^-$	SO$_4^-$	Cr(III)
Salt, NaCl	—	1.0	1.0	—	—	—	1.0	1.0	—	—
Acid	0.2	—	0.2	—	—	0.2	—	—	0.1	—
Na$_2$SO$_4$	—	0.5	—	0.25	—	—	0.5	—	0.25	—
Cr(OH)SO$_4$	—	—	—	0.25	0.25	—	—	—	0.25	0.25
Total molarity	*0.2*	*1.5*	*1.2*	*0.5*	*0.25*	*0.2*	*1.5*	*1.0*	*0.6*	*0.25*
Ionic strength before chrome	1.95					2.05				
Ionic strength after chrome	2.6					2.7				

[a]33% basic chrome powder contains 25% Cr$_2$O$_3$, empirically as Cr(OH)SO$_4$, and 50% sodium sulfate.

chrome tanning. If it is necessary to consider the ionic strength of the solution before and after chrome tanning, the only difference arises from the concentration difference in sulfate from sulfuric acid in comparison with hydrochloric acid.

9.6 Formic Acid

It is common for the acid in the pickle formulation to be split between sulfuric and formic acids. Since formic acid is more expensive than sulfuric acid, there has to be a distinct technological advantage for its inclusion.

As with sulfuric acid, formic acid must be diluted before adding it to the pelt. However, in this case, the reason is not connected to heat damage, because there is little heat of dilution. Nor is the danger concerned with a localised high concentration of hydrogen ion for hydrolysis catalysis, even though the acid is weak. Here the danger is associated with the lyotropic effect of the molecule, which can cause swelling damage by hydrogen bond disruption. The acid can be added as the concentrated acid or in the form of sodium formate, which can be converted to the acid by the addition of extra sulfuric acid, according to the following equation:

$$HCO_2Na + H_2SO_4 \rightleftharpoons HCO_2H + SO_4^{2-} + Na^+ + H^+$$

It is often said that 'formic acid penetrates faster than sulfuric acid'. Technologically, what this means is that a pickle formulation containing formic acid will acidify through the pelt cross-section faster than a formulation containing sulfuric acid alone. The assertion is not strictly true. Although it is true that formic acid penetrates, strictly sulfuric acid does not penetrate. In the case of any strong acid, the penetrating species is hydrated hydrogen ion, $[(H_3O^+)(H_2O)_n]$. The penetration rate is hindered by the ion carrying a full charge, because it can interact with the protein substrate. On the other hand, the weak formic acid penetrates predominantly in the form of an electrically neutral molecule and therefore can move through the protein relatively unhindered. In the production of hides, the penetration rate benefit of including formic acid may be sole reason to use it.

It is a common misconception that the inclusion of formate in the pickle constitutes a masked chrome tannage. The term 'masking' is defined as the creation of a complex between a ligand moiety, such as formate, and a metal tanning species, such as chromium(III): this traditionally is part of tanning technology, because the modification to chrome reactivity is designed to reduce the ability of the chrome species to complex further and thereby to reduce the rate of fixation with respect to the rate of penetration through the pelt. However, there are two problems associated with that view of chromium(III) complex chemistry.

1. The rate of reaction between chromium(III) and formate is the same as the rate of reaction between chromium(III) and collagen: they are both carboxylate complexation.[3] This has been demonstrated experimentally using acetate as the masking ligand. Therefore, the chromium must be unmasked at the beginning of the reaction and become masked as the tanning reaction proceeds; this runs counter to the requirement for increasing the chrome reactivity as the reaction proceeds.[4] If the tanning process calls for true formate masking, the complexation reaction must be undertaken prior to initiating the tanning reaction or an appropriately premasked tanning product must be purchased (see Chapter 11).
2. The assumption of reactivity reduction is flawed, since it is possible, indeed probable, that masking can increase the reactivity of chromium(III) complexes, depending on the chemistry of the masking agent and the masking ratio, defined as the ratio of the number of moles of applied ligand to the number of gram atoms of metal[5] (see Chapter 11).

9.7   Colour

The colour of chrome-tanned pelt, called 'wet blue', depends on how the tanning reaction is conducted: it is dependent on the nature of the chromium(III) complexes created during the process and the complexation of the tanning agents *in situ*. The colour of transition metal complexes depends on the ligand field around the central metal ion: in the case of chromium(III), there are six ligands in the octahedral complex. In a tanning context, the ligands are likely to be water, oxy bridges and carboxy compounds, including collagen (see Chapter 11).

The ligands in the octahedral ligand field of chromium(III) split the five degenerate 3d orbitals into non-degenerate orbitals and the energy difference between them determines the colour (see Figure 9.5).

If the chrome tannage is conducted in a sulfuric acid only pickle liquor, the leather is coloured green–blue, because the ligand field is modified by substituting an aquo ligand for a carboxylate group attached to protein. The additional inclusion of formate in the ligand field changes the colour to bright pale blue. The paler colouring is much better for dyeing, which is influenced by the base colour, especially pastel shades. In many cases, this is actually the sole valid reason for including formate in the system.

The addition of masking salt from the pickling process to the complex strongly influences the colour of the chrome complexes and hence modifies the colour of the leather, as indicated in Table 9.4.

Some processes employ a sulfuric acid pickle and introduce sodium formate either during the tannage or as part of the basifying system towards the end of the chrome tanning process. Both of these approaches result in a masked chrome process, yielding pale-blue leather. Sodium formate is a mildly alkaline salt and the equilibrium

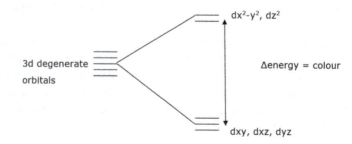

Figure 9.5   Effect of the ligand field on the energy of the d orbitals of chromium(III).

Table 9.4 Effect of complexation on the colour of chromium(III) ions.

Masking salt/ligand	Cr(III) complex colour	Leather colour – paler than the Cr complex itself
Water only	Violet	Pale blue–green
Hydroxyl	Dark green	Pale blue–green
Phosphate	Apple green	Apple green
Formate	Pale blue	Pale blue
Acetate	Royal blue	Royal blue
EDTA	Purple	Purple

pH of a solution of the salt can be calculated as follows, where the mass terms are given as concentrations, although strictly they should be activities:

$$HCO_2^- + H_2O \rightleftharpoons HCO_2H + OH^-$$

The hydrolysis constant is

$$K_h = \frac{[HCO_2H][OH^-]}{[HCO_2^-]}$$

The dissociation constant is

$$K_a = \frac{[HCO_2H][H^+]}{[HCO_2^-]}$$

$$K_h = \frac{[H^+][OH^-]}{K_a}$$

The ionic product of water is

$$K_w = [H^+][OH^-]$$

Then,

$$K_h = \frac{K_w}{K_a} = \frac{10^{-14}}{10^{-4}} = \frac{[OH^-]^2}{[HCO_2^-]}$$

If it is assumed that $[HCO_2^-] \simeq [HCO_2^-]_{total}$, since the salt is a relatively strong electrolyte, and that the total concentration of formate is 0.1 M, $[OH^-]^2 = 10^{-11}$ and therefore pOH = 5.5, *i.e.* pH = 8.5.

The colour of the chrome-tanned leather can be modified by the use of reagents capable of complexing in post-tanning steps, such as 'neutralisation', when the pH of the leather is raised by basic compounds. Although the colour of the chrome-tanned leather can be altered by the choice of masking agents, the effect on dried or 'crust' leather colour is slight and insufficiently deep in shade to constitute a colouring process. The influence of tanning on colouring is limited to modifying the colour struck in the conventional dyeing process.

9.8   Non-swelling Acids

So-called 'non-swelling acids' can be used in place of conventional pickling acids: they can be regarded as auxiliary syntans, synthetic tanning agents, in the acid form (see Chapter 15). These reagents may be characterised as possessing the following features:

- aromatic, typically phenyl or naphthyl, which may be substituted, *e.g.* with alkyl or other groups;
- low content of phenolic hydroxyl groups, including none at all;
- high content of sulfonic acid groups.

They have low tanning ability, due primarily to their lack of phenolic hydroxyl content, but they function as strong acids because of their sulfonic acid group content. Table 9.5 gives data for some examples of non-swelling acids: the range of possible structures is wide, as cam be seen in Heidemann's book.[6] It should be noted that the pH will influence the shrinkage temperature of the collagen; in the case of acidification to the values given in Table 9.5, the shrinkage temperature might be as low as 50 °C.

The properties of the non-swelling acids can be summarised as follows.

1. At similar pH values, the swelling effect is inversely related to the increase in shrinkage temperature.
2. The tanning effect, as measured by the shrinkage temperature elevation, is directly related to the complexity of chemical structure.

Table 9.5   Effects of some non-swelling acids.

Acid	pH	T_s/°C	Swelling/%	Tanning effect[a]
Water	6.3	66	100	0
Sulfuric acid	1.2	43	235	—
Benzenesulfonic acid	1.8	49	198	—
4-Hydroxybenzenesulfonic acid	1.8	58	193	—
Naphthalene-2,6-disulfonic acid	1.8	60	114	0
Naphthalene-2-sulfonic acid	2.3	61	101	0
3,4-Dicarboxybenzenesulfonic acid	1.7	62	112	0
Naphthalene-2-hydroxy-3,6-disulfonic acid	1.7	63	87	+
4-Hydroxybenzylsulfonic acid	2.1	65	62	++
Naphthalene-2-hydroxy-3-sulfonic acid	2.1	68	63	++

[a]Scored semiquantitatively positively or negatively.

3. Complexity of structure is a function of the presence of phenolic hydroxyl groups as the first priority, and the presence of sulfonic groups as the second priority.

Non-swelling acids perform three functions.

1. The sulfonic acid group is strong and is therefore effectively completely ionised in solution:

$$R - SO_3H \rightleftharpoons RSO_3^- + H^+$$

The situation can be understood by analogy with sulfuric acid:

$$(HO)_2SO_2 \rightleftharpoons [O_2SO_2]^{2-} + 2H^+$$

Therefore, the sulfonic acid equilibrium could be rewritten in the form

$$RS(OH)O_2 \rightleftharpoons RSO_3^- + H^-$$

This view of the equilibrium highlights the properties of sulfonate/sulfonic acid chemistry, which is a common feature of reagents in leather-making chemistry. In aspects other than non-swelling acids, the group is in the neutralised form, as the sulfonate group, and typically undergoes fixation under acid conditions (see Chapters 15 and 16). Just as sulfate ion is not converted to sulfuric acid by mild acidification (the reaction must be conducted in extremely acidic solution, when the solvent is effectively no longer water), sulfonate groups are not protonated. The reaction involves electrostatic interaction with liberated charged amino groups on the collagen.

2. The compounds interact with collagen: the degree to which they interact and how they interact create a tanning effect. Therefore, the term 'non-swelling acid' is relative, since not all members of the group are capable of overcoming the swelling effect of low pH. The degree of effectiveness depends on the extent to which the agent acts as a syntan, since conventional tanning confers resistance to osmotic swelling. Note that the only difference between a non-swelling acid and a syntan lies in the stage of production. This type of syntan is conventionally made by sulfonating an aromatic compound; the resulting compound is a sulfonic acid – a non-swelling acid. After neutralisation to the sulfonate salt, the auxiliary syntan is produced.

   It should be noted that the use of a non-swelling acid implies a degree of stabilisation of the pickled pelt, which can also be viewed as a pretannage. Therefore, it should be expected that the more effective a non-swelling acid is, the more the character of the leather will be altered by the pretanning process. However, the more effective the tanning is, the safer it is to conduct a salt-free pickling process.[7]

3. Like many other organic compounds, aromatic polysulfonic acids can interact with collagen in a lyotropic manner:[8]

$$P-NH_3^+ \cdots {}^-O_2C-P + PSA^- \rightleftharpoons P-NH_3^+ \cdots {}^- PSA + {}^-O_2C-P$$

As discussed in Chapter 6, with regard to the Hofmeister series, the effect is to liberate carboxyl groups, to assist the chrome tanning reaction by providing more reaction sites. However, this is probably at the expense of modifying the character of the leather (see Chapter 11).

9.9 Pickle Formulations

From the information set out above, it should be clear that there is an optimum formulation for pickling; it also implies that deviation from the optimum conditions will create problems. This is summarised in Figure 9.6, which applies both to conventional tannery pickling and to curing by storage pickling. The diagram, modified

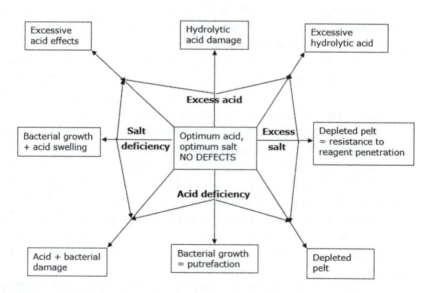

Figure 9.6 Effects of variations in pickling formulation compositions. Modified from ref. 9, courtesy of Eurofins BLC.

from Jordan Lloyd,[9] indicates that insufficient or excess components of the pickle solution can be equally counterproductive and result in problems in processing or in leather performance.

9.10 Implications of Pickling for Chrome Tanning

Because tanners routinely deal with a substrate with finite thickness, this implies that conditions may vary through the cross-section. The example of the effect of different pH profiles can be represented as shown in Figure 9.7, in which three variations of liming practice can yield (notionally) five patterns of pickling profile. Since chrome fixation is dependent on the pH conditions met by the chrome as it penetrates, the pattern of chrome fixation through the cross-section can also vary in those notional ways.

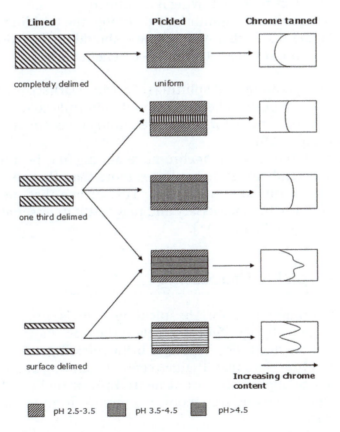

Figure 9.7   Implications of variations in deliming and pickling for pH profiles through a hide cross-section, modelling the consequences for chromium tanning.

The practice of partial deliming used to be common, but is less so nowadays, probably because of issues of inconsistency. Processing closer to equilibrium, at least to apparent pH uniformity over the cross-section, will yield more consistent results in subsequent steps, although the outcome may not be ideal, as discussed below.

Pickling to equilibrium will cause lower fixation in the centre than at the surfaces, because the continuing generation of acid during the tanning reaction will reduce fixation as penetration proceeds and there is always a higher concentration of chrome outside the pelt than inside. Pickling close to equilibrium will result in more uniform fixation, with the pattern dependent on the degree of pickling. As the pickling becomes more non-uniform, with a distinctly less acidified central region, the chrome fixation in the centre can be higher than at the surfaces. In a more extreme case, high chrome fixation at the surfaces creates a barrier to chrome penetration, resulting in much lower fixation in the centre, even leaving it untanned.

The outcome of non-equilibrium deliming and pickling, as illustrated in Figure 9.7, is that more chrome should be deposited in the centre than at the surfaces. This has two benefits.

1. Following tanning and splitting (of bovine leather), the split surface is more likely to be strongly and uniformly coloured, reducing the requirement for chrome retanning for cosmetic purposes (see Chapter 15).
2. Whatever the reasons for chrome retanning may be, the process typically tends to tan the surfaces more than the cross-section, because of the conditions used. This means that the grain is tanned more than necessary and may result in lesser affinity for acid dyes (see Chapter 11).

9.11   No-pickle Processing

Since the primary role for the pickling step is preparation for (chrome) tanning by acidifying the pelt, if the technology is changed there may be no need for pickling in brine. Pickling does not apply to tannages other than metal salts: vegetable tanning and the use of syntans and aldehydes all start at neutral pH or higher. Therefore, eliminating the pickling step requires a modification to the chrome tanning process.

In Section 11.7, the concept of a 'chaser' pickle is discussed. Here, a small amount of acid or even the acidic chrome tanning salt alone is allowed to penetrate pelt that has not undergone any pickling at all, for example, bated pelt at pH ≈ 9, that is, the chromium salt 'chases' the acid into the alkaline pelt. The theory is that the alkali within the pelt is balanced by the acidity of the tannage, such that the equilibrium conditions match those at the end of conventional pickling and tanning, thereby eliminating both the pickling and the basification steps. Clearly, this is a high-risk strategy if the mass balance does not work.

In addition, the different rates within the process have to be considered.

1. Fast penetration of chromium species means that the localised conditions around the chrome are dominated by the acidity of the chrome, *i.e.* the penetration rate is likely to be faster than the fixation rate, giving rise to uniform fixation throughout the cross-section.
2. Slow penetration of the chromium species means that the environment within the pelt is dominated by the alkalinity, so the chrome is likely to encounter conditions that are too astringent, *i.e.* fixation is faster than penetration and might include localised precipitation.

Depending on the specific processing conditions, control over the reactivity of the chrome species might be necessary, including reducing it by an appropriate degree of premasking (see also Chapter 11).

References

1. M. Komanowsky, *J. Am. Leather Chem. Assoc.*, 1989, **84**(12), 369.
2. (a) F. G. Donnan and H. E. Potts, *Kolloid-Z.*, 1910, 7, 208; (b) F. G. Donnan and A. B. Harris, *J. Chem. Soc.*, 1911, **99**, 1554.
3. S. G. Shuttleworth and A. E. Russell, *J. Soc. Leather Trades' Chem.*, 1965, **49**(6), 221.
4. A. D. Covington, *J. Am. Leather Chem. Assoc.*, 2001, **96**(12), 467.
5. A. D. Covington, *J. Am. Leather Chem. Assoc.*, 2008, **103**(1), 7.
6. E. Heidemann, *Fundamentals of Leather Manufacturing*, Eduard Roether, Darmstadt, 1993.
7. P. M. Pojer and C. P. Huynh, *Leather*, 1999, 12.
8. J. Christner, *et al.*, *J. Am. Leather Chem. Assoc.*, 2012, **107**(12), 409.
9. D. J. Lloyd, *Progress in Leather Science: 1920–45*, British Leather Manufacturers' Research Association, 1948, p. 151.

10 Tanning

10.1 Introduction

Before reviewing the options for tanning, it is first necessary to consider what constitutes a tanning reaction. More specifically, what is the definition of tanning? Most practitioners will offer some or all of the following features of tanning as the expected outcomes.

1. Appearance: dried raw pelt is translucent and horny, but tanned leather dries opaque and may change colour, *e.g.* in chrome tanning.
2. Handle: some degree of softness in comparison with dried raw pelt.
3. Smell: some tanning agents will introduce smell, *e.g.* cod oil for chamois leather, extracts of plant materials for vegetable tanned leather or aldehydic compounds.
4. Increase in denaturation temperature.
5. Resistance to putrefaction by microorganisms.
6. A degree of permanence to the changes.

In addition, those without a knowledge of leather science might assert that tanning increases the strength of the pelt. It is worth noting that every process step (with very few exceptions) weakens the pelt: it would be odd if Nature had arranged that skin could be improved in strength by some physical and chemical modification. Any process that is suggested to strengthen actually means less weakening.

Tanning Chemistry: The Science of Leather 2nd edition
By Anthony D. Covington and William R. Wise
© Anthony D. Covington and William R. Wise 2020
Published by the Royal Society of Chemistry, www.rsc.org

Table 10.1 Some observations on modified collagen.

Material	Appearance	$\Delta T_s/°C^a$	Effect of water
Solvent-dried pelt	White, opaque, soft	0	Reverts to raw
Aluminium tawed[b]	White, opaque, soft	0	Al(III) washed out, reverts to raw
Oil tanned	Beige, opaque, soft	0	No change, absorbs and holds water

[a]Quoted here as zero, but the reactions may stabilise collagen by a few degrees only, depending on the conditions of processing.
[b]The term 'tawing' is traditional, used in contrast to the term tanning, because it is relatively reversible.

The only strict definition of tanning is the conversion of a putrescible organic material into a stable material that resists putrefaction by spoilage bacteria. Tanners do expect other changes and the first three changes in the list, appearance, handle and smell, will typically accompany a tanning process. Most tanners, however, additionally expect an increase in hydrothermal stability, typically observed as an increase in shrinkage/denaturation temperature, ΔT_s. Furthermore, there is usually an expectation that the change is not transitory, but has some degree of permanence in conventional use. Consider the examples of modified collagen presented in Table 10.1.

In each case given in Table 10.1, the material looks and feels like leather but, in the absence of further information or access to some testing facilities, it would not be possible to demonstrate that it is actually otherwise. However, it can be seen from the table that the materials do not conform to the expected requirements of the definition of tanning, although these materials will not putrefy provided that they are kept dry. This begs the question: are the materials referred to in Table 10.1 really leather, even though they may have serviceable properties?

The solvent-dried pelt is not truly stabilised, except when dry, and remains raw: if it is wetted, it reverts to its previous raw, vulnerable state. However, the aluminium-treated material may also not be considered to be leather, hence the reaction is called tawing (see Chapter 12), because the reaction is reversible and there is no increase in hydrothermal stability. In contrast, oil tanning creates chamois leather (see Chapter 14), which is well known as a surface cleaning material, particularly for glass and automobile bodies; it meets all the criteria for tanning that might be expected, except for the lack of increase in hydrothermal stability.

Therefore, it is possible to distinguish two types of stabilising processes: tanning and leathering. Tanning achieves all the

expected changes to the properties of collagen. Leathering confers sufficient stability for the material to be useful, but it may
not exhibit all the changes expected from a tanning reaction, so
there will be some limitations to the range of uses. When asked
about the definition of tanning, most tanners immediately think
of the shrinkage temperature. That is understandable, because so
much of leather technology revolves around achieving a required
degree of hydrothermal stability and the measuring of it. However,
strictly, the fundamental requirement of tanning does not invoke
the notion of hydrothermal stability. Raw skin is vulnerable to
attack from the proteolytic enzymes produced by bacteria, which
can target sites in the protein to break bonds. Tanning agents interfere with the interaction between the enzymes and the protein and
thereby prevent the degradative reactions. It does not matter what
the tanning reactions are, which specific chemical changes are
made to the protein or whether the stabilising reaction is weak or
strong, they will still provide the required biochemical resistance.
This begs the question: if the stabilising reaction involves only
weak chemical interaction with the protein, does that still constitute a tanning process?

Returning to the question of what is expected from a tanning
process, along with a change in hydrothermal stability, it is generally assumed that the change has a degree of permanence. For
example, the reaction should not be reversed by simple wetting. It
was recognised hundreds of years ago that a stable material could
be made from a treatment with alum (aluminium sulfate or potassium aluminium sulfate), salt, flour and egg yolk, and the product
was a preferred leather to be used for gloving. However, the process could be reversed by washing, so it was referred to as 'tawing',
to distinguish it from tanning, and the producers were known as
whitawyers to distinguish them from the tanners. Modern tanners know that aluminium salts are poor tanning agents, even
under the best of conditions, and that the tawing process does
not alter the shrinkage temperature. Hence we can recognise two
levels of stabilisation, tanning and leathering. The former confers
at least some permanence to the enhanced hydrothermal stability
by strong reaction with the protein. The latter may have neither
permanence nor a strong chemical interaction between agent and
collagen, but nevertheless the product exhibits resistance to biochemical attack.

Alum tawing is not the only leathering process known in the art.
In addition, the following processes can all be called leathering,

characterised perhaps by some degree of stability/permanence, but with little or no change in shrinkage temperature. Some of these technologies are addressed in more detail in Section 10.3 and later chapters.

- Oil tanning for chamois leather or buckskin.
- Oil dressing, *e.g.* for Japanese traditional 'leather'.
- Low-astringency mineral tannage [except chromium(III)].
- Low-affinity polymers, *e.g.* acrylates.
- Brains 'tanning': a form of oil dressing. Post-dressing treatment with smoke is a true tanning process through reaction with acrolein, $CH_2=CH-CHO$.
- (Poly)phosphate 'tanning': this is merely a weak electrostatic interaction.
- Silica 'tanning': this is merely a weak electrostatic interaction with silicic acid.
- Sulfur 'tanning': this is the precipitation of elemental sulfur within the protein fibre structure.
- Solvent dehydrating.

Oil tanning is well known in the art: unsaturated oil, such as cod liver oil, can be made to polymerise by activating the reaction with heat, causing it to create a polymer within the matrix of the collagen structure. The lack of chemical interaction between the interpenetrating matrices means that the shrinkage temperature is not elevated, but the material structure is altered enough to confer resistance to biochemical attack. The unusual structure of oil-tanned leather gives rise to two remarkable properties, described in Chapter 14.

Oil tanning should not be confused with oil dressing, which is the principle of Japanese leather. The difference can be seen in the chemical makeup of the oils used in the two types of process. Oil tanning can be modelled by the structure of linoleic acic. In contrast, the composition of rapeseed oil, favoured in making Japanese leather, is as follows, although it does contain some linoleic acid:

- 45% erucic acid $CH_3(CH_2)_7CH=CH(CH_2)_{11}CO_2H$
- 16% oleic acid $CH_3(CH_2)_7CH=CH(CH_2)_7CO_2H$
- 13% linoleic acid $CH_3(CH_2)_4CH=CHCH_2CH=CH(CH_2)_7CO_2H$
- 8% linolenic acid $CH_3(CH_2CH=CH)_3(CH_2)_7CO_2H$
- 10% eicosenoic acid $CH_3(CH_2)_7CH=CH(CH_2)_9CO_2H$.

The high degree of unsaturation in linoleic acid allows the polymerisation to occur, but rapeseed oil, which contains about half of the content of double bonds, cannot undergo polymerisation and therefore functions only as a lubricant. The coating of the fibre structure with oil provides sufficient protection against biochemical attack to make it a useful material, provided that it does not get wet. When colleagues at the University of Northampton recently examined a sample of modern Japanese leather, it was confirmed that there was no increase in shrinkage temperature and no metal content, beyond trace amounts. It is noteworthy that the most abundant metal, at the parts per million level, was erbium – picked up during the river washing step in the manufacture! That does not count as a tanning process.

Brains tanning is an interesting process, used as a traditional technology by the plains dwellers of Asia and North America, where sources of plant extracts were scarce. It works because the chemical makeup of all brains is phospholipid: by partially cooking the brain and converting it into a paste, it can be worked into the fibre structure of the pelt, providing a lubricating effect. The stability of the product arises from the lubricant preventing access to the protein by enzymes just like a conventional oil dressing. However, if the material becomes wet, it is no more stable than raw pelt. The Sioux Indians have a saying: 'every animal has enough brains to tan its own hide'.

Often the brains treatment is followed by smoking, a process well known in food preservation, which allows the leather to be wetted without reverting to a raw state. This is a true tanning reaction because of the fixation of acrolein, an aliphatic aldehyde; the fact that it is incapable of crosslinking is immaterial, since the tanning effect depends on interfering with the shrinking transition and not on some notional creation of structure, a consequence of the link–lock mechanism (see Chapter 23).

Polyphosphate is a generic term, covering a range of degrees of polymerisation, all of which have been tried. In every case the outcome is the same: a weak electrostatic interaction with the cationic centres on the collagen, which does not constitute a tanning reaction. The main usefulness is as a pretreatment for vegetable tanning.

Silica tanning is a similar reaction, involving polymeric silicic acid. In contrast, sulfur tanning should be thought of as a type of lubrication, since the reducing conditions produce elemental sulfur. This was a side effect of using compounds such as thiosulfate for reducing chromium(VI) in the two-bath chrome tanning process.

The ultimate example of leathering is solvent drying, achieved by exchanging the water within the pelt by, for example, alcohol then applying several changes of acetone. In this way, the wetting medium is changed from water to organic solvent, so that when the solvent is evaporated the fibre structure has a lesser tendency to stick and the result is a white, soft, leather-like product. However, it cannot be considered to be a leather, because it has undergone no chemical change; indeed, it might be argued that it has not even been leathered, since the merest contact with water will cause it to revert to the raw state.

This review serves to highlight the relationship between the properties of a material, in terms of the physical and chemical characteristics, and the requirements of performance in use. Before the advent of chrome tanning and modern analytical methods, that relationship was pragmatic. Nowadays, leather is probably over-engineered; that is not necessarily a bad thing, but it does not hurt to recognise it as such.

10.2 Hydrothermal Stability

Hydrothermal stability is the resistance of a material to wet heat. In the case of collagenic materials, pelt or leather, it is the effect of heat on water-saturated material. Note that the thermal properties of collagenic materials are dependent on water content,[1,2] hence when they are tested in this context they are required to be wetted to equilibrium and therefore they are presented in a reproducible condition for testing.[3] The shrinkage temperature has been assumed to represent the breaking of hydrogen bonds in the triple helices, which would apply equally to raw and chemically modified (tanned) collagen, allowing the protein to undergo the transition from helix to random coil.[4,5] This point is developed in Chapter 23.

Weir[5] demonstrated that, in the early stages of the transition between the intact helix and random coil structures of collagen, shrinking is a rate process and the rate is influenced by the moisture content: shrinking is independent of the soaking period once the collagen is hydrated to equilibrium, but drier specimens shrink more slowly. The rate therefore affects the value assigned to the shrinkage temperature: the slower the rate of shrinking, the apparently higher is the quoted shrinkage temperature, as discussed in Chapter 1. Figure 10.1 illustrates the dependence of the apparent hydrothermal stability of different leathers on moisture content:

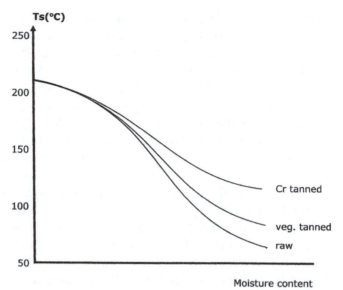

Figure 10.1 Dependence of apparent shrinkage temperature on moisture content.

Table 10.2 Thermodynamics of collagen shrinking at 60 °C where ‡ signifies a transition state.

pH	$\Delta H^{\ddagger}$/kJ mol^{-1}	$T\Delta S^{\ddagger}$/kJ °C^{-1} mol^{-1}	$\Delta G^{\ddagger}$/kJ mol^{-1}
1.8	280	210	73
3.0	330	243	86
3.6	590	506	83
4.2	610	512	94
6.8	700	600	103
7.2	635	532	104
10.8	550	455	98
12.5	320	246	77

at high moisture content the leathers exhibit the conventional shrinkage temperature but, at close to zero moisture content, all leathers tend to the same apparent shrinkage temperature, when the leather decomposes.

The kinetics of shrinking are affected by pH: it was demonstrated in Chapter 1 that the value decreases at pH extremes and this is reflected in the thermodynamics of the reaction, as shown in Table 10.2.

From Table 10.2, it is apparent that the thermodynamic parameters change depending on the relationship of the sample's condition relative to the isoelectric point: the further away from the point of highest stability, the lower are the activation parameter values, when the free energy is controlled by both the enthalpy

Table 10.3 Thermodynamics of shrinking for stabilised collagen at 60 °C where ‡ signifies a transition state.

Tannage	Offer/% Cr_2O_3	$\Delta H^{\ddagger}$/kJ mol^{-1}	$\Delta S^{\ddagger}$/J °C^{-1} mol^{-1}	$\Delta G^{\ddagger}$/kJ mol^{-1}
None		670	1705	104
Formaldehyde		470	1055	121
Zirconium(IV)		570	1285	147
Aluminium(III)		345	720	104
Chromium(III)	0.06	670	1690	111
	0.13	850	2180	119
	0.57	1350	3600	151
	1.02	1630	4350	184
	1.98	1360	3515	189

and entropy terms, the former indicating a breakdown of structure and the latter indicating the change in the state of disorder. From investigations of collagen in the form of tendon, it is more apparent than in skin that the shrinking reaction is not uniform: it initiates at certain points in the specimen, then propagates until the whole specimen is affected. This is consistent with the concept of the effect of zones in the protein that have fewer secondary amino acids, as discussed in Chapter 1.

Weir showed that when collagen is stabilised chemically, the thermodynamic parameters change, as illustrated in Table 10.3.[5]

Treatments that decrease the entropy term or increase the enthalpy term increase the free energy of activation and consequently increase the shrinkage temperature. Chromium(III) tans principally by the latter method, *i.e.* by increasing the structural component: this applies to offers up to 1% Cr_2O_3 (corresponding to half of the usual industrial offer), but at higher offers the free energy is still controlled by the enthalpy rather than the entropy. All other tannages presented in Table 10.3 are controlled by the entropy term: this may be pictured as resulting from an increased shrinkage temperature by steric hindrance to the shrinking process, when the uncoiling structure collapses into the interstices between the triple helixes. The amount of steric hindrance may be assumed to be proportional to the chemical impact of the tannin on collagen structure and its consequent ability to interfere with the collapsing process. The traditional way of thinking about tanning is based on the idea that the tanning agent confers stability to the collagen by changing the structure through crosslinking and thereby preventing the helices from unravelling: here is the first indication that a new concept of tanning is required, in which the function of the tanning agent is to prevent shrinking occurring by altering the thermodynamics of the process. This view of tanning is developed further in Chapter 23.

Hydrothermal stability can be measured in several ways, as set out in Sections 10.2.1–10.2.4.

10.2.1 Shrinkage Temperature (T_s)

The shrinkage temperature of pelt or leather is the most commonly quoted measurement of hydrothermal stability, certainly in technical publications and specifications. The principle of the method is to suspend the test piece in water, in the form of a strip (the dimensions are immaterial since the property does not depend on sample size), then to heat the water at a rate of 2 °C min^{-1}, according to the Official Method.[3] The shrinkage temperature is noted when the sample visibly shrinks. The extent of any reaction is a product of the rate and the duration of the reaction period. In this case, shrinking is observed when the rate becomes sufficiently fast to be seen. Since the observed effect at any point in the test is the product of temperature and time, it must also be dependent on the rate of heating: the faster the rate of heating, the higher the temperature of shrinking appears to be and, conversely, the slower the rate of heating, the lower the temperature of shrinking appears to be. Therefore, the test is somewhat arbitrary, but it is reproducible if the rate of heating is consistent. However, cases may be encountered when the nature of the material significantly alters the kinetics, making comparisons less reliable.

A feature of the Official Method[3] is that it specifies the use of water as the heating medium. This is not surprising, because the sample must be wetted with water before testing. However, the limiting factor is the boiling point of the solvent. Leathers with shrinkage temperature above 100 °C, *e.g.* chrome-tanned leathers, cannot be accurately tested. Two simple methods have commonly been used to resolve the problem. High-boiling paraffin can be used as the heating medium, but the results cannot be relied upon above 105 °C, because the water in the pelt boils off during the test, effectively increasing the shrinkage temperature of the piece under test. It is doubtful that the water within the pelt becomes so superheated that measurement of the temperature of wet material remains accurate. Nevertheless, it is common for values of >110 °C obtained in this way to be quoted in the technical literature. An alternative is to use mixtures of water and glycerol as the heating medium: the American Leather Chemists Association suggested a 75% glycerol–water mixture.[6] However, the same criticism applies, because although it might be argued that the pelt does become saturated with the higher boiling aqueous solvent, the shrinkage temperature is dependent on the amount of water in the solvent, increasing with decreasing water content, as might be expected.

10.2.2   Boil Test

Tanneries rarely possess shrinkage temperature testing equipment: conventional testing would be of arguable value in a chrome tanning context because of the high shrinkage temperatures above 100 °C, and typically they would not have a differential scanning calorimetry (DSC) instrument (see later), hence an alternative testing approach is needed. In most cases, tanners test chrome-tanned leather at the end of the process, prior to unloading the tanning vessel. The procedure is to cut a piece of leather and note the area, by drawing around it roughly on paper. The test piece is then dropped into boiling water and left for a specified period: this might be 1–5 min, sometimes longer, but commonly is 2 min. After the required time, the piece is retrieved, quickly cooled in cold water and the area is compared with that of the starting shape. The criterion is pass or fail: if there is no discernible change, the leather has passed, but if any shrinkage can be seen, the leather has failed. Often, in the case of failure, other changes can be seen in the texture of the leather, such as a rubbery feel and wrinkling of the grain. These observations can yield information regarding the uniformity of the tannage, particularly through the cross-section.

The rationale for the boil test can be expressed as follows. At any temperature, the rate of shrinking or the development of apparent thermal damage depends on the time, t, for which the sample is held at that temperature: at any temperature below the conventionally measured shrinkage temperature, where the difference is ΔT, the rate of shrinking is slower than the discernibly fast rate at the shrinkage temperature, so the accumulation of damage depends on the product of t and the reciprocal of ΔT. In the boil test, the temperature is fixed at 100 °C, when the rate is fixed at k_{100} for a particular leather. Therefore, shrinking depends on the product of rate multiplied by time, $k_{100} \times t$, so the longer the boil test period lasts, the more likely the pelt is to exhibit discernible shrinkage.

If the shrinkage temperature T_s is higher than 100 °C, then the difference between the point at which shrinking is observably fast and the test conditions is $T_s - 100 = \Delta T_{>100}$. Therefore,

$$k_{100} \propto 1/\Delta T_{>100}$$

The maximum period that the piece can resist the boil test without discernible shrinking, t_{max}, is inversely proportional to k_{100}, so

$$t_{max} \propto \Delta T$$

that is, the longer a leather can resist boiling water, the higher the shrinkage temperature is above 100 °C. However, the correlation between resistance time and T_s is typically not used commercially; the test time is normally fixed to give a pass/fail result that corresponds to the hydrothermal requirements/specifications of the product. Because the kinetics of the shrinking reaction may vary, depending on how the tanning reaction is conducted, comparisons are less reliable than shrinkage temperature measurements. However, the results should be consistent for any given process.

10.2.3   Differential Scanning Calorimetry

This is an instrumental technique in which the sample is placed in a metal capsule, which may be open or closed, and heated electrically. The amount of energy absorbed by the sample and capsule is compared with that for a control empty capsule and displayed as change in energy as a function of temperature. As in the shrinkage temperature test, the heating rate must be controlled and defined: here, a rate of 5 °C min^{-1} is common. An example of a thermograph is presented in Figure 10.2. The advantage of DSC is that wet samples can be sealed in the capsule, so the shrinking transition can be obtained under true hydrothermal conditions.

Thermal transitions can be characterised by an onset temperature (the peak maximum is less commonly quoted): this is generally assumed to be close to the conventionally measured shrinkage temperature. A feature of this technique is that it yields the enthalpy associated with the thermal transition: this is conventionally normalised to energy per unit dry weight of the material, but in this context it ideally should be adjusted to energy per unit weight of dry collagen. In the case of collagenic biomaterials, including leather, additional information can be obtained; the values of the parameters obtained can be summarised as follows.

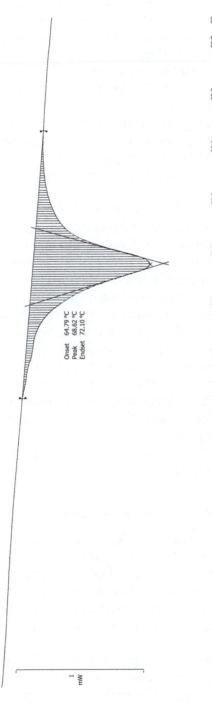

Onset 64.79 °C
Peak 68.62 °C
Endset 72.10 °C

Figure 10.2 DSC thermograph of wet bovine collagen.

- The shrinkage temperature is a reflection of the chemical status of the collagen. Chemical modifications alter the shrinkage temperature: stabilisation by tanning reactions increases T_s, detanning of tanned collagen decreases T_s.
- The enthalpy of shrinking/denaturation reflects the intactness of the collagen structure. In general, the energy of shrinking is relatively constant regardless of the tanning chemistry, but is lower than the energy associated with native collagen. Any process that results in damage to the collagen will lower the enthalpy of shrinking: the effect of damage is to decrease the amount of energy required to convert the intact helical structure to random coil.
- A change in either T_s or ΔH can be independent of the other parameter. In any case, an increase in shrinkage temperature is caused by chemical stabilisation of the structure (tanning) or measurements taken on dried material, but a decrease in shrinkage temperature is due to destabilisation of the newly stabilised collagen structure (detanning) or a damaging effect that might be described as a negative tanning effect, such as is observed when collagen is treated with hydrogen bond breakers, lyotropic agents. The enthalpy of shrinking is likely to increase only when the sample is drier than it should be or an error has been made in the moisture content measurement. The enthalpy can be decreased by damage to the collagen structure, which corresponds to a contribution to shrinking/denaturing of the protein.
- The entropy of shrinking can be calculated, but this is a less frequently used parameter.

10.2.4 Hydrothermal Isometric Tension

This is a much less common technique, borrowed from polymer science.[7] The method is similar to the Official Method of measuring shrinkage temperature,[3] except that here the sample is constrained from shrinking and changes in tension are recorded. Additional information regarding the molecular status of the piece is obtained, but there are significant differences between the temperature values of transition onsets.

The traces can be divided into three parts, as illustrated in Figure 10.3: the first is the tension-increasing process from zero to maximum tension, followed by a relatively constant-tension process, if present; finally, the tension is either constant or a relaxation process occurs,

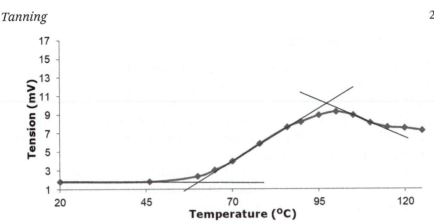

Figure 10.3 Hydrothermal isometric tension trace for bovine hide.

due to the gradual destruction of collagen structure or rupture of some crosslinking bonds. The slope of the curve in the tension-increasing process accounts for the rigidity of the collagen fibre, caused by crosslinks: the steeper the slope of the contraction curve, the more crosslinks should be present in the collagen materials. Relaxation represents the stability of these connecting elements (crosslinking bonds): the steeper the rate of the relaxation curve, the more unstable is the crosslinking.

To date, no mathematical or physical model has been proposed for a full analysis of hydrothermal isometric tension (HIT) curves obtained under linear heating conditions.[8] In 1987, Kopp and Bonnet proposed a tentative model for the development of isometric tension in collagen,[9] but their equations are probably suited only to very limited conditions (medium, pH, ionic strength, *etc.*). However, the value of HIT in the context of collagenic materials is that the shapes of the HIT curves can yield useful information about the relative crosslink density and stability in comparable specimens.

10.3   Historical Tannages

There are some chemistries in leather technology that are well known, referred to in the jargon as tannages, but that are no longer in regular current use. Some of these are addressed in Chapter 14 and are termed leathering processes. It is useful to recall these chemistries[10-16] and to review their effects with modern leather scientific eyes, to judge whether or not they might have some usefulness in modern leather technology.

10.3.1   Sulfur Tanning

Sulfur has featured in leather technology from the time of the two-bath chrome tanning process, where thiosulfate was the reducing agent.[11] Elemental sulfur can also be generated when thiosulfate is acidified. Although sulfur is known to have many allotropes, the stable form when precipitated at ambient temperature is a cyclic structure of eight atoms, known as 'crown' sulfur.

By carefully allowing pelt to absorb colloidal sulfur to equilibrium, a white, soft, leather-like material can be obtained. The interaction is almost certainly one of preventing fibre sticking, the usual function of fatliquoring/lubrication; here, the effect is to maximise the ability of the fibre structure to resist tearing, but at the same time not to cause weakening by disrupting the natural bonding when an emulsion of oil is deposited (see Chapter 17). The effect of coating the fibre structure is supported by the observation that the softening effect is reversed when the dry material is wetted, then dried. Also, the sulfur can be extracted with carbon disulfide, indicating that there is no chemical bonding between the sulfur and the protein. The chemistry offers an alternative approach to the lubrication of leather that may confer additionally useful properties.

10.3.2   Silica Tanning

Silicate has been mentioned already in the context of beamhouse processing, when the main effects are due to its alkaline state. Colloidal silica is referred to in the context of the wet white technology (see Chapter 14). Here, the effect of silicic acid as a tanning agent can be considered: it would seem to have some function in that regard, because it precipitates gelatine.[11,14]

It is necessary to add the silicate to acid, not the other way around, to avoid creating a gel by polymerisation: this allows the reaction to proceed at about the same rate as vegetable tanning. The product is pure white. In the earlier days of chromium tanning, this was suggested as an alternative. However, upon ageing, the leather became extremely weak, which might be a consequence of its acidic nature.

The 'tannage' can be used in combination with mineral agents, including chromium(III): the implication is that the silicic acid does not react with the sidechains on the collagen. Retanning with vegetable tannins is not hindered, but silica tannage after vegetable tanning is hindered, which indicates an interaction between the polymeric acid and the polyphenol groups. Since the acid has $pK_a = 10$, the reagent can function in the acidic state and therefore react *via* hydrogen bonds.

10.3.3 Phosphate Tanning

Treating pelt with polyphosphate has been known since the 1930s,[12,14] when the first experiments were conducted on Calgon, a commercial product of sodium hexametaphosphate, a linear salt with the structure $HO[PO_2NaO]_6H$ or $Na_4[Na_2(PO_3)_6]$. Higher molecular weight phosphates, with up to 20 phosphorus moieties, were shown to have a better tanning effect: applying the reagent at pH 2.4 for optimum fixation means that the collagen is in the acidic state, but because the phosphate is a strong electrolyte, it can only interact electrostatically. This is in contrast with the chemistry of silicate processing.

It was recognised at the time of the earliest trials that polyphosphate could function as a pretreatment for vegetable tanning. This was developed later in the twentieth century to become the LIRITAN process, from the Leather Industries Research Institute of South Africa.

10.3.4 Quinone Tanning

Quinone tannage has been known since the beginning of the twentieth century.[10,11,14-16] It has been of interest not only because of the powerful effect that is has on collagen but also because of the crossover to the chemistry of the condensed tannins (see Chapter 13).

As a single reagent, it is claimed that it can confer a shrinkage temperature of 90 °C, indicating that it is a true tanning agent. The process is suggested to be direct reaction at pH 6 by the sidechain amino groups at the quinone ring, followed by reduction of another quinone molecule. An additional crosslinking reaction has been postulated to occur at another amino group, followed by reduction of another quinone molecule, as illustrated in Figure 10.4. However, from the link–lock theory (see Chapter 23), this may not be a necessary feature of the mechanism.

It has been proposed that, because the mechanism depends on the presence of air, the tannage may involve the formation of polymerised species, as indicated in Figure 10.5. It has been reported that quinone-tanned leather exhibits exceptional strength; this observation implies analogies with some of the chemistries discussed in Chapter 24, such as that of nordihydroguaiaretic acid (NDGA).

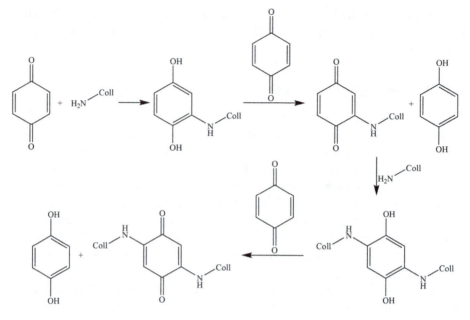

Figure 10.4 Mechanism of hydroquinone tannage.

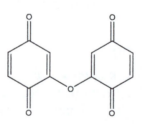

Figure 10.5 Proposed mode of polymerisation of quinone in tanning.

The effect of derivatives demonstrates the importance of structure, as discussed later in Chapters 13 and 24, and illustrated in Figure 10.6. Tanning with benzoquinone confers a shrinkage temperature of only 69 °C, but naphthaquinone can produce 85 °C.

10.3.5 Aldehydes

The reaction between polyphenols and aldehydes has long been exploited in the formation of phenol–formaldehyde resins – the ancestors of modern plastics. However, the reaction was also adopted for tanning when it was found that formaldehyde reacted with resorcinol at a molar ratio of 3:1. It was reported that pyrogallol worked as the polyphenol and the linking agent could also be acetaldehyde, crotonaldehyde, furfural or benzaldehyde.[14]

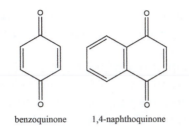

benzoquinone   1,4-naphthoquinone

Figure 10.6   Tanning reactions with quinone derivatives.

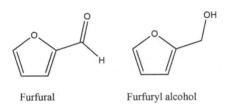

Furfural   Furfuryl alcohol

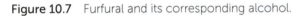

Figure 10.7   Furfural and its corresponding alcohol.

The use of monocyclic polyhydroxybenzene compounds together with aldehydic reagents has been applied as models for condensed vegetable tannins (see Chapters 13 and 24). The principle of combining aromatic phenolic species with the right stereochemistry with a crosslinking agent with appropriate affinity is an underdeveloped research area.

10.3.6   Furfuryl Alcohol

Furfural is an alicyclic aldehyde, already mentioned in Section 10.3.5. Its derivative as an alcohol has been reported to function as a tanning agent (Figure 10.7).[14] Merely contacting pelt with the compound under acidic conditions can initiate the reaction. Alternatively, impregnating pelt with the acid and exposing it to air can cause the reaction to proceed. This might be a principle for achieving aldehydic tannage without using aldehydes.

10.3.7   Sulfite Cellulose (Lignosulfonate)

The by-product of paper manufacture would appear to be a good candidate for contributing to tanners' options for reagents: there are analogous structural features in lignin that echo the structures of vegetable tannins.[11,13] Although the literature suggests that there are practical uses for the materials, such as pretreatment for vegetable tanning and blending with vegetable tannins or syntans, there has not been any real success in incorporating the material into mainstream tanning technology.

The subject is touched upon in Chapter 13, when the composition of lignin is discussed in terms of the major structural components and the possible reconstructing of them into a syntan. That was not successful, but the goal of making use of the by-product makes it worth returning to the chemistry.

References

1. S. S. Kremen and R. M. Lollar, *J. Am. Leather Chem. Assoc.*, 1951, **46**(1), 34.
2. C. A. Miles, *et al.*, *J. Mol. Biol.*, 2005, **346**, 551.
3. SLP18/IUP16, *Official Methods of Analysis*, Society of Leather Technologists and Chemists, 1996.
4. A. D. Covington, R. A. Hancock and I. A. Ioannidis, *J. Soc. Leather Technol. Chem.*, 1989, 73(1), 1.
5. C. E. Weir, *J. Am. Leather Chem. Assoc.*, 1949, **44**(3), 108.
6. J. Beek, *J. Am. Leather Chem. Assoc.*, 1941, **36**(12), 682.
7. B. M. Haines and S. G. Shirley, *J. Soc. Leather Technol. Chem.*, 1988, 72(5), 165.
8. A. J. Bailey and N. D. Light, *Connective Tissue in Meat and Meat Products*, Elsevier, 1989.
9. J. Kopp and M. Bonnet, *Advances in Meat Research: Collagen as Food*, 1987, vol. 4.
10. H. R. Procter, *The Principles of Leather Manufacture*, Spon, London, 1922.
11. J. A. Wilson, *The Chemistry of Leather Manufacture*, American Chemical Society, New York, 1929.
12. J. A. Wilson, *Modern Practice in Leather Manufacture*, Reinhold, New York, 1941.
13. D. J. Lloyd and M. P. Balfe, *Progress in Leather Science: 1920–41*, British Leather Manufacturers' Research Association, London, 1948, p. 487.
14. F. O'Flaherty, W. T. Roddy and R. M. Lollar, *The Chemistry and Technology of Leather*, American Chemical Society, Reinhold, New York, 1958.
15. K. Bienkiewicz, *Physical Chemistry of Leather Making*, Krieger, Florida, 1983.
16. E. Heidemann, *Fundamentals of Leather Manufacturing*, Eduard Roether KG, Darmstadt, 1993.

11 Mineral Tanning: Chromium(III)

11.1 Introduction

The use of chromium(III) salts is currently the commonest method of tanning: perhaps 90% of the world's output of leather in tanned in this way. Up to the end of the nineteenth century, virtually all leather was made by 'vegetable' tanning, using extracts of plant materials (see Chapter 13). The development of chrome tanning can be traced back to Knapp's treatise on tanning of 1858,[1] in which he described the use of chrome alum: this is referred to as the 'single-bath' process, because the steps of infusion and fixing of the chromium(III) species are conducted as consecutive procedures in the same vessel. It is usually accepted that chrome tanning started commercially in 1884, with the new process patented by Schultz:[2] this was the 'two-bath' process, in which chromic acid was the chemical infused through the hides or skins, conducted in one bath, then the pelt was removed to allow equilibration (but no fixation), and subsequently the chrome was simultaneously reduced and fixed in the second bath. The story is told that the invention was a solution to the problem of corset construction: iron bracing strips (an alternative to whale bone) in the garments would interact with the vegetable (plant polyphenol) tanned leather, under the conditions of moist heat around the body. Dampness caused the iron to rust and the liberated iron reacted with the plant polyphenols of the vegetable tannins, to produce dark colours. This reaction is similar to the basis of the chemistry of ink, and hence was an unacceptable change to the garment. The introduction of the new

Tanning Chemistry: The Science of Leather 2nd edition
By Anthony D. Covington and William R. Wise
© Anthony D. Covington and William R. Wise 2020
Published by the Royal Society of Chemistry, www.rsc.org

Table 11.1   Calculated basicity of linear chromium(III) aquo complexes.

No. of Cr atoms	No. of hydroxyl groups	Maximum No. of hydroxyl groups	Basicity/%	Approximate pH of solution of complexes
1	0	0	0	<2
2	2	6	33	2.7
3	4	9	42	3.3
4	6	12	50	3.9
n	$2n - 2$	$6n - 6$	67	>5

chrome tanning process eliminated this reaction. However, chrome tanning may have a longer history than that, as proposed by Thomson,[3] but the clearly observed change in tanning technology occurred around the turn of the nineteenth to the twentieth century. It was soon recognised in the industry that the new reaction was faster and more versatile than the then current vegetable tanning reactions, so after several thousands of years of using vegetable tannins, the global leather industry was converted to chrome tanning over a period of less than 50 years.

The basis of the chrome tanning reaction is the matching of the reactivity of the chromium(III) salt with the reactivity of the collagen. It was shown in Chapter 9 that the availability of ionised carboxyls varies over the range pH 2–6. This is also the reactivity range of collagen, since the metal salt reacts only with ionised carboxyls: the rate of reaction between chromium(III) and unionised carboxyls is so slow that it can be neglected.[4]

Chromium(III) salts are stable in the range pH 2–4, where the basicity changes, but at higher pH values they will precipitate. This might be modelled in the following way, using empirical formulae (see also later):

$$Cr^{3+} \xrightleftharpoons{OH^-} [Cr(OH)]^{2+} \xrightleftharpoons{OH^-} [Cr(OH)_2]^+ \xrightleftharpoons{OH^-} Cr(OH)_3$$

The parameter Schorlemmer 'basicity' (named after the chemist who defined it) is used a great deal to express the status of tanning salts' basicity: it is a way of expressing the degree to which a metal salt has been basified. The percentage basicity is defined by the following expression:

$$\frac{(\text{total number of hydroxyls}) \times 100}{(\text{total number of metal atoms}) \times (\text{maximum number possible per metal ion})}$$

The calculation of basicities is illustrated in Table 11.1, assuming that the polymerised complex ions are linear, as illustrated later in Figure 11.1.

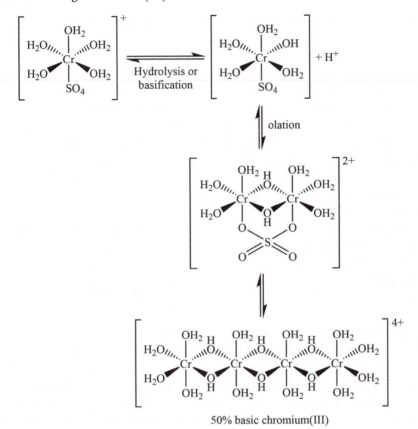

Figure 11.1 Structures of basic chromium(III) complexes.

It is interesting that the calculation indicates that the ultimate basic compound has the empirical formula $Cr(OH)_2^+$ rather than $Cr(OH)_3$, which might be expected from the valency.

11.2 Preparation of Chrome Tanning Salts

Chromium(III) salts are typically prepared from chromium(VI) compounds, which are commercially derived from chromite ore:

$$Cr(OH)_3 + OH^- \rightleftharpoons CrO_2^- + H_2O$$
$$\text{chromite salt}$$

The chromite ore is roasted in a rotary kiln at 1200 °C, in the presence of alkali and oxygen, to convert it into dichromate:

$$Cr_2O_3 + Na_2CO_3 + \tfrac{1}{2}O_2 \rightleftharpoons Na_2Cr_2O_7 + CO_2$$

Acidification converts the dichromate into chromic acid:

$$Na_2Cr_2O_7 + H_2SO_4 + H_2O \rightleftharpoons 2H_2CrO_4 + Na_2SO_4$$
chromic acid

The chemistry of chromite processing also highlights the potential dangers in burning or incinerating waste products from chrome tanning, such as leather dust and trimmings: the fly ash will contain Cr(VI) and therefore constitute an environmental danger. Cases have been reported from developing economies of people using chrome leather shavings as fuel for cooking and consequently developing lung cancers, because they had breathed in fly ash from the open fires.

Reduction converts the chromic acid or chromate or dichromate into basic chromium(III) salt. Reduction is illustrated with the following equation using the most commonly applied reducing agent, sulfur dioxide:

$$2H^+ + 2CrO_4^{2-} + 3SO_2 \rightleftharpoons 2Cr(OH)SO_4 + SO_4^{2-}$$
33% basic

The outcome of sulfur dioxide reduction is basic chromium(III) sulfate, accompanied by free sulfate salt: commercial tanning salt contains roughly 50% sodium sulfate. Almost any reducing agent can be used as an alternative and all the obvious organic materials have been used at some time, *e.g.* starch, cellulose, sawdust. The commonest organic reductant was, and still is, waste molasses or glucose, producing so-called sugar-reduced basic chromium(III) tanning salt. The relationship between the ratios of reactants and the outcomes is set out in Table 11.2, where

133 − (% basicity required) = kg acid per 100 kg dichromate

Other reducing agents have also been used industrially, particularly sulfite, metabisulfite and thiosulfate. The difference between organic and inorganic reduction is the presence of organic salts as residues of the breakdown of the organic molecules. These salts can act as complexing ligands, referred to as 'masking agents' (see later), which

Table 11.2 Preparation of basic chromium(III) tanning salts.[a]

	Reactant offers (parts by weight)			
Reactant	0% basic	33% basic	42% basic	50% basic
Dichromate	100	100	100	100
Acid	133	100	91	83
Glucose	25	25	25	25

[a]4 mol acid + 1 mol dichromate = 0% basicity; 3 mol acid + 1 mol dichromate = 33% basicity; 2 mol acid + 1 mol dichromate = 67% basicity.

change the composition of the ligand field around the chromium(III) and hence lead to the following perceived or assumed effects on the leather.

- The leather is brighter blue in colour, compared with the distinctly greener shade of blue from sulfur dioxide-reduced chrome tanning salt.
- The leather is softer and more full in handle.
- The tanning reaction is assumed to be less astringent, *i.e.* slower, so penetration is assumed to be better than with SO_2-reduced salt.
- The leather is more susceptible to mould growth in storage.

Sugar-reduced chrome salts are now relatively less common in the modern industry than they used to be, but may be found in those tanneries that still prepare their own chrome tanning salts. In these cases, it is crucial to ensure that the reducing reaction is complete, so that the leather is not contaminated with chromium(VI). Residual chromium(VI) is difficult to analyse unequivocally in the presence of chromium(III), see also Section 11.18.

11.3 Brief Review of the Development of Chrome Tanning

The development of modern chrome tanning went through three distinct phases.

1. The original single-bath process: The original process used chrome alum, $Cr_2(SO_4)_3 \cdot K_2SO_4 \cdot 24H_2O$, applied as the acidic salt, typically giving pH ~2 in solution. Following penetration at that pH, where the collagen is unreactive, the system is basified to pH ~4, with alkalis such as sodium hydroxide or sodium carbonate to fix the chrome to the collagen.
2. Two-bath process: The first commercial application was an alternative approach to the single-bath process: it was recognised that a more astringent, and consequently more efficient, tannage could be achieved if the technology of making chromium(III) tanning salts was conducted *in situ*. This means that the process was conducted in two steps. The pelt is saturated with chromic acid in the first bath, then it is removed, usually with standing overnight. At this time, there is no reaction, because Cr(VI) salts do not complex with protein. Next, the pelt is immersed in a second

bath, containing a solution of a reducing agent and sufficient alkali to ensure that the final pH reaches at least 4. At the same time, processes were also devised that combined both valencies of chromium, exemplified by the Ochs process.[5] However, the dangers of using chromium(VI) drove a change back to the single-bath process. Not least of these considerations was the incidence of damage to workers by chromium(VI) compounds: the highly oxidising nature of the reagents typically caused ulceration to the nasal septum, so for this reason notices warning of the dangers of chromium(VI) to health can still be found in UK tanneries, even though these compounds have long since ceased to be used.

3. The modern single-bath process: With the development of masking (see later) to modify the reactivity of the chromium(III) salt and hence its reactivity in tanning, the global industry universally reverted to versions of the single-bath process.

There have been attempts to reintroduce versions of the two-bath process. Notably, the Gf process was developed towards the end of the twentieth century in Copenhagen. However, European tanners were reluctant to handle chromium(VI), regardless of the advised measures to ensure complete reduction. For this reason, the method was never generally adopted.

11.4   Chromium(III) Chemistry

Chromium is an $[Ar]3d^54s^1$ element, so chromium(III) compounds have the electronic configuration $3d^3$, forming octahedral compounds, as illustrated in Figure 11.1. The hexaaquo ion is acidic, ionising as a weak acid, or may be made basic by adding alkali. The hydroxy species is unstable and dimerises, by creating bridging hydroxy compounds, because the oxygen of the hydroxyl can form a dative bond *via* a lone pair, as shown. This process is called olation: it is a rapid, but not immediate reaction. The possible presence of a bridging sulfato ligand is addressed in the following.

In Figure 11.1, the chromium complexes are presented as linear species: the shape of basic chromium(III) species is discussed in detail in the following.

A question that might arise is: how do we know that there are hydroxy bridges in these complex molecules? The answer is provided by the work of Bjerrum,[6] who titrated chromium(III) salts with alkali, until they precipitated, then immediately back-titrated with

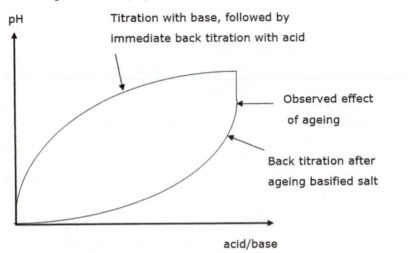

Figure 11.2 Model titration curve for chromium(III) salt.

acid. As illustrated in Figure 11.2, the pH trace for the alkaline reaction is retraced as acid is added. However, if there was an overnight delay between the alkali titration and the subsequent acid titration, the traces did not overlay. The difference indicates that some of the hydroxyls created by the alkali reaction are not available to react with acid, *i.e.* they have been changed in nature by their bridging function.

In 1974, Irving reviewed the chemistry of chromium(III) complexes from the point of view of the leather industry.[7] Some of his more important observations can be summarised as follows.

1. The half-life of water exchange in the Cr(III) ligand field, determined by isotopic exchange of ^{18}O, is 54 h at 27 °C: this is an associative interaction. This can be compared with other metals, including Al(III), which have half-lives of the order of 10^{-2} s: these are dissociative interactions. This difference between associative and dissociative complexation has implications not only for the stability of Cr(III) complexes but also for the role of Al(III) in tanning technology.
2. The diol complex, constructed from the two hydroxyl bridges, was assumed to be the preferred form of the chromium dimer. It was also argued that the trimer is the linear version of the bridged structure (see later).
3. The stability of transition metal complexes can be discussed in terms of the thermodynamic stability and the kinetic stability. In the case of Cr(III) complexes with carboxylates, they are

thermodynamically less stable than some other complexes, such as ammines, but they are kinetically more stable.

4. The mechanism of exchange between ligands into an octahedral complex depends on the stability of the intermediate crystal field, either five-coordinate square-pyramidal (S_N1 mechanism) or seven-coordinate pentagonal bipyramid (S_N2 mechanism). From calculations of the crystal field activation energies, both mechanisms exhibit high values, with a higher value for the mechanism involving seven-coordination. Whichever is the actual dominant mechanism, the high activation energies explain why the complexes are kinetically stable.

5. The stability of complexes between Cr(III) and carboxylates is inversely proportional to the dissociation constant of the carboxylic acid. This was first proposed by Shuttleworth.[8] A plot of log K_{CrL} against log K_{LH} indicates there is a correlation, as illustrated in Figure 11.3.[7,9–11] Note that a log–log plot is typically a way of

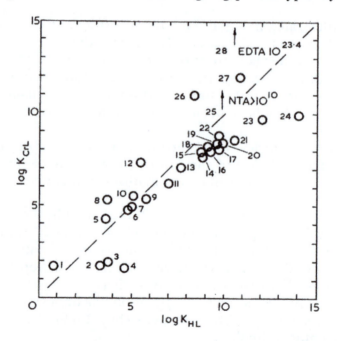

Figure 11.3　Relationship between chromium(III) complex stability and carboxylate dissociation constant, mainly at 25 °C and I = 0.1 M. 1, Thiocyanate; 2, nitrite; 3, formate; 4, azide; 5, fluoride; 6, acetate; 7, propionate; 8, oxalate; 9, maleate; 10, phthalate; 11, succinate; 12, malonate; 13, citrate; 14, asparagine; 15, arginine; 16, serine; 17, lysine; 18, methionine; 19, valine; 20, glycine; 21, alanine; 22, leucine; 23, sulfosalicylate; 24, hydroxide; 25, nitrilotriacetate (NTA); 26, EDTA. Courtesy of the *Journal of the Society of Leather Technologists and Chemists*.

flattening a curve, and so does not necessarily mean that there is a chemical relationship of that nature.

6. The formation of chelate complexes is favoured compared with complexes with monobasic carboxylates (Table 11.3).[12] The im Although there is not a simple stepwise relationship between the successive stability constants, particularly for the monobasic carboxylates, the relative values do show the enhanced stability of chelates.[13] where Interestingly, these results for chelation do not clearly reflect the well-known stability of five- and six-membered rings in organic chemistry, where there is preference for those ring sizes.[14]

7. The olation reaction occurs at the *trans* positions because the rate of ionisation of the aquo ligand is faster (Figure 11.4). This suggests that the polymerisation of basic chromium(III) complexes is linear: structure I in Figure 11.5. It should be pointed out that there is an alternative structure for polymers larger than the dimer, the three-dimensional Orgel structures, II and III in Figure 11.5.[15]

Several studies have elucidated the kinetics of formation of the Orgel structures and their stabilities,[16–18] where it has been suggested that the three-dimensional structures are the preferred

Table 11.3 Stability constants for carboxylate complexes with chromium(III): 25 °C, 0.1 M KClO$_4$ for dibasic acids, 0.3 M HClO$_4$ for monobasic acids.

Ligand	Log K_1	Log K_2	Log K_3
Oxalate	5.3	5.2	4.9
Malonate	7.1	5.8	3.3
Succinate	6.4	4.6	2.9
Maleate	5.4	3.0	1.9
Formate	1.9	0.7	1.3
Acetate	4.6	2.4	2.5
Propionate	4.7	2.4	2.7

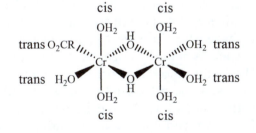

Figure 11.4 Stereochemistry of chromium(III) complexes.

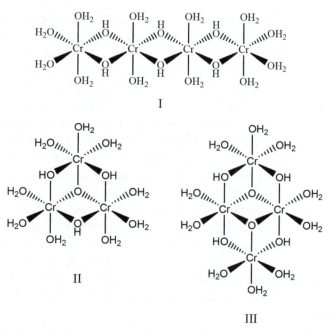

Figure 11.5   Polymeric chromium(III) aquo structures: linear and non-linear (Orgel).

conformations and the trimer is the fundamental unit of structure of basic chromium(III) oligomers. The structures as represented in Figure 11.5 are controversial because of the three-centre
bonding,[19] but the unusual lability of the molecules may require
unusual structural features to explain it.

It was reported by Ramasami and co-workers[20] that structures
similar to the Orgel structures are found accumulated in recycled
tanning solutions, indicating that they are formed as minor polymer components, but also that they have poorer tanning power
than the linear complexes. In contrast, Montgomery and co-
workers reported that the Orgel trimer is a better tanning agent
that the dimer, particularly in the presence of sulfate.[21] These
and other studies of the effect of structure on tanning power and
ability were based on the investigation of species in solution,
rather than species actually engaged in the tanning process on
collagen. This is discussed in more detail later.

8. Irving[7] addressed the relationship between sulfate and chro-
mium(III), the rate of exchange between ionic and coordinated sulfate. Infrared spectroscopy indicated that sulfate is
unidentate in the pentaaquo chromium(III) ion,[22] but in the
dichromium species the sulfate is bidentate. Using ^{35}S-labelled

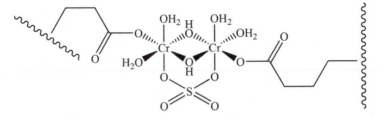

Figure 11.6 Indicative model of bridging sulfate in the dichromium aquo complex.

sulfate, measurement of the stability of the monosulfatope-ntaaquo complex gave a $t_{1/2}$ value of 20 h at 20 °C, but for the bidentate complex gave $t_{1/2} = 30$ h at 25 °C,[23] indicating greater stability.

However, questions remain regarding the stability of the sulfato complex of polymeric chromium(III) under tanning conditions and in a tanning reaction, particularly whether or not the sulfate acts as a bridging ligand, as indicated in Figures 11.1 and 11.6. In the latter case, the model is the reaction mechanism assumed over the last half century or more to summarise the chrome tanning reaction.[24,25] The implications of this model and its accuracy are discussed in detail in the following.

11.5 Chrome Tanning Reaction

Although the modern process is conventionally referred to as 'chrome tanning', it is important to remember that the reaction is most commonly conducted with basic chromium(III) sulfate, as the commonest reagent used in the global leather industry. The importance of that caveat lies in understanding the roles of each component of the salt and in reviewing the alternative options.

The popularity of the process is clear when the features of the process are compared with vegetable tanning (see Chapter 13).

- The process time for the chrome tanning reaction itself is typically less than 24 h, whereas the vegetable tanning reaction takes several weeks, even in a modern process.
- Chrome tanning confers high hydrothermal stability; a shrinkage temperature of 110 °C is easily attainable. This opens up new applications, compared with vegetable-tanned leather, where the maximum achievable shrinkage temperature is 85 °C, depending on which vegetable tannin type is used.

- Chrome tanning alters the structure of the collagen in only a small way. The usual chrome content of fully tanned leather is 4% Cr_2O_3, whereas vegetable-tanned leather may contain up to 30% tannin and hence the handle and physical properties are inevitably modified, restricting applications of the leather.
- Vegetable tanning creates hydrophilic leather, because of the chemical nature of the plant polyphenols that constitute the tanning material, but chrome tanning makes collagen more hydrophobic, so the tannage allows water resistance to be built into the leather.
- Chromium(III) can act as a mordant (fixing agent for dyes) and its pale colour allows bright, deep and pastel shades (even though the base colour of the leather is pale blue). Tanning with plant polyphenols has the effect of making the dyeing effect dull, whichever vegetable tannin or dye types are used – the leather is said to be 'saddened'.
- Vegetable-tanned leather may exhibit poor light fastness, depending on the type of vegetable tannin, whereas chrome-tanned leather is lightfast. Hence dyed chrome tanned leather will retain its colour better.

In summary, chrome tanning is faster, better in many different ways and offers greater versatility to the tanner with regard to the leather that can be made from wet blue, the name given to leather after the chrome tanning reaction is complete. Versatility is a key characteristic of the process. It is theoretically possible to create any type of leather from any wet blue hide: men's weight or ladies' weight shoe upper, combat upper, soling, clothing, gloving, upholstery, *etc.* It is true that the leathers would not be ideal, since in this example the wet blue is not created specifically for every application, but the principle of the general assertion is true.

In essence, the chrome tanning reaction is the creation of covalent complexes between collagen carboxyl groups, specifically the ionised carboxylate groups and the chromium(III) molecular ions. Molecular modelling studies have indicated that reaction at glutarate is favoured over aspartate: reaction is faster at aspartate, but the complexation at glutarate is calculated to be more stable.[26] It is noteworthy that the calculations also indicate that sulfate as a ligand has no stabilising or accelerating effect on complexation at collagen carboxylates. This is in line with advanced considerations of the tanning mechanism (see Chapter 23).

Table 11.4   Calculation of the relative availability of chrome complexation reaction sites on collagen.

	$[CO_2H]:[CO_2^-]$		Anions per 1000 residues		Anionic ratio
pH	Asp	Glu	Asp	Glu	[Asp]:[Glu]
2.5	20	50	1.7	1.2	1.4
3.0	6.3	16	4.8	3.6	1.3
3.5	2.0	5.0	12	10	1.2
4.0	0.6	1.7	21	23	0.9
4.5	0.3	0.7	27	35	0.8

Tanning studies by Ioannidis and co-workers,[27] using aluminium(III) as a model for the mineral reaction and NMR spectrometric analysis of its interaction with polyaspartate and polyglutamate, showed that the former reaction is stronger than the latter. Although this reaction differs from that with chromium(III), because its nature is much more electrostatic, the difference between the sidechains was attributed to the additional entropy penalty of reaction at the longer sidechain of glutamate. In extended instrumental analysis of the chrome tanning reaction, Brown *et al.* concluded that aspartate is favoured over glutamate;[28] the indication from Table 11.4 is that glutamic acid should be favoured because of its greater availability, although aspartic acid has a slight advantage with regard to its dissociation constant. In an attempt to isolate chrome crosslinks on collagen and obtain direct evidence for the chemistry of the reaction, Covington *et al.* produced somewhat inconclusive evidence, although it was not inconsistent with preferential reaction at aspartate.[29]

The relative amounts of aspartic and glutamic acids in collagen are about 26 and 45 per 1000 residues, respectively, increased to about 35 and about 60 per 1000 residues, respectively, after liming to the typical commercial extent, owing to hydrolysis of asparagine and glutamine. The pK_a values of the aspartic and glutamic sidechain carboxyls are 3.8 and 4.2, respectively. Hence the availability of the carboxylate at any pH value can be readily estimated, as shown in Table 11.4, taking into account that reaction *via* the unionised carboxylic acid group is so slow that it can be neglected.[4] The analysis in Table 11.4 indicates that the availability of reaction sites is similar in number for aspartate and glutamate, although the trend with pH is clear.

Regardless of the competition between aspartate and glutarate, the reaction is no different to making any other carboxylate complex, such as acetate or oxalate: although tanners tend to think of the reaction as fixation of chrome onto collagen, they might equally think of fixing collagen onto chromium compounds. Either way, the reaction is the same, since the way of thinking does not alter the product. The one difference between simple complexation, such as acetate and chromium(III), and the typical tanning reaction is the ability of the reactants to come together. In the case of the simple reaction, the reactants are in solution and can come together without hindrance, limited only by diffusion. In the case of the tanning reaction, the substrate has finite thickness, so the additional parameter of penetration through the cross-section comes into play. The tanner must balance competing process rates, so there are three possible conditions:

1. rate of penetration > rate of reaction;
2. rate of penetration = rate of reaction;
3. rate of penetration < rate of reaction.

In the first case, the rate of penetration is faster than the rate of fixation, so the chrome salt is likely to colour all the way through the pelt, but may not be fixed, depending on the degree to which the system is made unreactive by the pH conditions in pickling. This is the usual condition created at the beginning of practical tanning.

In the second case, this is the ideal situation: following colouring through the cross-section, the rates should be the same or similar, because that will lead to the most uniform fixation through the pelt, although the actual concentration profile through the cross-section will depend on the initial pH profile, as described in Chapter 9. Completely uniform fixation is clearly difficult to control, but is only one of the various processing problems facing the tanner.

In the third case, the system would be too reactive (astringent), resulting in excessive fixation on or at the surfaces, potentially leading to a raw centre in the pelt. This is the usual condition aimed for at the end of practical tanning, when the remaining chrome in solution becomes less of a damaging threat even if fixation occurs at the surface.

These rates are controlled in large part by the pH and temperature profiles, so the conditions at the beginning and the end of the reaction are adjusted accordingly.

The term 'ambient' as applied to temperature is, of course, relative: in considering the initial conditions of tanning, account must be taken of local conditions, where the temperature might be lower or more likely higher than the range quoted in Table 11.5. For example,

Table 11.5 Conditions at the beginning and end of the chrome tanning process.

	pH	Temperature	Reaction rate	Penetration rate
Start	2.5–3.0	Ambient, *e.g.* 15–25 °C	Slow	Fast
Finish	3.5–4.0	Elevated, *e.g.* 40–60 °C	Fast	Slow

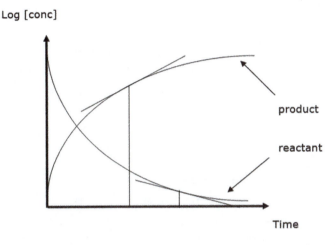

Figure 11.7 Representation of the kinetic plot for a first-order reaction: the rate at any point is a tangent to the curve.

in the height of summer in some parts of the world, the ambient and municipal water temperature may be significantly higher than 25 °C, then tanners sometimes have to resort to the addition of ice before tanning can start. From a kinetics point of view, the reaction must be accelerated during its progress, because the rate will fall as the chrome is fixed. The rate follows a pseudo-first-order path, as illustrated in Figure 11.7.

The tanner cannot afford to wait until the reaction reaches equilibrium, so the conditions of reaction are constantly varied, to allow the reaction to fit into the available process time. It should be noted that attempts to define the kinetics of the chrome tanning reaction have met with little success, primarily because the rate is dependent on the finite thickness of the substrate. Therefore, only empirical equations have been developed of the following kind:[30]

$$[Cr]^{-n} = kt + 1 \text{ or } [Cr] = kt^m$$

Such expressions have limited use, since they do not eliminate the effect of the solid substrate, they merely emphasise that the rate depends on the chromium concentration. The solution kinetics of carboxylate complexation with chromium(III) have been analysed according to the scheme set out in Figure 11.8.[31]

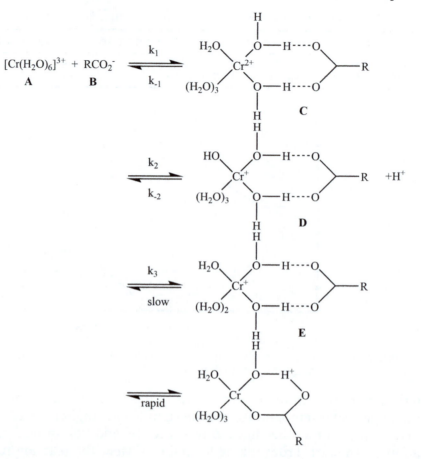

Figure 11.8 Reaction mechanism of carboxylate complexation with chromium(III).

The rate equation for the rate-determining step leading to species E is as follows:

$$\frac{d[Cr_t]}{dt} = -k_3[D]$$

where $[Cr_t] = [A] + [C] + [D]$. If

$$\frac{[C]}{[A][B]} = \frac{k_1}{k_{-1}} = K_1 \text{ and } \frac{[D][H^+]}{[C]} = \frac{k_2}{k_{-2}} = K_2$$

then

$$\frac{d[Cr]}{dt} = \frac{k_3 K_2[A^-][Cr]}{[H^+] + ([H^+] + K_2)K_1[A^-]}$$

From the rate equation developed from the proposed mechanism, some assumptions can be made regarding the parameters of the reaction, in order to simplify the kinetics, to be able to draw some conclusions about their reliance on solution conditions. The argument is as follows:

If $[H^+] > K_2$: lower pH

$$\frac{d[Cr]}{dt} = \frac{-k_1[A^-][Cr]}{[H^+](1 + K_1[A^-])}$$

If $K_1[A^-] < 1$:

$$\frac{d[Cr]}{dt} = \frac{-k_1[A^-][Cr]}{[H^+]}$$

If $K_1[A^-] > 1$:

$$\frac{d[Cr]}{dt} = \frac{-k_2[Cr]}{[H^+]}$$

$$= \frac{-k_2[Cr][A^-]}{K_a[HA]_r}$$

$$= -K_5[Cr][A^-]$$

At lower pH, the rate depends on chrome concentration and number/concentration of available reaction sites.

If $[H^+] < K_2$: higher pH

$$\frac{d[Cr]}{dt} = \frac{-K_3[A^-][Cr]}{K_2 K_1[A^-]}$$

$$= K_4[Cr]$$

At higher pH, the rate depends only on chrome concentration.

The relative quantities of the reactants can be calculated (Table 11.6). In the case of the collagen, the total content of carboxyl groups is 1 mol kg^{-1} dry weight, from which can be assumed 0.2 mol of carboxyls kg^{-1} wet weight during the tanning reaction and that the pK_a of the carboxyls is 4. The chrome offer in 100% float is 2% Cr_2O_3 and the fixation rate can be assumed to terminate at 90%. To correct the amount of carboxylate available for reaction, it can be assumed each carboxylate reacts with a Cr_4 species.[32] At the beginning of the reaction, the chrome is in excess, so the rate depends on the substrate's reactant concentration, the carboxylate groups: by the end of the reaction, the substrate reactant is in excess and the rate depends on the chromium concentration.

It is interesting to calculate the relative quantities of reactants, to check the possible impact on the kinetics: 1 kg contains 1 equiv. of carboxyl groups; at pH 4 that corresponds to 0.5 equiv. of carboxylate. If tanning finishes with 4% Cr_2O_3 fixed and it is assumed

Table 11.6 Conditions during conventional chrome tanning.

pH	Buffer ratio $[A^-]:[HA]$	$[CO_2^-]$/M	Chrome uptake/%	$[Cr]$/M	$[CO_2^-]_{corr}$/M	$[Cr]:[CO_2]$
2.5	0.032	0.006	0	0.26	0.006	43
3.0	0.10	0.018	40	0.16	0.016	10
3.5	0.32	0.048	70	0.08	0.037	2
4.0	1.0	0.100	90	0.03	0.071	0.4

that the chrome is bound as a Cr_4 species, so that the empirical formula, in similar style to the conventional representation of the standard, is Cr_4O_6, with molecular weight 304, 1 kg of leather contains 0.13 mol of tetrachromium species bound by unipoint fixation.

The following inferences can be drawn.

1. By that analysis, the total carboxylate is in excess at 0.37 mol kg^{-1}, corresponding to the reaction being theoretically 26% of completion if all carboxylates were to be reacted.
2. If we assume multipoint fixation of the metal complex, *i.e.* in some form of crosslinking, the excess carboxylate is reduced to 0.24 mol kg^{-1}, corresponding to the reaction being 52% complete.
3. The results in (1) and (2) indicate that the maximum chrome content would be 16 or 8% Cr_2O_3, respectively. Since the maximum has been calculated to be around 16%, the analysis is consistent with a unipoint complexation reaction, *i.e.* no crosslinking.

Bowes *et al.*[33] showed that the shrinkage temperature of chrome-tanned leather is dependent on the amount of chromium fixed in the leather. In Figure 11.9a, reproduced from their paper, the relationship is presented: the curve is predictable, indicating that the equilibrium maximum chrome content lies somewhere around 15–20% Cr_2O_3. Interestingly, the curve is logarithmic, shown in Figure 11.9b. As the authors reflected: 'The linear relationship probably has no simple physical significance, but it allows the relationship between shrinkage temperature and Cr_2O_3 content to be expressed in a simple manner'.

Hence increasing the chrome content by equal percentage increments increases the shrinkage temperature by the same rise in degrees. For example, increasing the chrome content by 50% increases the shrinkage temperature by $31.81 \times \log 1.5 = 5.6 \,°C$, or double would be $31.81 \times \log 2 = 9.6 \,°C$. In this way, chrome contents of 20 and 30% Cr_2O_3 would correspond to shrinkage temperatures of 127 and 133 °C, respectively. Since it is known how high shrinkage temperatures in chrome tannage can be achieved, and it is nothing like the conditions predicted by this artificial, mathematical curiosity, the results remain just that, an interesting curiosity.

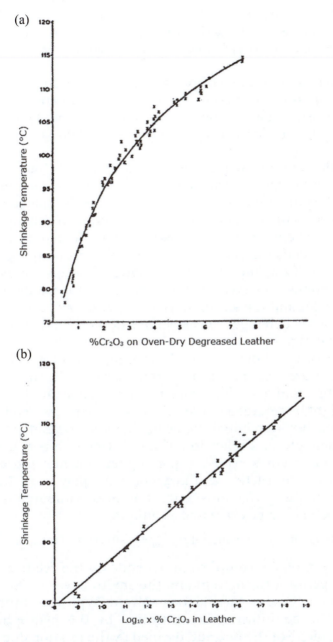

Figure 11.9 (a) Relationship between shrinkage temperature and chrome oxide content (Courtesy of the *Journal of the Society of Leather Technologists and Chemists*). (b) Relationship between shrinkage temperature and $\log(Cr_2O_3$ content).

In the chrome tanning reaction, there is another factor that may be taken into account, namely the mechanics of the process. The steps involved in any general reaction in which a reagent is fixed onto a substrate may be defined as follows (see also Chapter 15):

1. transfer of the reagent from solution into the substrate;
2. hydrophobic interaction between the reagent and the substrate;
3. electrostatic interaction between the reagent and the substrate;
4. covalent reaction between the reagent and the substrate.

The consideration of step 1 proposed here is analogous to the measurable free energy of transfer between two solvents;[34] in this case, the solvating environment for the reacting species is changed from the conventionally understood aquation in water to an environment that reflects the chemistry of the substrate, typically more hydrophobic than the aqueous solvent itself. Alternatively, this is analogous to the partitioning of a solute between two solvents, a concept used in solvent extractions. However, an empirical view of the relative properties of the solvent and the substrate is a powerful way of understanding and predicting heterogeneous reactions of the type routinely used in leather making.

The role of step 2 depends on the chemistry involved: hydrophobic interactions are thought to be important in the initial stages of vegetable tanning[35] and may play a part in other organic tanning reactions. Step 2 might be regarded as a special case of step 3, which is the start of most reactions, an initial charge interaction; however, it might also be considered to be a reflection of step 1, when transfer is driven by affinity. The extent to which step 4 applies depends entirely on the chemistry of the reaction and may not even apply at all. In the case of chrome tanning, the transfer reaction is very important. The controlling factors can be expressed as follows:

$$[\text{reagent}]_{\text{solvated}} + [\text{substrate}]_{\text{solvated}} \rightleftharpoons [\text{substrate} - \text{reagent}]_{\text{solvated}}$$

The position of the equilibrium depends on the relative affinities of the reagent for the solvent and the environment within the substrate. In the case of an aqueous solvent, typical in tannery processing, the equilibrium is controlled by the hydrophilicity or hydrophobicity of the reagent, defined by the relationship between the reagent and the solvent water, i.e. the equilibrium in the following equation:

$$[\text{reagent}] + \text{water} \rightleftharpoons [\text{reagent}]_{\text{hydrated}}$$

A hydrophilic reagent will tend to remain in solution, because it interacts favourably with water. On the other hand, a hydrophobic reagent will not be as soluble in water and will tend to move into a more hydrophobic environment, into the substrate.

In the case of chrome tanning, all the steps of the reaction mechanism operate. Transfer depends on the relationship between the chrome species and the solvent, in this case water: the more hydrophobic the species, the faster the transfer will be. This effect operates when chromium(III) is basified, when the chrome species are polymerised and become less soluble in water. Ultimately, if the chromium salt is on the point of precipitation, reaction will be very fast on the surfaces, the origin of 'chrome staining', the appearance of green marks on the leather. Staining is a serious defect, because the chrome confers colour and also takes up reaction sites on the leather required for dyeing: each of these will adversely affect the colour of the dyed leather. The electrostatic interaction depends on the charge on the chrome complex, which can be adjusted by altering the ligand field: this is the effect of 'masking', addressed in detail later. Clearly, the charge will also contribute to the partitioning of the solute between the water and the more hydrophobic environment. Ultimately, covalent complexes are formed between the chromium species and the collagen carboxyl groups.

Sykes[36] demonstrated that the primary chrome tanning reaction occurs at the carboxyl groups and this is the reaction responsible for raising the shrinkage temperature of collagen. Minor reactions also occur at the sidechain amino groups and the peptide links, where the interaction is hydrogen bonding between the charged sites on the collagen and the aquo ligands and hydroxyl bridges of the chromium(III) molecular ions: these reactions account for only a few percent of the total amount of fixed chrome, and they have no effect on raising the shrinkage temperature of the leather. This provides us with an understanding of the consequences of chrome staining, which ultimately cause 'dye resist', when the dyeing reaction cannot take place where the precipitated chrome species is fixed to the leather surface. The highly basic chrome fixes very rapidly to the surface by hydrogen bonding, a much faster reaction than complexation. Therefore, the hydrogen bondable sites on the collagen are no longer available for reaction with acid dyes, so no colouring is possible and the surface remains stained green.

It is both implicit and explicit in publications in tanning science that the fundamental element of the mechanism of collagen stabilisation is crosslinking – as may appear to be the case in the foregoing exposition. Without pre-empting the argument set out in Chapter 23 regarding the general mechanism of tanning, in which a theoretical model of tanning is derived to explain all tanning reactions, the term 'crosslinking' will be significant by its absence from this treatise. Once that theory has been developed, it might be useful to return to the data presented here and in the literature, to review the significance of the information in the light of newer fundamental thinking.

11.6 Basification

From the figures given in Table 11.4, it can be seen that raising the pH of the solution has the effect of ionising the carboxyl groups on the collagen: this is the mechanism by which the reactivity of the system is increased. At the same time, two things happen to the other reactant, the chromium(III) salt.

1. The basicity is increased, so the molecules are polymerised by olation. As the molecular weight increases, the hydrophobicity of the species increases: this makes the species more reactive towards the collagen. This is part of the explanation of the tanners' 'rule of thumb': *the tanning power of a metal is greatest just before the point of precipitation.*
2. The polymerisation reaction increases the number of chromium atoms per molecular ion; therefore, if the reaction rate remains the same as measured by the reaction of the carboxyl groups, the fixation rate of the chromium is increased. This means that for every complexation reaction between a chrome species and a carboxyl group, the number of chromium atoms fixed to the collagen increases. Since the reaction kinetics involve complexation at the *trans* positions, Figures 11.3 and 11.5 illustrate that the kinetics of linear polymeric ions will be unaffected when the reaction pH is increased. This is additional rationale for that 'rule of thumb'.

Any compound that is capable of raising the pH of the tanning solution to a value higher than 4.0 can be considered as a candidate to be a basifying agent. Examples of basifying agents are set out in Table 11.7. In addition to the basifying effect, the other effect that must be considered is reaction between the basifying

Table 11.7 Approximate pH values of solutions of industrial basifying agents, as 10% solutions or maximum values for less soluble salts.

Basifier	Concentration/%	pH
Sodium carbonate	10	11
Sodium aluminium silicate	1	10
Magnesium oxide (magnesia)	Maximum $(9 \times 10^{-5} \text{ g L}^{-1})$	10
Sodium tetraborate (borax)	10	9.2
Sodium formate	10	8.5–8.7
Sodium bicarbonate	10	8.0
Sodium sulfite	10	7.8–8.0
Ammonium bicarbonate	10	7.5–7.8
Dicarboxylates:	10	
Mono		6–8
Disodium salts		8–11

species and the chromium species, because any complexation reaction will influence the reactivity of the chrome to tanning complexation (see later).

In all cases, the basifying reaction must be related to the pH in solution at any time during the basifying process. Above the typical endpoint of basification, usually pH 3.8–4.0, there is a danger zone comprising the maximum allowed pH leading to the precipitation point. The danger arises from the olation reaction causing polymerisation and hence elevated reactivity of the chrome species: this reaction is fast, and so is the ionisation of the carboxyl groups on the collagen, so even localised high pH can create conditions for enhanced reaction in that vicinity. Here, it is not necessary to have complexation to cause a fast reaction – hydrogen bonding is all that in needed. The mechanism of allowing chrome to fix on the collagen can be regarded as a one-way reaction, since the bonding is effectively irreversible. A mild effect would be enhanced reaction on the surface, leading to chrome 'shadowing', and a more serious effect would be clearly uneven colouring of the surface; both of these would lead to uneven colouring in dyeing, but the ultimate problem of chrome precipitation would cause dye resist. In all cases, the solution to the defect may involve pigment finishing, resulting in loss of added value for the final leather. Therefore, any basification procedure that involves approaching or exceeding the maximum allowed pH value must cause concern. It is for this reason that basifying chrome tannage is often regarded as the most dangerous step in leather making, because the effect on uniformity of dyeing can be so great. Any adverse effect on dyeing the leather is of ultimate importance, since the appearance of the surface is the first perception by a (potential) purchaser.

11.6.1   Soluble Alkaline Salts

Examples of soluble basifying agents include sodium carbonate, sodium bicarbonate and sodium hydroxide. The only difference between these salts is the degree of their solubility in water and their ability to buffer, giving rise to different pH values in solution. In all cases, the type of effect of their use in basification is the same, illustrated in Figure 11.10. In the figure, it is assumed that the danger point is not far above the ideal final value, when precipitation of the chrome in solution is likely to occur. In practice, the critical value may be just above pH 4, but certainly under typical tanning conditions precipitation will occur at pH < 5.

It is worth recalling that the following relationship between bicarbonate and carbonate is temperature dependent:

$$2NaHCO_3 \rightleftharpoons Na_2CO_3 + CO_2 + H_2O$$

Above 40 °C, which may apply in practical tanning processes, bicarbonate loses carbon dioxide, thereby becoming a more alkaline salt, hence this can affect the basification process in terms of the extent and outcome.

Figure 11.10 contains an idealised basification curve and a representation of the impact of adding all the required base at once.

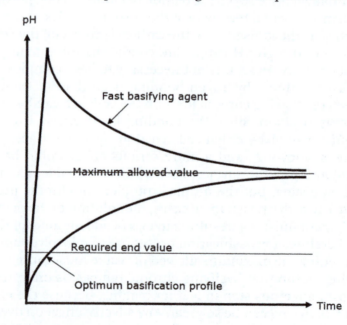

Figure 11.10   Basification curves for chrome tannage.

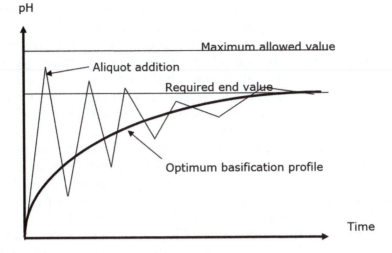

Figure 11.11 Basification by aliquots of base.

The effect of a single addition of base will be the same, the only difference will be the maximum value achieved, dependent on the solubility and buffering ability. The initial rate of increase in pH depends on the solubility, so it will be higher if the base is soluble and strong.

The effect of adding the base in aliquots is shown in Figure 11.11, where the ideal curve is tracked by multiple additions, but the critical point is never exceeded. The number of aliquots to be added is determined by the maximum concentration allowed to remain below the critical pH value; the minimum number is usually arrived at by experience.

11.6.2 Carboxylate Salts

It is not common for carboxylates, such as sodium formate, to be used alone as basifying agents, mainly owing to the cost and their side reaction of masking the chrome in solution. However, as part of a basifying system, they can contribute a buffering effect, useful in controlling the pH during the reaction. For example, formate will buffer around its pK_a value of 3.75, which is ideal for chrome tanning. Other options are succinate, glutarate, adipate and phthalate, of which only phthalate (pK_a = 4.0) is commonly used in industry.

The side reaction of masking is a critical feature of the chrome tanning reaction, since the effects are not always as expected. The impact of a masking salt on the chrome tanning reaction depends on the chemistry of the carboxylate, which can have effects beyond merely occupying reaction sites on the chromium species: the structure of the carboxylate can have a marked influence on the solubility of the complex.[37] This in turn can impose limitations on the amount that can be used in practice, best illustrated by phthalate, which will precipitate chrome at low masking ratios. The situation is analysed in detail in Section 11.11 on masking.

A less common outcome is the formation of chrome soaps, ionic compounds of chromium(III) and carboxylate: such compounds may form, but do not necessarily result in any particular effect, either desirable or undesirable. The exceptions are chrome soaps of higher molecular weight carboxylates, classified as anions of free fatty acids: these soaps sometimes form in wet blue, observed as solvent-soluble pink coloration. The mechanism of their formation is not obvious, because it is actually difficult to make them in the laboratory.

11.6.3   Other Basic Salts

There are not many other reagents commonly available and used in the global industry. Among the less common reagents for basification is sodium aluminium silicate (marketed as Coratyl G), promoted as a solo agent or in combination with sodium glutarate (marketed as Coratyl S) for masking, where it is claimed that the modification can lead to an improvement in chrome uptake.

11.6.4   Self-basifying Salts

The commonest drawback to basification by the aliquot method is the need to stop the tanning drum to add the base – unless automatic or external dosing is possible. Therefore, the concept of self-basification was developed. Here, the basifying agent is a water-insoluble compound, capable of reacting with acid: since the reaction cannot occur in solution, it must take place on the surface of the solid, so the rate of reaction depends not only on the chemistry but also on the available surface area, *i.e.* the finer the particle dimensions, the faster is the reaction with acid.

11.6.4.1   Dolomite

Dolomite is an equimolar mixed complex of calcium and magnesium carbonates, available naturally as dolomite rock and

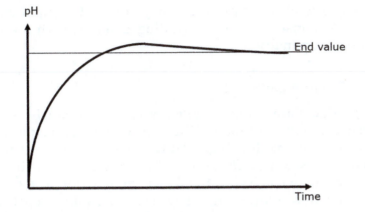

pH

End value

Time

Figure 11.12 Illustration of the basifying reaction of dolomite.

patented by the Bayer Company as a basifying agent for chrome tannage. The term self-basifying in this context means that the powdered dolomite is mixed with 33% basic chrome tanning powder (Baychrom D), so that at the end of the reaction the pH is slightly below 4.0, *i.e.* there is no need for the tanner to add base separately, because the system basifies itself (Figure 11.12). The reactions are as follows:

$$MCO_3(\text{solid}) + H^+(\text{solution}) \rightleftharpoons M^{2+}(\text{solution}) + HCO_3^-(\text{solution})$$

In the Bayer product, the $CaCO_3 : MgCO_3$ molar ratio is typically $1:1$ or smaller, *i.e.* the calcium component never dominates the compound. However, other commercial versions of dolomite exist, in which the molar ratio is $>1:1$. It might be tempting to use such products, since they are typically offered at a lower price, but there is a potential problem associated with the cheaper products that is not found with the Bayer product. Consider the progress of the reaction with acid. The magnesium carbonate in the complex is more reactive than the calcium carbonate, so it is preferentially leached from the particle, from the surface inwards. If the structure is equimolar in Mg(II) and Ca(II), as the magnesium component is removed there is insufficient residual calcium carbonate for the structure to remain intact as a solid, so it rapidly crumbles and the calcium salt continues to react as a finely divided solid. Conversely, if the compound contains a higher molar ratio of calcium carbonate to magnesium carbonate, the reaction of the magnesium component leaves a strong structure of calcium carbonate. The particle is intact, but the surface is deeply pitted, leaving sharp edges that can cut into the surface of the leather.

In this way, the cheaper products may abrade the grain surface, which would constitute a downgrading damage, with an associated economic penalty.

11.6.4.2 Calcium Carbonate

Several products are commercially available that consist solely of powdered calcium carbonate. These products are actually safer than the products with a high $CaCO_3 : MgCO_3$ ratio, because the salt decomposes uniformly in acid. The rate at which they react depends on the particle size: the faster the reaction, the higher is the pH reached in solution. In many cases, the solution pH is high enough to require the product to be added in aliquots – defeating the idea of self-basification. The effect of these products on the pH profile is the same as for strong base, illustrated in Figure 11.10. In addition, the high concentration of calcium ions in solution may lead to the formation of calcium sulfate, by reaction with the sulfate present from the chrome tanning salt and (perhaps) from the pickling acid. This may result in the formation of sparingly soluble crystals within the grain surface, giving rise to an effect analogous to limeblast.

11.6.4.3 Magnesium Oxide

Laboratory-grade magnesium oxide (magnesia) is available in different particle sizes, often distinguished as 'light' or 'heavy'. These chemicals are essentially insoluble in water, but they can be hydrated on the surface of the solid, giving rise to the hydroxide, also a very low-solubility compound, so reaction with acid occurs on the surface:

$$MgO + H_2O \rightleftharpoons Mg(OH)_2$$

$$Mg(OH)_2 + 2H^+ \rightleftharpoons Mg^{2+} + 2H_2O$$

In this way, they function in precisely the same way as the calcium carbonate products, giving rise to the same effects, *i.e.* too high pH if a high-reactivity species is added as a single dose (see Figure 11.10).

Industrial versions of this compound can be obtained in lightly or heavily calcined (heated) forms. The effect of heating is to drive off water from the hydroxide (magnesium hydroxide is obtained by precipitation from sea water) to greater or lesser extents and thereby reduce the chemical reactivity to greater or lesser extents. The lightly calcined products are offered in different particle size distributions and are no better for chrome tannage basification than laboratory-grade

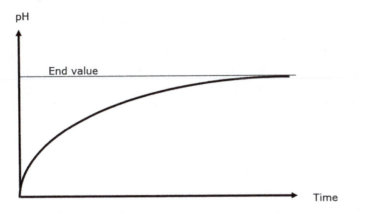

Figure 11.13 Illustration of the basifying effect of magnesium oxide (Tanbase).

reagents. On the other hand, heavily calcined magnesia is rendered so unreactive that it follows an idealised basification trace, as illustrated in Figure 11.13.

The first product based on magnesium oxide available to the industry was Tanbase, *ex* ICI, in which the particle size distribution is uniform, in order to ensure uniform reaction.[38] It is safe to add this product at the beginning of tanning, when the offer can be calculated as 5% of the offer of 33% basic chrome tanning salt (25% Cr_2O_3 content). This will typically produce a final pH of about 3.8, but the precise value will depend on the specific conditions of pickling.

There is one caveat for users of this technology: the reaction must be heated at the end of basification, in other words, the temperature at the end of the tanning plus basification time period must have reached at least 40, preferably 50 °C, and been held at the elevated temperature for at least 2 h. The reason is to ensure complete reaction and dissolution of the magnesia particles. If the reaction is not completed, for example because the process was conducted at lower temperatures, such as ambient conditions, minute particles of magnesia may remain, invisible to the naked eye but large enough to continue to react with the available acid in the pelt. The result can be chrome stains, created during the period of piling after unloading from the tanning vessel.

11.7 Avoiding Basification

It is recognised in the industry that completing chrome tanning by basification is a high-risk operation. Therefore, an alternative to the automation of basification by self-basification is the avoidance of applying basifying agents. This can be achieved by controlling the pickling step, to leave sufficient residual alkali in the pelt to act as the basifying mechanism. This approach is not new and was originally called the 'chaser' pickle process, when the offer of acid on bated pelt was designed only to acidify the surface, then the chrome salt 'chased' the mild pickle through the cross-section. Cooper and Davidson suggested an alternative approach, applying a formaldehyde pretannage to limed pelt, squeezing out much of the moisture and then treating with chrome liquor: the technology combined rapid uptake of the tanning liquor and rapid fixation by the alkaline pelt.[39]

In New Zealand, DasGupta redeveloped the chaser pickle technology into the 'ThruBlu' process.[40] The principle is to avoid the conventional low-pH pickle, with its attendant requirement for salt, and to decrease the pH after deliming to a suitable value so that there is sufficient capacity to basify the chrome without a separate step. For sheepskins or lime split hides, pH 7 may be sufficient; for thicker hides, pH 5 may be required. The reaction involves elevated astringency, which can increase the chrome uptake to nearly 99%. A slight variation on the chaser pickle is to pretreat the pelt with acid protease at pH 4.5 for up to 90 min, then tan with basic chromium(III) sulfate to achieve 97% chrome uptake and a shrinkage temperature of 115 °C.[41]

11.8   Reactivity at High Basicity

It has been assumed that the use of high-basicity chromium salts means that the salt is more reactive than lower-basicity salts. The assumption is based on the following generalised view of the relationship between process rates, as applied to the binding of a reagent to a substrate of finite thickness.

- High reactivity means fast reaction, which also means slower penetration, leading to surface reaction, with the consequence of non-uniform reaction.
- Low reactivity means slow reaction, which also means faster penetration, leading to reaction through the cross-section, with the consequence of even surface reaction.

However, although high reactivity causes surface reaction, the observation of surface reaction does not necessarily mean high reactivity. The latter can be due merely to the molecular size of the reactant, a known effect of basification, since:

- higher molecular weight means larger molecules, leading to slower penetration, causing more surface reaction.

In common with every other fixation reaction, the first step of the chrome tanning reaction is transfer from the solution environment into the substrate environment. The thermodynamics of the equilibrium are determined by the solvating power of the solvent on the solute, where greater affinity between the solvent and a solute results in a tendency for the solute to remain in the solvent. Basification makes the chrome species larger and hence more hydrophobic, therefore facilitating or promoting transfer of the species from the solvent into the more hydrophobic environment of the substrate, which will indeed have a positive effect on chrome uptake. However, those are second-order effects, relative to the role of the complexation in the reactivity; that is, the effect of basicity on available reaction sites for complexation by carboxyl groups on collagen. It is typically assumed that polymerisation in basification results in the creation of more reaction sites within the higher molecular weight chromium ion. However, that assumption is based on the further assumption that all the available reaction sites on the chromium atoms within the molecular ion are equivalent with regard to complexation with carboxylate species. The literature evidence indicates that this is not true, because reaction is preferred at the *trans* positions, *i.e.* in the plane of the hydroxy bridges, as shown in Figure 11.14.[42] These positions are favoured by the presence of both aquo ligands and the bridging hydroxyl groups.

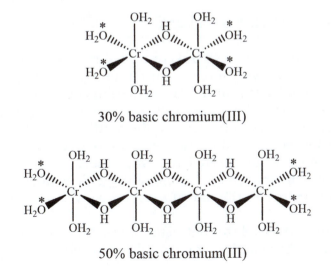

30% basic chromium(III)

50% basic chromium(III)

Figure 11.14   Reactions (asterisked) in chromium(III) tanning complexes.

Table 11.8   Influences of chromium(III) salt basicity and solution pH on the chrome tanning reaction.

System conditions			Consequences		
Solution pH	Cr(III) salt basicity	Size of Cr(III) species	Reaction rate as Cr uptake	Cr reaction	T_s
Low	Low	Small	Low	Penetrating	Low
Low	High	Large	Low	Penetrating	Low
High	Low	Increasing	High	Penetrating	High
High	High	Large	High	Surface	Low

Figure 11.14 demonstrates that, if only the *trans* positions are reactive for complexation, the polymerisation process does not increase the availability of reaction sites. Therefore, the more basic chrome salt is essentially no more reactive than the less basic salt. However, the buffering effect of the more basic salt will affect the state of ionisation of the collagen, which will make the chemical system more reactive accordingly.

It is important to note that the rate of olation, *i.e.* rate of polymerisation of basic chrome, is fast, but the reverse reaction is slow, an observation well known in the chemistry of chrome recovery.[43] This means that the addition of high-basicity chrome salt into a tanning bath whose pH corresponds to lower basicity will result in equilibration of pH. However, the distribution of chrome species sizes will remain unaltered. The effect of these variables on the chrome tanning reaction is summarised semiquantitatively in Table 11.8.

11.9   Role of Temperature

The rates of most reactions are controlled by temperature; usually the rate increases with increase in temperature. The relationship can be expressed in the form of the Arrhenius equation, as discussed in Chapter 3. The rate of ionisation of collagen carboxyls is very fast and therefore can be considered to be hardly affected by temperature. The speciation of chromium(III) compounds is also relatively fast and hence little affected by temperature change. Therefore, raising the temperature of chrome tannage does not alter the system, it only increases the rate of reaction between the chrome and the collagen.

Although it is possible to control the temperature in a tanning vessel by circulating the float through a so-called 'lab. box', it is common for tanners to rely on the development of frictional energy within the tanning vessel, to raise the temperature. This is relatively consistent for a given vessel containing similar loads and water to rawstock ratios, but tends to be inconsistent between vessels. An option is to inject steam or add hot water. The latter is referred to as 'hot water ageing', often used in conjunction with basification and hence sometimes regarded as a basification process itself. That is strictly true only when the solution is diluted to the extent that hydrolysis occurs, likely to result in chrome precipitation.

11.10 Relative Effects of pH and Temperature

The chrome tanning reaction is characterised by the type of curve shown in Figure 11.15, which can represent the effect of time on chrome uptake and the effect of chrome content on shrinkage temperature. It also is the shape of the effect of pH or temperature on chrome uptake or shrinkage temperature. Of course, the lines of those dependencies are quite different. However, this raises the question as to whether pH and temperature have equivalent effects on chrome uptake and the resulting shrinkage temperature.

At this point, it is important to recognise that there are two aspects to the chrome tanning reaction. Fixation is important, in terms of the economics of the process and its environmental impact. However, more important is the introduction of hydrothermal stability, which reflects and determines the properties of the leather in terms of the physical, chemical and biochemical resistance to external conditions. The tanner's function strictly is to make stable leather, not

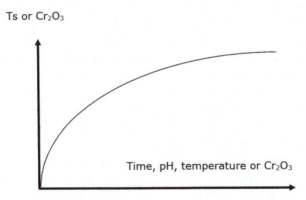

Figure 11.15 Generalised effect of controlling parameters in chrome tanning.

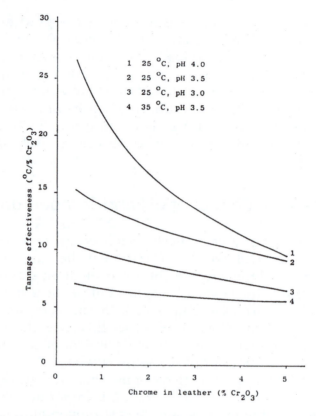

Figure 11.16 Relative effects of pH and temperature on the effectiveness of chrome tanning under constant conditions. Courtesy of the *Journal of the American Leather Chemists Association*.

merely to absorb chromium(III) from solution – in other words, the parameter of actual importance is the effectiveness of the reaction, which can be defined as the increase in shrinkage temperature per unit of bound chrome. This can easily be calculated as the tanned collagen shrinkage temperature minus the shrinkage temperature of the collagen before tanning (normalised to the same pH) divided by the percentage of chrome in the leather, as Cr_2O_3 on dry weight. The influence of pH and temperature on tanning effectiveness is presented in Figure 11.16, based on model studies in which the tanning conditions were held constant.[44]

It is clear that chrome fixation is dominated by temperature, whereas rise in shrinkage temperature is dominated by pH. This can be understood with reference to the direct roles of each parameter. The effect of increased temperature is to drive the reaction under the particular pH conditions that apply at any point in the process;

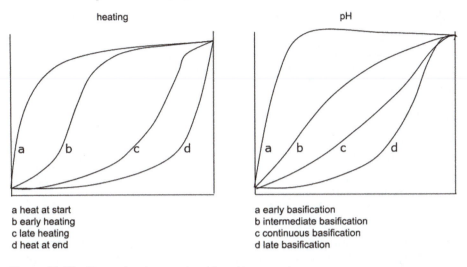

a heat at start
b early heating
c late heating
d heat at end

a early basification
b intermediate basification
c continuous basification
d late basification

Figure 11.17 Dynamic changes in pH and temperature.

in that way, the constitutions of the reactants are unaffected, only their interaction is speeded up. In the case of increasing pH, both reactants are affected. The collagen is most affected by increasing pH, which causes ionisation of carboxyls, creating reaction sites for chrome complexation, resulting in faster reaction, but there is no obvious link between the kinetics of the reaction and the effect of the reaction. The chromium species are increased in size by olation, although it has been argued that this process does not influence the reactivity. However, the change in molecular dimensions introduces a change in the outcome with regard to the nature of the product, and this is the reason why pH affects the shrinkage temperature in a different way to temperature change. This idea is developed further in Chapter 23, with regard to the basis of hydrothermal stability.

The typical chrome tanning reaction is not conducted under constant conditions, but requires dynamic changes of pH and temperature. This begs the question: if the starting and finishing values of pH and temperature are known, does the route taken affect the outcome? A model of the options is presented in Figure 11.17.[45]

The conditions have their equivalents in industrial processing. Heating at the start is similar to starting at elevated temperature, common in hotter countries. Early and late heating could be the result of efficient and inefficient frictional heating, respectively. Heating at the end is accomplished by so-called hot water ageing, where the drum is floated up with hot water just prior to unloading. Early basification

Table 11.9 Effects of dynamic pH and temperature changes on hydrothermal stability: 5 min boil tests conducted on wet samples (designated 'wet boil test') or on dry, conditioned samples (designated 'dry boil test').[a]

	Heat at start	Early heating	Late heating	Heat at end
Early basification	X[b]	X	X	X
Intermediate basification	0	0	0	0
Continuous basification	0	0	0	0
Late basification	0	0	XX	XX

[a]0 means no shrinking in either test, X means shrinking failure in the dry boil test and XX means failure in both tests.
[b]The only anomaly to remain after ageing at 60 °C.

applies when the tannage is basified right at the start of the process: the solution of the basifying agent is added immediately after adding the chrome in solution. Intermediate and continuous basification refers to variations in self-basification. Late basification means basifying after prolonged drumming to allow the chrome to penetrate to completion.

A detailed study of hydrothermal stability conducted on chrome-tanned leather with high chrome content and high shrinkage temperature, >115 °C, revealed some inconsistent behaviour when the leathers were subjected to relatively prolonged boil testing, as shown in Table 11.9.[46] Purely on the basis of the shrinkage temperature, which is consistent with the high chrome content in the leathers, it would be assumed that the leathers should pass the conventional wet boil test and there would be no reason to suppose that the leathers would fail the same test but starting from the dry condition.

From Table 11.9, the way in which pH and temperature changes are managed produces an anomaly in the hydrothermal stability, in which performance in boil testing is apparently dependent on the way in which the shrinkage temperature was produced, but not on the value of the shrinkage temperature. Although it is still not clear why the extremes of processing conditions should influence the hydrothermal stability of wet leather, an explanation for the difference between boil testing wet leather and boil testing dry leather has been offered as follows.[46] If the dry leather is hydrophilic, then the wetting process at the beginning of the test would be rapid, thereby releasing the heat of hydration quickly; this energy will contribute to the heating of the leather and cause faster shrinking. Therefore, it appears that the relatively extreme processing conditions give rise to the anomalous

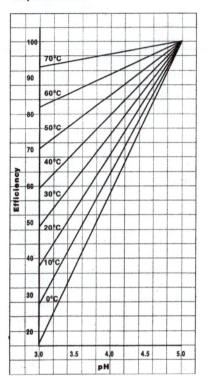

Regression of efficiency on pH. at different temperature levels

Figure 11.18 Combined effects of pH and temperature on chrome fixation efficiency. Courtesy of *World Leather*.

results from the dry boil test, because they create more hydrophilic leather. This is not a large effect in practice because the test is unusually stringent, but it is worth considering that it may have some influence on the production of water-resistant leather.

One approach to making the best use of chrome salts in the tannery is to achieve extreme conditions of pH and temperature. Some data taken from the industrial development work of Daniels are presented in Figure 11.18;[47] although the observations were made under scientifically uncontrolled conditions, they do represent the final conditions in industrial processes.

It can be seen that the efficiency tends to 100% at pH 6 and 70 °C. It is possible to basify to pH 6, but only under the following conditions.

1. The tannage must be highly efficient before raising the pH above the conventional finishing point, pH < 4.0. The chrome concentration in solution needs to be low for two reasons.

(i) The rate of reaction should be as low as possible. This is necessary to control the relative rates of penetration and fixation.

(ii) The likelihood is for the residual chrome to fix more on the surface than within the pelt. The lower the amount of residual chrome in solution, the less likely surface fixation is to cause a problem, particularly in subsequent dyeing.

2. The rate of pH increase must be as slow as possible. As already indicated, the chrome must be allowed to penetrate as much as possible during the process of enhancing the reactivity of the pelt.

3. The basification should not proceed beyond pH 6, because then the reaction begins to show signs of reversal.[48] The reason for this is not clear, because reversal by hydrolysis to chromite might not be expected to apply much below pH 9. However, polymerisation of the chrome species and interference with the establishment of the tanning matrix may be a contributor to the high pH effect. High-pH basification was proposed by Gauglhofer,[49] to achieve very high chrome uptake efficiency; however, he was obliged to admit that the leathers could not be dyed owing to excessive surface fixation.

High-temperature tanning is technologically possible, feasible and highly desirable. In conventional processing, a final temperature of 40–60 °C should be the target. The value achieved depends on the method of heating, usually obtained solely by frictional energy within the process vessel, which limits the extent of heating. Higher temperatures can be achieved by heating the recirculated float or by injecting steam into the vessel. There are limitations to this approach, which relate to the kinetics of shrinking. Consider the dynamic conditions in the tanning process.

- As the reaction proceeds, temperature rises, chrome fixation increases and shrinkage temperature rises.
- Shrinking, either as gross denaturation or incipient heat damage, is a kinetic process. The closer the process temperature approaches the shrinkage temperature, the faster damage occurs, *i.e.* the sooner damage occurs.

From this analysis, the shrinkage temperature must always be higher than the temperature in the vessel, and preferably much higher. In the case of chrome tanning, assuming that the temperature rise is not fast in the early stages of tanning, that condition will be met. If we consider reaching the extreme condition indicated in Figure 11.17, is it feasible? The benefit is clear, but what are the contraindications?

1. Under conventional industrial conditions, the shrinkage temperature of the chrome-tanned leather is not likely to exceed 120 °C, so the rule about the temperature of the solution not approaching closer than 30 °C lower than the shrinkage temperature might be broken if the processing temperature is abnormally high. Therefore, the safety margin will probably apply, but equally may not under extreme conditions.
2. The energy cost may be prohibitive.
3. The workers will not be able to handle the leather when it is dropped from the process vessel, because it will be too hot. This may not seem to be a problem, since they merely have to wait for it to cool, but that in itself creates additional problems.
   (i) Cooling will take a long time, whether within the process vessel or after unloading, so delaying production.
   (ii) Reaction between the residual chrome and the pelt will continue. This can cause shadowing on the surfaces, depending on the amount of residual chrome in the float.
   (iii) The leather, whether in or out of the process vessel, is always creased: if further tanning reaction occurs, the creases in the leather can be made permanent.

11.11   Masking

Masking is conventionally defined as:

the modification of metal complexes, by replacing aquo ligands with other less labile ligands – the purpose is to render the complex less susceptible to additional complexing reactions, including precipitation.

In this way, it has been assumed that the function of masking is to aid chrome penetration at the start of tanning, by reducing the reactivity of the complexes. Although such masking can in principle be achieved with any ligand capable of creating a complex, in the case of chrome tanning the commonest practical masking agent is formate, usually derived from the pickling formulation, but it may be added after a sulfuric acid pickle or the masking reaction might be conducted prior to the tanning reaction.

The traditionally accepted purpose of masking is threefold.

1. Since the chrome tanning reaction depends on creating carboxyl complexes with collagen carboxyl groups, modifying the complexes by including other carboxyls as ligands reduces the number of reaction sites available to the collagen, so the reactivity of the chromium(III) is reduced. This is the rationale for using masking at the beginning of the tanning process, when the masking reaction is completed before tanning starts: to decrease the rate of fixation relative to the rate of penetration.

   The ratio of the number of moles of carboxylic masking agent to the number of gram atoms of chromium is known as the masking ratio. Typical masking ratios for industrial processes lie in the range 0.5–1.0. The reaction is illustrated in Figure 11.19, where R in the structure would typically be a group from an aliphatic carboxylate salt.

   Since the masking agents are usually anionic, they reduce the charge on the complex, thereby reducing the initial affinity of the chrome complex for collagen carboxyls. In the limit, the chrome complex can be made anionic, for example in the trioxalato complex; this does not necessarily prevent further reaction, since ligands can be exchanged, but clearly the affinity for additional complexation with anions would be markedly adversely affected. The reactive sites for complexation are the *trans* positions (see Figure 11.3); it is assumed that multiple substitutions will minimise steric interactions.

2. An additional consequence of reducing the number of complexation reaction sites is that the concentration of hydroxyl ions required to precipitate the salt is increased; in other words, the pH of precipitation is raised. This means that the system can be made more reactive by raising the pH, to increase the availability of reaction sites on collagen, avoiding precipitating the

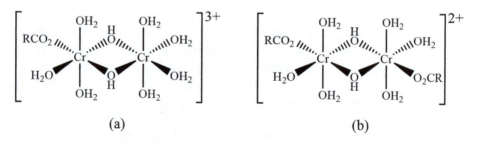

(a)   (b)

Figure 11.19   Masking ratios (a) 0.5 and (b) 1.0.

chrome. Although this argument seems reasonable, the effect is not often used, because the precipitation point is raised significantly only by high masking ratios. In this way, the tanner sets up a competition between the increase in substrate astringency caused by increasing the reactivity of the collagen and the decrease in the astringency of the chromium caused by the loss of reaction sites.

3. The colour of the chrome complex is altered by the ligands in the molecule and this affects the colour struck in the wet blue leather, as discussed earlier.

It should be recalled that the complexation reaction between collagen and chromium(III) is the same as the rate of any other carboxylate complexation reaction, *i.e.* the masking rate is the same as the chrome fixation rate, as shown by Shuttleworth and Russell in Figure 11.20.[30,50] Therefore, it is important to recognise that, if a masked tannage is required, the masking reaction needs to be completed prior to commencing the tanning reaction.

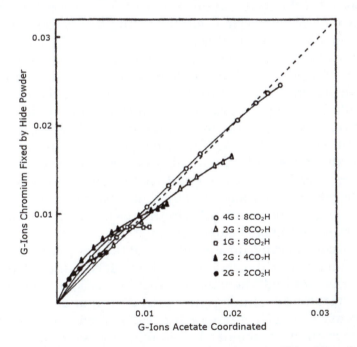

Figure 11.20 Comparison of reaction rates for chromium(III) with collagen and with acetate: reactions were for $[Cr]:[CO_2H]_{total}$ ratios of 1:1 and 1:8 and the results presented refer to comparable reaction times. Courtesy of the *Journal of the Society of Leather Technologists and Chemists*.

However, it is difficult to obtain a truly accurate comparison of collagen complexation *versus* carboxylate salt complexation, because of diffusion effects with the protein substrate and uncertainties in the pK_a values of the reaction sites on collagen. Nevertheless, a simple calculation is useful.

If the offer of formic acid in the pickle is 1% on limed weight, this corresponds to about 5% on dry weight, which is the same as 1 mol formate kg^{-1} dry protein. Dry limed collagen contains about 1 mol of carboxylic acid groups. The pK_a of formic acid is 3.75 and the pK_a values of the carboxylic acid groups on collagen lie in the range 3.8–4.2. Therefore, allowing for errors in numbers of carboxyl groups, differences in pK_a and sites of reaction on collagen, the concentrations of formate and collagen carboxylate groups are certainly within the same order of magnitude.

By that argument, the two sources of complexing carboxylate groups compete on roughly equal terms. Therefore, if the masking and tanning kinetics are comparable, fixation of formate and collagen ligands proceeds at the same rate. This means that the tannage is unmasked at the start, but masked to some extent by the end. It is not known if the chrome in solution and the chrome bound to collagen are equally reactive in this regard. If they are, then the masking ratio is less than assumed and masking competes with collagen complexation. If they are not, this would mean that chrome in solution is more reactive and therefore the masking ratio for unbound chrome will increase during the process, making the chrome increasingly unreactive. It is reasonable to assume that the bound chrome can react with more masking agent, although the reaction will be hindered by the decrease in the number of available reaction sites on the chrome species: if a chromium(III) molecular ion is complexed with collagen at a *trans* site, the adjacent *trans* site will be sterically hindered and the only available sites for easy access are situated at the other end of the molecular ion.

There is considerable uncertainty about the way in which masking operates under typical reaction conditions. If certainty of the masking effect is required, the chrome must be masked to equilibrium prior to starting the tanning process and, to that end, commercial products are available, otherwise the reaction must be conducted in the tannery. In most commercial tanning processes, when masking is taking place at the same time as chrome is being fixed onto collagen, it is probable that the dynamic masking effect plays little part in the tanning process, for example in its role of affecting chrome penetration.

Table 11.10   Effects of masking agents on relative chrome uptake.

Masking salt	Chelate ring size	Complex reaction	Relative chrome content[51]	Masking effect
None	—	—	1.00	None
Formate	—	—	1.06	Hydrophobic
Acetate	—	—	1.18	Hydrophobic
Oxalate	5	Chelate	0.97	Complexation
Malonate	6	Chelate	1.05	Hydrophobic
Maleate	7	Chelate	1.40	Hydrophobic
Succinate	7	Chelate/ crosslinking	1.85	Hydrophobic/ molecular size
Phthalate	7	Chelate	1.93	Hydrophobic
Adipate	11	Crosslinking	2.03	Hydrophobic/ molecular size

The technology of masking must be reconsidered, because there is much misunderstanding regarding its effects on chrome tanning. Introducing a ligand into the chrome complex can be considered to have two competing outcomes: reduction in reactivity due to the occupation of complexation reactions sites, but alternatively the reactivity of the masked chrome species can be enhanced, because the complex becomes more hydrophobic by the introduction of the carboxylate into the ligand sphere. It is typically the case that all carboxylates are more hydrophobic than an aquo ligand – at least those commonly used in the art. An exception might be sugar-reduced chrome salt, but that is a different tanning technology, where conventional masking technology is not relied upon.

The effects of masking with formate and other masking salts, which can react in different ways, are shown in Table 11.10. Here the masking ratio is 1 mol mol^{-1} Cr_2O_3: the effect of the number of equivalents of carboxyl per mole should be taken into account. In each case, the effect of the masking on the reactivity of the chrome species is designated, based on the known chemistry of the masking agent and the observation of the relative extent of reaction.

The designation 'hydrophobic masking' indicates the influence of the masking agent on the properties of the complex and hence the element of the reaction mechanism that is affected: increasing the hydrophobicity of the complex encourages transfer from aqueous solution.

The designation 'complexation masking' indicates that the net effect is to modify the ability of the complex to react further, *i.e.* this conforms more to the conventional view of masking. Comparing the effects of formate and oxalate masking in Figure 11.21, the notional

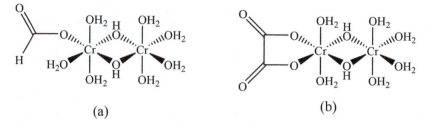

(a)   (b)

Figure 11.21   Masking with (left) formate and (right) oxalate.

structures indicate that formate is likely to be less able to interact with water than an aquo ligand, thereby conferring a degree of hydrophobicity. On the other hand, oxalate has the potential for effective hydrogen bonding, because it offers only carbonyl groups for interaction with the solvent, so the net effect is a reduction in the number of available reaction sites and hence reduced affinity for collagen carboxyls, driven by the chelating mechanism.

Table 11.10 illustrates the following important points.

1. If it is assumed that chrome content in pelt is a reflection of the overall reaction rate or reactivity, there is no indication that masking with formate at this level reduces the reactivity of the chrome salt; indeed, from Table 11.10, there is an increase in reactivity. This runs contrary to conventional thinking about the masking effect of formate. It must be recognised that the relative effects of masking are dependent on the masking ratio. Those apparently contradictory effects are the reduction in reactivity by using up reaction sites *versus* the increase in reactivity by increasing the hydrophobicity. Taking the example of formate, as the simplest and commonest masking agent, at higher masking ratios, $[RCO_2]:[Cr_2O_3] > 1$, the reactivity-reducing effect will begin to dominate, typically resulting in a reduction in tanning efficiency.[51] At lower masking ratios, which is more typically related to industrial practice (the masking ratio of 1 mol of formate per mole of chrome oxide corresponds to a formic acid offer of 0.6% if the chrome offer is 2% Cr_2O_3), the outcome is an enhancement of chrome reactivity.

The effect of incorporating formate has the same hydrophobic effect as masking with malonate; the latter has a more predictable impact on the complex properties, because of the presence of the methylene group, but the outcome is similar for formate and malonate masking.

2. In recognising the role of masking ratio, the combination of the chemical outcome with the kinetics of the process must also be taken into account. Unless the chrome is masked by conducting the reaction separately, prior to starting the tanning process, the masking ratio will necessarily be low at the beginning of the chrome tanning reaction and then increase as the tanning reaction proceeds; the progress and change of masking ratio will depend on the chrome fixation kinetics, controlled by the pH profile, and on the offers of chrome and masking agent.

The influence of masking ratio on chrome reactivity is modelled in Figure 11.22, where the ideal masking situation is presented in terms of reduced reactivity at the beginning of the reaction, gradually increasing by the end of the reaction. It is clear that this bears no resemblance to any real situation. The assumed effect is to reduce the reactivity by eliminating reaction sites, but the actual effect is to include some enhancement of reactivity by increasing the hydrophobicity or reducing the hydrophilicity of the ion. In this way, the actual effect of masking is directly opposite to the effect required. If more control is required, then the chrome must be premasked by prior complexation and the effect will be the same as the assumed effect presented in Figure 11.22, in terms of the effect of masking ratio: at low masking ratios the chrome will be made more reactive and only at unusually high ratios will the classical masking effect come into play. In Figure 11.22, the effect of premasking is modelled as a function of

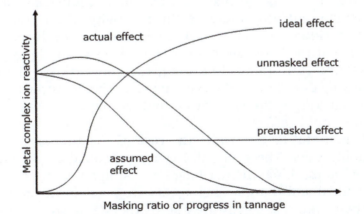

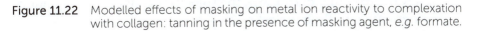

Figure 11.22   Modelled effects of masking on metal ion reactivity to complexation with collagen: tanning in the presence of masking agent, *e.g.* formate.

progress in tannage: the reactivity of the chrome complex is fixed by the masking ratio, so the reactivity will not change; the position of the premasked line on the reactivity axis will depend on the masking ratio and therefore may even lie above the value of the unmasked chrome, *i.e.* low ratios of premasking can produce chrome salt of enhanced reactivity.

The observation that low masking ratios will usually increase rather than decrease chrome astringency is interesting, because it means that the impact of the hydrophobicity of the complex can be more powerful than the effect of a reduction in charge. In other words, it is confirmation of the stepwise mechanism of reaction, where transfer from solution precedes electrostatic interaction of solute with substrate.

The premasked effect is assumed here to be constant, if there is no additional masking agent available. What is predictable is that the reactivity will not decrease by reversal of the masking reaction as the tannage proceeds; the contrary is another unaccountable assumption by some tanners. The stability of chrome-tanned collagen is well known and applies equally to masking complexation.

In Figure 11.21, the relative effects of the two contributing parameters of hydrophobicity and reaction site occupation will depend strongly on the chemistry of the masking agent. The role of masking in practical tanning depends on the tanner understanding the nature of the actual effect, even if only qualitatively.

3. From the analysis of typical tanning conditions presented above, it is likely that conventional, industrial formate masking will only increase the reactivity of chrome, since the masking ratio is unlikely to exceed the condition represented at the beginning of the 'actual' curve in Figure 11.21. Therefore, in practical tanning, the conventional notion of masking does not apply: even in premasked chrome tannage, the masking ratio will not be high enough to overcome the hydrophobic effect.

It is useful to reconsider the case of sugar-reduced chrome tanning salts. The properties can be rationalised in terms of classical masking: the masking ratio is high (assumed, but not specified) and the masking agents are breakdown products of carbohydrate, *i.e.* polyhydroxy polycarboxylates, so they are relatively hydrophilic. In this way, the salts have reduced astringency, but enhanced filling power.

4. The role of masking as a general aspect of chrome tanning is an important feature that can be exploited technologically. The use of specific masking agents, which gradually increase the astringency of the chrome species, by increasing their tendency to transfer from solution to substrate, is an aspect of the reaction that has not attracted scientific attention. Here, the requirement is to match the rate of diminishing concentration of chrome in solution with the rate of masking complexation, to maintain or increase the rate of chrome uptake. Since the complexation interaction cannot easily be reversed, changing the reactivity in the way indicated by the ideal masking trace in Figure 11.21 would require changes within the masking ligand, altering the ligand from hydrophilic to hydrophobic within the timescale of conventional tannage and under the restrictions of the typical chrome tanning conditions. Potential exploitable commercial options are not difficult to imagine, such as the following:
   (i) hydrolysis of esters;
  (ii) reaction at keto groups;
 (iii) aldehydic reactions;
 (iv) solvent modification (see later).

11.11.1 Polycarboxylates

The use of dicarboxylates can lead to different kinds of masking: chelation, when the dicarboxylate compound reacts with a single metal ion, or crosslinking, when the dicarboxylate reacts with two metal ions, illustrated in Figure 11.23. The reaction adopted depends on the size of the ring formed if the salt can act as a chelate.

The relationship between ring size and chelate or chain-forming masking is set out in Table 11.9. The formation of the ring type of complex is dependent on the steric distortion in the ring and the loss of entropy that results from the two ends of the chelating molecule being constrained on the same atom. The loss is allowable only if the ring is five- or six-membered: smaller rings are enthalpically highly strained, so less favoured, whereas larger rings lose too much entropy and are not formed in favour of a crosslinking structure, which can adopt any conformation. Therefore, for ring sizes of 5–7 members chelation is generally favoured, but for larger ring sizes crosslinking is generally preferred.[52]

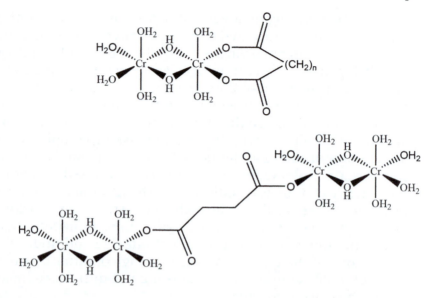

Figure 11.23   Reactions of dicarboxylate masking agents: (top) chelation and (bottom) crosslinking.

Table 11.11   Effects of bidentate masking agents on chromium(III) and aluminium(III) tanning.

Masking salt	Relative chrome uptake	T_s of Al(III)-tanned collagen/°C
None (sulfate)	1.00	60 (estimated)
Fumarate	—	65
Maleate	1.40	80
Phthalate	1.93	92

Although crosslinking increases the size of the molecular ion, the number of reaction sites is effectively not increased. In the case of chelation, the reactivity of the species is significantly reduced, in comparison with monodentate masking, since the model in Figure 11.23 indicates blocking of potential crosslinking power. Some effects of masking are presented in Table 11.11, as the relative chrome uptake[51] and the shrinkage temperature of aluminium-tanned collagen, using decomposed zeolite.[53]

Table 11.11 introduces an interesting example of commercial masking technology, the use of phthalate, benzene-1,2-dicarboxylate. The use of this salt is well known, typically applied late in the process, to increase the reactivity of the chrome. It is widely assumed that the reactivity increase stems from crosslinking and Vernali and Amado

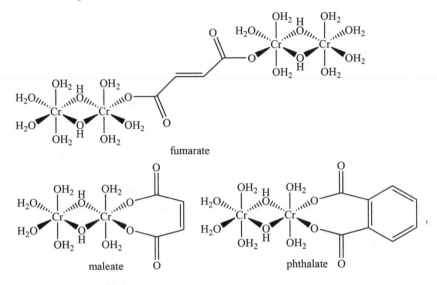

Figure 11.24 Bidentate masking of chromium(III).

calculated the energetics of crosslinking by molecular modelling.[54] However, crosslinking is sterically less likely, so the masking reaction is more likely to be chelation, like the reaction with maleate, creating a seven-membered ring, because the carboxyls are sterically fixed: here, only fumarate is sterically capable of crosslinking. If that analysis is correct, the effect of masking must be to alter the nature of the interaction with the solvent, due to the hydrophilic/hydrophobic nature of the solute. From considerations of the structures of the masking salts, the relative effects on the hydrophobicity of the salts can be understood, as illustrated in Figure 11.24.

The extent of increasing the reactivity of the masked chrome is predictable, based on the evident chemical nature of the masking ligand, *i.e.* the CH=CH π-bond of the maleate can interact with water more than the saturated linked methylene groups of succinate, which are in turn less hydrophobic than the large benzene ring of phthalate. The relative effects of the ligands are based on their interaction with the solvent. This is clearly a mechanism for manipulating chrome reactivity that can be exploited. The relative effects of the chelating and crosslinking reactions can be compared as shown in Table 11.12.

Crosslinking masking depends on the calculated chelate ring size exceeding seven: for aliphatic dicarboxylates, it starts with glutarate and the effect of the four methylene groups in the chain of adipate is

Table 11.12   Comparison of chelation and chain-forming masking reactions.

Chelate	Chain
No change in size of chrome species	Polymerisation of chrome species: more surface reactive, not more active towards complex formation
Strong masking effect	Weaker masking effect
Less effect on solubility of complex	Complexes less soluble: more surface reactive

clear in the reactivity illustrated in Table 11.10. Therefore, all cross-linking masking agents are likely to increase the reactivity of chrome; the effect is not concerned with the inherent reactivity towards complex formation, but is dependent on the enhancement of fixation rate by polymerisation.

The relative affinities of potential masking agents for reacting in the chrome complex sphere are as follows:

hydroxide > oxalate > citrate > lactate > malonate > maleate > phthalate > glycolate > tartrate > succinate > adipate > acetate > formate > sulfite > sulfate > chloride > nitrate > chlorate

This list is not exhaustive, but in practical terms only formate and phthalate are conventionally important.

There is another effect of masking, which is to change the colour of the chromium complex, discussed in Chapter 9. As indicated in Table 9.4, the colour of the complex and hence the colour of the tanned leather depend on chrome complexation. If chrome tanning is conducted in the presence of only sulfate and chloride as counterions, the tannage is unmasked. Therefore, the only complexes formed are with collagen carboxyls; the result is wet blue that has a distinctly green cast. If formate is present, whether truly acting as a masking agent or merely present in the system, the resultant wet blue is bright pale blue.

11.11.2   Other Chemistries

For many years, the supply sector has tried to exploit the availability of lignosulfate, a 1 million tonne per annum by-product of the paper industry, as the basis of syntans, but with little success. In more recent research, Shi and co-workers[55] used chromic acid to oxidise lignosulfonate to create a masked chromium(III) product: using a lignosulfonate-to-dichromate molar ratio of >0.5 and then spray drying, the product was apparently free from chromium(VI). The product was reported to make the leather more full, but there were no other advantages.

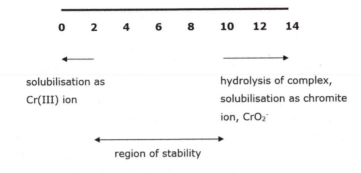

Figure 11.25 Guideline limits of the stability of chrome-tanned leather over the pH scale.

11.12 Stability of Chrome-tanned Leather

The range and limits of stability of chrome-tanned leather over the pH scale can be illustrated as shown in Figure 11.25.

The chromium complex can be broken by the presence of other complexing agents. The following conditions are combined in the accelerated artificial perspiration test, capable of stripping all chrome out of leather overnight:

- high concentration of an alternative complexing ligand, *e.g.* 8% sodium lactate;
- elevated pH, at least pH 8, to assist in the hydrolysis reaction;
- elevated temperature, *e.g.* 50 °C, to accelerate the rate of reaction.

11.13 Role of Sulfate in the Chrome Tanning Mechanism

In the literature, it is suggested that sulfate participates in the chrome tanning process because it is included in the reactive chromium(III) crosslinking species. The frequently presented model of the chrome tanning crosslink has sulfate acting as a bidentate ligand in a dichromium species, as shown in Figure 11.6. Conventional thinking in the latter half of the twentieth century had the sulfate acting in the following roles:

- as a counterion (or gegenion) for cationic Cr(III);
- as a counterion for protonated amino groups on collagen;
- as a ligand in the Cr(III) complex.

This was the model used by Gustavson[24,25] in his analysis of the chrome tanning reaction, in which wet chemistry was used to distinguish between those different roles of sulfate. From an analysis of wet blue, he concluded that the dominant chrome species was of the form indicated in Figure 11.6, a dichromium complex bridged by sulfate. The important conclusion from his chemical argument is that only 10% of the bound chrome is acting as a crosslinker.

This argument has been challenged in the following way.[56] The simple association of the anion with the cationic centres in the collagen would not appear to offer a contribution to the stabilising effect. It is also unlikely that a complexation reaction would be part of the stabilising effect; indeed, because sulfate is a weak ligand, it might be expected to be hydrolysed out of the complex rapidly, although the studies in solution tend to disagree.[7] Gustavson[57] found that adding excess sulfate to chrome tannage maximised the shrinkage temperature, indicating that it does play some role in the tanning mechanism, and it has been shown that sulfate does exchange between the coordinate chrome and the solution in tanning.[58] On the other hand, Grace and Spiccia demonstrated that the μ-sulfato dichromium species loses the sulfate during reaction with collagen carboxyl groups.[59]

The relationship between sulfate and the chrome complexes bound to collagen was directly addressed in the first published study of the species using extended X-ray absorption fine structure (EXAFS) analysis by Covington *et al.*[56] This advanced analytical technique is based on the absorption of a spectrum of X-rays from a synchrotron source by the target atom, in this case chromium, also described in a similar study by Reich *et al.*[60] The way in which energy quanta are reflected within the molecule is seen as the absorption data, shown in Figure 11.26.

The analysis depends on predicting the structure of the molecule, in terms of the nearest neighbours and next-nearest neighbours and so on: a computer program generates an absorption spectrum, which is compared with the acquired data and coincidence confirms the predicted structure. The EXAFS study yielded data for the ratios of neighbouring atoms and the experimental results can be compared with the theoretical values for numbers of nearest neighbours, given in Tables 11.13 and 11.14.

Comparison of the calculated ratios from the non-linear structures with the linear structure, given in Figure 11.5, supports the linear nature of the average chrome species bound to collagen and confirms that the average molecular ion contains four chromium atoms.

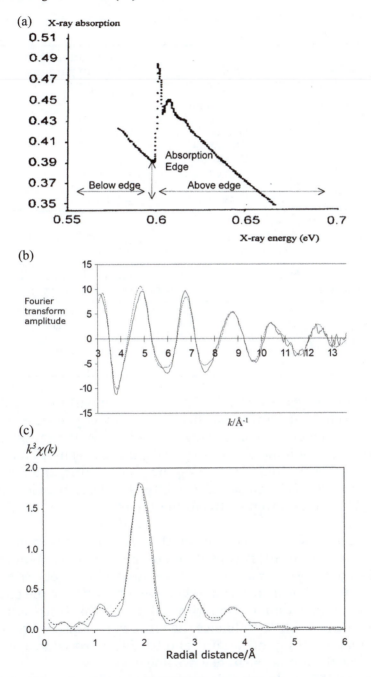

Figure 11.26 Basis of chromium(III) EXAFS in tanned collagen: experimental data are solid lines, best-fit lines are broken lines. (a) The K-edge absorption of X-rays by chromium in leather. (b) Cr K-edge spectrum for hide powder tanned with 33% basic chromium(III) sulfate, basified to pH 4.0. (c) The corresponding Fourier transform, corrected with the phase shift of the first shell. Reproduced from ref. 56 with permission from Elsevier, Copyright 2001.

Table 11.13   EXAFS data for nearest neighbours in chrome species bound to collagen.

Basicity/%	Shell	Coordination number (error in parentheses)
33	Oxygen	6.29 (0.14)
	Chromium	1.45 (0.30)
	Oxygen	7.43 (1.37)
42	Oxygen	6.46 (0.18)
	Chromium	1.42 (0.34)
	Oxygen	5.55 (1.46)
50	Oxygen	6.54 (0.16)
	Chromium	1.45 (0.32)
	Oxygen	5.85 (1.46)

Table 11.14   Calculated values for nearest neighbours in linear and non-linear chrome species (structures given in Figure 11.5).

Chrome species	No. of first-shell oxygens	No. of second-shell chromiums	No. of third-shell oxygens
Non-linear Cr_3 (II)	6.0	2.0	4.0
Non-linear Cr_4 (III)	6.0	2.5	3.25
Linear Cr_4 (I)	6.0	1.5	6.0
Linear Cr_4 (I) measured	6.5	1.4	6.2

It has been suggested that sulfate ion plays a part in the polymerisation mechanism during basification.[56] This is plausible because the transient bidentate complexing effect of sulfate ion at the *cis* positions in the chromium(III) species (see Figure 11.3) causes labilisation of all the other positions, including the *cis* positions, and facilitates exchange reactions by an associative interchange, I_a, reaction mechanism.[61] This would promote the formation of non-linear species (see Figure 11.27).

These data lead to the conclusion that the polymerisation mechanism for chromium(III) may be different in solution from that *in situ* on collagen, the former favouring three-dimensional species and the latter favouring linear species and hence contributing to the effectiveness of the primary interaction with collagen. If this truly is the case, it should be possible to influence the formation of basic chrome species by manipulating the tanning and basifying procedures, in order to optimise the formation of reactive linear species.

In the EXAFS analysis of chrome-tanned collagen, when a sulfur atom was introduced into the complex it produced a theoretical spectrum totally unlike the experimental data. Therefore, it was concluded that sulfate was not detectably close to the chromium atoms in the ionic complexes. This means that the structure of the tanning species

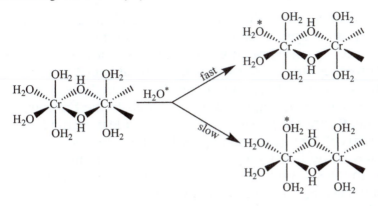

Figure 11.27 Kinetics of water exchange at chromium(III).

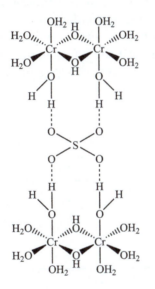

Figure 11.28 Possible interaction between chromium(III) species and sulfate, which may involve more water molecules between the aquo ligand and the sulfate ion.

is likely to be as illustrated in Figure 11.28, in which the sulfate ion can link chrome species *via* the aquo ligands, but does not approach closer.

11.14 Role of the Counterion in Chrome Tanning

In the conventional understanding of the typical range of tanning chemistries, it is frequently asserted that chromium tanning is unique, because it is the only solo tannage capable of raising the shrinkage temperature of collagen above 100 °C. However, if we examine some observations of the effects of chromium(III) salts in this regard, that assertion must be modified. If collagen is tanned with basic chromium(III) chloride[62] or perchlorate[63] in the conventional way, the shrinkage temperature of the leather reaches only 80–85 °C, illustrated in Table 11.15.

In the commonest version of chrome tanning, the shrinkage temperature reaches as high as 120 °C, which is the typically expected outcome. Furthermore, rigorous washing of leather made from chromium(III) sulfate in the usual way will reduce the high shrinkage temperature to the lower range.[64] In each case, simply adding sodium sulfate to the solution restores the expected high shrinkage temperature; interestingly, orthophosphate does not have the same effect – indeed, it actually destabilises the tannage.

DasGupta[65] demonstrated that the practical limit to the shrinkage temperature of chrome-tanned collagen can be reached by including polycarboxylate in the process, as shown in Figure 11.29. In these cases, the anions can interact with the aquo ligands of the chromium atoms, but in a strong way since each carboxyl groups carries a whole negative charge.

Table 11.15 Effect of anions on the shrinkage temperature of leather tanned with basic chromium(III) perchlorate (T_s = 80 °C).

Anion	$\Delta T_s/°C$
Sulfate	32
Formate	24
Acetate	16
Oxalate	30
Maleate	32
Fumarate	37

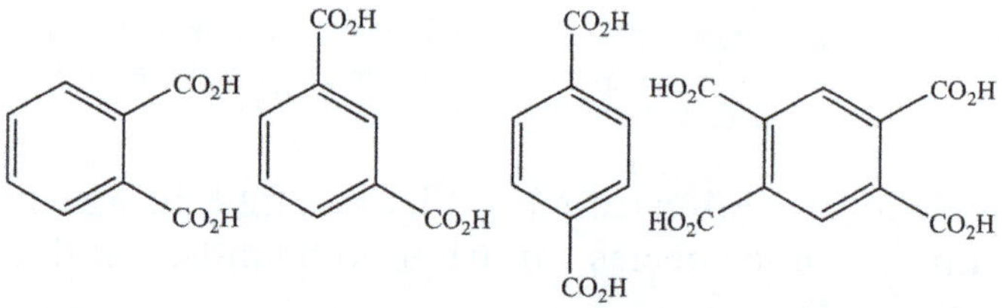

	Phthalate	Isophthalate	Terephthalate	Pyromellitate
Treated	115	121	121	129
Rechromed	121	132	129	133

Figure 11.29 Shrinkage temperature of wet blue sheepskin treated with polycarboxylate for 1 h, then rechromed.

Therefore, it can be concluded that the chromium(III) molecular ions themselves have only a moderate effect on the shrinkage temperature. This observation means that chromium(III) is not unique, but apparently conforms to the pattern of performance of other mineral tannages, which typically confer limited hydrothermal stability. However, unlike the other mineral tannages, the introduction of the weak reaction of sulfate ion and other anions produces a synergistic effect on the shrinkage temperature, giving rise to the familiar high hydrothermal stability. Therefore, the conventional chrome tanning effect should not be regarded as a solo tanning reaction: it is, in fact, a synergistic combination of the effects of the chromium(III) molecular ions and their counterions. The conclusion raises the possibility of improving the effect of chrome tanning by manipulating the type and amount of reactive counterions. In solution chemistry, sulfate acts as a structure maker and it has a positive effect on the stability of raw collagen, in a way that might be interpreted as a weak tanning effect.[66] It was shown that a component of the stability of raw and processed collagen is the presence of a sheath of supramolecular water surrounding the triple helix, centred on the hydroxyproline.[67] It was therefore postulated that the role of sulfate in stabilising collagen and also in chrome tanning is to contribute to the stabilising effect of the supramolecular water. This may work by direct incorporation within that layer or by acting as a link between the solvent layer and the collagen or the chromium, or by all three mechanisms.[68] This view of tanning is further developed in the discussion of tanning theory in Chapter 23.

11.15 Role of the Solvent

The mechanism of reaction between a reagent and a substrate was introduced earlier. There, the primary interest was in the interaction between the reagent and the aqueous solvent; however, discussion of the role of the solvent can be widened, to include all modes of delivering reagents to the substrate. This is developed in Chapter 19.

11.16 Role of Ethanolamine

Rohm and Haas Co. supplied ethanolamine hydrochloride as an auxiliary agent for pickling, to improve chrome tanning efficiency.[69] In this way, the reagent is applied as a solid salt: as the pH increases, the amine is neutralised and the medium is converted to a mixed aqueous and organic solvent. However, the effect was also observed using ethylene glycol and ethylenediamine, with only marginal superiority with ethanolamine in terms of additional hydrothermal stability.[48] Therefore, a more likely rationalisation is that the effect is based on the changes to the solvent system. The controlling parameter for the solute–solvent interaction is the ability of the solvent to solvate. For hydration reactions, a parameter that reflects the electrostatic nature of the solvent is the dielectric constant: the values for water, ethylene glycol, ethanolamine and ethylenediamine are 78, 38, 34 and 14, respectively. Therefore, the addition of even small amounts of these solvents will markedly change the properties of water, reducing the solvating effect of the primarily aqueous mixed solvent. Indeed, it was observed that an acceleration is obtained when the offer of organic solvent was as little as 3×10^{-4} mole fraction, corresponding to 1% w/v.[48] Many other solvents also exhibit this effect. However, it is not known to what extent solvation itself influences the effect. For example, dimethyl sulfoxide (DMSO) in the aqueous float markedly increases the chrome penetration rate,[46] but here it is known that in such solvent mixtures, surprisingly, the chromium(III) ions are preferentially solvated by DMSO.[70]

This is a general principle: a change in the solvating power of the reaction solvent will alter the affinity of any solute for the solvent. Examples are known of reactions in the leather-making process which already exploit this effect, although often without recognition of the scientific principles involved. There is clearly scope for further exploitation, whether to accelerate solute uptake onto the substrate, by reducing the solvating energy, or to promote penetration through the substrate, by increasing the solvating energy.

The concept of atom economy (AE) has been introduced as a way of comparing chemical processes in analysing the ecological impact or sustainability. It is defined as follows:

$$AE = \frac{\text{mol. wt of reaction product}}{\sum_{i}(\text{mol. wt of reactants})_i}$$

that is, the molecular weight of the reaction of interest divided by the sum of the molecular weights of all the reactants, expressed as a percentage.

For chrome tanning, the AE is 98% for the simple reaction, but if chloride, sulfate and water are included, the AE falls to 30%. In a comparison of options for chrome tanning, the order of AE is no pickle > ethanolamine assisted > liquor recycling > conventional.[71]

11.17 Nature and State of the Substrate

11.17.1 Modifying the Substrate

It has been shown that the reaction rate between chrome and collagen depends on the availability of reaction sites on the collagen, at least at the beginning of the process, so a simple method of enhancing the chrome tanning reaction, in terms of the fixation efficiency, is to increase the availability of reaction sites. Although that can be achieved by increasing the pH, to change the degree of ionisation of the natural carboxylic acid groups there is a limit to the use of this approach because of the effect on the chrome species. The alternative is to introduce more carboxyl groups into the collagen. One way is to hydrolyse more sidechain amide groups in liming; however, extending the liming period is likely to be counterproductive with regard to the physical properties of the leather. The use of amidases in this context might be a future practical biotechnology. The remaining option is to attach carboxyl-bearing groups to the collagen, usually at the amino groups of lysine. The reaction is exemplified by the use of glycine attached *via* formaldehyde, shown in the following reaction; this is a technology developed by Joan Bowes[72] that has been used as an industrial process in the UK:

$$Coll-NH_2 + HCHO + H_2NCH_2CO_2H \rightleftharpoons Coll-NH-CH_2-HNCH_2CO_2H$$

This principle was adapted and extended by Feairheller and coworkers to include the Mannich[73] and Michael[74] reactions, illustrated as follows, respectively:

$$P-NH_2 + HCHO + CH(CO_2H)_2 \rightarrow P-NH-CH_2-CH(CO_2H)_2$$
$$\text{malonic acid}$$

$$P-NH_2 + CH_2 = CH-CO_2CH_2CH_2CO_2{}^- \rightarrow P-NH-CH_2CH_2CO_2CH_2CH_2CO_2H$$
$$\text{β-carboxyethyl acrylate}$$
$$\rightarrow P-N-(CH_2CH_2CO_2CH_2CH_2CO_2H)_2$$

Holmes[75] demonstrated the technology of improving complex metal ion fixation by linking carboxylate moieties to collagen *via* halogenated azo heterocycles, in reactions analogous to reactive dyeing (see Chapter 16). This modification to collagen has more important application to tannages weaker than chromium(III), discussed in Chapter 12. More recently, Indian workers[76] reported the application of an acrylate polymer for increasing chrome tanning efficiency, shown in Figure 11.30, where the gains were marginal.

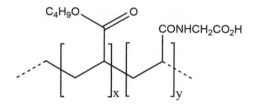

Figure 11.30 Chrome tanning auxiliary.

Since it has been shown that in unmodified collagen reaction sites are in excess for the usual levels of chrome fixation, the requirement for more is doubtful. There will be an effect of rate increases at lower pH values, at the beginning of tanning, but that is a diminishing effect and, anyway, there are simpler ways of influencing the kinetics of fixation.

11.18   Chromium(VI) and Leather

Since the turn of the century, increasing attention has been paid to the possibility of the presence of chromium(VI) in leather. It is also worth noting that there is continuing controversy regarding the validity of the test for Cr(VI), especially as a contaminant formed in chrome leather by ambient oxidation of Cr(III).[77] Criticisms of methodology are based on the possibility of obtaining false-positive results from the wet chemical procedures involved. However, until the validity of analyses of low concentrations of chromium(VI) in the presence of high concentrations of chromium(III) is unequivocally confirmed, the industry has to accept the application of the accepted standard method of analysis.

However, as Moretto pointed out,[78] even though the toxic properties of chromium(VI) are well known, it would be expected that its presence in leather in contact with skin would give rise to widespread reaction in the form of irritation and allergic contact dermatitis. However, it seems that the problem arises only with people who are already sensitised to such allergens: '10% of chromium-allergic patients will react to 7–44 ppm Cr(VI) when patch tested'.[79] Moretto concludes:[78]

> ... *there is no evidence that skin reactions in all those (sensitised patients) are due to the Cr(vi) and/or Cr(vi) plus Cr(iii) released from leather. ... this issue involves a few subjects in the whole population and appears to be of minor relevance if the leather industry avoids the use of Cr(vi) and properly uses Cr(iii) for tanning.*

Tegtmeyer and Kleban[80] reviewed the formation of chromium(VI) in the context of chrome-tanned leather. Under normal conditions, *i.e.* pH 3.5–5.0, extractable Cr(III) 50–500 ppm, < 100 °C, the ratio of Cr(III) to Cr(VI) is estimated to be higher than $10^4:1$, corresponding to < 2 ppm Cr(VI), *i.e.* below the assumed detection limit of 3 ppm. Furthermore, if it is assumed that the extractable chrome is the fraction available for the equilibrium between Cr(III) and Cr(VI), then the Cr(VI) content of the leather would be estimated as 0.005–0.05 ppm. In the absence of reagents that can contribute to the oxidation of Cr(III), such as unsaturated oils, leather is a reducing environment. Tegtmeyer and Kleban[80] offered some rules for preventing Cr(VI) formation during processing.

1. Use premium chromium(III) tanning agents.
2. No oxidising agents to be used, *e.g.* bleach after tanning.
3. Finish processing at pH 3.5–4.0.
4. Finish with a final wash.
5. Avoid excess ammonia prior to dyeing.
6. Use high-performance fatliquors – no unsaturated lipids or waxes.
7. Avoid chromate pigments – inorganic, yellow and orange.
8. 1–3% vegetable tannin can provide antioxidant protection.
9. Alternatively, use synthetic antioxidants

It is widely accepted that the availability of oxidisable lubricating oils, through the intermediates of epoxides, can lead to the presence of Cr(VI),[81] although the mechanism is actually at the site of the hydrogen α to the double bond where the epoxide forms.[82] The reaction is based on the Fenton redox mechanisms:

$$Fe^{2+} + H_2O_2 \rightleftharpoons Fe^{3+} + OH + OH^- \quad k_1 = 76 \text{ L mol}^{-1} \text{ s}^{-1}$$

$$F^{3+} + H_2O_2 \rightleftharpoons Fe^{2+} + OOH + H^+ \quad k_2 = 0.01 \text{ L mol}^{-1} \text{ s}^{-1}$$

$$Fe^{2+} + OH \rightleftharpoons Fe^{3+} + OH^- \quad k_3 = 3.0 \times 10^8 \text{ L mol}^{-1} \text{ s}^{-1}$$

The mechanism of oxidation is a chain reaction, as follows, where R represents the unsaturated lipid and an accompanying H is the hydrogen α to the double bond:

Chain initiation:

$$RH + X^\cdot \rightleftharpoons R^\cdot + XH$$

Propagation:

$$R^{\bullet} + O_2 \rightleftharpoons ROO^{\bullet}$$
$$ROO^{\bullet} + RH \rightleftharpoons ROOH + R^{\bullet}$$
$$ROOH \rightleftharpoons RO^{\bullet} + OH^{-}$$
$$2ROOH \rightleftharpoons ROO^{\bullet} + RO^{\bullet} + H_2O$$
$$RO^{\bullet} + RH \rightleftharpoons ROH + R^{\bullet}$$
$$RH + OH^{\bullet} \rightleftharpoons R^{\bullet} + H_2O$$

Chain stop:

$$ROO^{\bullet} + ROO^{\bullet} \rightleftharpoons ROOR + O_2$$
$$ROO^{\bullet} + R^{\bullet} \rightleftharpoons ROOR$$
$$R^{\bullet} + R^{\bullet} \rightleftharpoons R-R$$

Benzoic acid can act as a free-radical scavenger: 3% on the weight of chrome can reduce any chromium(VI) content by 99% in 90 min.[83] The more complex aromatic structures of vegetable tannins are better known as reducing agents; the antioxidant properties are variable, as demonstrated by Ozgunay and co-workers.[84] Using two different analytical methods, the reduction of ferric tripyridinyl-triazine or the inhibition of chromogen radical cation, the average effect can be calculated, also correct for tannin content, as shown in Table 11.16.

The results are consistent in terms of tannin type, insofar as they exhibit a pattern of reactivity in contrast to their lightfastness properties: tara is very lightfast and mimosa is known to redden readily in light. Although the effect of light on polyphenols is characterised as free-radical oxidative changes, the results show that the more immediate chemical oxidative reactions do not necessarily correlate. The difference in effectiveness at stopping the free-radical reactions also indicates that there are subtle effects of the details of the chemical structures, *i.e.* not all hydrolysable tannins are the same in this context.

Table 11.16   Mean relative antioxidant effect of plant polyphenols on chromium(VI), derived from Ozgunay and co-workers.[84]

Tannin	Tannin type	Relative antioxidant effect
Tara	Hydrolysable	100
Sumac	Hydrolysable	87
Valonia	Hydrolysable	76
Chestnut	Hydrolysable	63
Quebracho	Condensed (catechol)	59
Mimosa	Condensed (pyrogallol)	57

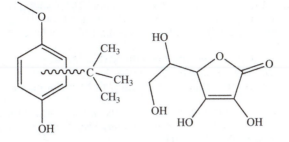

Figure 11.31   Components of an antioxidant mixture.

Ogata *et al.*[85] demonstrated that conversion of chromium(III) to chromium(VI) can occur during ageing for 24 h at 80 °C. Storing the treated leathers at 20 °C for 6–12 months at different relative humidities caused the Cr(VI) content to diminish, with faster conversion at higher moisture content; the mechanism is thought to be due to the water providing the redox reaction field. Experiments with a three-component antioxidant mixture (Figure 11.31) consisting of 2- or 3-*tert*-butyl-4-hydroxyanisole plus ascorbic acid (vitamin C) plus a collagen peptide with molecular weight of 9400 Da in a molar ratio of 1:0.01:0.001 resulted in complete inhibition of chromium(VI) formation.

The global leather industry contends that chromium(III) salts are non-toxic and therefore they have little environmental impact, particularly because the complexes cannot pass through cell walls and interfere with the DNA in the nucleus. Indeed, the world has had a century of experience of people wearing chrome-tanned leather next to the skin, with incidences of allergic response, *e.g.* chrome eczema, being rare. However, the industry does agree with the regulatory authorities that chromium(VI) is toxic, carcinogenic and teratogenic, because its compounds can pass through the cell wall, whereupon it is reduced and able to react with DNA. Hence the validity of testing is paramount, but currently under scrutiny.

References

1. F. Knapp, *Nature und Wesen der Gerberei und des Leders*, ed. J. G. Cotta, Jena, 1858.
2. A. Schultz, *US Pat.*, 291,784, 1884.
3. R. S. Thomson, *J. Soc. Leather Technol. Chem.*, 1985, **69**(4), 93.
4. J. Geher-Glucklich and M. T. Beck, *Acta Chim. Acad. Sci. Hung.*, 1971, **70**(3), 235.
5. E. E. Ochs, *et al.*, *J. Am. Leather Chem. Assoc.*, 1953, **48**(12), 730.
6. N. Bjerrum, *Z. Phys. Chem.*, 1910, **73**, 724.

7.   H. M. N. H. Irving, *J. Soc. Leather Technol. Chem.*, 1974, **58**(3), 51.
8.   S. G. Shuttleworth, *J. Soc. Leather Trades' Chem.*, 1954, **38**, 419.
9.   L. A. Sillel, Chemical Society, *Stability Constants of Metal Ion Complexes*, Chemical Society, 1971.
10. O. Mitsuo, *et al.*, *Bull. Chem. Soc. Jpn.*, 1964, **37**, 692.
11. K. Nagata, *et al.*, *Bull. Chem. Soc. Jpn.*, 1965, **38**, 1059.
12. P. H. Tedesco, V. B. de Rumi, and J. A. Gonzalez Quintana, *J. Inorg. Nucl. Chem.*, 1973, **35**(1), 285.
13. H. M. N. H. Irving and R. J. P. Williams, *J. Chem. Soc.*, 1953, 3192.
14. R. P. Bell and M. A. D. Fluendy, *Trans. Faraday Soc.*, 1963, **59**, 1623.
15. L. E. Orgel, *Nature*, 1960, **187**, 504.
16. H. Stunzi and W. Marty, *Inorg. Chem.*, 1983, **22**, 2145.
17. H. Stunzi, W. Marty and F. P. Rotzinger, *Inorg. Chem.*, 1984, **23**, 2160.
18. L. Spiccia, W. Marty and R. Giovanoli, *Inorg. Chem.*, 1988, **27**, 2660.
19. D. A. House, *Adv. Inorg. Chem.*, 1997, **44**, 341.
20. J. Raghava, *et al.*, *J. Soc. Leather Technol. Chem.*, 1997, **81**(6), 234.
21. T. Gotsis, L. Spiccia and K. C. Montgomery, *J. Soc. Leather Technol. Chem.*, 1992, **76**(6), 195.
22. J. E. Finholt, *et al.*, *Inorg. Chem.*, 1965, **4**, 43.
23. A. Kawamura and K. Wada, *J. Am. Leather Chem. Assoc.*, 1967, **62**(9), 612.
24. K. H. Gustavson, *J. Am. Leather Chem. Assoc.*, 1953, **48**(9), 559.
25. K. H. Gustavson, *J. Soc. Leather Trades' Chem.*, 1962, **46**(2), 46.
26. W. Han, *et al.*, *J. Am. Leather Chem. Assoc.*, 2016, **111**(3), 101.
27. A. D. Covington, R. A. Hancock and I. A. Ioannidis, *J. Soc. Leather Technol. Chem.*, 1989, **73**(1), 1.
28. E. M. Brown, R. L. Dudley and A. R. Elsetinow, *J. Am. Leather Chem. Assoc.*, 1997, **92**(2), 225.
29. A. D. Covington and O. Menderes, *et al.*, *Proceedings of IULTCS Congress*, Capetown, South Africa, 2001.
30. S. G. Shuttleworth and A. E. Russell, *J. Soc. Leather Trades' Chem.*, 1965, **49**(6), 221.
31. R. E. Hamm, *et. al.*, *J. Am. Chem. Soc.*, 1958, **80**, 4469.
32. A. D. Covington, *J. Am. Leather Chem. Assoc.*, 2001, **96**(12), 461.
33. J. Bowes, *et al.*, *J. Soc. Leather Technol. Chem.*, 1947, **31**(7), 236.
34. A. K. Covington and K. E. Newman, *Pure Appl. Chem.*, 1979, **51**, 2041.
35. E. Haslam, *J. Soc. Leather Technol. Chem.*, 1988, **72**(2), 45.
36. R. L. Sykes, *J. Am. Leather Chem. Assoc.*, 1956, **51**(5), 235.
37. A. D. Covington, *J. Am. Leather Chem. Assoc.*, 2008, **103**(1), 7.
38. A. D. Covington and R. D. Allen, *J. Soc. Leather Technol. Chem.*, 1982, **66**(1), 1.
39. D. R. Cooper and R. J. Davidson, *J. Soc. Leather Technol. Chem.*, 1969, **53**(11), 428.
40. S. DasGupta, *J. Soc. Leather Technol. Chem.*, 1998, **82**(1), 15.
41. R. Venba, *et al.*, *J. Am. Leather Chem. Assoc.*, 2008, **103**(12), 401.
42. S. J. Crimp, *et al.*, *Inorg. Chem.*, 1994, **33**, 465.
43. A. D. Covington, *et al.*, *J. Soc. Leather Technol. Chem.*, 1983, **67**(1), 5.
44. A. D. Covington, *J. Am. Leather Chem. Assoc.*, 1991, **86**(10), 376.
45. A. D. Covington, *J. Am. Leather Chem. Assoc.*, 1991, **86**(11), 456.
46. A. D. Covington, *J. Soc. Leather Technol. Chem.*, 1988, **72**(1), 30.
47. R. P. Daniels, *World Leather*, 1994, 7(2), 73.
48. A. D. Covington, unpublished results.
49. J. Gauglhofer, *J. Soc. Leather Technol. Chem.*, 1991, **75**(3), 103.
50. D. L. Huizenga and H. H. Patterson, *Anal. Chim. Acta*, 1986, **206**, 263.
51. H. C. Holland, *J. Int. Leather Trades' Chem.*, 1940, **24**(5), 152.
52. F. A. Cotton and G. Wilkinson, *Advanced Inorganic Chemistry*, John Wiley and Son, 1988.

53. A. D. Covington and N. Costantini, *J. Am. Leather Chem. Assoc.*, 2000, **95**(4), 125.
54. T. Vernali and A. Amado, *J. Am. Leather Chem. Assoc.*, 2002, **97**(5), 167.
55. Z. Peng, *et al.*, *J. Soc. Leather Technol. Chem.*, 2011, **95**(4), 158.
56. A. D. Covington, *et al.*, *Polyhedron*, 2001, **20**, 461.
57. K. H. Gustavson, *J. Amer. Leather Chem. Assoc.*, 1950, **45**(8), 536.
58. A. Kawamura and K. Wada, *J. Am. Leather Chem. Assoc.*, 1967, **62**(9), 612.
59. M. R. Grace and L. Spiccia, *Inorg. Chim. Acta*, 1993, **213**, 103.
60. T. Reich, *et al.*, *J. Am. Leather Chem. Assoc.*, 2001, **96**(4), 133.
61. T. W. Swaddle, *Coord. Chem. Rev.*, 1974, **14**, 217.
62. K. Shirai, *et al.*, *Hikaku Kagaku*, 1975, **21**(3), 128.
63. K. Shirai, K. Takahashi and K. Wada, *Hikaku Kagaku*, 1984, **30**(2), 91.
64. D. E. Peters, *J. Am. Leather Chem. Assoc.*, 1978, **73**(7), 333.
65. S. DasGupta, *J. Soc. Leather Technol. Chem.*, 1979, **63**(4), 69.
66. W. P. Jencks, *Catalysis in Chemistry and Enzymology*, McGraw-Hill, 1969.
67. H. M. Berman, J. Bella and B. Brodsky, *Structure*, 1995, **3**(9), 893.
68. A. D. Covington, *J. Soc. Leather Technol. Chem.*, 2001, **85**(1), 24.
69. W. E. Prentiss and I. V. Prasad, *J. Am. Leather Chem. Assoc.*, 1981, **76**(10), 395.
70. A. D. Covington and A. K. Covington, unpublished results.
71. M. Kumar, *et al.*, *J. Am. Leather Chem. Assoc.*, 2011, **106**(4), 113.
72. J. H. Bowes and R. G. H. Elliott, *J. Am. Leather Chem. Assoc.*, 1962, **57**(8), 374.
73. S. H. Feairheller, M. M. Taylor and E. M. Filachione, *J. Am. Leather Chem. Assoc.*, 1967, **62**(6), 408.
74. S. H. Feairheller, M. M. Taylor and E. H. Harris, *J. Am. Leather Chem. Assoc.*, 1988, **83**(11), 363.
75. J. M. Holmes, *J. Soc. Leather Technol. Chem.*, 1996, **80**(5), 133.
76. J. Kanagaraj, *et al.*, *J. Am. Leather Chem. Assoc.*, 2008, **103**(1), 36.
77. R. M. Cuadros, *et al.*, *J. Soc. Leather Technol. Chem.*, 1999, **83**(6), 300.
78. A. Moretto, *Regul. Toxicol. Pharmacol.*, 2015, **73**, 681.
79. M. B. Hansen, *et al.*, *Contact Dermatitis*, 2002, **47**, 127.
80. D. Tegtmeyer and M. Kleban, *IULTCS, IUR-1*, 2013.
81. K. Nakagawa, *Hikaku Kagaku*, 1999, **45**(3), 210.
82. P. Liu, *et al.*, *J. Soc. Leather Technol. Chem.*, 2007, **91**(3), 116.
83. L. Zhang, *et al.*, *J. Soc. Leather Technol. Chem.*, 2010, **94**(4), 156.
84. C. K. Ozcan, *et al.*, *J. Soc. Leather Technol. Chem.*, 2015, **99**(5), 245.
85. K. Ogata, *et al.*, *J. Soc. Leather Technol. Chem.*, 2018, **102**(2), 53.

12   Mineral Tanning

12.1   Introduction

The recent history of chromium tanning has been centred on its alleged environmental impact. Therefore, attention has turned to the possibility of finding an alternative tanning metal for the global leather industry, ideally to confer the same properties as chromium. The availability of candidates to be alternative mineral tanning agents can be assessed by reviewing the Periodic Table (Figure 12.1), in terms of the known chemistries of the elements and their compounds and their likely reactivity towards collagen.

The following criteria should be applied.

1. Does the reaction work?
   It is assumed that an alternative to chromium(III) tanning should be able to confer high shrinkage temperature to collagen. For this to happen, the metal ion must be able to complex with carboxyl groups, ideally covalently. Using the analogy of chromium(III), it is desirable that the reactions of an alternative should be similar, in order to have a similar outcome. It is implied in Chapter 11 that the reaction may not have to rely on the mechanism of crosslinking, but the defined and understood elements of chrome tanning should be present. Kuntzel and Dröschler[1] concluded that a satisfactory mineral tannage involved aggregation or polymerisation, to allow multiple interactions with collagen, ideally complexation reactions. The role of the counterion was recognised and Gustavson[2] summarised the contribution of the counterion in the following order.

Tanning Chemistry: The Science of Leather 2nd edition
By Anthony D. Covington and William R. Wise
© Anthony D. Covington and William R. Wise 2020
Published by the Royal Society of Chemistry, www.rsc.org

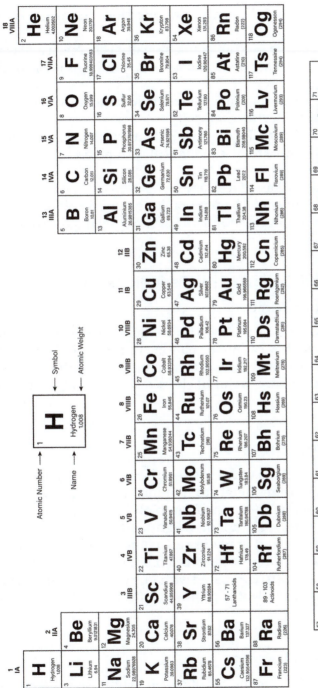

Figure 12.1   Periodic Table of the elements. © Shutterstock.

$$NO_3^- < ClO_4^- < Cl^- < SO_4^{2-}$$

2. Are there toxicity or other hazard implications in the use of the metal salts?
   Could the reagent be used as a bulk industrial chemical and what would be the environmental impact of the industrial use of the element?
3. Is the tanning reagent available in sufficient quantity to be practical for the global leather industry?
   A chemical is industrially useful only if it is available in sufficient quantity for the price to be economical with respect to the value of the final product.
4. Does it produce coloured leather?
   Given a choice, it would be preferable to have a colourless tannage, to assist dyeing, but tanners have managed quite well with a blue substrate.

12.1.1 Blocks and Groups of the Periodic Table

12.1.1.1 The s-Block Elements

These comprise Groups IA to IIA. The elements are metallic, with chemistries dominated by cation formation. They have little complex chemistry; in particular they do not form complexes with carboxylic or amino groups and hence are unlikely to react covalently with protein sidechains.

12.1.1.2 The p-Block Elements

These include Groups IIIA to VIIA and Group 0 or VIIIA. These elements range from metals to semi-metals to non-metals, going towards the higher numbered groups and going from high to low atomic weight. Only the elements with high atomic weight and low Group number have chemistries that could apply in this context. At higher atomic weight the lower oxidation state becomes more stable, *e.g.* Tl(I) compared with Tl(III) and Pb(II) compared with Pb(IV): the oxidation state becomes of importance only if it positively affects the complexation chemistry.

12.1.1.3 The d-Block Elements

These are the transition metals, all of which exhibit some degree of complex formation. Hence they are candidates as tanning agents.

12.1.1.4 The f-Block Elements

The lanthanides have a complex chemistry, hence they might work. The actinides exhibit radioactivity and toxicity or they may even be synthetic. Therefore, they are unlikely to be of use in this context.

In reviewing the properties of the elements in a tanning context, the elements in Row 7, the synthetic or predicted elements, will be excluded to avoid repetition in the text.

12.1.1.5 Group IA: the Alkali Metals – (Hydrogen), Lithium, Sodium, Potassium, Caesium, Francium

These form monovalent, highly electrovalent, cationic monomeric species: there is no useful complex chemistry with carboxylate groups.

12.1.1.6 Group IIA: the Alkaline Earth Metals – Beryllium, Magnesium, Calcium, Strontium, Barium, Radium

These form divalent, cationic monomeric species.

The lighter elements form oxyions: beryllium in the form of the sulfate or the phosphate can raise the shrinkage temperature to 93 °C,[3] but the salts are toxic.

The heavier metals can create precipitates with collagen by electrostatic interaction, but this is not tanning. Indeed, magnesium, calcium and barium chlorides have a destabilising effect due to lyotropy.[4]

12.1.1.7 Group IIIA: Boron, Aluminium, Gallium, Indium, Thallium, Nihonium

These elements are on the borderline between the metals and the non-metals; they form trivalent species.

Boron forms oxy complexes and is toxic. Aluminium forms cationic basic ions, capable of reacting with collagen (see later).

Gallium and indium do not have a complex chemistry. Additionally, they are expensive. Thallium has very limited complex chemistry in the 1+ and 3+ states. It is toxic.

12.1.1.8 Group IVA: Carbon, Silicon, Germanium, Tin, Lead, Flerovium

Carbon creates organic chemistry, which is quite another topic in leather making.

Silicon, as a non-metal, does not form cations. It has limited use in the industry in the form of silicates,[5] where it is claimed to be a useful substitute for lime in opening up. Perhaps its best application is as a pretreatment prior to tanning; this was discussed in Chapter 10. However, it is clear that tanning is not a likely option: although the oxyanions can interact with collagen *via* hydrogen bonds and electrostatic attraction, this has no effect on hydrothermal stability. However, as silicates or as silicic acid ($pK_a \approx 10$) they can function as auxiliaries, where they have a minor application as a pretreatment for vegetable tannage.[6] They can occupy reaction sites on collagen and hence slow the reaction rate between the polyphenols and collagen; this is similar to the function of polyphosphates (see Chapter 10 and later in this chapter).

Germanium, tin and lead have chemistries with oxygen and so form complex anions. In the lower valence state they tend to hydrolyse in solution and can form complex cations. However, they are expensive, toxic and have limited tanning power.[7]

12.1.1.9   Group VA: Nitrogen, Phosphorus, Arsenic, Antimony, Bismuth, Moscovium

Nitrogen has a chemistry that is not applicable to this situation.

Phosphorus chemistry is dominated by oxy compounds. The use of phosphates is limited to a pretreatment for vegetable tanning. In this context, phosphate in the form of polyphosphate or pyrophosphate can interact *via* hydrogen bonds and electrostatic interactions, to mask reaction sites, making collagen less reactive towards vegetable tannins, promoting penetration,[6] as discussed in Chapter 10.

Arsenic, antimony and bismuth are semi-metals with limited chemistries as complex anions, so they have no useful chemistry for tanning.

12.1.1.10   Group VIA: Oxygen, Sulfur, Selenium, Tellurium, Polonium, Livermorium

These elements form anions and oxyanions, which have no tanning use. Furthermore, as the atomic weight increases, the compounds become more toxic and much more smelly in comparison with sulfur compounds.

The concept of 'sulfur tannage' is well known: it involves the precipitation of elemental sulfur by, for example, acidifying thiosulfate *in situ*,[8] discussed in Chapter 10:

$$S_2O_3{}^{2-} + 2H^+ \rightleftharpoons S + SO_2 + H_2O$$

The chemistry does offer an alternative approach to the lubrication of leather, which may confer additionally useful properties.

12.1.1.11 Group VIIA: the Halides – (Hydrogen), Fluorine, Chlorine, Bromine, Iodine, Astatine, Tennessium

The chemistry is dominated by anion formation, which has no tanning value.

12.1.1.12 Group VIIIA or 0: the Noble or Inert Elements – Helium, Neon, Argon, Krypton, Xenon, Radon, Oganesson

The elements are certainly inert in a tanning context.

12.1.1.13 Group IIIB: Transition Elements – Scandium, Yttrium, Lanthanum, Actinium

Scandium(III) tends to behave like a main group element, with similarity to the lanthanides.

Yttrium can be considered with the lanthanides, lanthanum is the first in the lanthanide series and actinium is the first in the actinide series (see later).

12.1.1.14 Group IVB: Transition Elements – Titanium, Zirconium, Hafnium, Rutherfordium

Titanium forms oxy complexes, as linear chains, in the 4+ oxidation state. The 3+ oxidation state is metastable: it does form complexes, but they revert spontaneously to the higher oxidation state in air.[9] Titanium(IV) salts have long been used in the leather industry, discussed later.

Zirconium(IV) has a well-known application in tanning (see later). Hafnium is too expensive and rare to be considered.

12.1.1.15 Group VB: Transition Elements – Vanadium, Niobium, Tantalum, Dubnium

Vanadium has several oxidation states, all capable of forming some complexes, *e.g.* the 3+ oxidation state can form covalent carboxy complexes, creating lemon-yellow leather with a shrinkage temperature of 69 °C.[10] However, it is an expensive and relatively rare metal.

Niobium and tantalum are stable in the 5+ oxidation state. They tend to form complex anions, resembling the chemistry of phosphorus or arsenic, therefore behaving like non-metals.

12.1.1.16   Group VIB: Transition Elements – Chromium, Molybdenum, Tungsten, Seaborgium

Molybdenum and tungsten have some history of involvement with the leather industry in Victorian times. However, this was confined to the use of the oxy compounds, molybdates, phosphomolybdates, tungstates, silicotungstates and phosphotungstates: Casaburi and Simoncini[11] demonstrated that these complex anions have no effect on hydrothermal stability, but might have some application in combination with vegetable tannins. Their use is retained in modern histological staining, which relies on electrostatic interaction with protein and other biological materials; they are particularly useful for incorporating heavy atoms into specimens for electron microscopy.

12.1.1.17   Group VIIB: Transition Elements – Manganese, Technetium, Rhenium, Bohrium

The most stable oxidation state of manganese, Mn(II), forms some relatively unstable complexes. Other oxidation states are unstable. Technetium and rhenium have little cationic aqueous chemistry, and cost and availability also rule them out.

12.1.1.18   Group VIIIB: Transition Elements – Iron, Cobalt, Nickel, Ruthenium, Rhodium, Palladium, Osmium, Iridium, Platinum, Hassium, Meitnerium, Darmstadtium

Iron forms complexes in the +2 and +3 oxidation states. Fe(II) forms octahedral complexes with a tendency to oxidise in air. Fe(III) also forms octahedral complexes, which tend to hydrolyse. In this way, Fe(III) functions more like Al(III) than Cr(III). However, iron has a history of application in the leather industry (see later).

Cobalt salts are most stable in the 3+ oxidation state, which has an extensive complex chemistry. The octahedral complexes are stable, with low lability. There is a preference for nitrogen donors, but oxy ligands may be reactive. Cost and availability are limitations.

Nickel salts are effectively available only in the 2+ oxidation state. Nickel(II) can form oxy complexes, but considerations of cost, allergenicity and toxicity rule it out.

Ruthenium(III) and osmium(III) have limited complex chemistries, dominated by ammines. Osmium is particularly toxic, so even though some compounds (*e.g.* osmium tetraoxide) are highly reactive in organic chemistry, they are unsuitable for tanning applications.

Rhodium(III) and iridium(III) form kinetically stable cationic complexes.

Platinum has some interesting complex chemistry, but application is confined to high-value products *e.g.* anti-cancer drugs such as cisplatin.

Ru, Rh, Pa, Ir and Pt are expensive; they are used as jewellery metals, so are ruled out for leather applications.

12.1.1.19   Group IB: Copper, Silver, Gold, Roentgenium

Copper(II) has a complex chemistry with oxy ligands, but with a tendency to form complex anions, which are not useful for tanning, with the exception of semi-metal.

Silver and gold do not have a suitable complex chemistry and these metals are more commonly used in jewellery or high-value applications.

12.1.1.20   Group IIB: Zinc, Cadmium, Mercury, Copernicium

Zinc and cadmium have chemistries similar to that of magnesium, but do exhibit some complex formation. They are toxic.

Mercury can form complexes: linear two-coordinate, tetrahedral four-coordinate and octahedral six-coordinate. The bonding can have significant covalent character, so it does have some tanning power:[7] acetate-buffered mercury(II) acetate can raise the shrinkage temperature by 11 °C.[12] Cost and toxicity rule out its use as a tanning agent.

12.1.1.21   The Lanthanides or Rare Earth Elements – Lanthanum, Cerium, Praseodymium, Neodymium, Promethium, Samarium, Europium, Gadolinium, Terbium, Dysprosium, Holmium, Erbium, Thulium, Ytterbium, Lutetium

These elements have very similar properties, so they are often encountered as in Nature as mixtures. An exception in chemical applications is cerium, routinely used as an analytical oxidising agent in the Ce(IV) oxidation state.

All of these elements usually function as trivalent compounds. They do not seem to have a useful complex chemistry for tanning, although it was claimed in 1910 that a mixture of the chlorides is a better tanning agent than aluminium(III).[13] There has been information in the technical literature regarding the application of lanthanides in the leather industry, mostly by the Chinese, who have substantial amounts of lanthanides in their geology. However, with the possible exception of semi-metal tanning as retanning agents for vegetable tanned leather (see Chapter 13), their use is limited because they do not have a large enough effect on raising the shrinkage temperature of collagen: typically the shrinkage temperature outcome is ~80 °C.

12.1.1.22   The Actinides – Thorium, Palmium, Uranium and the Transuranic Elements

This group strictly includes only thorium, palmium and uranium, since the transuranic elements are synthetic. Of the remaining elements, ^{238}U, the non-fissile isotope, is the least radioactive, but it is toxic. Its use is confined to salts as the uranyl ion, UO_2^{2+}, where it used as a histological stain in the form of the acetate.

12.2   Experimental Tanning Reviews

From the brief survey presented in the previous section, it is clear that the number of mineral options open to the tanner is limited. The remaining question is whether any of those limited options can rival chromium(III) as an industrial tanning agent. A few wide-ranging reviews of tanning abilities of inorganic species have been undertaken, all characterised by being somewhat selective, *i.e.* they were not completely comprehensive, nor were the salts all related.

In a study of the thermodynamics of tanning, Weir and Carter[4] made several leathers, summarised in Table 12.1.

In 1957, Borasky[14] reviewed mineral tanning, including plutonium nitrate, as shown in Table 12.2.

The most comprehensive study to date was undertaken by Chakravorty and Nursten in 1958;[7] some of their results are presented in Table 12.3.

Although Tables 12.1–12.3 do not cover all of the options, it is apparent that there has not been much success in finding an alternative to chromium. Indeed, it is discussed in Chapter 23 that there can be no alternative. The only generalisation that comes out of reviewing tanning effects is that all metals exhibit their greatest tanning power at the point of precipitation.[7] This is understandable because at that point the following conditions apply.

Table 12.1 Outcomes of some unrelated mineral tannages.

Salt	pH of tannage	Maximum T_s (°C)
Diamminemercury(II) chloride: $Hg^{II}(NH_3)_2Cl_2$	7.5	67
Lead(II) nitrate: $Pb(NO_3)_2$	7	68
Thorium(IV) sulfate: $Th(SO_4)_2$	4	68
Magnesium sulfate: $MgSO_4$	7.5	68
Titanium(IV) chloride: $TiCl_4$	6	70
Beryllium(II) sulfate: $BeSO_4$	5.5	71
Copper(II) sulfate: $CuSO_4$	5.5	73
Mercury(II) acetate: $Hg^{II}(O_2CCH_3)_2$[a]	7	91

[a]Hg(II) removed by washing.

Table 12.2 Outcomes of some unrelated mineral tannages.

Salt	pH of tannage	Maximum T_s (°C)
None		63
Strontium(II) chloride: $SrCl_2$	6.0	59
Silver(I) nitrate: $AgNO_3$	6.0	63
Barium(II) chloride: $BaCl_2$	6.0	63
Sodium tungstate: Na_2WO_4	6.0	63
Mercury(II) chloride: $HgCl_2$	4.2	63
Lead(II) acetate: $Pb(O_2CCH_3)_2$	6.0	60
Lead(II) nitrate: $Pb(NO_3)_2$	5.3	46
Uranyl nitrate: $UO_2(NO_3)_2$	3.8	43
Chrome alum	4.0	98
Phosphotungstic acid: $H_3PW_{12}O_{40}$	3.0	63
Osmium tetraoxide: OsO_4	4.9	78
Uranyl acetate: $UO_2(O_2CCH_3)_2$	4.6	73
Plutonium(IV) nitrate: $Pu(NO_3)_4$	1.0	78

Table 12.3 Outcomes of some unrelated mineral tannages.

Salt	pH of tannage	Maximum T_s (°C)
Tin(II) chloride: $SnCl_2$	5	60
Manganese(II) chloride: $MnCl_2$	7.5	62
Yttrium(III) chloride: YCl_3	7	62
Silver(I) nitrate: $AgNO_3$	8	63
Cerium(III) chloride: $CeCl_3$	7.5	63
Thorium(IV) chloride: $ThCl_4$	4.5	63
Lead(II) nitrate: $Pb(NO_3)_2$	7	63
Lanthanum(III) nitrate: $La(NO_3)_3$	8.5	63
Cuprammonium hydroxide: $[Cu(NH_3)_4(H_2O)_2](OH)_2$	11	63
Thorium(IV) nitrate: $Th(NO_3)_4$	4.5	64
Potassium titanyl oxalate: $KTiO(O_2C)_2$	5.5	64
Zinc(II) sulfate: $ZnSO_4$	7	65
Neodymium(III) nitrate: $Nd(NO_3)_3$	7.5	65
Cuprammonium sulfate: $[Cu(NH_3)_4(H_2O)_2]SO_4$	10.5	65
Lithium chloride: LiCl	8	66
Tin(IV) chloride: $SnCl_4$	6.5	66
Cadmium(II) sulfate: $CdSO_4$	6.5	67

1. Collagen is at its most reactive for that metal. The solution pH is at the highest value consistent with still keeping the metal ions in solution.
2. The molecular metal ions are in their most highly polymerised state, consistent with remaining solubilised. This gives the species maximum hydrophobicity and hence the greatest affinity for the environment of the collagen substrate.

The only practical options as alternatives to chrome, which have some history of use in the leather industry, are aluminium(III), titanium(IV) and zirconium(IV) [and to a lesser extent iron(II) and zinc(II); see later], and some of their properties in the context of tanning are summarised in Table 12.4.[13]

Generalising the information in Table 12.4, aluminium, titanium and zirconium cannot match the tanning power of chromium. The tanning power of titanium(IV) falls between those of aluminium(III) and zirconium(IV). From their positions in the Periodic Table in Group IVa, similarities between titanium and zirconium chemistry would be expected. If aluminium in Group IIIa is viewed as a pre-transition element, then any similarities between aluminium and titanium constitute an example of a diagonal relationship, more

Table 12.4 Comparisons between chromium(III), aluminium(III), titanium(IV) and zirconium(IV).

Property	Tanning metal ion			
	Cr(III)	Al(III)	Ti(IV)	Zr(IV)
Electronic configuration	$[Ar]3d^3$	$[Ne]$	$[Ar]3d^0$	$[Kr]4d^0$
Species in aqueous acid	$[Cr(H_2O)_6]^{3+}$	$[Al(H_2O)_6]^{3+}$	$[TiO]_n^{2n+a}$	$[ZrO]_m^{2m+a}$
Species in aqueous alkali	Chromite	Aluminate	Titanate	Insoluble
Coordination number	6	6	6	6–8
Effect of complexation:[b]				
Oxy complexes	**	****	***	*
Carboxy complexes	****	**	**	***
Amino complexes	**	*	**	**
Metal oxide	Cr_2O_3	Al_2O_3	TiO_2	ZrO_2
Equivalent basicity for tanning/%	33	33	50	50
Equivalent weight of oxide	38	25	40	61
Relative equivalent tanning offer/% metal oxide	6	4	6	10
Maximum T_s (°C)[c]	>120	90	96	97

[a] $n > m$.
[b] More asterisks mean a greater tanning effect.
[c] Extreme conditions of astringency and magnitude of offers.

commonly associated with the first and second periods. These relationships depend on the polarising power of the ions (charge-to-radius ratio, *i.e.* charge density), which determines the nature of the bonding in compounds.

12.3 Aluminium in Leather Making

Potash alum, a mixed salt of aluminium and potassium sulfate, occurs native, *i.e.* it can be dug out of the ground. It has been used for thousands of years in leather making; this is known because written recipes exist.[15] In combination with traditional vegetable tanning, it makes semi-alum leather (see Chapter 13): this combination of reagents yields unusually high hydrothermal stability, but that was almost certainly unknown to the ancient Egyptians and earlier civilisations. The earliest mention of this effect was from Procter,[16] who pointed out that heating linseed oil-treated vegetable tanned leather, to make patent leather, could be conducted at higher temperatures if the leather had been treated with an aluminium salt.

A more common reason for including aluminium(III) in a tanning formulation is for colouring. In traditional dyeing processes, the dye is better fixed if there is a mordant present in the leather, because natural dyes have a greater affinity for metal mordants than for collagen (see Chapter 16). In his review of the history of the role of alum in leather, Thomson[17] argued that, at least from the second millennium BCE, it was not used for tanning, but for its effect in dyeing. The role in tanning came later, but it was not until Gioanventura Rosetti's *The Plictho*, published in Venice in 1548, that the weak tanning effect was recognised and hence differentiated as tawing.

Another possible use of aluminium salts is to precipitate the salt as the hydroxide (hydroxide is spontaneously turned into hydrated oxide), which allows dye to be adsorbed on the oxide. This creates a densely coloured type of pigment, called a 'lake', providing colour to cover the leather surface.

A role for aluminium salt, originally potash alum, used to be in making white gloving leather. Strictly, this process is not tanning, and is known as 'tawing'. Traditionally, this process was operated in England by the whitawyers, in times when leather-making operations were separated and controlled by the London Guilds in the earlier part of the last millennium. The formulation has been described as a 'Yorkshire pudding' mix, with the following ingredients (Yorkshire pudding omits only the alum).[6]

For 100 kg of pelt:

50–60 L of water at 35 °C, to create a paste;

6–8 kg of potash alum, to react with the collagen;

8–12 kg of flour, to provide starch, which acts as a masking agent for the Al(III);

1–2 kg of salt, to prevent swelling under the acidic conditions;

8–15 kg of egg yolk, to act as a lubricant, due to the content of phospholipids.

The paste is introduced into pelt, which is then dried and softened. The white leather is functional, but the effect may be reversed if it gets wet.

Aluminium(III), $[Al(H_2O)_6]^{3+}$, is an acidic complex, with a greater tendency than chromium(III) to hydrolyse to basic salts, especially at low concentration. It also has a lesser tendency to olate and thereby to form discrete basic salts: it readily polymerises to an Al_{13} species and precipitates.[19] The high molecular weight species is likely to be the most reactive species in a tanning context. The formation of Al_{13} is dependent on the conditions in solution, as shown in Table 12.5.[19]

Sulfate does not complex with Al(III) in the dry salt, as it does in basic chromium(III) sulfate; the interaction is much more electrostatic. Aluminium(III) is stable at pH 4, where the basicity is 30–35%, helped by the presence of a masking agent, otherwise it is likely to precipitate at pH < 4. Masking effects can be summarised by the following series:[20–24]

Table 12.5 Amount of Al_{13} in basic aluminium salts, as determined by ^{27}Al NMR spectroscopy.

Al(III) compound	Al_{13} (%)
$AlCl_3$	57
$Al_2(SO_4)_3$	60
Al:citrate 4 : 1	
pH 2.0	71
pH 3.0	74
pH 4.0	92

glucoheptonate > gluconate > oxalate > citrate > malonate > lactate > tartrate > succinate > acetate > glycolate > formate > sulfate

Masking may be applied to aluminium(III) sulfate, to allow the creation of a more reactive salt at pH approaching 4. Alternatively, basic chloride salts are available. They can be more basic because chloride ion is one of the few ligands that can create true covalent complexes. For this reason, the chloro complexes are typically less reactive than those based on the sulfate and additional sulfate in the tannage is beneficial.[25]

Aluminium(III) interacts with collagen carboxyls, but in a reaction that is much more electrostatic than covalent. The consequence is that the leather retains a significantly cationic character, and in this state the fibre structure is vulnerable to collapsing and sticking together, producing thin, hard leather or, in the case of pickled pelt, a tendency to form irreversible adhesions.[26] The cationic character also poses difficulties for applying anionic reagents, such as fatliquors, but this can be overcome to a large extent by treating the leather with polyphosphate, to sequester the charge.

The weak interaction between aluminium(III) and collagen means that the shrinkage temperature from a tawing process is barely raised. By the use of masking, the aluminium tanning process can be conducted at pH 4, yielding leather with a shrinkage temperature of about 80 °C. Despite the obvious inadequacies of aluminium(III) as a tanning agent, there are some modern uses of aluminium(III) salts, summarised in the following.

12.3.1 Alum Pickle

The presence of aluminium in the tannage, whether alone or in conjunction with chromium(III), confers firmness, as mentioned. For some applications, this is desirable: a well-known example is for lightweight leathers for shoe uppers, such as ladies' weight calf leather, which is stiff leather capable of being shaped on the shoemaker's last. One way of incorporating Al(III) into the tannage is to include an aluminium(III) salt such as the sulfate in the pickle formulation. In this way, the acidity of the salt contributes to the pickling process and the conditions of chrome tannage are sufficient to fix some aluminium into the pelt.

12.3.2   Suede

The stiffening effect of aluminium(III) is also useful for making suede leather. Here, one of the technological requirements is to cut the fibres cleanly in order to obtain a fine suede nap. If the fibres are not cut cleanly, they may be pulled from the surface, creating a 'raggy' suede. If the fibres are stiffened they are cut by the sueding paper more easily. The aluminium(III) may be introduced in the pickle, but more usually is applied as part of the retannage of chrome leather.

12.3.3   Furskins

Furskin processing is more concerned with stabilising the pelt than conferring high stability to elevated temperature, *i.e.* the processing can be characterised as leathering rather than tanning. Aluminium(III) salts play a role in this technology.

   Furskin technology and associated science are traditionally separated from leather, both technically and ethically. Consequently, fur is not addressed in this treatise.

12.3.4   Semi-alum Tannage

The relationship between plant polyphenols and metal salts, to create semi-metal-tanned leather, is addressed in Chapter 13. In that context, aluminium(III) is the preferred metal salt, because it has greatest affinity for phenolic hydroxyls.

12.3.5   Aluminium Silicate

Aluminium silicate has been available as a tanning auxiliary in the form of the product Coratyl G, where it was offered as a basifying agent for chrome tanning. Its effect was probably a mixture of aluminium and silica tanning, along with the mildly basic reaction.

   The structure of intact zeolite is tetrahedrally coordinated aluminium and silicon atoms connected by bridging to coordinated oxygen atoms. Costantini *et al.*[27] showed that aluminosilicate in the form of zeolite breaks down under acidic conditions, as indicated in Table 12.6.

   The impact of the decomposition on the tanning effect is shown in Table 12.7.

Table 12.6 Stepwise decomposition of zeolite in acid for 17 h.

$[H^+]:[Al^{3+}]$	Aluminosilicate oligomers (%)	$[Al(H_2O)_6]^{3+}$ (%)
1.0	50	0
2.0	50	50
4.0	0	100

Table 12.7 Tanning sheepskin with 11% Al_2O_3 as decomposed zeolite masked with phthalate, starting at pH 6.

$[H^+]:[Al^{3+}]$	Final pH	T_s (°C)
1:1	5.9	85
2:1	5.0	92
4:1	3.8	65

The most recent product of this type is Tanfor T,[28] described as an aluminium–silicon complex with polycarboxylate, sold as a wet white agent and marketed as 'wet bright', on the basis that the dyeing effect is enhanced.

12.3.6 Mixed Complexes with Aluminium

Aluminium(III) has often been incorporated into complexes with other metal salts, usually with polycarboxylate masking, for tanning purposes. In all cases, the effects of the metals can be considered to be additive with regard to the contribution of each to the physical properties.[29,30]

Organic complexes of aluminium syntans are less common. Chinese workers reported making a complex of aluminium(III) sulfate with modified calcium lignosulfonate, by enhancing the reactivity of the lignosulfonate through a Mannich reaction to bind glycine *via* formaldehyde.[31] Tanning with 2% Al_2O_3 + 10% modified lignosulfonate resulted in a shrinkage temperature of 83 °C.

12.3.7 Chrome Uptake

The presence of aluminium(III) in chrome tanning has been shown to enhance the rate of chrome fixation, by an effect analogous to catalysis.[20] The results of laboratory studies are presented in Figure 12.2.

It can be seen that the pretreatment with aluminium has a positive effect on both chrome content and shrinkage temperature, maximised

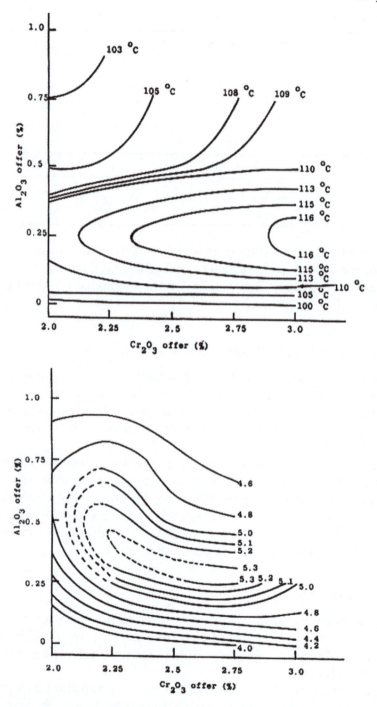

Figure 12.2   Effect of aluminium and chromium offers on the shrinkage temperature and chrome content (% on dry weight) of bovine leather. Courtesy of the *Journal of the Society of Leather Technologist and Chemists*.

at up to 0.25% Al_2O_3 offer; at higher offers the aluminium seems to compete with the chrome and the benefit is diminished. Part of the overall reaction is the displacement of fixed aluminium(III) by incoming chromium(III), as shown in Figure 12.3.[32]

The aluminium effect appears to arise from the difference between aluminium(III) complexation and chromium complexation reactions: in the comparison presented in Table 12.8, the rate of complexation by aluminium(III) is favoured by the entropy of activation. Therefore, it can be postulated that the reaction between collagen carboxyls and aluminium(III) is fast and results in loose complex formation. Hence the sidechains can adopt a suitable conformation for reaction with metal ions without a high energy penalty and the consequent reaction with chromium(III) is energetically favoured.

It is useful to compare the stability of aluminium-tanned leather with that of chrome-tanned leather and some of the properties of the chemical species are set out in Table 12.8.[33-35] There are some superficial similarities between aluminium and chromium chemistry, due to their preferred valence state and the amphoteric nature of the hydroxides (capable of acting as acids or bases). However, a closer examination of the properties of ions in solution reveals they have very different relationships with solvating species: this can be used as a reflection of the ability of the ions to complex with collagen. The reactions may be summarised as associative for chromium and dissociative for aluminium: the rates of exchange of solvate species at aluminium(III) are 10^5–10^6 times faster than those of exchange at chromium(III).

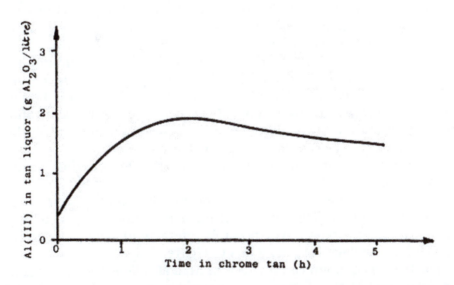

Figure 12.3   Displacement of aluminium(III) by chromium(III) at pH 3.8–4.0. Courtesy of the *Journal of the Society of Leather Technologist and Chemists.*

Table 12.8 Comparisons between the chemistries of chromium(III) and aluminium(III).

Property	Cr(III)	Al(III)
Hydration number:		
By NMR	6	6
By conductivity	12	11
Solvation of M^{3+} ion at 298 K:		
$\Delta H°$ (kJ mol^{-1})	−4400	−4660
$\Delta S°$ (J K^{-1} mol^{-1})	−405	+155
Transfer from crystal lattice to solution, $T\Delta S_s$ (kJ mol^{-1})	−109	−113
Transfer from unit molality to dilute solution, ΔS_{tr} (J K^{-1} mol^{-1})	−405	−426
Partial molar hydration volume, relative to zero for H^+ at 298 K (cm^3 mol^{-1})	−40	−42
$\left[M(H_2O)_6\right]^{3+} \rightleftharpoons \left[M(H_2O)_5 OH\right]^{2+} + H^+$:		
pK_1 ($I = 0$)	3.9	5.0
ΔH_1 (kJ mol^{-1})	38	79
$qM^{3+} + pH_2O \rightleftharpoons M_q(OH)_p^{(3q-p)+} + pH^+$:		
Log $\beta_{2,2}$	−3.4	−7.3

	Cr(III)			Al(III)		
Solvent exchange[a]	H$_2$O	DMSO	DMF	H$_2$O	DMSO	DMF
pK (298 K)	6.3	7.5	7.3	0.8	1.2	0.8
$\Delta H^{\ddagger}$ (kJ mol^{-1})	109	97	100	113	84	75
$\Delta S^{\ddagger}$ (J K^{-1} mol^{-1})	0	−50	−42	+118	+17	+2
Complex formation	Associative			Dissociative		

[a]DMSO, dimethyl sulfoxide; DMF, dimethylformamide.

The basic ions in solution have similar structures and interact with the same sites on collagen, but the outcomes of tanning reactions are very different because the bonding is very different. This difference is addressed further in Chapter 23.

The comparison between the tanning powers of aluminium and chromium is illustrated diagrammatically in Figure 12.4. The nature of the bonding confers the range of stability: the covalently bound chromium(III) is stable over eight pH units, but the electrostatic bonding of aluminium(III) is stable over only two pH units. In contrast, the semi-alum reaction produces covalent bonding and the semi-alum bond is actually more stable than chrome tanning bonding at the alkali end of the scale.

It has been shown by ^{27}Al NMR spectroscopy that the magnetic environment of the aluminium(III) nucleus is not changed during the shrinking reaction.[18] The possible interactions between aluminium(III) and collagen are modelled in Figure 12.5.

The (basic) aquo ion can interact electrostatically *via* a water ligand or a complex might be formed, which would be more electrovalent than

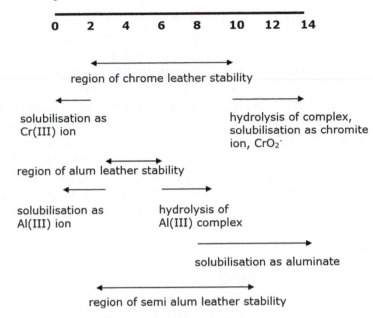

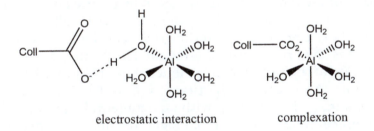

Figure 12.4 Guideline limits of tanned leather stability within the pH scale.

electrostatic interaction   complexation

Figure 12.5 Interactions between collagen and aluminium.

covalent. Since it is known that aluminium(III) does not form stable complexes, it may be surmised that the former case is more likely. Therefore, the bonding between the collagen and the matrix could break down hydrothermally, but the environment of the aluminium nucleus would not be substantially altered. Therefore, it can be concluded that the aluminium-based matrix involves water that can break down, to allow shrinking, *i.e.* the electrostatic interaction with collagen carboxyls is sufficiently distant to allow this to happen, but the aluminium(III) nucleus experiences no change in the shielding of its magnetic field. In contrast, covalent complexation between collagen carboxyls and chromium(III) is a direct interaction, which cannot break down under the conditions of shrinking. Therefore, the shrinking transition cannot involve breakdown of the metal–collagen bonds, so the process must involve the breaking of the hydrogen bonding in the triple helices, causing them to unravel, and in the associated matrix around the triple helices.

An alternative approach was proposed by Holmes,[36] in which chelating groups were bound to lysine sidechains using dihalogenated azo heterocycles, *i.e.* exploiting the chemistry used in reactive dyeing (see Chapter 16). There is an additional side effect to metal fixation, *i.e.* the shrinkage temperature is higher than expected. This effect, which applies to all metal tannages, chromium(III) and less effective reactions, demonstrates that the tanning effect of metals depends in part on the way in which the metal species interacts directly with the collagen. Here, the firmer interaction of chelate bonding is sufficient to increase the shrinkage temperature of aluminium-tanned leather, as shown in Table 12.9. This aspect of the tanning reaction is developed further in Chapter 23.

12.4   Titanium Tanning

Titanium(IV) salts have a similar affinity for collagen as aluminium(III), in part due to some similarities in properties,[37] such as the acidity of the ion and the tendency to hydrolyse and precipitate as the pH is raised above 3. The interaction with collagen carboxyls is similarly electrovalent, rather than covalent. However, one difference is the greater filling effect of Ti(IV) salts, due the polymeric nature of the salts, which produces softer leather. The empirical formula of the cation, the titanyl ion, is TiO^{2+}; in reality, the ion is a chain, $(-Ti-O-Ti-O-)_n$ with an octahedral ligand field. The weak chemical interaction with collagen results in shrinkage temperatures of 75–80 °C. The resultant leather is initially colourless, although it tends to turn pale yellow on ageing, owing to the creation of charge-transfer bonding; this is a change from hydrogen bonding to electrostatic bonding.

Table 12.9   Effect of chelation on the shrinkage temperature of metal-tanned collagen.

	T_s (°C)		
Chelating group	No Al(III)	With Al(III)	$\Delta T_{s\,Al}$ (°C)
None	63	79	16
Salicylate: 1-Hydroxy-2-carboxybenzene	70	90	20
Mandelate: $PhCH(OH)CO_2H$	72	94	22
Phthalate: 1,2-Dicarboxybenzene	74	101	25
Isophthalate: 1,3-Dicarboxybenzene	55	77	22

The only widespread industrial use of Ti(IV) salts was the traditional process for gentlemen's hat banding leather: this is the leather that is sewn on the inside of the rim, the area in contact with the head. The process was semi-titan: starting with vegetable-tanned leather, it was retanned with potassium titanyl oxalate, to give a very hydrothermally stable leather, which is also capable of resisting the effects of perspiration because the elements of the combination tannage are rendered non-labile.

An application for titanium(IV) chemistry was suggested by Russian workers,[38] in which sole leather is made from large offers of ammonium titanyl sulfate, $(NH_4)_2TiO(SO_4)_2$ (a salt that gives pH 4 in solution), basified with hexamethylenetetramine (HMT). This has not met with commercial success. Using excessive offers of ammonium titanyl sulfate (30% salt containing 6%TiO_2) on sheepskin and basifying with sodium sulfite and HMT, shrinkage temperatures up to 96 °C could be achieved and acceptable leather was made, although clearly not like chrome-tanned leather.

Chinese workers reviewed the effects of masking on titanium tannage, when they tried 13 hydroxyl carboxylates and 22 other agents; they could achieve shrinkage temperatures of only up to 92 °C.[39,40]

It is possible to prepare metastable salts of titanium(III): these violet salts can be easily made by dissolving titanium metal in sulfuric acid and can be held for many days in this valence state, but only if air is excluded. Attempts to tan with these salts produced no additional benefits; although it can be initially bound as Ti(III), the tanning agent rapidly reverts *in situ* to the 4+ oxidation state in the leather,[9] with no improvement over tanning with titanium(IV).

12.5   Zirconium Tanning

Zirconium(IV) salts are expected to perform in a similar way to titanium(IV) salts, because of the position of Zr in the Periodic Table. They have been available commercially from the middle of the twentieth century; their chemistry and tanning properties were comprehensively reviewed by Hock.[41] The tanning effect is similar to that with Ti(IV), yielding slightly higher shrinkage temperatures, but the colour is stable and makes white leather with a full handle.

The chemistry of the complexes is more like that of chromium(III) than titanium(IV), illustrated in Figure 12.6. It can be seen that the 50% basic ion is a tetrameric species, where the metal is eight-coordinated, with μ-hydroxy bridges between metal ions.

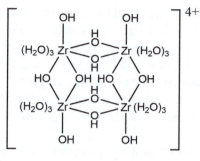

Figure 12.6 Structure of 50% basic zirconium(ɪᴠ) ion.

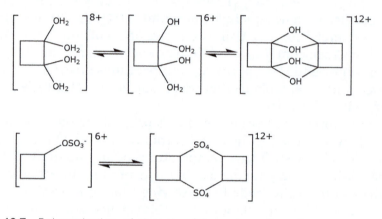

Figure 12.7 Polymerisation of zirconium(ɪᴠ) species, where the squares represent the unit of structure illustrated in Figure 12.6.

Like titanium(ɪᴠ), the unmasked zirconyl ion is acidic and unstable to hydrolysis. The conventional way of applying it is to use as little water as possible for the penetration stage, then add water (float up, in the jargon) to cause polymerisation by hydrolysis, in the same way as chromium(ɪɪɪ) polymerises, shown in Figure 12.7.

The increase in molecular weight increases the astringency of the species, driving it into the substrate, which results in fixation. It has been claimed that, by deactivating the carboxyl groups and amino groups of collagen, zirconium(ɪᴠ) complexes do not interact with the carboxyls and that they interact significantly with the amino groups.[42] However, the mechanism is unclear, since zirconium(ɪᴠ) can coexist as cationic, neutral and anionic species, by interaction with counterions, particularly sulfate, and masking agents such as citrate. It has been suggested[43] that part of the mechanism is based on interaction between the zirconium species and tyrosine residues: UV irradiation studies have shown that there is a change in conformation at those sites and the presence of the zirconium quenches fluorescence. In reacting with collagen, it is significant that counterions are displaced from the zirconium complexes.

Zirconium(IV) tanning has been referred to as the mineral equivalent of vegetable tanning, because of the presence of so many hydrogen bonding sites on the complex. This is a significant contributor to the binding mechanism in tanning. The large size of the complexes produces a filling tannage (a characteristic of vegetable tanning) and the commonest use for this chemistry is as a retannage, particularly to stiffen (tighten) the grain. As in vegetable tanning, zirconium(IV)-tanned leather is hydrophilic because of the hydrophilic nature of the tanning species.

12.6   Iron Tanning

This mineral tannage does not have a long history: initial developments took place in Germany in the mid-part of the twentieth century, when the availability of chromium was restricted. Tanning with iron(II) salts is preferred to that with iron(III), primarily due to the oxidising power of Fe(III) with consequent damage to the collagen: this is a feature of both oxidation states, so there is a progressive weakening due to the catalytic redox effect from the iron species. Even when the iron complex is stabilised by gluconate, the tannage is reversed over time; for example, over 3 months the shrinkage temperature of iron(II)-tanned collagen decreased from 88 to 82 °C, whereas that of the gluconate-masked leather changed from 96 to 94 °C over the same period.[44]

The applications are limited, because the lack of true complexation with collagen carboxyls means that the resulting shrinkage temperature is much the same as that for aluminium(III). A disadvantage for workers is that they suffer from a metallic taste in the mouth when they handle wet leather, due to absorption of the metal salt through the skin.

Iron(II) sulfate complexed with ethylenediamine and masked with dicarboxylates can produce a shrinkage temperature of 76 °C,[45] but tartrate masking alone can give a shrinkage temperature of 85 °C.[46]

Iron salts, especially Fe(III), have high affinity for plant polyphenols, although this may be a disadvantage in tanning because the resulting colour is typically intensely blue–black. Hence these reactions have been exploited for thousands of years for making inks. The effect is often observed in factories using vegetable-tanned leather, where the leather becomes affected by black spots from the action of sparks from, for example, the continuous sharpening of splitting band knives. The treatment or 'clearing' action is to use oxalic acid or EDTA (ethylenediaminetetraacetic acid disodium salt) to complex the ion and solubilise it. However, the reaction can be exploited, for example to obtain a stable black colour on leather, by combining iron(II) with gallnut and sappan wood tannins.[47]

12.7   Zinc(II)

It is known that that zinc(II) can interact with plant polyphenols, in the way that most of the transition elements can, as mentioned in Chapter 13. However, it has been suggested that it might be regarded as a tanning agent in its own right: citrate-masked zinc(II) is claimed to complex at histidine, to give a shrinkage temperature of 81 °C.[48] Also, as with other metal salts in a tanning context, zinc(II) can be part of a mixed complex, for example with titanium(IV). Studies on mixed complexes containing Zn:Ti ratios of 9:1, 8:2, 7:3 and 6:4 found that the last ratio was optimal, masked with carboxylate, to give a maximum shrinkage temperature of 77 °C.[49] The reaction makes a wet white that can be shaved, although the toxic nature of zinc limits its application.

12.8   Mixed Mineral Tannages

All metal salts can be used together in all proportions. The benefits of introducing another metal salt can be summarised as follows.

- The efficiency of chrome uptake is improved by reducing the offer.[50]
- The rate of chrome uptake can be improved by the presence of another, weaker mineral tanning agent, applied as a pretreatment.[24]

The contraindications can be summarised as follows.

- In no case is the shrinkage temperature positively affected; often the opposite applies. If it is required that the leather should be boilfast, the chrome offer must be not less than 0.75% Cr_2O_3.
- The properties of the leather are modified, changed by the presence of another mineral tanning agent, maximised if the substitution is total.

- The handle or feel of the leather may be changed, in terms of collapsing [as with aluminium(III) salts] or filling [as with titanium(IV) or zirconium(IV) salts] the fibre structure.
- The leather will become more cationic and therefore more surface reactive towards anionic reagents.
- The colour may be changed (as with iron salts). At all proportions in a mixture with a colourless substitute, chrome-tanned leather remains distinctly blue, even at very low chrome offers, *i.e.* the effect of blue to whiten, as applied to fabric washing, does not apply to chrome-tanned leather.
- The hydrothermal stability is adversely affected, in terms of the pH window within which the mineral–collagen bonding remains intact.

Aluminium(III) with chromium(III) has been discussed earlier, in the context of assisting the chrome tanning reaction.[32] Iron(II) masked with tartrate or citrate has been suggested as a suitable accompaniment to chrome tanning,[51] although there appears to be no obvious advantage.

Aluminium(III) with titanium(IV) has been suggested as a suitable mixture for colourless tannage for sheepskin rugs.[13] A mixture of aluminium(III) sulfate and ammonium titanyl sulfate was masked with polyhydroxycarboxylate, $HO(CHOH)_nCO_2^-$, preferably gluconate, where $n = 5$, or glucoheptonate, where $n = 6$. The tannage was adequately fast to washing. A formulation of tanning salts and masking salt was briefly commercialised by ICI as Synektan TAL, but it could never achieve the aspirations of the producer to replace chromium(III).

Development work was conducted on combination tannages of chromium(III) with the aluminium(III) + titanium(III) complex.[13] Tannages with offers as little 0.75% Cr_2O_3 with 1.3% total metal oxide as Al_2O_3 + TiO_2 could produce a shrinkage temperature of 106 °C. In addition, the leathers were both softer and more full and less cationic than similar leathers tanned with chrome and aluminium.

12.9 Overview

... the periodic table holds out little hope of the discovery of new commercially useful tanning agents, based on elements not previously considered.

Chakravorty and Nursten[7]

The value of a tanning process depends on two main parameters: the reversibility of the tannage and the shrinkage temperature of the leather. For those salts that might be candidates to compete with chromium(III) on cost and availability, both of these requirements are impossible because of the nature of the bonding between the metal ions and the collagen carboxyls. The stability of mineral-tanned leather can be enhanced if the approach demonstrated by Holmes is adopted, in which he grafted multifunctional complexing sites onto collagen.[36] In this way, the linking part of the tanning mechanism is improved, fixing the mineral matrix to the collagen more firmly. This is a chemical modification to the substrate that could be exploited commercially, but it clearly complicates the tanning operation and is likely to be impractical. However, for some specialised applications, it could be appropriate.

Consequently, further searching for a viable single mineral tanning alternative to chromium(III) for the global leather industry is futile. The future of mineral tanning remains with chromium(III) – either in its present form or modified in ways discussed in this treatise.

References

1. A. Kuntzel and T. Dröschler, *Collegium*, 1940, **839**, 104.
2. K. H. Gustavson, *The Chemistry of the Tanning Process*, Academic Press Inc., New York, 1956.
3. *Rev. Tech. Ind. Cuir*, 1947, **39**, 238.
4. C. E. Weir and J. Carter, *J. Am. Leather Chem. Assoc.*, 1950, **44**(7), 421.
5. K. H. Munz and R. Sonnleitner, *J. Am. Leather Chem. Assoc.*, 2005, **100**(2), 45.
6. P. Chambard, in *The Chemistry and Technology of Leather*, ed. F. O'Flaherty, W. T. Roddy and R. M. Lollar, Reinhold Publishing Corp., New York, 1958.
7. H. P. Chakravorty and H. E. Nursten, *J. Soc. Leather Technol. Chem.*, 1958, **42**(1), 2.
8. J. A. Wilson, *The Chemistry of Leather Manufacture*, Chemical Catalog Co. Inc., New York, 1929.
9. G. S. Lampard, PhD thesis, University of Leicester, 2000.
10. A. Kuntzel and H. Erdmann, *Collegium*, 1938, **824**, 630.
11. V. Casaburi and E. Simoncini, *J. Int. Soc. Leather Trades' Chem.*, 1936, **20**(1), 2.
12. M. Parenzo, *Collegium*, 1910, **403**, 121.
13. A. D. Covington, *J. Am. Leather Chem. Assoc.*, 1987, **82**(1), 1.
14. R. Borasky, *J. Am. Leather Chem. Assoc.*, 1957, **52**(11), 596.
15. R. Reed, *Ancient Skins, Parchments and Leathers*, Seminar Press, 1972.
16. H. R. Procter, *Textbook of Tanning*, Spon, London, 1885.
17. R. Thomson, *J. Soc. Leather Technol. Chem.*, 2009, **93**(4), 125.
18. A. D. Covington, R. A. Hancock and I. A. Ioannidis, *J. Soc. Leather Technol. Chem.*, 1989, **73**(1), 1.
19. K. Ding, *et al.*, *J. Am. Leather Chem. Assoc.*, 2006, **101**(11), 381.
20. A. D. Covington and R. L. Sykes, *J. Am. Leather Chem. Assoc.*, 1984, **79**(3), 72.

21. A. Kuntzel, *et al.*, *Collegium*, 1935, **786**(10), 484.
22. T. C. Thorstensen and E. R. Theis, *J. Int. Soc. Leather Trades' Chem.*, 1950, **34**(6), 230.
23. A. Kuntzel and S. Rizk, *Das Leder*, 1962, **13**(5), 101.
24. A. Simoncini, *et al.*, *Cuoio*, 1978, **54**(4), 439.
25. C. Pal and B. M. Das, *J. Indian Leather Technol. Assoc.*, 1955, **3**, 67.
26. M. Komanowsky, *J. Am. Leather Chem. Assoc.*, 1989, **84**(12), 369.
27. F. Ciardelli, *et al.*, *J. Am. Leather Chem. Assoc.*, 2000, **95**(4), 125.
28. A. Bacardit, *et al.*, *J. Am. Leather Chem. Assoc.*, 2014, **109**(4), 117.
29. D. Chen, *et al.*, *J. Soc. Leather Technol. Chem.*, 2013, **97**(3), 116.
30. S. Cai, *et al.*, *J. Am. Leather Chem. Assoc.*, 2015, **110**(4), 114.
31. Q. He, *et al.*, *J. Soc. Leather Technol. Chem.*, 2011, **95**(5), 204.
32. A. D. Covington, *J. Soc. Leather Technol. Chem.*, 1986, **70**(2), 33.
33. F. A. Cotton and G. Wilkinson, *Advanced Inorganic Chemistry*, Wiley, 1962.
34. J. Burgess, *Metal Ions in Solution*, Wiley, 1978.
35. C. S. G. Phillips and R. J. P. Williams, *Inorganic Chemistry Vol. 2*, OUP, 1966.
36. J. M. Holmes, *J. Soc. Leather Technol. Chem.*, 1996, **80**(5), 133.
37. M. P. Swamy, *et al.*, *Leather Sci.*, 1983, **30**(10), 291.
38. V. Yakutin *et al.*, *USSR Pat.*, 234598, 1972.
39. B. Y. Peng, *et al.*, *J. Am. Leather Chem. Assoc.*, 2007, **102**(9), 261.
40. B. Y. Peng, *et al.*, *J. Am. Leather Chem. Assoc.*, 2007, **102**(10), 297.
41. A. L. Hock, *J. Soc. Leather Technol. Chem.*, 1975, **59**(6), 181.
42. T. S. Raganathan and R. Reed, *J. Soc. Leather Trades' Chem.*, 1958, **42**(2), 59.
43. N. Nishad, *et al.*, *J. Am. Leather Chem. Assoc.*, 2008, **103**(12), 422.
44. H. Chen, *et al.*, *J. Am. Leather Chem. Assoc.*, 2013, **108**(7), 257.
45. S. Balasubramanian and R. Gayathri, *J. Am. Leather Chem. Assoc.*, 1997, **92**(2), 218.
46. E. L. Tavani and N. A. Lacour, *J. Soc. Leather Technol. Chem.*, 1994, **78**(2), 50.
47. S. C. Lee, *et al.*, *J. Am. Leather Chem. Assoc.*, 2012, **107**(2), 33.
48. S. Cao, *et al.*, *J. Am. Leather Chem. Assoc.*, 2013, **108**(11), 428.
49. S. Cao, *et al.*, *J. Soc. Leather Technol. Chem.*, 2015, **99**(3), 120.
50. A. D. Covington, *Future Tanning Chemistries*, UNIDO Workshop, Casablanca, 2000.
51. R. Karthikeyan, *et al.*, *J. Soc. Leather Technol. Chem.*, 2007, **102**(12), 383.

13   Vegetable Tanning

13.1   Introduction

The relationship of humans with plant polyphenols is ancient, since they are components of plant materials and hence are a traditional feature of our diet. Modern interest usually centres on the antioxidant properties, which means that they can scavenge carcinogenic and mutagenic oxygen free radicals in the body. However, polyphenols are also associated with other physiological reactions of importance, such as accelerating blood clotting, reducing blood pressure and lowering blood serum lipid concentration.[1]

The oldest exploitation of plant polyphenols in technology is their ability to stabilise collagen in skin against putrefaction. It is not difficult to understand how vegetable tanning may have come about: prehistoric humans perhaps observed changes in hide or skin after it had lain in a puddle with plant material. The plant polyphenols that can be leached out of vegetable matter or plant parts have the power to react with the collagen, rendering it resistant to biochemical degradation, but more importantly remaining soft after a wetting–drying cycle. All manner of plant parts may contain tannins. They are present as a defence mechanism against insects, because they impart a stringent taste to the plant material. An example is unripe fruit, such as green apples, which can dry up the inside of the mouth, because the polyphenols react with protein in the saliva: this effect is designed to deter consumption until the seeds are mature.

Tanning Chemistry: The Science of Leather 2nd edition
By Anthony D. Covington and William R. Wise
© Anthony D. Covington and William R. Wise 2020
Published by the Royal Society of Chemistry, www.rsc.org

The reactivity of polyphenols is referred to as 'astringency', because they can react with protein, but also implying that the reactivity is high. The traditional test for astringency is to precipitate gelatine from solution with the active agent under investigation, in a tanning type of reaction.

It was said that for every type of leather the tanner wished to tan, there is a suitable plant tannin. Few are commonly used in the modern leather industry, but to get some idea of the range of tannins that were employed, a useful source is Procter's seminal work *The Principles of Leather Manufacture*, where he lists exploited plant materials in use at the turn of the nineteenth to twentieth century.[2]

Tannins are extracted from plant materials that contain commercially viable concentrations: this may be done with water at ambient or elevated temperature. Alternatively, organic solvents, such as ethanol, can be used, where it is claimed that with wattle the tans to non-tans ratio is higher.[3]

The extracts typically contain three fractions of materials, as follows.

1. *Non-tans* are the low molecular weight fraction, with molecular weight <500.[4]
   These compounds are designated non-tans to indicate that they have low tanning power; however, their presence in the extract is useful, because they contribute to solubilising the tans and hence assist penetration into the pelt. They have an additional use in their own right, as components in organic tanning reactions (see Chapter 24).
2. *Tans* are the medium molecular weight fraction, with molecular weight 500–3000.[4]
   The tanning function depends primarily on the phenolic hydroxyl content, to provide the astringency. Lower molecular weight compounds lack astringency, higher molecular weight impedes penetration into the pelt.
3. *Gums* are the high molecular weight fraction, with molecular weight >3000.
   These compounds are tans and higher molecular weight polyphenols complexed with carbohydrate species. The astringency and high molecular weight prevent penetration into pelt, causing surface reaction only. Hence their presence is undesirable. It is a common experience that higher yields of extract can be

obtained by using more aggressive conditions, such as higher temperatures, but this is actually counterproductive, because the higher yield of the extract contains higher amounts of gums rather than tannins and so the product can be damaging to the quality of the leather.

The mechanism of fixation of tannins onto collagen follows the stepwise mechanism introduced in Chapter 11. It has been proposed that the preliminary interaction between polyphenols and collagen starts with hydrophobic interactions, because these molecules are relatively insoluble in water, solubilised by the presence of the non-tans. Reaction then proceeds to hydrogen bonding, as the electrostatic element of the process. The details of the mechanism and structure–reactivity relationships of the vegetable tannins have been comprehensively reviewed by Haslam,[5] whose paper contains the following important discussions.

1. Other than the leather industry, there was no major industry exploiting vegetable tannins, hence the development of the understanding of the structural chemistry of these secondary metabolites was slow in comparison with other groups of natural products.
2. The roles of the components of plant extracts were not and still are not well understood, particularly the effects of the low molecular weight species on the solution properties of the higher molecular weight species.
3. The hydrolysable tannins are freely soluble components of plant tissue and thereby have dual roles of protection by astringency and contribution to taste.
4. The condensed tannins are chemically associated with the plant tissue. The structural relationships between the procyanidins indicate that there are preferred conformations, although the significance is not known (see later).
5. The role of hydrophobicity in tanning is illustrated by the observation that more tannin is required to precipitate a dilute solution of protein than a concentrated solution. This is because the tannin creates a hydrophobic layer over the surface of concentrated protein, causing precipitation. In the case of dilute protein, the tannin reacts with multiple sites, making it less hydrophilic, but more tannin is required to make it drop out of solution.
6. Polyphenols inhibit enzymes due to interaction with their protein structure. This has important implications in leather

processing, particularly leather area, as discussed in Chapter 2, and in leather recycling, discussed later. In the former case, it is worth recalling that the biotechnology of area gain by elastin degradation in chrome leather by elastase does not work in the presence of vegetable tannins: they interact with protein *via* hydrophobic bonding, therefore they can effectively tan elastin, rendering it resistant to elastolytic biochemical action, in the same way chrome stabilises collagen to microbiological attack.

7. Polyphenols react with proteins *via* multiple bonding and in this way they are better suited to reaction with structured proteins such as collagen, rather than unstructured species such as the globular proteins.

The family of plant polyphenols is characterised by the presence of phenolic hydroxyls, capable of reacting with collagen, at the basic groups on sidechains and at the partially charged peptide links *via* hydrogen bonding.[5] This is illustrated in Figure 13.1.

Haslam[6] proposed that a reaction of importance occurs in the gap regions of collagen, between the ends of the triple helices. Whereas the apparent ability of galloylated glucose molecules, the hydrolysable gallotannins, to fit into the available spaces might indicate that tannage can take place there, so that four catechin molecules could be fitted into a location around the hydroxyl of serine, the same interaction could not be modelled in the overlap regions. However, in general, molecular modelling could not confirm affinity between catechin and sidechains in the gap zone as the primary site of the vegetable tanning mechanism.[7] Therefore, in the absence of any contradictory information, it may be assumed that reaction can take place all along the helical structure.

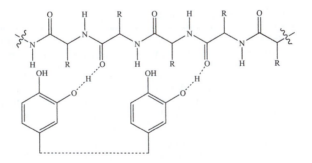

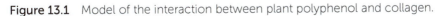

Figure 13.1   Model of the interaction between plant polyphenol and collagen.

Modelling the reaction between gallotannin and collagen does confirm the hydrogen bonding reaction at hydroxyproline, but also interaction at proline. In addition, it has demonstrated hydrophobic interaction at alanine and isoleucine.[8]

13.2 Vegetable Tannin Classification

Plant polyphenols can be classified into the following four groups, based on their structural characteristics: the hydrolysable tannins, with subgroups gallotannins and ellagitannins, the condensed tannins and the complex tannins.[9] Some examples of commercial tannins are given in Table 13.1.

The chemistry of the plant polyphenols is highly complex, as might be expected in a class of natural products. Tanners take a simplistic attitude to the structures of available tanning products, based on the technological outcome of the process rather than the details of the fixation mechanism, which is a neglected area of research. The ways in which the polyphenol types engage in tanning reactions depend on the primary structure rather than on the precise stereochemistry or shapes of the molecules and how they fit into the collagen structure. Tanners take account of the primary structure, but not the secondary structure. For many purposes, that is sufficient, but if plant polyphenols and their derivatives are to be used with more precision, for example in new organic tannages, a deeper understanding of molecular structure–reactivity–outcome relationships is required.

13.2.1 Hydrolysable Tannins

The so-called 'hydrolysable' or 'pyrogallol' tannins are saccharide-based compounds, in which the aliphatic hydroxyls are esterified by carboxylate species carrying the pyrogallol group, 1,2,3-trihydroxybenzene.

Table 13.1 Some sources of tannins and the composition of the tannin source.

Plant	Source of tannin	Tannin class	Tannin content (%)	Non-tans content (%)
Chestnut	Wood	Hydrolysable	5–15	1–2
Myrobalan	Nuts	Hydrolysable	25–48	14–17
Sumac	Leaves	Hydrolysable	22–35	14–15
Oak	Wood	Hydrolysable	4–10	1–2
Oak	Bark	Condensed	6–17	5–6
Quebracho	Wood	Condensed	14–26	1–2
Mimosa	Bark	Condensed	22–48	7–8
Mangrove	Bark	Condensed	16–50	9–15

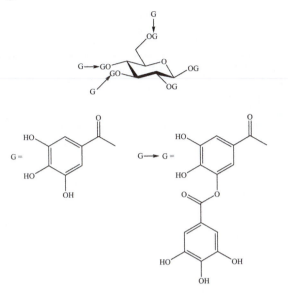

Figure 13.2 Structures of the gallotannins, hydrolysable plant polyphenols, illustrated by Chinese gallotannin.

The tannins are typically based on glucose, but there are variations in which derivatives of glucose can be the central moiety, such as sucrose.[9] They can be separated into two groups: the gallotannins and the ellagitannins. In each case, the properties are dominated by the ease of hydrolysis of the ester linkages and the reactivity of the phenolic hydroxyl groups.

13.2.1.1 Gallotannins

The smaller group within the hydrolysable tannins, *i.e.* less common polyphenols, is the gallotannins, in which the glucose core is esterified only with gallic acid. In addition, the bound gallate groups can be esterified themselves, at their phenolic hydroxyls: this is called depside esterification. The structures are illustrated in Figure 13.2.

Variations in structures come from the degree of esterification of the glucose centre and the degree of depside esterification. The astringency of any polyphenol is dependent on the effective concentration of reaction sites within the molecule, *i.e.* the number of phenolic hydroxyls. This is demonstrated in Table 13.2, in which the direct interaction between protein and polyphenol can be quantified: the more negative the free energy of transfer, the greater is the interaction.

Table 13.2   The astringency of model hydrolysable tannins, measured by the free energy of transfer of the protein bovine serum albumin from aqueous solution to an aqueous solution containing the polyphenol.[10]

Polyphenol: D-glucose esters	Molecular weight	ΔG (kJ mol^{-1})
β-1,2,6-Tri-O-galloyl	636	−0.9
β-1,2,3,6-Tetra-O-galloyl	788	−9.1
β-1,2,3,4,6-Penta-O-galloyl	940	−26.9

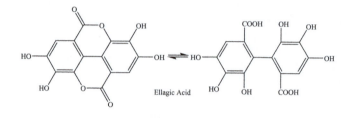

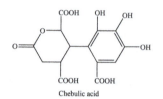

Figure 13.3   Structures of ellagic and chebulic acids.

13.2.1.2   Ellagitannins

In these compounds, the esterifying moieties include gallic acid, ellagic acid and chebulic acid: the structures are given in Figure 13.3 and examples of combination esterification are given in Figure 13.4.

Some of the properties of the hydrolysable tannins can be discerned from their structures.

1. They are highly astringent, owing to the large number of phenolic hydroxyls groups so close together; in this regard, the gallotannins are more astringent than the ellagitannins. A contributor to the reactivity of the hydrolysable tannins is the acidity due to the presence of carboxylic acid groups, *e.g.* the pHs of 50° Barkometer solutions of valonia, sumac and chestnut (Figure 13.5) are 3.2, 3.7–4.2 and 2.6–2.8 respectively; the less acidic tannins are therefore self-buffering.

2. Reaction with collagen is exclusively *via* hydrogen bonding. This means that the reaction can be reversed by the action of

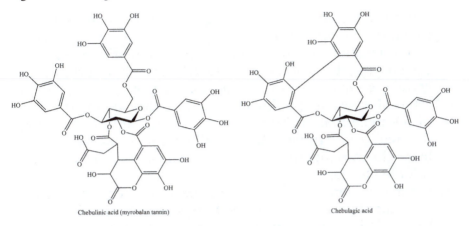

Chebulinic acid (myrobalan tannin)

Chebulagic acid

Figure 13.4 Structures of chebulinic and chebulagic acids.

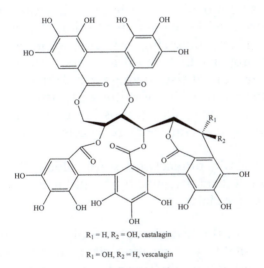

R_1 = H, R_2 = OH, castalagin

R_1 = OH, R_2 = H, vescalagin

Figure 13.5 Major components of chestnut tannin.

hydrogen bond-breaking compounds, *e.g.* urea, calcium and chloride, or an organic solvent; in this way, complete detanning is possible.

3. The shrinkage temperature conferred by these tannins is 75–80 °C; this is characteristic, *i.e.* if a leather is known to be fully tanned and the shrinkage temperature is <80 °C, it is probably tanned with hydrolysable tannin(s).

4. The ability of phenols to discolour (in the case of vegetable tannins, the darkening effect is to redden) depends on the formation of phenyl radicals by the loss of hydrogen to atmospheric

oxygen. The formation of free radicals causes bond shifts and oxidative coupling, which means polymerisation; if this results in the creation or the linking of chromophoric groups (see Chapter 16), then colour is developed. In the case of the hydrolysable tannins, the chromophores, the benzene rings of the ester moieties, are not linkable because they are too far apart in the molecule; in this way, they are resistant to reddening, referred to as 'lightfast'. Tara, a gallotannin, is often quoted as being the most lightfast vegetable tannin and makes very pale-coloured leather.

5. From their name, these tannins can be hydrolysed like any other ester and this can occur both within the tan bath and within the leather. The former causes a loss of tanning material and silts up the tanning pits; the latter is no bad thing, because the deposited carboxylic acid moieties contribute to the water resistance and wear properties of the leather. In addition, the carboxylates have a buffering effect, conferring resistance to pH change; this is an important property in archival leathers, such as bookbinding, where leathers are at risk of long-term damage by the action of acidic atmospheric oxides of sulfur and nitrogen.

6. The presence of pyrogallol groups in these tannins means that they can undergo semi-metal reactions, discussed in Section 13.7.

7. They are generally unreactive towards aldehydic compounds at the aromatic nuclei, because of the inductive effect of the carboxyl groups (see later).

The astringency of the hydrolysable tannins seems to depend primarily on the availability of reaction sites, the phenolic hydroxyls. There is no information on the second-order effect of the conformation of the molecules. This is in part because the offers are so large in pure vegetable tanning, which might be as high as 30% on pelt weight in order to fill up the pelt structure. In processes involving less vegetable tannin, such as semi-metal (see later), the influence of detailed polyphenol structure and conformation has not been studied.

13.2.2 Condensed Tannins

The so-called 'condensed' or 'catechol' tannins have a flavonoid ring structure, as shown in Figure 13.6; the more common structures have a catechol group, 3,4-dihydroxybenzene, as the B-ring. The A- and B-rings are aromatic and the C-ring is alicyclic.

The differences between tannins lie in the pattern of hydroxyls: position 3 in the C-ring is always occupied, in the A-ring position 7 is always occupied, but position 5 may or may not be occupied, and in the B-ring positions 3' and 4' are always occupied, but position 5' may or may not be occupied, as shown in Figure 13.7.

The hydroxyl in the 3-position in the central C-ring may be *cis* or *trans* with the B-ring, as exemplified by gallocatechin in Figure 13.8.

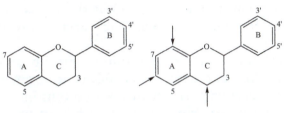

Hydroxylation positions   Polymerisation positions

Figure 13.6 Flavonoid ring system, showing the positions for hydroxyls and the positions for polymerisation.

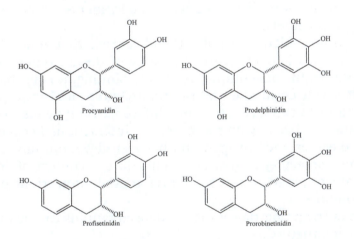

Procyanidin   Prodelphinidin

Profisetinidin   Prorobinetinidin

Figure 13.7 Variations in hydroxyl patterns in the flavonoid polyphenols.

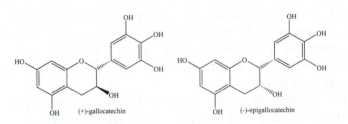

(+)-gallocatechin   (-)-epigallocatechin

Figure 13.8 Stereochemistry in the flavonoid ring system.

Because the linking point for polymerisation is the alicyclic C-ring, there is a variation in conformation, shown in Figure 13.9:[5] B-1 to B-4 are the principal conformations; B-5 to B-8 are known are but less common.

The hydroxylation pattern determines the shape of the polymers of flavonoids, because the presence of a 5-hydroxyl group restricts the ability of the ring system to react at the 6-position; therefore, these tannins tend to be linear. If the 5-hydroxyl group is absent, the ring system can involve reaction at the 6-position; therefore, these tannins can be branched. The variation is illustrated in Figure 13.10.

The procyanidins and prodelphinidins are linear polyphenols. The profisetinidins and prorobinetinidins are branched polyphenols. As is the case with the hydrolysable tannins, the effects of polymeric structure on tanning reactions have not been investigated. However, it can be postulated that the shape of the molecule will probably not have much effect on conventional vegetable tanning, where the pelt structure is filled with excess tannin. However, shape is more likely to have some influence on combination tannages (see later).

There is no evidence available to indicate whether or not these stereochemistries impact on the reactivity of these isomers in a tanning context. In conventional vegetable tanning, it is probable that any effect will be masked by the high content and multiple weak interactions in collagen, so subtle differences will be lost. However, any tannages that rely on more precise interactions between flavonoid molecules and collagen and between flavonoid molecule and flavonoid molecule, as discussed later in the context of combination tannages, may exhibit some preferential interactions based on conformation.

Some of the properties of the condensed tannins can be discerned from their structures.

1. Although the condensed tannins may contain as many aromatic hydroxyl groups as the hydrolysable tannins, they are dispersed over larger molecules, hence the flavonoid compounds are typically less astringent. The molecular weight of catechin is 290, so the lower limit for inclusion in the tans fraction is the dimer.
2. Like the hydrolysable tannins, the condensed tannins react with collagen *via* hydrogen bonds. In addition, they are capable of reacting at the 5- and 7-positions *via* quinoid species,

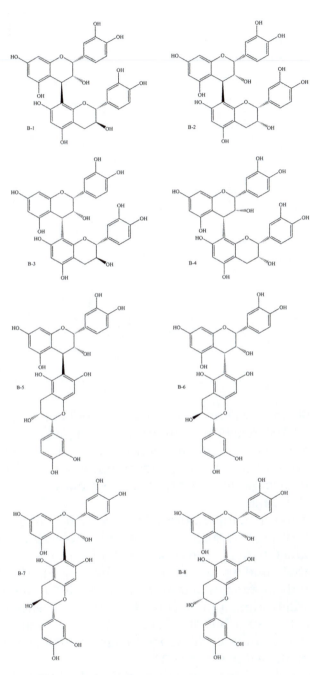

Figure 13.9 Conformations of procyanidins.

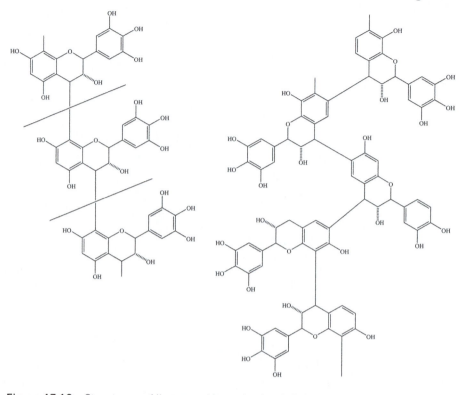

Figure 13.10 Structures of linear and branched polyflavonoid tannins.

resulting in covalent reaction at the collagen lysine amino groups;[11] the reaction is modelled in Figure 13.11. See also Chapter 10. Because of this contribution to the fixation reaction, 5–10% of the tannin is not removable by hydrogen bond breakers or organic solvent, so the pelt does not revert to raw after detanning.[12]

3. The shrinkage temperature (T_s) of leather tanned with condensed tannin(s) is 80–85 °C; as with hydrolysable tannins, the T_s is characteristic, *i.e.* an observed T_s of >80 °C is a strong indication that condensed tannins have been used. The difference between the outcomes of tanning with the different types of tannins is interesting: it must come from the difference in reaction type, *i.e.* the contribution of covalency to the similar reactions makes a measurable difference to the shrinkage temperature. This is explained in Chapter 23.

4. The proximity of the aromatic nuclei in the flavonoid structure means that the free radical oxidative bond rearrangements can

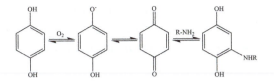

Figure 13.11 The quinoid reaction with collagen, modelled by hydroquinone.

take place easily. Therefore, these tannins redden, creating a rapid colour change on the leather surface: mimosa-tanned leather exhibits this effect, darkening visibly after only a few minutes of exposure to sunlight in air.

5. The flat structure of these tannin molecules means that they can aggregate together as layers, linked by van der Waals forces and electrostatic interactions; the aggregates are referred to as 'reds'. The formation of reds occurs particularly at lower temperatures, resulting in the insoluble particles reacting on the pelt surface because they are astringent but too large to penetrate into the pelt. The attractive forces are weak and are readily reversed by warming the solution or including a solubilising agent in solution, such as an auxiliary syntan (see Chapter 14). The formation of reds is so important in practical tanning that the tendency is quoted in product specifications. Under acidic conditions, reds may take the form of phlobaphenes, which are insoluble, red pigments that will function like reds in their tanning action, but are not solubilised merely by warming the solution.

6. Condensed tannins typically do not engage in semi-metal reactions, except for the less common prodelphinidins and prorobinetinidins, which have a pyrogallol B-ring. However, they can undergo analogous polymerisation reactions with aldehydic reagents, discussed later in the context of combination tannages.

One of the clear differences between tannins known in the art is the reaction to light: the lightfastness of some tannins is indicated in Table 13.3 by the change in colour in the CIELAB colour space after 24 h of exposure to UV light in the standard Xenotest. In this limited group, there is a clear distinction between the relatively lightfast hydrolysable tannins and the condensed tannins the redden in light.[13]

The interaction of plant polyphenols with UV light in a tanning context can be regarded as a mechanism for protecting the integrity of the collagen substrate. It has been reported that collagen tanned with Chinese bayberry extract (*Myrica rubra*, a prodelphinidin) confers resistance to UV damage at collagen backbone sites where there is tyrosine.[14]

13.2.3 Complex Tannins

These are mixtures of tannin types in which a hydrolysable gallotannin or ellagitannin moiety is bound glycosidically to a condensed tannin moiety.[4] Such tannins demonstrate the properties of both types of polyphenol, illustrated by acutissimin A, shown in Figure 13.12. Such tannins are not uncommon in Nature, but they are traditionally uncommon in industrial vegetable tanning.

Table 13.3 CIELAB colour change after 24 h in the Xenotest: more negative ΔL means darker on the scale of black to white, more positive Δa means more yellow and less blue and more positive Δb means more red and less green.

Tannin	Type	Structure	ΔL	Δa	Δb
Mimosa	Condensed	Profisetinidin	−19.9	10.19	10.54
Quebracho	Condensed	Procyanidin	−18.2	6.98	9.89
Gambier	Condensed	Procyanidin	−14.6	8.33	8.33
Valonia	Hydrolysable	Ellagitannin	−8.27	1.54	5.58
Chestnut	Hydrolysable	Ellagitannin	−8.05	0.11	4.24
Tara	Hydrolysable	Gallotannin	−6.80	3.88	5.19

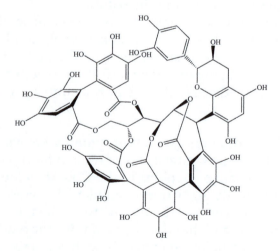

Figure 13.12 Structure of a complex tannin, acutissimin A.

13.3 General Properties of Vegetable Tannins

Vegetable tannins exhibit two general properties.

1. The reaction of plant polyphenols at the basic sidechains lowers the isoelectric point of collagen by 1–2 pH units.[12] The effect is to reduce the need to neutralise before post-tanning, in contrast to the necessity for neutralising chrome-tanned leather, in which the isoelectric point is raised by a similar amount (see Chapter 15).
2. The reactivity of vegetable tannins can be modified in two ways. First, the pH can be raised, *e.g.* in the case of chestnut extract it is said to be 'sweetened' when the pH is adjusted from about 2.7 to 3.5–4.5. Second, the water solubility can be increased by sulfonation using sodium bisulfite; this is more commonly applied to the condensed tannins, which are less water soluble naturally and readily undergo such reactions at their A-rings.

13.4 Practical Vegetable Tanning

The history of vegetable tanning concerns the development of both the tanning agents employed and the equipment used for the process. In the former, changes involved the nature of the agent, gradually changing from the use of the plant material itself, to the availability of extracts. In the latter, pits have been the preferred vessel, but the way in which they were used changed. Historically, tanning was conducted in pits in a process referred to as 'layering'. A layer of the appropriate plant material, for example oak bark, was placed in the bottom of the pit, followed by a layer of hides or skins, then another layer of plant material was placed on the hides, followed by another layer of plant material, and so on. The alternate layering continued until the pit was full, as indicated in Figure 13.13. The pit was then filled with water.

The water leached the polyphenols out of the plant material and the dilute tannin diffused into the hides, converting them into leather. The dilute nature of the solution limits the reactivity of the tannins, allowing it to penetrate through the pelt cross-section. The reaction is slow, in part due to the static nature of the process, so an early form of quality assurance applied more than half a millennium ago in England was the requirement for hides to stay in the pit for 'a year and a day'.

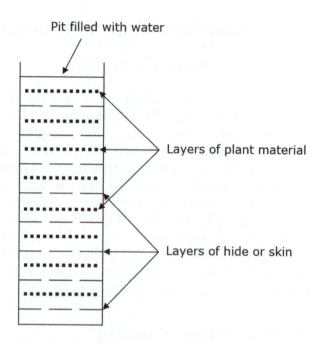

Figure 13.13 Principle of layering as a tanning technique.

13.5 Modern Vegetable Tanning

Once the principles of extracting and concentrating vegetable tannins had been established, their application meant that the process could be accelerated to reaction times of weeks. This was made faster by introducing heating to the pits, so-called 'hot pitting'. However, the use of extracts created a problem in itself. The astringency of vegetable tannins means that there is a strong likelihood of overtanning the surfaces. This is an example of 'case hardening', where the structure at the surface becomes so overtanned or filled with tannin that a barrier to tannin penetration is created and the tanning reaction within the cross-section stops. In order to avoid this happening, corrective action must be taken. It was recognised that one approach to mitigate the effects of tannin astringency is to reduce the reaction rate, and the easiest way to control the rate is to control the concentration of the tannin: the lower the concentration, the slower is the rate of reaction. This is another example of the requirement for the tanner to balance the rate of penetration, k_p, and the rate of fixation, k_f.

- If $k_p > k_f$, the tannin colours through the pelt, but does not react – this is the requirement at the outset of the reaction.

- If $k_p = k_f$, the tannin reacts uniformly through the cross-section of the pelt – this is the desired effect during the majority of the tanning process.
- If $k_p < k_f$, the tannin will tend to react more on the surfaces than within the pelt – this is an acceptable situation in chrome tanning, when the concentration of the tanning agent diminishes towards the end of the reaction, but this does not apply in vegetable tanning and hence this particular balance of rates is undesirable.

The vegetable tanning process has been modelled as a diffusion-controlled process,[15] which conforms to Fick's Second Law, expressed here for one dimension:

$$\frac{\partial C}{\partial t} = D_{eff} \frac{\partial C^2}{\partial z^2}$$

where C = concentration of tannin free to diffuse, expressed as the amount of tannin per unit volume of hide, D_{eff} = effective diffusion coefficient and z = longitudinal coordinate.

It has also been modelled using modified versions of the Cegarra–Puente equation:[16]

$$Q_e = K\sqrt{t} \tag{13.1}$$

$$\ln[1 - (Q_t^2/Q_\infty^2)] = -Kt \tag{13.2}$$

$$\ln\{-\ln[1 - (Q_t^2/Q_\infty^2)]\} = a \ln t + a \ln K \tag{13.3}$$

where Q is the quantity of fixed tannin, K is the rate constant and t is time. Using eqn (13.3), the rates of reaction for wattle tannage were determined at different temperatures, with or without a pretreatment with proteolytic enzyme to open up the hide structure, as shown in Table 13.4.[17] The kinetic studies followed experiments on enzyme treatment to assist the vegetable tanning process, where it was shown that 99% uptake could be achieved in 1 h after treatment with 0.2% of protease, compared with 85% in the control.

Table 13.4 Kinetics of wattle tanning assisted by enzyme pretreatment.

	Rate constant, K (min^{-1})	
Temperature (°C)	Control	Enzyme pretreated, pH 7.2
25	0.0088	0.0482
30	0.0214	0.0728
35	0.0582	0.1484
40	0.0718	0.1562

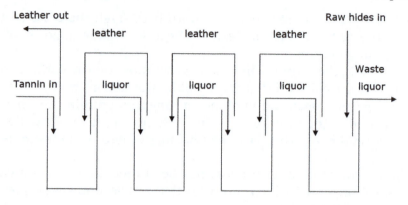

Figure 13.14   Countercurrent system of pit tanning.

13.5.1   Countercurrent Pit Tanning

The modern countercurrent system of vegetable tanning is illustrated in Figure 13.14: four pits are used in the illustration, but in reality more might be used. The method is called countercurrent, because the tannin concentration moves one way and the hides move the other way, *i.e.* counter to the current of tannin movement.[18]

The hides start in the first pit, where the tannin concentration is low. Therefore, the reaction is slow and the tannin can penetrate through the cross-section. After a few days, the hides are ready to go into a more concentrated solution: faster reaction is allowed, because tannin fixation from the first bath has reduced the reactivity of the substrate somewhat. Some of the depleted solution is discharged to waste and the concentration is brought back up with some of the solution from the second pit, ready to receive the second pack of rawstock. The process is repeated: the hides are moved to a third pit, and the first and second pits are 'mended' with higher-concentration tannin solutions, the second pack is moved to the second pit and a third pack is introduced into the first pit, and so on. This process is continued until the leather is fully tanned and taken from the pits.

13.5.2   Rockers

Figure 13.15 represents what happens at the pelt surface when tannage is conducted with the pelt suspended in a static manner. Because tannin in solution is reacting at the surface, the concentration at the surface is lower than the bulk concentration; consequently, the reaction rate is lower than it might be if the concentration at the surface

Tannin content of pelt or solution

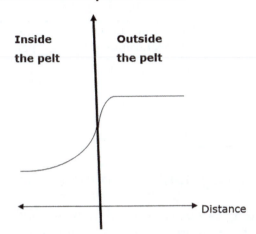

Figure 13.15 Representation of tanning reactions at the pelt surface.

were the same as the bulk concentration. Within the pelt, there is a concentration gradient, because fixation is always greater nearer the surface.

The concentration gradient outside the pelt is easily broken by rocking the pelt in the solution, creating movement at the surface, so that the mixing rate is faster than the reaction rate.

13.5.3 Ultrasound

Power ultrasound is a newer way to introduce energy into a system, where it can have the effect of breaking chemical bonds and dispersing components in solution. It has been investigated within leather technology as a way to improve the diffusion of chemicals into pelt by reversing aggregation reactions of the type exhibited by condensed tannins. For example, the effect on particle size is similar for chestnut (hydrolysable) and mimosa or quebracho (condensed), but greater for gambier (condensed).[19] In this way, the reaction rate can be improved, as indicated in Table 13.5. The effect of ultrasound is dependent on the power applied, but less dependent on the vegetable tannin.[20]

It is generally agreed that ultrasound can influence the rate of processes, particularly vegetable tanning, for example, doubling the rate of diffusion of wattle compared with stirring or static reaction.[21] It has not been applied in industry, where it might aid the transition from pits to drums.[22]

Table 13.5　Use of ultrasound to assist vegetable tanning with mimosa, quebracho and chestnut: maximum power 1000 W at 27 kHz.

Power (%)	Vegetable tannin concentration (%)	Time to complete penetration (h)
100	30	24
100	60	12
75	30	24
75	60	14
50	30	24
50	60	24

13.6　Other Vegetable Tanning Technologies

Vegetable tanning can be conducted in drums, although it is uncommon. In order to facilitate rapid penetration, higher offers of pretanning agents are used: this includes non-swelling acids (see Chapter 9), syntans (see Chapter 14) and other hydrogen-bonding auxiliaries such as polyphosphates. The reaction is typically conducted at about pH 3.5 and up to 38 °C, but in zero float.[18]

It has been reported that a more effective and efficient wattle tannage can be achieved from an ethanolic extract, followed by tannage in ethanol.[23] A more radical suggestion is to tan in an apolar solvent.[24] This is technology discussed in Chapter 19: the principle is the same as applied in chrome tanning, but in the work reported on vegetable tanning the solvents were dichloromethane, 1,1,1-trichloroethane, 1,1,2-trichloroethene and 1,1,2,2-tetrachloroethene. Experiments on 20% offers of chestnut, mimosa and sulfited quebracho extracts in a 300% volume of solvent for sheepskins showed that 99–100% uptake of tannin was achieved in 1–2 h.

13.7　Combination Tanning

Strictly, the term 'combination' just refers to the use of more than one tanning agent. The retanning of chrome-tanned leather could be viewed as a combination tannage, but there is no assumption of interaction between the components. A useful basis for several combination tannages in which the components interact positively is the polyphenols and some of those options are discussed here. A more general analysis of combination tanning is presented in Chapter 24.

13.7.1 Semi-metal Tanning

The combination of vegetable tannins with metal salts has been used for thousands of years. Almost certainly, the purpose was concerned with the mordanting effect on dyeing, rather than a stabilising process. What was not known for most of that time was how the two reagents actually combine and what the outcome was in terms of leather properties and performance.

It is known now that the shrinkage temperature of semi-metal-tanned leather is much higher than expected. This is an example of synergy, when the total effect is greater than the sum of the parts. If the two components of the tannage can be separated for the purposes of this argument, there are three possibilities:

1. $\Delta T_s(\text{observed}) < [\Delta T_s(\text{vegetable tannage}) + \Delta T_s(\text{aluminium tannage})]$
2. $\Delta T_s(\text{observed}) = [\Delta T_s(\text{vegetable tannage}) + \Delta T_s(\text{aluminium tannage})]$
3. $\Delta T_s(\text{observed}) > [\Delta T_s(\text{vegetable tannage}) + \Delta T_s(\text{aluminium tannage})]$

If the first case applies, the observed change in stability is less than the calculated value; therefore, the tannages are antagonistic, creating unexpected destabilisation. If the second case applies, the effects of the two tannages are simply additive, so it would indicate independent reactions. If the third case applies, the observed change in stability is greater than the calculated value; therefore, the tannages are synergistic, creating a new stabilising reaction. The degree of difference, from the equation, is an indication of the extent of the potency of the new reaction. We can put some actual numbers to the equation, where the T_s of raw pelt is typically 65 °C:

$$\Delta T_s(\text{observed}) > [\Delta T_s(\text{vegetable tannage}) + \Delta T_s(\text{aluminium tannage})]$$
$$\Delta T_s/°C \quad 115-65 \quad 75-65 \quad 75-65$$
$$= 50 \quad\quad = 10 \quad\quad = 10$$

More than half a century after Procter's observation about stoving patent leather, Frey and Beebe demonstrated the useful properties of semi-alum leather[25–27] and, about a quarter of a century later, Rao and Nayudamma studied the reaction using citrate-masked aluminium(III) sulfate, illustrated in Table 13.6.[28,29] Here individual tannages were compared with semi-alum, when aluminium(III) followed vegetable tannin, and vegetable retan, when the reverse order was used.

Table 13.6 Semi-alum tannages.

Tannin	T_s (°C)			
	Vegetable tannage	Al tannage	Semi-alum tannage	Vegetable retan
None		80		
Divi–divi	68		115	85
Myrobalan	68		120	84
Quebracho	76		88	85
Wattle	78		98	102

Divi-divi and myrobalan are typical hydrolysable tannins, producing high hydrothermal stability in the semi-alum reaction. However, when the vegetable tannin is applied after the aluminium, the outcome appears to be more like an additive combination. This can be understood in terms of the high affinity of the interaction: in the semi-metal reaction, the metal creates a network by linking the polyphenol molecules together, but the vegetable retan reaction merely creates metal complexes at the collagen reaction sites, so no network is formed. The procyanidin molecules of quebracho do not complex with metals to any significant extent, unlike the prorobinetinidin component of wattle (mimosa). It is interesting that the semi-metal and vegetable retan reactions are almost the same: the lesser reactivity of the wattle even allows the combination reaction to take place if the reagents are offered together.[30] Kallenberger and Hernandez reviewed the scope of the semi-metal reaction using vegetable-tanned leather (the composition was not disclosed, but is likely to have included mimosa), indicated in Tables 13.7 and 13.8.[31,32]

It can be seen that the outcome is variable and that masking is more likely to reduce the reactivity of transition metal salts in this reaction. It was concluded that the preferred metal salts are of aluminium(III)[33] and that the reaction is based on polymerisation of the polyphenol by the metal, moreover suggesting that the carboxyls of collagen play no significant part in the reaction. On this basis, they proposed the formation of a stiff network within the collagen fibre structure.

Sykes and Cater[34] proposed that the critical requirements of the semi-metal reaction are the presence of a polyhydroxy aryl ring and more than one such moiety in the molecule. With the use of aromatic

Table 13.7 Semi-metal tannages.

Metal salt	T_s (°C)	
	Unmasked	Citrate masked
None	86	
Co(II) chloride	112	95
Pb(II) thiocyanate	90	85
Mn(II) sulfate	90	90
Ni(II) sulfate	107	89
Fe(III) chloride	101	111

Table 13.8 Semi-titan tannages, using citrate-masked Ti(IV).

Vegetable tannin	T_s (°C)
None	84
Quebracho	99
Chestnut	90
Wattle	106

Table 13.9 Model vegetable tannin species for the semi-alum tanning reaction.

OH positions in hexahydroxydiphenyl sulfone, $(HO)_3C_6H_2-SO_2-C_6H_2(OH)_3$	T_s (°C)
2,2′,3,3′,4,4′	98
2,3,3′,4,4′,5′	104
3,3′,4,4′,5,5′	105

sulfones as models, Orszulik *et al.*[35] demonstrated more specifically that the reaction requires the presence of pyrogallol moieties in the polyphenol compounds (see Table 13.9).

From Table 13.10, the presence of the third hydroxyl group in the pyrogallol system lowers the second ionisation constant compared with the dihydroxyphenyl system. Since the complexation reaction occurs at pH values much lower than the range in which the phenol groups are ionised, the reaction involves dative complexation, indicated in Figure 13.16. Therefore, the inductive effect of the third hydroxyl weakens the O–H bonds, increasing the ability of the other hydroxyls to engage in complexation with the aluminium(III) species.

Table 13.10 Properties of model polyphenol compounds.

	Catechol	Pyrogallol	3,3',4,4'-Tetrahydroxydiphenyl sulfone	3,3',4,4',5,5'-Hexahydro xydiphenyl sulfone
Ionisation				
pK_{a1}	9.5	9.2	7.8	7.5
pK_{a2}	12.5	11.6	12.1	10.4
pK_{a3}		12.7		12.6
Metal–ligand equilibrium				
pK_1	6.6	6.0	5.0	4.7
pK_2	9.6	8.6	7.2	6.6
pK_3	14.3	14.6	10.7	9.6

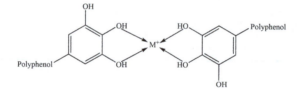

Figure 13.16 Basis of the semi-metal tanning reaction.

It has been shown that the semi-metal tannage can be conducted with a range of metals salts, including zinc(II)[36] and lanthanide(III) compounds,[37] so it has been proposed that semi-metal tannage with aluminium(III) could be an alternative to chrome tanning.[38] Using a pretannage of 1% glutaraldehyde or 12% syntan, then tannage with 20% tara extract and 8.5% Al_2O_3 as citrate-masked sulfate, the respective shrinkage temperatures were 113 and 108 °C. It is interesting that the enthalpies of denaturation show an inverse relationship between T_s and ΔH, indicating that there is more disruption to the natural collagen structure by the more effective tanning process. The quantities of reagents employed in this study are unlikely to be practical with regard to a commercially viable and versatile product.

In a modern review of semi-metal tanning, Covington and Lampard[39] retanned myrobalan-tanned pelt with vanadium(IV) oxysulfate or basic titanium(IV) sulfate, isoelectronic d[1] transition metal salts; the respective shrinkage temperatures were 104 and 100 °C. After storing the leathers for 1 year under ambient conditions, the shrinkage temperature of the semi-vanadyl leather had fallen to

Table 13.11 Properties of some semi-metal leathers.

Vegetable tannin	Metal salt retannage	Rate of shrinking (10^2 s^{-1})	T_s (°C)
Mimosa	Al(III)	4.06	94
	Zr(IV)	3.65	91
Myrobalan	Ti(III)	0.51	98
	V(IV)	0.57	95

60 °C, equivalent to raw pelt (a similar observation was made by Kallenberger and Hernandez in a semi-molybdenum leather).[31] At the same time, the tear strength decreased to one-third of the original value. Since the decrease in hydrothermal stability includes a complete loss of the tanning effect of the vegetable tannin, the mechanism must be based on the degradative redox effect of the metal salt. This is an uncommon example of reversal of tannage under ambient conditions, extending to the collagen structure itself, with the seeds of reversal built into the leather. In the context of a 'cradle-to-cradle' approach to materials (the reuse of materials rather than disposal at the end of the working life – a 'cradle-to-grave' approach), the reactions are interesting. This is discussed further in Chapter 24.

From Table 13.11, measurement of the rate of shrinking in the denaturation transition shows an order of magnitude difference between the ellagitannin myrobalan and the profisetinidin/prorobinetinidin mimosa: the indication is that the pyrogallol and gallic acid groups are not equivalent, so a more stable complex is formed by the pyrogallol group when it is unaffected by the inductive influence of the carboxyl group.[38]

The ability of the polyphenol moieties in vegetable tannins to complex metals is widespread across the Periodic Table, allowing the chemistry to be exploited as a scavenging mechanism. Shi[40] reported on the effectiveness of *Myrica rubra* immobilised on collagen to bind several metallic species, as set out in Table 13.12.

In comparison,[41] a more unusual semi-metal reaction has been reported for tannic acid with laponite clay, a synthetic compound that forms a clear thixotropic gel in solution, when it is thought to decompose according to the following equation:

$$Na_{0.7}Si_8Mg_{5.5}Li_{0.3}O_{20}(OH)_4 + 12H^+ + 8H_2O \rightarrow 0.7Na^+ + 8Si(OH)_4 + 5.3Mg^+ + 0.4Li^+$$

Tannage with 20% tannic acid produced a shrinkage temperature of 67 °C, then treatment with 3% laponite at pH 3.5 changed the shrinkage

Table 13.12 Absorption capacity, Q_e, of *Myrica rubra* for metal ions at 30 °C for 24 h.

Metal ion	Q_e (mg g^{-1})	Mechanism(s)
Au^{3+}	1400	Redox/complexing
Hg^{2+}	198	Chemical
UO$_2$$^{2+}$	107	Chemical
Mo^{6+}	82	Varies with Ph
Pd^{2+}	80	Chemical
Pb^{2+}	79	Physical/chemical
Cr^{6+}	78	Redox
Bi^{3+}	73	Multiple
Pt^{2+}	73	Chemical
Th^{4+}	55	Chemical
Cd^{2+}	24	Physical
Cu^{2+}	7	Chemical

temperature to 89 °C.[42] The chemistry may or may not be useful for making wet white, but it does illustrate the reactivity of polyphenols and the versatility of the semi-metal reaction.

13.7.2 General Properties of Semi-metal Leathers

These can be summarised as follows.

1. The most marked property of semi-metal leathers is the elevated shrinkage temperature; like other tannages, the value of the shrinkage temperature depends on the nature of the complexation reaction.
2. The vegetable tannin is immobilised because it is cross-linked. This means that the tendency for the polyphenol to migrate when the leather is wetted is practically eliminated. Therefore, the semi-metal-tanned leather is useful in applications where the leather is in contact with skin, *e.g.* wristwatch straps.
3. The complexation reaction reduces the cationic nature of the metal species; the charge on the cation is dispersed to some extent over the polyphenol complex but not completely, because the bonding is not made with ionised hydroxyls. Although the problem of interaction with anionic species is not eliminated, complexation *in situ*, *e.g.* sequestration by polyphosphate, is effective in neutralising the cationic nature.
4. The leathers are significantly fuller and more hydrophilic than, for example, chrome-tanned leather, *i.e.* they have the

Table 13.13 Observed differences between semi-chrome and semi-alum tannages.

Reaction parameter	Semi-chrome	Semi-alum
Rates of reaction:		
Metal salt uptake	Slow	Fast
Increase in T_s	Slow	Fast
Effect of heating tannage	Apparent increase in both rates of reaction	No apparent increase in either rate of reaction
T_s achieved/°C	<110	Up to 120

characteristics of vegetable-tanned leather, because they conform to the tanners' 'rule of thumb' which asserts that the dominant character of the leather comes from the tannage applied first.

13.7.3 Semi-chrome Tanning

In practical semi-metal tanning, the semi-chrome process appears to be anomalous, because it is observed to be less easily accomplished than the corresponding tannage with aluminium(III) or other successful metal salts. Generalised observations of semi-metal tanning processes are summarised in Table 13.13, which reveals important differences between chromium(III) reactions and those involving other metal salts, exemplified by aluminium(III).[43]

The reaction mechanism of semi-metal tanning has been explained in terms of the metal species crosslinking the pyrogallol moieties, to create a matrix that interacts with the collagen in a concerted manner,[44] so it has been assumed that the less effective reaction involving chromium(III) is due to the lack of affinity between chromium(III) and phenolic hydroxyls, causing a slow reaction and hence a slow rise in shrinkage temperature.

A powerful piece of evidence regarding the nature of the reaction comes from dynamic mechanical thermal analysis.[45] The response of the sample, illustrated in Figure 13.16, provides information about the viscoelastic properties of the material; in this case, the properties are affected by the chemistry of the tanning processes. The technique is based on measuring the way in which the response or strain of the material lags behind a deforming stress. When the stress is sinusoidal, the phase lag, δ, will lie between 0° for a perfectly elastic material and 90° for a perfectly viscous material. Tan δ is measured as a function of temperature, which changes the viscoelastic response. Collagen has been likened to a copolymer,

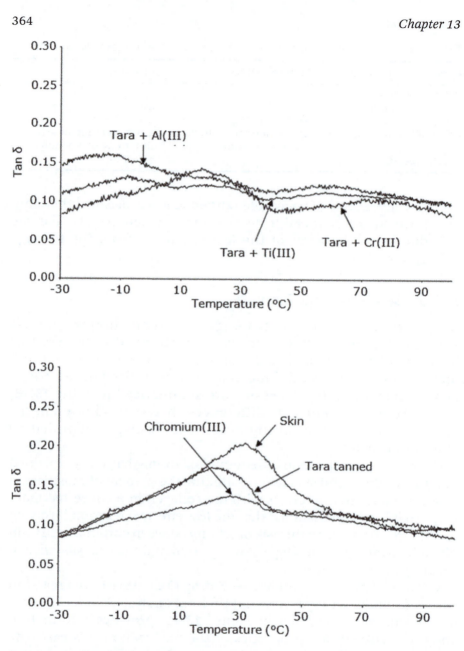

Figure 13.17   Dynamic mechanical thermal analysis (DMTA) of leathers. Courtesy of the *International Union of Leather Technologists and Chemists Societies*.

comprised of so-called soft and hard blocks.[46] In Figure 13.17,[47] the region studied corresponds to the β or soft block transition, which has been shown to be dependent on shrinkage temperature and is assumed to relate to the mobility of the α-amino acids, as opposed to the imino acids, which determine the restricted motion of the α or hard blocks.

It can be seen that all tannages affect the trace with respect to untanned skin. It is apparent that tara (hydrolysable gallotannin) and chromium(III) have markedly different effects. However, there are similarities between the traces for semi-alum(III) and semi-titan(III) tannages, as might be expected. What is not expected is that the graph for semi-chrome-tanned leather is unlike the graphs for the other two semi-metal tannages and is more like a combination of the tara-tanned result and the chromium-tanned result. It is this observation that leads to the conclusion that the semi-chrome tanning reaction is a combination of the two separate reactions, so the high shrinkage temperature is due to chrome tanning, although modified by the interfering or blocking effect of the polyphenol. This does not rule out the possibility of some complexation between chromium(III) and the polyphenol hydroxyl groups, which is established in other semi-metal reactions,[34] but these latest observations and conclusions are consistent with the qualitative observations set out in Table 13.13.

13.8 Condensed Tannins and Aldehydic Crosslinkers

The interaction between condensed tannins and formaldehyde is well known and had been used as the basis for making materials analogous to plastic long before the chemistry of polymerisation was developed.[48] The reaction might be modelled as shown in Figure 13.18, while recognising that formaldehyde is actually a polymeric compound, depending on the conditions (see Chapter 14).

The application of a combination of condensed tannin and aldehyde to tanning was first proposed by DasGupta.[49] The specific requirements of the flavonoid ring system in the reaction were defined in Covington and Ma's patent of 1996, in which the parameters of the reaction were specified, as illustrated in Tables 13.14 and 13.15.[50] In Table 13.14, the tanning method for East Indian leather is typically not specified – usually whatever comes to hand is used – but in this case, it is clear that at least one component of the tanning mixture was reactive condensed tannin.

The reaction conditions to fix condensed tannin and aldehyde are mutually exclusive: vegetable tannin is fixed at relatively low pH, 3–5, but aldehydic compounds are fixed at relatively high pH, 7–9. Therefore, the best approach for the combination process is to conduct the crosslinking retannage of vegetable tanned leather at the intermediate condition of pH 6 and drive the reaction with elevated temperature.

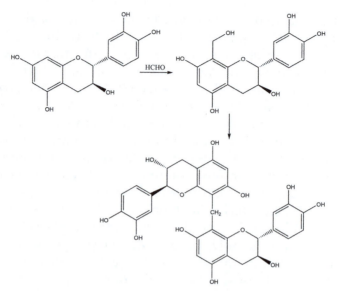

Figure 13.18   Reaction between catechin and formaldehyde.

Table 13.14   Effect on shrinkage temperature (°C) of retanning East Indian buffalo calf leather with 10% of aldehydic agents at different temperatures.

		Reaction temperature (°C)		
Aldehydic crosslinker[a]		**20**	**40**	**60**
None	84			
Phosphonium salt[b]		94	93	93
Glutaraldehyde derivative		90	91	92
Oxazolidine		100	105	110

[a]See Chapter 14.
[b]Tetrakis(hydroxymethyl)phosphonium sulfate (THPS).

Table 13.15   Shrinkage temperature (°C) from aldehydic combination tannage with different vegetable tannins: 5% syntan pretan, 20% vegetable tannage, 5% oxazolidine retannage.

Vegetable tannage		Vegetable tannage only	After oxazolidine
Condensed	Mimosa	84	111
	Quebracho	84	105
Hydrolysable	Chestnut	73	87
	Tara	75	86

The reaction does not apply to all aldehydic agents: here, the use of tetrakis(hydroxymethyl)phosphonium sulfate (THPS) or a glutaralde-hyde derivative merely makes an additive tannage, but the oxazolidine creates a synergistic tannage.

Aldehydic retannage is additive only following hydrolysable polyphe-nol tannage. The comparison between the condensed tannins suggests that the presence of a pyrogallol B-ring is advantageous in this reac-tion. The reaction was further studied by Covington and Shi,[51] using the crosslinker oxazolidine, shown in Figure 13.19: it is an alicyclic ring system, capable of opening to create two N-methylol reactive sites.

Table 13.16 shows the dependence of the reaction on different tan-nins and the crosslinker offer. The reaction is almost independent of the offer of the aldehydic retanning agent, indicating that the synergy requires only a few linkages between the polyphenol molecules to create the effective network. In the case of the hydrolysable tannins, there is evidence of antagonism at higher retan offers.

The synergy of the reactions can be easily calculated from the fol-lowing relationship:

$$\Delta T_{s\ synergy} = \Delta T_{s\ observed} - \Delta T_{s\ vegtan} - \Delta T_{s\ oxaz}$$

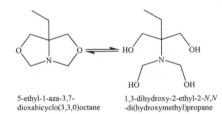

5-ethyl-1-aza-3,7-
dioxabicyclo(3,3,0)octane

1,3-dihydroxy-2-ethyl-2-*N,N*
-di(hydroxymethyl)propane

Figure 13.19 Structure of oxazolidine II.

Table 13.16 Shrinkage temperatures (°C) of leathers tanned with 20% vegetable tannin and retanned with oxazolidine.

Vegetable tannage	Oxazolidine offer (%)						
	0	2	4	6	8	10	12
None	65	76	78	80	80	79	79
Mimosa	84	108	114	114	114	114	113
Quebracho	84	95	101	101	101	101	101
Gambier	80	104	103	101	101	101	101
Chestnut	78	86	85	84	85	84	84
Valonia	75	88	88	87	82	82	82
Sumac	75	91	90	88	90	89	89
Myrobalan	72	88	88	86	86	86	84

Table 13.17   Shrinkage temperatures (°C) of leathers tanned with oxazolidine and retanned with 20% vegetable tannin.[a]

Vegetable tannage	Oxazolidine offer (%)							
	2		4		6		8	
	T_s	ΔT_s	T_s	ΔT_s	T_s	ΔT_s	T_s	ΔT_s
Mimosa	96	12	97	14	100	14	100	14
Quebracho	91	4	92	9	93	8	96	5
Chestnut	81	5	81	4	83	1	82	3
Myrobalan	78	10	78	10	81	5	80	6

[a]ΔT_s is the difference between the reactions with polyphenol then oxazolidine and with oxazolidine then polyphenol.

Table 13.18   Relative contributions of mimosa and oxazolidine offer to the outcome of the reaction: shrinkage temperature (°C).

Mimosa offer/%	Oxazolidine offer (%)					
	2	4	6	8	10	12
20	108	114	114	114	114	113
15	101	102	112	112	112	112
10	99	100	103	103	103	103
7.5	94	96	98	99	99	100
5	92	95	95	95	95	95

For comparison, Table 13.17 shows the results of changing the order of tanning reactions. The values of the difference between the alternative combinations is in part a measure of the synergy of the oxazolidine retanning process, but is modified by the antagonistic contribution.[50]

From Tables 13.18 and 13.19, the dependence of T_s on the polyphenol offer is confirmed.[51] It is apparent that the hydroxylation pattern of the condensed tannin plays an important role in the eventual hydrothermal stability conferred by the combination tannage. An NMR spectroscopic study of the interaction between catechin and oxazolidine demonstrated that reaction occurs at the 6- and 8-positions in the A-ring, and that the B-ring remains unchanged.[52] Molecular modelling of catechin and gallocatechin and their reaction with formaldehyde indicated that the 6- and 8-positions in catechin are equally reactive towards aldehydes, but the 8-position is more reactive than the 6-position in gallocatechin. Also, the B-ring of gallocatechin is more reactive than the B-ring of catechin, as shown in Table 13.20.[53]

Table 13.19 Calculated synergy (°C) of mimosa–oxazolidine combination tannages.

Mimosa offer/%	Oxazolidine offer (%)					
	2	4	6	8	10	12
20	13	17	15	15	16	15
15	6	5	13	13	14	14
10	7	6	7	7	8	8
7.5	5	5	6	5	7	8
5	6	6	5	5	6	6

Table 13.20 Calculated charges associated with positions within the flavonoid ring system.

Position	(+)-Catechin	(+)-Gallocatechin
C-6	−0.163	−0.162
C-8	−0.197	−0.193
C-2′	−0.042	−0.074
C-5′	−0.155	—
C-6′	−0.133	−0.161

Table 13.21 Characteristics of some condensed tannins.

Tannin	Ratios of main components[a]				Molecular weight	Degree of polymerisation
	PC	PD	PF	PR		
Pecan shell pith	16	84			3000	10
Myrica esculenta		100			2500	7–8
Quebracho			100		1230	4–5
Mimosa		5	25	70	1250	4–5
Gambier	100				570	2

[a]PC, procyanidin; PD, prodelphinidin; PF, profisetinidin; PR, prorobinetinidin.

In the reaction of condensed tannins and oxazolidine, the preferred polyphenols are prodelphinidin or prorobinetinidin, but they are uncommon in terms of availability: pecan and *Myrica esculenta* are less well known commercial options; see also Table 13.21.[54]

The relative reactivity of the components in the flavonoids can be gauged from the effect of polymerisation of the individual components, as presented in Table 13.22;[55] interactions may be in the form of positive synergy or negative synergy, where the latter indicates antagonism between the components of the process.

Table 13.22 Tanning effects of polyhydroxybenzene compounds (0.18 mol% offer, equivalent to 20% dihydroxybenzene) crosslinked with oxazolidine (equimolar offer).

Hydroxybenzene	T_s (°C)[a]	ΔT_{s1} (°C)[b]	ΔT_{s1+2} (°C)[c]	Predicted T_s (°C)	Measured T_s (°C)	Synergy (°C)
Hydroquinone (1,4-dihydroxy)	69	16	43	96	79	−17
Catechol (1,2-dihydroxy)	60	7	34	87	77	−10
Phloroglucinol (1,3,5-trihydroxy)	63	10	37	90	97	+7
Pyrogallol (1,2,3-trihydroxy)	58	5	32	85	96	+11
Resorcinol (1,3-dihydroxy)	56	3	30	83	115	+32

[a]The effect of the polyphenol alone.
[b]Compared with raw pelt, T_s 53 °C.
[c]$\Delta T_{s\,oxaz} = 27$ °C.

Table 13.23 Rates of reaction (L mol^{-1} s^{-1}) at pH 6.0 and 20 °C for equimolar concentrations of polyphenol and crosslinker. Rates are not corrected for the availability of reaction sites on the polyphenols.

	Crosslinker	
Polyphenol	Oxazolidine	Formaldehyde
Phloroglucinol	0.433	0.234
Resorcinol	0.024	0.008
Pyrogallol	0.018	—
Catechol	<0.001	<0.001

The effects of different aldehydic crosslinkers on polyphenols are not the same, as shown in Table 13.23.[54] The flavonoids themselves can be modelled by mixtures of polyphenols (Table 13.24).[54]

The reaction rates of resorcinol and pyrogallol are consistent, whether they are alone or mixed with another polyphenol. Catechol is consistently slow, indicating the lack of participation in crosslinking reactions of this type. Pyrogallol is reactive, so its presence in the flavonoid system contributes an effect of about an additional quarter of an A-ring; this explains the enhanced effect in combination tanning processes, because the pyrogallol B-ring provides a measurable added facility to the creation of a crosslinked network.

The role of molecular weight on combination tannages is presented in Table 13.25,[54] in which conventional plant extracts are compared

Table 13.24 Modelling the crosslinking reactions between flavonoid tannins and oxazolidine: total phenol-to-oxazolidine molar ratio 1:1, pH 6.0, 20 °C.

Phenol	Reaction rate ($L\ mol^{-1}\ s^{-2}$)	Crosslinking reaction
Resorcinol	0.048	Re-4 to Re-4
Phloroglucinol	0.433	Ph-2 to Ph-2
Pyrogallol	0.036	Py-4 to Py-4
Phloroglucinol + catechol	Phloroglucinol 0.396	Ph-2 to Ph-2
(model procyanidin)	Catechol <0.001	
Catechin (procyanidin)	A-ring 0.217	A to A
	B-ring <0.001	
Phloroglucinol + pyrogallol	Phloroglucinol 0.393	Ph-2 to Ph-2
(model prodelphinidin)	Pyrogallol 0.056	Ph-2 to Ph-4
Gallocatechin	A-ring 0.217	A to A
(prodelphinidin)	B-ring 0.058	A to B
		B to B
Resorcinol + catechol	Resorcinol 0.048	Re-4 to Re-4
(model profisetinidin	Catechol <0.001	
Resorcinol + pyrogallol	Resorcinol 0.046	Re-4 to Re-4
(model prorobinetinidin	Pyrogallol 0.050	Re-4 to Py-5

Table 13.25 Effects of crosslinking polyphenol tannages and effect of washing with acetone (hydrogen bond breaker).

	T_s (°C)		
Tannin offer	Phenol alone	Crosslinked	Washed
8% oxazolidine		83	0
5% phloroglucinol	60	95	−2
5% green tea extract	68	101	−5
20% pecan	83	112	−8
20% *Myrica*	85	113	−8
20% mimosa	82	110	−6

with phloroglucinol and green tea extract.[56] The low molecular weight green tea extract is relatively effective in this combination tannage; this is discussed further in Chapter 24. The components of green tea are monomeric catechin, epicatechin, gallocatechin and epigallocatechin, half not esterified and half esterified with gallic acid at the aliphatic 3-position.[56] The effect of the isomeric structures presented in Figure 13.9 on the rate of reaction with oxazolidine is shown in Table 13.26.

The following conclusions may be drawn from the rates of reaction in Table 13.26.

Table 13.26 Rates of reaction between flavonoid species and oxazolidine.

Flavonoid	Rate $(L\ mol^{-1}\ s^{-1})$
Catechin	0.18
Epicatechin	0.20
Epigallocatechin	0.22
Epigallocatechin gallate	0.24

- There is a clear indication that the stereochemistry of the polyphenol can influence the reactivity, at least as far as the locking aspect of tanning is concerned. This has not been further investigated scientifically.
- In the only comparison possible from these results, the reactions for catechin and epicatechin, in which the locking reaction takes place only at the A-ring, the *cis* isomer favours the aldehydic crosslinking reaction.
- The presence of a pyrogallol B-ring increases the reactivity of the polyphenol to the aldehydic locking reaction; effectively, it adds about 10% to the rate of reaction.
- The presence of a gallate group (esterified at the hydroxyl group of the C-ring) also increases the rate of reaction.

Unlike the semi-metal reaction, in which the crosslinking agent has little affinity for the collagen but a high affinity for the polyphenol, the aldehydic crosslinker has a high affinity for both the collagen and the polyphenol. The result is that the crosslinked matrix of polyphenol is not only hydrogen bonded (with some covalent bonding) to the collagen, but also it is covalently linked to the collagen *via* oxazolidine crosslinks. This confers more resistance to reversing the tanning effect and extracting the tanning species.[56–58] This is discussed further in Chapter 23.

References

1. K.-T. Chung, *et al.*, *Crit. Rev. Food Sci. Nutr.*, 1998, **38**(6), 421.
2. H. R. Procter, *The Principles of Leather Manufacture*, Spon Ltd, London, 1922.
3. K. J. Sreeram, *J. Am. Leather Chem. Assoc.*, 2015, **110**(4), 97.
4. T. White, *The Chemistry of Vegetable Tannins*, Society of Leather Trade's Chemists, 1956.
5. E. Haslam, *J. Soc. Leather Technol. Chem.*, 1988, **72**(2), 45.
6. E. Haslam, *J. Soc. Leather Technol. Chem.*, 1997, **81**(2), 45.

7. E. M. Brown and D. C. Shelly, *J. Am. Leather Chem. Assoc.*, 2011, **106**(5), 145.
8. E. Brown and P. Qi, *J. Am. Leather Chem. Assoc.*, 2008, **103**(9), 290.
9. K. Khanbabaee and T. van Ree, *Nat. Prod. Rep.*, 2001, **18**, 841.
10. Z. Shen, *et al.*, *Phytochemistry*, 1986, **25**(11), 2629.
11. E. Heidemann, *Fundamentals of Leather Manufacturing*, Eduard Roether, Darmstadt, 1993.
12. K. H. Gustavson, *J. Soc. Leather Technol. Chem.*, 1966, **50**(12), 845.
13. H. Ozgunay, *J. Am. Leather Chem. Assoc.*, 2008, **103**(10), 345.
14. Z. Shan and Z. Liang, *J. Soc. Leather Technol. Chem.*, 2012, **96**(5), 210.
15. D. Sannino, *et al.*, *J. Soc. Leather Technol. Chem.*, 2013, **97**(4), 139.
16. S. V. Kanth, *et al.*, *J. Am. Leather Chem. Assoc.*, 2010, **105**(1), 16.
17. S. V. Kanth, *et al.*, *J. Am. Leather Chem. Assoc.*, 2009, **104**(11), 405.
18. C. Jones, *World Leather*, 2006, 37.
19. J. M. Morera, *et al.*, *J. Am. Leather Chem. Assoc.*, 2008, **103**(4), 151.
20. J. M. Morera, *et al.*, *J. Am. Leather Chem. Assoc.*, 2008, **103**(4), 15.
21. V. Sivakumar, *J. Am. Leather Chem. Assoc.*, 2008, **103**(10), 33.
22. J. M. Morera, *et al.*, *J. Am. Leather Chem. Assoc.*, 2010, **105**(11), 369.
23. N. Reddy, *et al.*, *J. Am. Leather Chem. Assoc.*, 2015, **110**(4), 97.
24. B. Teliba, *et al.*, *J. Soc. Leather Technol. Chem.*, 1993, **77**(6), 174.
25. R. W. Frey and C. W. Beebe, *J. Am. Leather Chem. Assoc.*, 1940, **35**(7), 440.
26. R. W. Frey and C. W. Beebe, *J. Am. Leather Chem. Assoc.*, 1942, **37**(10), 478.
27. R. W. Frey and C. W. Beebe, *J. Am. Leather Chem. Assoc.*, 1942, **37**(11), 539.
28. K. P. Rao and Y. Nayudamma, *Leather Sci.*, 1963, **10**, 433.
29. K. P. Rao and Y. Nayudamma, *Leather Sci.*, 1964, **11**, 6; K. P. Rao and Y. Nayudamma, *Leather Sci.*, 1964, **11**, 39; K. P. Rao and Y. Nayudamma, *Leather Sci.*, 1964, **11**, 84; K. P. Rao and Y. Nayudamma, *Leather Sci.*, 1964, **11**, 89.
30. E. Heidemann and B. Balatsos, *Das Leder*, 1984, **35**, 186.
31. W. E. Kallenberger and J. F. Hernandez, *J. Am. Leather Chem. Assoc.*, 1983, **78**(8), 217.
32. J. J. van Benschoten, *et al.*, *J. Am. Leather Chem. Assoc.*, 1985, **80**(10), 237.
33. W. E. Kallenberger and J. F. Hernandez, *J. Am. Leather Chem. Assoc.*, 1984, **79**(5), 182.
34. R. L. Sykes and C. W. Cater, *J. Soc. Leather Technol. Chem.*, 1980, **64**(2), 29.
35. S. T. Orszulik, R. A. Hancock and R. L. Sykes, *J. Soc. Leather Technol. Chem.*, 1980, **64**(2), 32.
36. J. M. Morera, *et al.*, *J. Soc. Leather Technol. Chem.*, 2004, **80**(4), 120.
37. Z. Shan and G. Wang, *J. Soc. Leather Technol. Chem.*, 2004, **8**(2), 72.
38. S. Vitolo, *et al.*, *J. Am. Leather Chem. Assoc.*, 2003, **98**(4), 123.
39. A. D. Covington and G. S. Lampard, *J. Am. Leather Chem. Assoc.*, 2004, **99**(12), 502.
40. B. Shi, *J. Am. Leather Chem. Assoc.*, 2008, **103**(8), 270.
41. S. Jatay and Y. M. Soshi, *Appl. Clay Sci.*, 2014, **97–98**, 72.
42. J. Shi, *et al.*, *J. Soc. Leather Technol. Chem.*, 2016, **100**(1), 25.
43. A. D. Covington, unpublished results.
44. A. D. Covington, *J. Soc. Leather Technol. Chem.*, 2001, **85**(1), 24.
45. S. Jeyapalina, PhD thesis, University of Northampton, 2004.
46. I. V. Yannas, *J. Macromol. Sci., Rev. Macromol. Chem.*, 1972, **7**(1), 49.
47. G. E. Attenburrow, A. D. Covington and S. Jeyapalina, *Proceedings of IULTCS Congress*, Cape Town, South Africa, 2001.
48. D. T. Rossouw, *et al.*, *J. Polym. Sci., Polym. Chem. Ed.*, 1980, **18**, 323.
49. S. DasGupta, *J. Soc. Leather Technol. Chem.*, 1977, **61**(5), 97.
50. A. D. Covington and S. Ma, *UK Pat.*, 2287953, 1996.

51. A. D. Covington and B. Shi, *J. Soc. Leather Technol. Chem.*, 1998, **82**(2), 64.
52. B. Shi, *et al.*, *J. Soc. Leather Technol. Chem.*, 1999, **83**(1), 8.
53. Z. Li, *et al.*, *J. Am. Leather Chem. Assoc.*, 2005, **100**(11), 432.
54. A. D. Covington, C. S. Evans, T. H. Lilley and L. Song, *J. Am. Leather Chem. Assoc.*, 2005, **100**(9), 325.
55. S. Phillips, MSc thesis, University of Northampton, 1998.
56. L. Song, PhD thesis, The University of Northampton, 2003.
57. A. D. Covington and L. Song, *Proceedings of IULTCS Congress*, Cancun, Mexico, 2003.
58. Z. Lu, *et al.*, *J. Soc. Leather Technol. Chem.*, 2003, **87**(5), 173.

14 Other Tannages

14.1 Oil Tanning

Oil tanning is the method for making chamois leather and a seminal review was published by Sharphouse.[1] This leather is best known for its properties of holding water, useful for cleaning and drying washed surfaces, such as windows. More traditionally, it was the tannage used for making buckskin clothing leather. It is an example of a leathering type of process because, although it resists microbial attack, the shrinkage temperature is not raised significantly above the value for raw pelt. In essence, the process involves filling wet pelt with unsaturated oil, typically cod liver oil, then polymerising the oil *in situ* by oxidation with hot air.

The principal steps are as follows, with comments in italics.

1. Start with pickled sheepskin pelt. (*The pelt needs to be acidic to allow acid swelling.*)
2. Wash to swell. (*Use fresh water – swelling of the pelt occurs only if water is available and if the swelling is not suppressed by neutral electrolyte. The effect of swelling will also extend the opening up or loosening of the fibre structure, which will assist the penetration of the oil.*)
3. Split, to remove the grain and leave the flesh split. (*This minimises the difficulties of getting oil to penetrate into pelt wet with water.*)

Tanning Chemistry: The Science of Leather 2nd edition
By Anthony D. Covington and William R. Wise
© Anthony D. Covington and William R. Wise 2020
Published by the Royal Society of Chemistry, www.rsc.org

4. Repickle, using 10% salt solution. (*Acidification in the presence of salt depletes the pelt, to optimise oil penetration, preferably by adjusting the pH to the isoelectric point, ~5.0, where swelling is minimal.*)

5. Optionally apply aldehydic pretannage, traditionally with form-aldehyde,[2] but now glutaraldehyde is most common: other aldehydic agents will become more common, *e.g.* phosphonium salt; see later. (*The change in the hydrophilic–hydrophobic balance (HHB) of the pelt aids oil penetration and confers 'dry soft' properties, avoiding the hardening effect when wetted oil-tanned leather dries. The tanning process is conducted under acidic conditions, under which the aldehyde reaction is very slow – this indicates the lack of reliance on fixing the polymerised oil to the collagen via aldehydic reactions.*)

6. Squeeze. (*Minimising the water content to allow oil penetration.*)

7. Add cod liver oil. (*The requirement is for unsaturated oil, which must be forced into the pelt, e.g. by drumming – the traditional equipment is 'fulling stocks' in which the oil was effectively hammered into the pelt by the action of a bank of large wooden mallets.*)

8. Oxidise the oil with air. (*Blow in hot humid air or use frictional heat, to raise the temperature to 40–50 °C to initiate auto-oxidation; thereafter cool to prevent hydrothermal damage by the rapidly rising temperature: the apparent shrinkage temperature rises as the moisture content is lowered, starting at 65–70 °C for wet pelt/leather, but the reaction temperature must not approach within 30 °C of the shrinkage temperature. The total time for the chamoising effect to run to completion may be up to 2 weeks.*)

9. Degrease with sodium carbonate solution or solvent. (*The residual reacted oil extracted at this point is called 'degras' or 'moellon' and is used to make lubricants for other leathers.*)

10. Buff the flesh surface, and also the split surface if necessary. (*This is for cosmetic reasons; use a rotating gritted wheel or automatic lightning buffer.*)

11. Condition, typically at 20 °C and 65% relative humidity. (*The area of any leather depends on moisture content; this applies to chamois more than most leathers.*)

12. Measure the area. (*The area is highly dependent on the measuring method, particularly if the leather is stretched during area measuring, as it may be going through the mechanical pinwheel instrument.*)

13. Grading for area, faults, handle and function. (*The guideline property of water retention is that it must be able to hold 800% water on pelt weight.*)

The reactions in the oil tanning process are not completely clear. The active agent is unsaturated oil, preferably cod oil, which can be modelled by linoleic acid, $CH_3(CH_2)_4CH=CHCH_2CH=CH(CH_2)_7CO_2H$, which is known to polymerise but not to form epoxides. It has been generally accepted that the reaction is based on the formation of aldehydic compounds, particularly since the process is accompanied by the release of acrolein, $CH_2=CHCHO$, which has been used as an element of quality control.[2] However, acrolein alone does not make acceptable chamois leather.[1]

Sharphouse[1] summarised the tannage in the following terms:

> *... fixation of oily or resinous auto-oxidation products to the protein fibre in some intimate sheath-like form. These may be in polymer form and resist removal by alkaline wash waters and common solvents. It is presumed that they account for the differences between aldehyde tannages and 'full oil' chamois tannage.*

In this way, he defined the outcome of the tannage as a polymer matrix within the collagen matrix: there is no certainty of reaction between the polymer and the collagen, unlike the product of aldehydic tanning. Therefore, the system can be pictured as a matrix of polymerised hydrocarbon chains, holding the collagen fibre structure apart, as an extreme form of lubrication to prevent the fibre structure coming together and sticking (see Chapter 17). This model does, however, provide a rationale to explain the three most important features of oil-tanned leather.

1. *Hydrothermal stability:* The shrinkage temperature of oil-tanned leather is only a few degrees higher than that of raw pelt. Therefore, the conventional view of tanning does not hold, *i.e.* there is little interaction between the tanning agent and the collagen. This is an example of the (tanning) agent having more affinity for itself than for the substrate. This aspect of collagen stabilisation is discussed further in Chapter 23.

2. *Water retention:* The effect of keeping the fibre structure apart means that the hydrophilic collagen can be hydrated and hold excess water within the hydrophobic polymerised oil matrix. In this way, there is the apparently contradictory situation of a hydrophobic tanning chemistry producing the most hydrophilic leather.

3. *The 'Ewald' effect:* Oil-tanned leather is one of a few cases where the leather exhibits reversibility of hydrothermal shrinking. If the leather is held in hot water, at or above the shrinkage

temperature, it will shrink as expected; however, if the leather is immediately placed in cold water, it regains about 90% of its original area. The phenomenon can be used to mould the leather into shapes, a shrink-fitting process called 'tucking'. When leather in general shrinks under hydrothermal conditions, the triple helix structure begins to unravel, hence reversibility applies only in the earliest stages, when the structure can reregister by itself. However, the oil matrix provides a scaffold that mirrors the collagen structure, so the denatured protein can regain much of its original structure by reregistering the helical structure, but not quite all of it.

In recent years, there have been several studies on alternative oils to fish oil to make chamois leather. Suparno *et al.*[3] led research into the application of rubber seed oil: this oil has an iodine value of 146, compared with 148 for fish oil, where the iodine value is a measure of the unsaturation in the oil. They showed that the results are similar to those for tanning with fish oil, but with less colour and odour, and the water absorption is 380%. Furthermore, by including 6% by weight of hydrogen peroxide in the oil, the chamoising period can be shortened to 3 days.[4,5] Linseed oil is another option:[6] with an iodine value of 160, it yields leather that exhibits water absorption of 303%, but it is weaker than conventional fish oil-tanned leather.

Studies on tanning with oil derived from Japanese anchovy (*Engraulis japonicus*) were based on peroxidising the oil before applying it to pelt. The lack of a chamoising effect led the authors to cast doubt on the accepted mechanism, but they did not take into account the role of *in situ* oxidation of the oil.[7]

14.2   Sulfonyl Chloride

Alkyl sulfonyl chloride, available in the early part of the twentieth century under the trade name Immergan, has been used in a sort of synthetic oil tanning, an alternative chamoising process. The chemistry of this reaction can be summarised in the following way:

$$\text{Prot} - \text{NH}_2 + \text{C}_{16}\text{H}_{33} - \text{SO}_2 - \text{Cl} \rightleftharpoons \text{Prot} - \text{NH} - \text{SO}_2 - \text{C}_{16}\text{H}_{33}$$

The leather is less coloured and the shrinkage temperature rises to 80 °C, giving similar properties to oil tanning. However, the reaction raises the following points of interest.

- The tanning agent cannot create a matrix, so the effect relates more to covering the fibre structure to produce a hydrophobic surface, like conventional oil dressing (see Chapter 10).
- The reagent can only react through the sulfone group *via* a single bond, *i.e.* it cannot crosslink the collagen, yet the shrinkage temperature is elevated. This is discussed further in Chapter 23, where an explanation is offered.

14.3 Syntans

The term 'syntan' refers to the range of synthetic tanning agents.[2,8] Their history starts with the work of Stiasny, who patented the synthesis of polymeric species capable of tanning.[9] This class of reagents is wide ranging, but typically they are aromatic, as hydroxy and sulfonate derivatives, so some general principles regarding their functions can be derived. This group does not strictly include other resins and polymers, which might also be used in leather making, *i.e.* acrylates, urethanes, *etc.*, are excluded (see later).

Usually, there are two steps in the synthesis of syntans: sulfonation and polymerisation. These steps can be conducted in either order: these are the nerodol and novolak (novolac) syntheses, typically leading to different molecular constitutions and hence properties. The reactions are represented in Figures 14.1 and 14.2, using naphthalene or phenol and monomeric formaldehyde as models. When the sulfonation step comes before the polymerisation step, typically the sulfonate content is relatively high. This means that the solubility in water is increased, which reduces the astringency. These effects can be balanced by the choice of precursor, to make the syntan more or less reactive. In contrast, when the sulfonation step comes after the polymerisation step, the syntan typically contains less sulfonate,

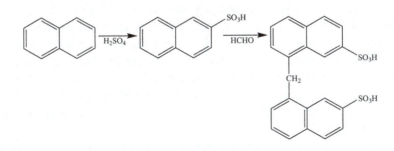

Figure 14.1   The nerodol synthesis: sulfonation then polymerisation.

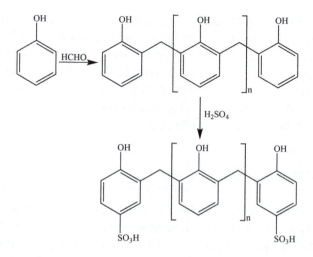

Figure 14.2 The novolak synthesis: polymerisation then sulfonation.

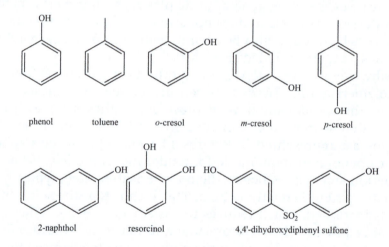

phenol   toluene   o-cresol   m-cresol   p-cresol

2-naphthol   resorcinol   4,4'-dihydroxydiphenyl sulfone

Figure 14.3 Some syntan precursors.

making the molecule less soluble in water and hence more reactive towards the substrate The overall properties of the syntans, with regard to the affinity for collagen or leather, depend on the choices of aromatic precursor and crosslinker, order of reactions in synthesis, the degree to which the reaction steps are applied, the solubility of the syntan and its molecular weight. Although the reaction conditions may be precisely controlled, the constitution of the syntans, their chemical structures, are relatively non-specific and are rarely analysed in detail.[10]

The properties of the syntan depend on the nature of the monomeric precursors, some of which are presented in Figure 14.3.

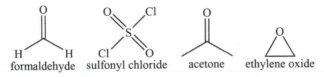

formaldehyde   sulfonyl chloride   acetone   ethylene oxide

Figure 14.4 Crosslinkers for syntans.

In addition, the crosslinker can have an effect on the properties, particularly the lightfastness, which depends on the ability of the aromatic moieties to interact to create colour (see Chapters 13 and 16); some of the crosslinkers used to make commercial syntans are presented in Figure 14.4.

Ammenn *et al.*[11] referred to the nerodol/novolak syntans as the first generation of syntans. The second generation is characterised by the more complex crosslinking mechanisms, including combination systems such as urea–formaldehyde, where the products made softer, fuller leather with no yellowing and better lightfastness.

The range of syntan properties and functions can be summarised in the notional model of trends in reactivity and properties, shown in Figure 14.5. Here, the reactivity of the syntans depends primarily on the availability of phenolic hydroxyl groups for hydrogen bonding, like the plant polyphenols. A secondary reaction is fixation *via* the sulfonate groups, as discussed in Chapter 15. However, the impact of the sulfonate groups is more concerned with its effect on the HHB value of the syntan and the consequent effect on the solubility in water, acting against its astringency properties. The range of syntans is typically separated into groups of syntans with names reflecting their properties, but the ranges are neither specific nor defined, so they overlap. This model does not include allowance for variations in the crosslinker, which is an additional variable in syntan constitution.

The size of the syntan molecules and hence their properties, depend on the ratio of crosslinker to aromatic nuclei, shown in Table 14.1, according to Thorstensen.[12]

The effects of the syntans with regard to their tanning power depend primarily on the combination of molecular weight and availability of reactive sites, phenolic hydroxyls and sulfonate groups, together with the secondary influence of the crosslinking chemistry, illustrated in Figure 14.6.[13]

The positions within the aromatic nuclei and the relationship between the phenolic hydroxyl groups and the sulfonate groups can influence the tanning power due to inductive effects;[9] for example,

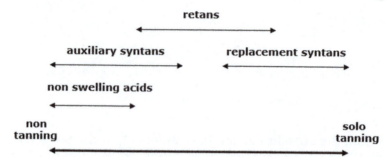

Range of syntan constitution and properties

Increasing molecular weight   ──────────▶

Increasing molecular complexity   ──────────▶

Increasing phenolic hydroxyl content   ──────────▶

Increasing sulfonate content (solubility)   ◀──────────

Increasing [OH⁻]:[SO₃⁻] ratio   ──────────▶

Decreasing solubility in water (hydrophobicity)   ──────────▶

Increasing solubility in water (hydrophilicity)   ◀──────────

Increasing tanning power = increase in Ts   ──────────▶

Increasing hydrophilicity of leather   ──────────▶

Trend of Nerodol to Novolac synthesis   ──────────▶

Increasing effectiveness of non swelling acids
in pickling, see Chapter 9   ──────────▶

Figure 14.5 Trends in syntan properties.

Table 14.1 Effect of crosslinking ratio on molecular weight.

Crosslinking ratio[a]	Average no. of aromatic nuclei per molecule	Average molecular weight
0.50	2	300–350
0.66	3	450–500
0.75	4	600–700
0.80	5	750–900
0.90	10	1500–2000
>1.0	Theoretically infinite	Theoretically infinite

[a]No. of moles of crosslinker to no. of moles of aromatic nuclei.

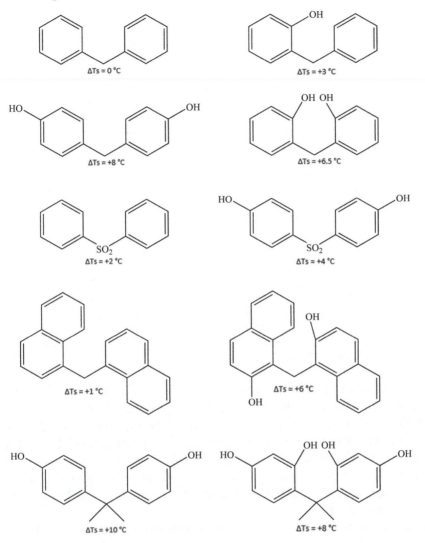

Figure 14.6 Effect of simple syntan structures on tanning properties.

syntans based on β-naphthols are better tanning agents than syntans based on α-naphthols[14] and the relative *meta* positions of the oxygens in xanthene species allow more effective tanning than an *ortho* or *para* arrangement (Figure 14.7), analogous to the arrangement in the condensed tannins.[15]

The effects of the relative positions of phenolic hydroxyl and sulfonate groups rely more on the combination of inductive effects; here,

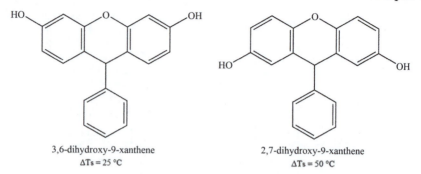

3,6-dihydroxy-9-xanthene
ΔTs = 25 °C

2,7-dihydroxy-9-xanthene
ΔTs = 50 °C

Figure 14.7   Effect of relative positions of phenolic hydroxyls on tanning power.

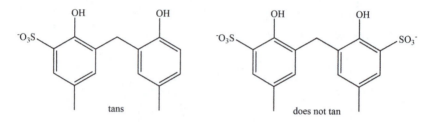

tans

does not tan

Figure 14.8   Effect of relative positions of hydroxyl and sulfonate groups on the ability to react with protein.

the inductive effect of a sulfonate adjacent to a phenolic hydroxyl is to reduce its reactivity by weakening the O–H bond, thereby weakening its ability to engage in hydrogen bonding. The relative tanning powers of the species in Figure 14.8 are attributed to the ability of the tanning species to interact *via* one hydrogen-bonded hydroxyl and one sulfonate ionic link, but the non-tanning species interacts only *via* sulfonate links. The effect of an HHB difference will contribute to the observed difference in properties.

The following conclusions can be drawn from Figures 14.6–14.8.

- The effectiveness of the syntan depends on its ability to engage in hydrogen bonding, from both the phenolic hydroxyls and the crosslinks.
- The positions of the hydrogen bonding groups determine the tanning effect.
- The tanning effect depends on a combination of inductive effects from all the substituents involved.
- The steric effects of additional groups can influence the tanning ability by reducing the free rotation and hence introducing some degree of fixed geometry.

14.3.1 Auxiliary Syntans

As the name implies, these low molecular weight syntans are relatively unreactive (in a tanning sense), so their performance functions are limited to aiding other reactions in the following ways.

- Solubilising higher molecular weight syntans or vegetable tannins, to allow more uniform surface reaction and promoting penetration (see Chapter 13).
- Dispersing reds and dyes, to allow more level tannin fixation and colouring (see Chapter 16).
- Non-swelling acids fall into this category, since they rely on a mild tanning effect after giving up their acid content to collagen in the pickling step.
- So-called 'bleaching' syntans may fall into this or the next group. When applied to chrome-tanned leather, the reaction is to strip some of the chrome from the leather or displace it into the leather and hence reduce the colouring on the surface, a detanning effect.
- 'Neutralising' syntans serve the function indicated by the name: the alkalinity in the product neutralises chrome-tanned leather, raising the pH from 3–4 to 5–7, depending on the post-tanning requirements with regard to charge on the leather (see Chapter 15). At the same time, they provide a retanning effect that may be useful for the production of certain kinds of leathers; the tanning/filling effect required is determined by the required properties/handle of the leather. Therefore, these syntans may also fall into any of the other categories.

14.3.2 Retans

As the name implies, the medium molecular weight, mildly astringent syntans are typically used in post-tanning as retanning agents. They will confer some filling effect, dependent on the molecular weight, which will also limit the degree of hydrophilicity conferred to the leather.

They also aid vegetable tanning and dyeing, if they are reacted with the collagen or leather before applying vegetable tannin or dye, thereby reducing the reactivity of the substrate. In this way, tanning and dyeing are hindered at the surface, not only causing even reaction on the surface, but also promoting penetration through the cross-section.

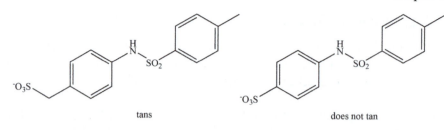

tans   does not tan

Figure 14.9   Functionality of Supra syntan structures.

14.3.3   Replacement Syntans

As the name implies, the high molecular weight, astringent syntans are designed to replace vegetable tannins and act as solo tanning agents. They can match the astringency of vegetable tannins and provide the same degree of filling. However, their structures can be created to minimise colour change under the influence of light, so they can offer improved lightfastness compared with the condensed tannins.

The effect of a syntan on collagen is dependent on its structure. This is demonstrated in Figure 14.5, where the effect of structure on the shrinkage temperature rise of collagen depends on both the cross-linker and the precursor, illustrated by dimeric syntan species.[14]

14.3.4   Other Syntans

The 'Supra' syntans may not contain any aromatic hydroxyl groups, so rely for their effect on hydrogen bonding reactions *via* their amino sulfonyl groups, $-NHSO_2-$, as illustrated in Figure 14.9, which also indicates the importance of the inductive effects.

The filling power of syntans depends on the molecular weight and the chemistry of the polymer; in order to increase the filling power, syntans have been made with additional chemical species, notably lignosulfonic acid, a by-product from the paper industry. These chemical species do not possess tanning power, as may be deduced from the indicative structures in Chapter 2 and Figure 14.10 and the break-down products shown in Figure 14.11.

Approximate relative yields of degradation products are as follows:[16] the ratio of vanillin to vanillic acid to 4-hydroxybenzaldehyde to 2-methoxyphenol is 1:1:5:10. Formaldehyde polymerisation of these breakdown products does not create effective syntans.[17] Therefore, it is understandable that syntans based on by-products from the paper manufacturing industry have never been a major success in the leather manufacturing marketplace.

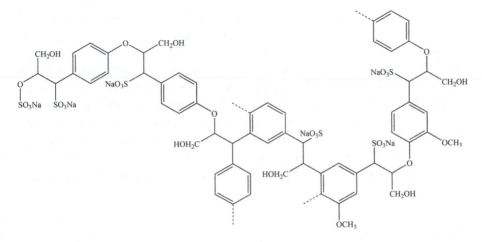

Figure 14.10 Intermediate breakdown product of Kraft lignin, lignosulfonic acid, where the structure is notional.

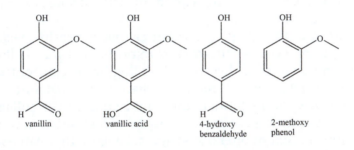

vanillin — vanillic acid — 4-hydroxy benzaldehyde — 2-methoxy phenol

Figure 14.11 Ultimate breakdown products of Kraft lignin.

Bacardit and co-workers[18,19] suggested an alternative, analogous source of compounds, humic acid and derivatives. In descending order of reactivity they are lignin, natural humic acids (extracted from lignite coal at pH 10–12), huminas (obtained by the oxidation of coal, huminas, bitumen) and bitumen. Like sulfited derivatives of lignin, these additional materials can be sulfited. Similarly, they can function as retanning and filling agents, with the same degree of success.

All the syntans discussed so far are anionic (or neutral in the case of the Supra syntans). However, by altering the chemistry of the precursor, it is possible to create amphoteric syntans, containing both acidic and basic groups. These syntans can be useful, because they are less reliant on pH variation for fixation. The same would be true for extreme nerodol syntans with no sulfonate groups, but these would be relatively insoluble in water.

It is apparent that any tanning reaction exhibited by natural polyphenols can be matched by appropriate choices of syntan precursor, crosslinker and production conditions. However, with the exception of Reich's work,[13,15] there is limited information regarding the structural parameters that contribute to syntan reactivity; these are necessary to allow the leather scientist to model the requirements for new tannins. Nevertheless, informed choices should allow improvements in tanning technology. In making those choices, reference should be made to the analyses presented in Chapters 23 and 24, which will form the basis of developments in this field.

14.4 Resins

The term 'resin' refers to synthetic polymers, some of which have applications in leather making. The chemistries of some examples are presented in Figure 14.12.

It is clear from the chemical structures that the resin species have low astringency, because the opportunity for fixing to collagen or leather is limited to some hydrogen bonding; some can contribute to the hydrothermal stability,[20] but a positive effect is uncommon, weak and not relied upon by tanners for stabilisation of the collagen. An exception is the melamine–formaldehyde resins, which have reactivity analogous to that of phenolic syntans, because they can hydrogen bond to collagen *via* their triazinyl amino groups. This functionality can be used in combination reactions with aldehydic crosslinkers, as indicated in Table 14.2.[21]

The reaction is analogous to the combination of condensed polyphenol and aldehydic crosslinker, discussed in Chapter 13.

- The reaction is driven by elevated temperature.
- The outcome depends more on the amount of the polymer than it does on the amount of the crosslinker.
- The aldehydic agents are not equally good at achieving high shrinkage temperature; the preferred crosslinker is tetrakis(hydroxymethyl)phosphonium sulfate (THPS).
- In this case, the reaction does not work with every commercial melamine resin: there appears to be a dimensional requirement; the preferred mean particle size is 80 nm.

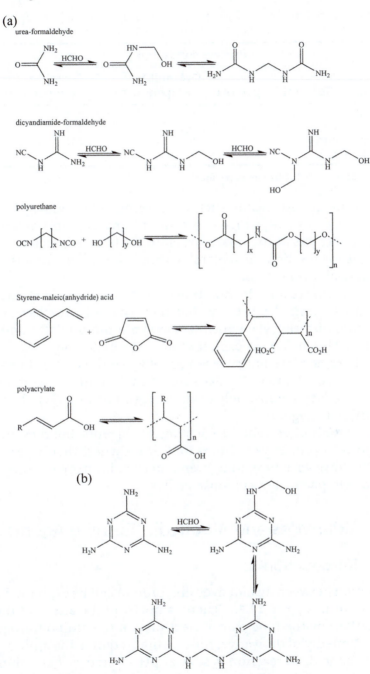

Figure 14.12 (a) The chemistries of resins. (b) Melamine–formaldehyde structure.

Table 14.2   Combination tannages of 10% commercial melamine–formaldehyde resin and aldehydic crosslinkers, applied to sheepskin.

	Process conditions		
Crosslinker offer	Initial pelt pH	Overnight temperature (°C)	Shrinkage temperature (°C)
2% glyoxal	7	50	106
5% oxazolidine	6	50	106
4% phosphonium salt (THPS[a])	5	50	112

[a]Tetrakis(hydroxymethyl)phosphonium sulfate.

One problem with resins such as melamine–formaldehyde is the formaldehyde content, which can result in unacceptable amounts of free formaldehyde in the final leather. It has been suggested that the solution is to crosslink the melamine with glyoxal,[22] but that is not yet a commercial proposition.

Typically, the resins do not function in combination reactions, owing to the limited reactivity that they possess. One exception is the reaction of carboxylate polymers with aluminium(III)[20] and chromium(III). This has been exploited in reagents for high water resistance, where acrylate polymer is partially esterified with long-chain alcohols[23] (see Chapter 17). Because of their ability to complex *via* their carboxylate groups, polyacrylates have been marketed as chromium(III) fixing agents.[24]

Resins have limited value as substantive reagents, but they are used as structure or property modifiers for leather, where they have value in filling the fibre structure, to tighten and even improve the properties over the pelt, particularly in loose bellies.

14.5   Aldehydes and Aldehydic Tanning Agents

14.5.1   Introduction

The reaction between an aldehyde and protein can be expressed in the form set out in Figure 14.13. There is nucleophilic attack at the carbonyl carbon by the lone pair of electrons on the amino nitrogen, to form an *N*-methylol derivative; the reaction requires the amino group to be uncharged, so reaction is slow at pH < 7 and fast at around pH 9. It has been shown that the reaction is dominated by the involvement of lysine amino groups,[25] with little involvement of other side-chains with active hydrogens. This hydroxy compound can lose water, to form the corresponding unsaturated Schiff base.

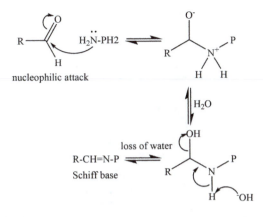

Figure 14.13 General aldehyde reaction with protein amino group.

It is commonly held that the tanning reaction mechanism involves further reaction between the *N*-methylol derivative and another basic group in the protein system, as illustrated in the following model reaction equations, using monomeric formaldehyde as the example:

$$P-NH_2 + HCHO \rightleftharpoons P-NH-CH_2OH \overset{P-NH_2}{\rightleftharpoons} P-NH-CH_2-HN-P$$
$$N\text{-methylol derivative}$$

The presence of the nitrogen in the *N*-methylol group with its lone pair of electrons makes the hydroxyl a good leaving group, *i.e.* it is reactive to groups with an active hydrogen. This can be illustrated in the extremes of structure as follows:

$$P-NH-CH_2-OH \rightleftharpoons P-N^+H=CH_2\cdots OH$$

It is clear that crosslinking reactions can occur and the role in tanning will be discussed further in Chapter 23. What is not in doubt is that the reaction product is covalently bonded, hence reversal can only be accomplished by hydrolysis, general acid and general base catalysed.

14.5.2 Formaldehyde

It is useful to use formaldehyde in the example of aldehyde tanning, because the structure can be regarded as the simplest. As illustrated in Figure 14.14, it is more than likely to be present in a polymerised form, depending on the history of the sample; it is common to see a white precipitate in laboratory bottles of aqueous formalin solution.

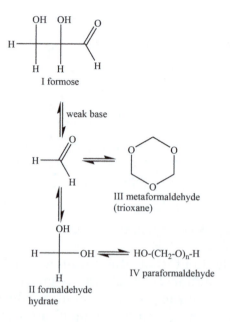

Figure 14.14 Formaldehyde structures.

Apart from the hydrate (II), formaldehyde species are dominated by paraformaldehyde (IV). Some insight has been gained into the reactions of formaldehyde using NMR spectroscopy.[26] Reaction with gelatine showed that lysine was the preferred site and then arginine. Studies with model amino acids showed that they are difficult to crosslink. In studies using polymeric amino acids, no crosslinking could be achieved in either polylysine or polyarginine, although links to the polymers could be formed. Formation of the methylol derivative and Schiff base has been demonstrated in reactions between formaldehyde and protein, but the subsequent reactions are complicated.[27] Preferred reaction occurs at lysine, arginine, histidine and cysteine. Schiff base adducts react preferentially with arginine and tyrosine: in the former, reaction is with the terminal amino groups, and in the latter, reaction is with the aromatic nucleus of the sidechain (see later). From subsequent reactions, the crosslinking chains can be variable in length. Therefore, although crosslinking of protein is theoretically possible and sensible paper chemistry can illustrate the reactions, it seems to be the case that simple crosslinking is not necessarily the natural consequence of these reactions and the overall reaction is complicated.

Formaldehyde has long held a place in leather-making technology, in the following applications.

- Non-mineral tannages, *e.g.* woolskins for nursing purposes.
- Applications where high perspiration resistance is required, *e.g.* gloving leather.
- Wool treatment in woolskins, *i.e.* straightening, in a reaction analogous to permanent waving of straight hair or fixing relaxed naturally kinked hair.

Tanning with formaldehyde produces white leather, which is characterised by plumpness and hydrophilicity. These features are the effect of the polymeric nature of the tanning agent, pushing the fibre structure apart, and the presence of hydrogen bondable groups in the polymer. The tannage yields leather with shrinkage temperature up to 80 °C.

It should be noted that, because of formaldehyde's toxicity hazard, the restrictions on the allowable concentration in the atmosphere of the workplace have effectively created a ban on its use as an industrial reagent for leather making. Its use in preserving biological specimens remains: although formaldehyde is effective as a bactericide, its mode of preservation of specimens in prolonged contact with it is to tan the protein. It is useful to recognise that such biological specimens have been transformed chemically and artefacts can be introduced because of the filling effect of polymerised formaldehyde.

The toxicity and tanning properties of formaldehyde call into question the benefits of its continued use in biology and histology. Some of the other aldehydic agents discussed in the following may be better options, because they are less toxic and react as monomers.

14.5.3 Glutaraldehyde

As the use of formaldehyde declined, the use of glutaraldehyde as a replacement grew. The structure of this aliphatic dialdehyde in solution is given in Figure 14.15.

Consideration of the structure of the monomer might lead to the conclusion that the tanning reaction involves the creation of a four-centre crosslink, since each end can theoretically link two amino groups. Intuitively, this is doubtful, because the coincidence of four adjacent amino groups is unlikely and the entropy penalty of such a bond would be high. However, it is not necessary to invoke a mechanism that is chemically unsound: the polymer structure, shown in Figure 14.15, based on condensation of the hydrate, is a linking structure between two reacted glutaraldehyde molecules, as shown in Figure 14.16.[28]

Olde Damink *et al.*[28] drew the following conclusions.

1. Within a complex scheme of reactions, glutaraldehyde forms Schiff bases with protein and these are stabilised by other glutaraldehyde molecules.
2. There is no evidence that crosslinks are formed.
3. Three glutaraldehyde molecules are fixed per lysine amino group: there is no evidence for a polymerised matrix.

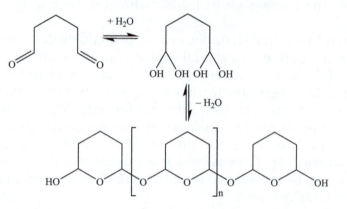

Figure 14.15   Polymerisation of glutaraldehyde hydrate.

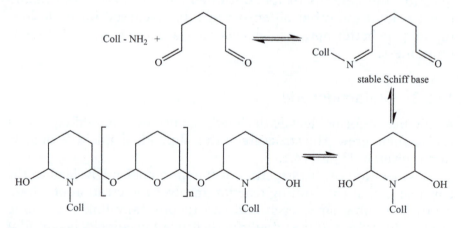

Figure 14.16   Reaction between glutaraldehyde and protein.

Therefore, although crosslinking is clearly possible, there is no evidence that the tanning reaction actually does rely on such a reaction (see Chapter 23).

As with formaldehyde, the leather tanned with glutaraldehyde is plump and hydrophilic, for the same reasons: shrinkage temperatures are similar. However, the colour is different, having a distinct yellow cast, which can appear quite orange. The reason for the colour development is unclear, but has something to do with changes in structure upon reaction with collagen. The colour of normally tanned glutaraldehyde creates a problem in dyeing such leathers, especially as it is a dynamic effect, and is unacceptable in leathers designed to be white. Consequently, derivatives of glutaraldehyde have been offered to the industry. The original option was Relugan GTW, the bisulfite addition derivative. This produces paler leather, but there is always a degree of yellow in the leather since the addition compound must be reversed for the reaction with collagen to occur. Other derivatives have included glyptal compounds with alcohols and aldol addition and condensates with other aldehydes, including formaldehyde. Unsurprisingly, none of these has successfully overcome the problem.

Glutaraldehyde is currently widely used. However, it is probable that it will experience the same regulatory fate as formaldehyde and eventually its use will have to be discontinued.

14.5.4   Other Aliphatic Aldehydes

Other aldehydes are not commonly used in the industry. This is partly because there is no environmental advantage, but mainly because they are not effective tanning agents. Other aldehydes in the aliphatic homologous series, R–CHO, can tan, but they are not effective and they are not cheap. Glyoxal, $(CHO)_2$, the lowest dialdehyde in the homologous series, is an industrial chemical, but not widely used in this context.

Acrolein, $CH_2{=}CH{-}CHO$, has both aldehyde and unsaturation functionality, allowing polymerisation. It is a toxic chemical, with consequent associated problems for industrial use. Its roles in leather making are incidental.

A more commercial agent is dialdehyde starch (or starch dialdehyde), made by the oxidation of starch with periodic acid (HIO_4), illustrated in Figure 14.17. The value of n, the degree of depolymerisation, depends on the ratio of starch to periodic acid.[29]

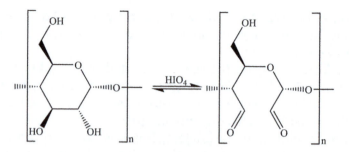

Figure 14.17 Preparation of starch dialdehyde.

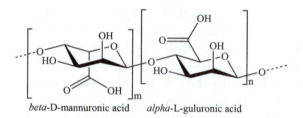

beta-D-mannuronic acid   *alpha*-L-guluronic acid

Figure 14.18 Structural features of alginates.

The tanning power is relatively weak, but it has potential value as a component in combination organic tanning. It produces white, hydrophilic, plump leather, where the last property depends on the size of the molecule.

Dialdehyde formation is not confined to starch: any other carbohydrate may be used, *e.g.* dextrin, a lower molecular weight version of the starch structure.[30] Alginates can be oxidised to dialdehydes in the same way; the structure is illustrated in Figure 14.18, where the polymer contains blocks of either mannuronic acid (M) or guluronic acid (G) or alternating G and M or random M and G. Alginate dialdehydes have been shown to be effective tanning agents, although the tensile strength of the leathers is inferior to that of glutaraldehyde-tanned leather.[31]

Because there is reactivity within the polymeric aldehyde structure from the carbohydrate base, it might be assumed that the alginate species might usefully perform in a combination process. One study did suggest that a small degree of synergy is possible, as indicated in Table 14.3.[32] All the leathers exhibited poorer performance with respect to all strength tests compared with chrome-tanned leather.

Table 14.3 Tanning with 10% dialdehyde alginic acid then 3% aldehydic agent.

Aldehydic agent	T_s (°C)
None, control	84
Tetrakis(hydroxymethylol)phosphonium sulfate	90
Oxazolidine	92
Glutaraldehyde	97

14.6 Aldehydic Tanning Agents

Aldehydic agents are distinguished from conventional aldehydes because they are not aldehydes themselves, but either they undergo some aspect of the typical aldehyde reaction or they exhibit analogous reactions.

14.6.1 Oxazolidines

Oxazolidines, illustrated in Figure 14.19, were introduced to the leather industry by DasGupta.[33] The chemistry of their reaction with protein is clearly aldehydic, although they are not themselves aldehydes.

The aldehydic character can be seen from the way in which they are synthesised, creating an aliphatic heterocyclic ring system, which can open to form *N*-methylol groups. Since they are capable of reproducing the intermediate in formaldehyde tanning, it might be assumed that they function in the same way. DasGupta *et al.*[34] proposed that the reaction is concerned in part with the production of a simple Schiff adduct, analogous to the formaldehyde reaction; the electronic changes are illustrated in Figure 14.20. Reaction scheme A is the route

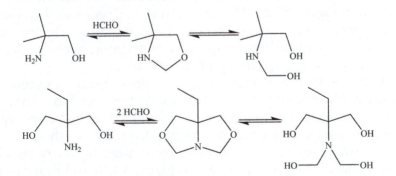

Figure 14.19 Examples of structures of mono- and bicyclic oxazolidines.

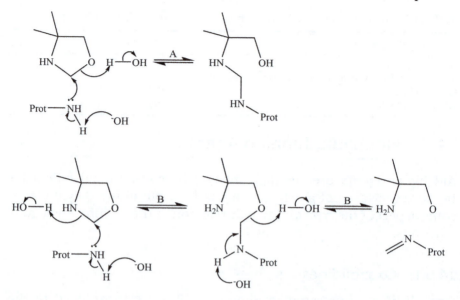

Figure 14.20   Reaction between oxazolidines and protein sidechain amino groups.

by which the oxazolidine is bound to the protein. Reaction scheme B is the route by which the oxazolidine could break down to create the Schiff base. The latter reaction appears less plausible, but there is evidence that the reaction can work in this way, adding a methylene group to amino sidechains.[27] However, considering the mechanistic options, it is not certain whether the reaction works because of scheme B or because there is formaldehyde in solution from the breakdown of the oxazolidine back to the components from which it was made.

The reaction between the Schiff base and the tyrosine sidechain is illustrated in Figure 14.21. This reaction is analogous to the reaction with polyphenol derivatives already discussed. Similarly, the Schiff base can react with other groups that contain an active hydrogen, such as the amino groups in arginine and lysine.

However, because oxazolidines are known to react as monomers, the leather is less filled than aldehyde-tanned leather. Since they contain no chromophores, they produce white leather. They show typical aldehyde-tanned shrinkage temperatures, up to 85 °C; there is a difference between the reactions depending on whether the oxazolidine is monocyclic or bicyclic, as shown in Figure 14.19 and illustrated in Table 14.4.[33]

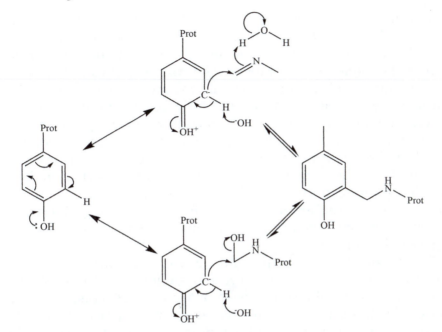

Figure 14.21 Reaction between a Schiff base or methylol derivative and a tyrosine sidechain.

Table 14.4 Comparison of the tanning reactions of mono- and bicyclic oxazolidines.

		Shrinkage temperature (°C)	
Oxazolidine offer (%)	Initial pH	Monocyclic	Bicyclic
5	2.0	84	65
5	4.0	85	72
5	7.4	84	82
10	2.0	84	69
10	4.0	85	74
10	7.4	84	82

The following points of interest arise from the results in Table 14.4.

- The outcome of the reaction, in terms of the hydrothermal stability, is not improved by offering excess reagent.
- The monocyclic oxazolidine confers higher shrinkage temperatures.
- The monocyclic oxazolidine reacts independently of pH, unlike the bicyclic oxazolidine, which reacts like a conventional aldehyde.

DasGupta *et al.*[34] proposed that the main tanning outcome from the oxazolidines is *via* the formation of the methylene derivative of the substrate amino groups. This suggests that the products of reaction by the monocyclic and bicyclic derivatives would be the same. It is a contrary observation that the monocyclic derivative yields shrinkage temperatures below 100 °C when retanning flavonoid polyphenol-tanned leather, but the bicyclic derivative yields shrinkage temperatures above 100 °C; in the former case, the outcome is merely additive tanning, in the latter case there is synergistic crosslinking. From the chemistry of the oxazolidines, it is clear that the bicyclic species can theoretically form crosslinks with protein, but the monocyclic species cannot; therefore, the tanning abilities of the latter versions become divorced from the necessity for crosslinking (see Chapter 23).

The monocyclic oxazolidine tends to react too quickly for a commercial tanning process; the bicyclic oxazolidine in Figure 14.19 is available commercially as Neosyn TX or Mirosan E. Consequently, the oxazolidine has featured in the literature in many processes, usually in combinations with most of the tanning agents well known in the art – usually with some predictable additive benefit, but no significant overall advantage, with the exception of condensed plant polyphenols (see Chapter 13).[35,36]

The chemistry can be extended, as shown in Figure 14.22, in which the ring closure adds an extra point of reactivity.[37]

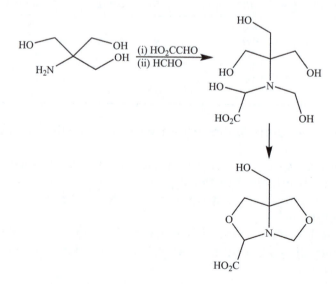

Figure 14.22   Synthesis of a modified oxazolidine.

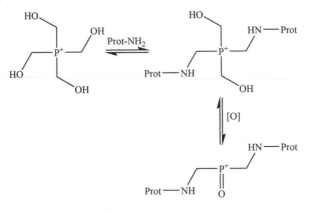

Figure 14.23 Reactions of phosphonium salts.

14.6.2 Phosphonium Salts

Methylol phosphonium salts have been known for many years, but it is only recently that the sulfate, tetrakis(hydroxymethyl)phosphonium sulfate (THPS), has become available commercially.[38] Before their potency as tanning salts became known, they were used as biocides. The chemistry is presented in Figure 14.23.

The reaction mechanism appears aldehydic, with the notable difference that it depends on the reactivity of the *P*-methylol group; considering the position of phosphorus relative to nitrogen in the Periodic Table, the analogy with *N*-methylol chemistry is apparent. However, the nature of the mechanism of the reaction with protein is not clear; it has not been extensively investigated, so it is still somewhat conjectural. It has been assumed that reaction occurs with at least one protein amino group – it is unlikely that a four-centre bond would be created – but it is known that each molecule reacts with 2.2 amino groups.[38] It has been argued that subsequent oxidation is advantageous, presumably to convert the residual methylol groups to an oxy moiety.

It is known that the phosphonium salt is hydrolysed at pH > 8 in the following way, to create the oxide, which is unreactive as a tanning agent:[39]

$$(HOCH_2)_4P^+ \xrightleftharpoons{OH^-} (HOCH_2)_4POH$$
$$\rightleftharpoons (HOCH_2)_3P + H_2O + HCHO$$
$$\xrightleftharpoons{[O]} (HOCH_2)_3PO$$

As an organic tanning agent, it appears to be less resistant to general biochemical attack than chromium(III)-tanned leather,; a study of the rate of degradation of a THPS–Al(III) combination

Table 14.5 Rate of degradation in soil of chrome-tanned and THPS–aluminium-tanned leathers.

	Weight loss (%)	
Time/days	Cr(III)	THPS–Al(III)
3	1.7	4.4
7	2.0	6.5
14	3.3	9.1
21	5.3	15.5
30	5.6	19.6

tanned leather in simulated landfill showed that it decomposes faster than chrome-tanned leather, as shown in Table 14.5.[40] Such studies are useful with regard to considerations of 'end of life' for leather goods.

THPS is a potent tanning agent and is likely to be a useful addition not only to the tanning industry but also to all other sectors that rely on the chemical stabilisation of protein by covalent organic means.

14.7 Other Tanning Applications

14.7.1 Wet White

In recent years, the concept of wet white has changed. It used to refer to a temporary condition, sufficiently stable to allow splitting and shaving prior to chrome tanning, but ideally allowing reversal of the chemical treatment. It is no longer concerned with a stage prior to chrome tanning, so although the options offered at the time might be of technological interest, they are solutions to a problem that is currently being ignored.[41]

However, there is one technology that comes under the heading of traditional wet white, which it is useful to recall, in case splitting and shaving in the raw state should be desired. Silica, in the form of colloidal silica, hydrogels or silicic acid, has been shown to have applications in leather processing:[42–44] the reagent binds to collagen *via* hydrogen bonding, but there is no increase in shrinkage temperature or even a leathering effect. However, the presence of silica within the fibre structure has the effect of reducing the ability of the fibre structure to slip over itself when subjected to distorting pressure. When the shaving blade applies pressure to untreated pelt, the surface can ripple ahead of the blade, until the pressure becomes too much and the blade skids over the surface, generating frictional heat. If the pelt is treated with silica, it is stiffened, so the shaving blade cuts rather than distorts the surface. The treated waste/by-product can be used as a fertiliser, instead of being disposed of to landfill.

Wet white is now regarded as a tannage in its own right, an approach to organic tanning, to allow, for example, recycling at the end of life, such as is increasingly required for automotive leathers. Several versions of wet white are made and sold – all are based on syntan plus glutaraldehyde. There is typically no rationale for this choice, especially the choice of syntan, and no attention is paid to the interactions between the components of the process: the tannage is viewed as the sum of the additive features of the individual contributions. A better approach would be to apply the principles set out Chapters 15 and 23, in order to achieve a more powerful tanning effect and achieve better properties and performance from the product.

14.8 Miscellaneous Tannages

It was demonstrated earlier that polymeric resins can be applied to collagen or leather to confer physical properties, such as filling up the fibre structure, but the chemistry does not allow high-energy bonding. Consequently, they cannot be considered to be tanning agents of any importance. The alternative approach is to create the polymeric species *in situ*.

14.8.1 Epoxide Tannage

Epoxides are currently used in the leather industry as crosslinking agents for polymeric finishes on the surface of grain leather; because of their instability to hydrolysis, they were typically applied in solvents. However, because of the requirement for aqueous processing, some epoxides have been developed that are water soluble and capable of being applied to wet collagen, and some of these are illustrated in Figure 14.24.[45,46]

It has been shown that epoxides can be used as tanning agents, but the shrinkage temperature is typically limited to 82 °C, with a maximum of 85 °C under extreme conditions, so in this way they

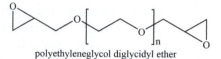

polyethyleneglycol diglycidyl ether

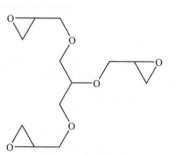

glycerol triglycidyl ether

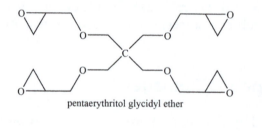

pentaerythritol glycidyl ether

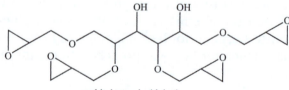

sorbitol tetraglycidyl ether

Figure 14.24 Some epoxides for tanning.

function like any other single-component tannage, without offering any advantage. They were compared with aldehydic reagents (glutaraldehyde, oxazolidine, THPS), but in all cases exhibited weaker tanning effects.

14.8.2 Isocyanate Tannage

In 2005, Traubel suggested that isocyanates and derivatives can offer an alternative approach to organic tanning.[47] They work by reacting with an active hydrogen in the following way:

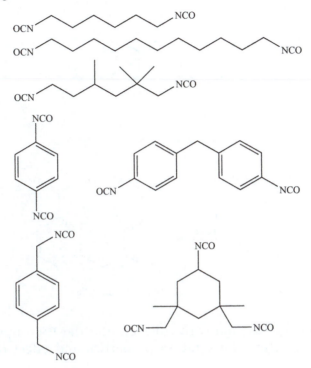

Figure 14.25 Diisocyanates for crosslinking tanning.

$$R-NCO + NaSO_3 \rightleftharpoons R-NH-CO-SO_3^- \ Na^+ \xrightarrow{\text{Protein}-H} R-NH-CO-Protein$$
$$\text{isocyanate} \qquad \text{carbamoyl sulfonate}$$

Diisocyanates can also be used to attempt crosslinking, such as those shown in Figure 14.25.

Although the compounds are capable of reacting with several types of sites on the protein and their functionality indicates the potential for crosslinking, the shrinkage temperature of the products is limited to 82 °C, just like the epoxides. The chemistry has been commercialised by Lanxess AG as X-Tan for making wet white.

14.8.3 Aromatic Heterocycles

The chemistry of aromatic heterocycles has been exploited for making 'reactive' dyes, so that the colouring can be fixed covalently to the leather. This same chemistry has been commercialised for tanning by Clariant as EasyWhite, as shown in Figure 14.26.[48] The nitrogen heterocycle (triazine) ring has two halogen leaving species that create

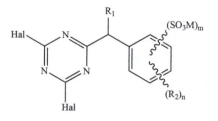

Figure 14.26 Basis of the chemistry of Clariant's EasyWhite.[48]

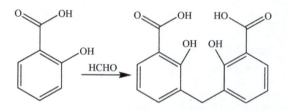

Figure 14.27 Salicylic acid polymerised by formaldehyde.

covalent sites for reaction with collagen and the R groups and sulfonate groups define the solution properties and effect on leather properties.

The Clariant patent notwithstanding, there is potential to create a new class of syntans based on the established chemistry of reactive dyes. However, the problem of simultaneous hydrolysis remains and must be addressed by the nature of the wider structure, as indicated in the example in Figure 14.26.

14.8.4 Multifunctional Reagents

Some reagents can offer a variety of functions: an example is the reaction product of salicylic acid and formaldehyde, reported by Evans and Montgomery[49] (Thiotan GDC 7, *ex* Sandoz), illustrated in Figure 14.27. The product has hydrogen bonding and metal complexation potential; by itself it raises the shrinkage temperature of pelt by only 7 °C, so its function as a tanning auxiliary is more important, as shown in Table 14.6.

A useful by-product of the use of this auxiliary is that the heat resistance is increased: in trials,[50] commercially chrome-tanned ovine leather with a shrinkage temperature of 110 °C exhibited 13% linear shrinkage after being boiled in water for 15 min, but the same leather treated with 5% of the salicylic acid–formaldehyde-treated product relaxed by 2% in the same boil test.

Table 14.6 Tanning processes with 5% salicylic acid–formaldehyde polymer as pretreatment.

Mineral offer	Mineral salt	Shrinkage temperature (°C)
2% Cr_2O_3	33% basic chromium(III) sulfate	121 ± 7
5% Al_2O_3	Basic formate-masked sulfate at pH 4	81 ± 2
5% TiO_2	Titanyl ammonium sulfate, solution at pH 4	79 ± 2
3% Al_2O_3 + 1.5% TiO_2	Basic gluconate-masked mixed sulfate, pH 4	89 ± 2

References

1. J. H. Sharphouse, *J. Soc. Leather Technol. Chem.*, 1985, **69**(2), 29.
2. M. P. Balfe, *Progress in Leather Science: 1920–45*, British Leather Manufacturers' Research Association, 1948, p. 493.
3. O. Suparno, *et al.*, *J. Soc. Leather Technol. Chem.*, 2009, **93**(4), 158.
4. O. Suparno, *et al.*, *J. Am. Leather Chem. Assoc.*, 2011, **106**(12), 360.
5. O. Suparno, *et al.*, *J. Am. Leather Chem. Assoc.*, 2013, **108**(5), 180.
6. K. V. Sandhya, *et al.*, *J. Am. Leather Chem. Assoc.*, 2015, **110**(7), 221.
7. H. Zhou, *et al.*, *J. Soc. Leather Technol. Chem.*, 2011, **95**(6), 250.
8. J. A. Wilson, *The Chemistry of Leather Manufacture*, American Chemical Society, The Chemical Catalog Co. Inc., New York, 1929.
9. E. Stiasny, *Austrian Pat.*, 58,405, 1913.
10. J. Guthrie, *et al.*, *J. Soc. Leather Technol. Chem.*, 1982, **66**(5), 107.
11. J. Ammenn, *et al.*, *J. Am. Leather Chem. Assoc.*, 2015, **110**(11), 349.
12. T. C. Thorstensen, *Practical Leather Technology*, Van Nostrand Reinhold Co., New York, 1969.
13. G. Reich, *Das Leder*, 1958, **9**, 6.
14. K. Bienkiewicz, *Physical Chemistry of Leather Making*, Robert E. Krieger Publishing Co., Malabar, Florida, 1983.
15. G. Reich, *Ges. Abh. Inst. Freiberg*, 1960, **16**, 40.
16. O. Suparno, PhD thesis, University of Northampton, 2005.
17. O. Suparno, *et al.*, *Resour., Conserv. Recycl.*, 2005, **45**, 114.
18. A. Bacardit, *et al.*, *J. Soc. Leather Technol. Chem.*, 2011, **95**(6), 259.
19. A. Bacardit, *et al.*, *J. Soc. Leather Technol. Chem.*, 2012, **96**(2), 64.
20. A. D. Covington and R. L. Sykes, *J. Am. Leather Chem. Assoc.*, 1981, **65**(2), 21.
21. A. D. Covington and M. Song, *UK Pat. Appl.*, GB 2287953A, 1994.
22. X. Heng, *et al.*, *J. Soc. Leather Technol. Chem.*, 2014, **98**(1), 17.
23. A. G. el Amma, *European Pat.*, EP 0372746A2, 1989.
24. R. Schmidt, *Das Leder*, 1989, **40**, 41.
25. J. H. Bowes and R. G. H. Elliott, *J. Am. Leather Chem. Assoc.*, 1962, **57**(8), 374.
26. S. K. Taylor, F. Davidson and D. W. Overall, *Photogr. Sci. Eng.*, 1978, **22**(3), 134.
27. B. Metz, *et al.*, *J. Biol. Chem.*, 2004, **279**(8), 6235.
28. L. H. H. Olde Damink, *et al.*, *J. Mater. Sci.: Mater. Med.*, 1995, **6**, 460.
29. E. M. Filchione, *et al.*, *J. Am. Leather Chem. Assoc.*, 1958, **53**(2), 77.
30. R. Sugiyama, Y. Chonan and H. Okamura, *Hikaku Kagaku*, 1997, **43**(1), 55.
31. J. R. Rao, S. V. Kanth and B. U. Nair, *J. Soc. Leather Technol. Chem.*, 2008, **92**(2), 65.
32. G. Jaya Kumar, *et al.*, *J. Am. Leather Chem. Assoc.*, 2011, **106**(2), 50.
33. S. DasGupta, *J. Soc. Leather Technol. Chem.*, 1977, **61**(5), 97.
34. S. D. Choudhury, *et al.*, *Int. J. Biol. Macromol.*, 2007, **40**(4), 351.
35. T.-T. Qiang, *et al.*, *J. Soc. Leather Technol. Chem.*, 2008, **92**(5), 192.

36. S. DasGupta, *J. Soc. Leather Technol. Chem.*, 2010, **94**(4), 167.
37. Z. Luo, *et al.*, *J. Am. Leather Chem. Assoc.*, 2009, **104**(4), 149.
38. Y. Li, *et al.*, *J. Soc. Leather Technol. Chem.*, 2006, **90**(5), 214.
39. Y. Wu and Q. Wu, *J. Soc. Print Dye.*, 1990, **54**(4), 11.
40. L. F. Ren, *et al.*, *J. Am. Leather Chem. Assoc.*, 2009, **104**(6), 218.
41. A. D. Covington, *Tanning Chemistry: The Science of Leather*, RSC Publishing, Cambridge, UK, 2009.
42. K. H. Fuchs, *et al.*, *J. Am. Leather Chem. Assoc.*, 1995, **90**(6), 164.
43. K. H. Munz, *et al.*, *J. Am. Leather Chem. Assoc.*, 2003, **98**(5), 159.
44. K. H. Munz and R. Sonnleitner, *J. Am. Leather Chem. Assoc.*, 2005, **100**(2), 66.
45. R. J. Heath, *et al.*, *J. Soc. Leather Technol. Chem.*, 2005, **89**(5), 186.
46. Y. Di, *et al.*, *J. Soc. Leather Technol. Chem.*, 2006, **90**(3), 93.
47. H. Traubel, *J. Am. Leather Chem. Assoc.*, 2005, **100**(7), 304.
48. Clariant Int. Ltd., *European Pat.*, PCT/EP2010/001334, 2010.
49. N. A. Evans and K. C. Montgomery, *Proceedings of IULTCS Congress*, Melbourne, Australia, 1987.
50. A. D. Covington, unpublished results.

15 Post-tanning

15.1 Definition

The term 'post-tanning' refers to the wet processing steps that follow the primary tanning reaction. This might refer to following tannage with chromium(III), as is usually the case in industry, but equally it applies to vegetable tanning or indeed any other tannage used to confer the primary stabilisation to pelt. The combination of post-tanning processes may not always be the same for all tannages: the choice of post-tanning processes depends on the primary tannage and the type of leather that the tanner is attempting to make. In all cases, post-tanning can be separated into four generic processes.

- *Retanning:* This may be a single chemical process or may be a combination of reactions applied together or more usually consecutively. The purpose is to modify the properties and performance of the leather. These changes include the handle, the chemical and hydrothermal stability and the appearance of the leather. The effects are dependent on both the primary tanning chemistry and the retanning reactions.

  Retanning can involve many different types of chemical reactions. These include mineral tanning with metal salts [including chromium(III) applied to chrome-tanned leather], aldehydic reagents, hydrogen-bondable polymers, electrostatic reactions with polymers or resins or any other type of synthetic tanning agent (syntan).

Tanning Chemistry: The Science of Leather 2nd edition
By Anthony D. Covington and William R. Wise
© Anthony D. Covington and William R. Wise 2020
Published by the Royal Society of Chemistry, www.rsc.org

- *Dyeing*: This is the colouring step. Almost any colour can be struck on any type of leather, regardless of the background colour, although the final effect is influenced by the previous processes.

  Colouring almost invariably means dyeing. Applying dye in solution or pigment suspended in solution, to confer dense, opaque colour, can be performed in the drum or colouring agents may be sprayed or spread by hand (padding) onto the surface of the leather.

- *Fatliquoring*: This step is primarily applied to prevent fibre sticking when the leather is dried after completing the wet processes. A secondary effect is to control the degree of softness conferred to the leather. One of the consequences of lubrication is an effect on the strength of the leather (see Chapter 17).

  Fatliquoring is usually conducted with self-emulsifying, partially sulfated or sulfonated (sulfited) oils, which might be animal, vegetable, mineral or synthetic. This step may also include processing to confer to the leather a required degree of water resistance.

- *Special treatments:* These are treatments to confer specific properties to the leather, but which do not fall into the three categories above. An example is conferring fire resistance.

15.2   Relationship Between Tanning and Post-tanning

In the whole of leather making, each step in the process affects all subsequent steps and none more so than the impact of tanning on the post-tanning reactions. The nature of tanning reactions and their possible effects are illustrated in Table 15.1.

Modification of collagen by the chemistry of the tanning agent(s) affects the following features of the properties of the new material.

1. The hydrophilic–hydrophobic balance (HHB) of the leather is markedly affected by the chemistry of the tanning agent, because it is likely to change the relationship between the leather and the solvent, which in turn will affect the equilibrium or partitioning of any reagent between the solvent and the substrate.

2. The site of reaction between the reagent and the collagen will affect the isoelectric point (IEP) of the collagen and consequently there will be a different relationship between pH and charge on the leather. The lower the IEP, the more anionic or less cationic

Table 15.1 Categorisation of tanning reactions.

Tannage	Effect on IEP	Tanning agent	Leather	HHB Charge	Reaction sites	Bonding
Mineral	Higher	Hydrophilic	Hydrophilic	Cationic	CO_2^-	Electrovalent
Chromium(III)	Higher	Hydrophilic	Hydrophobic	Cationic	CO_2^-	Covalent
Aldehydic	Lower	Hydrophilic	Hydrophilic	Neutral	NH_3^+	Covalent
Syntan	Lower	Hydrophilic/ hydrophobic	Hydrophilic	Neutral	NH_3^+, amide	Electrovalent, H-bonding
Vegetable	Lower	Hydrophobic	Hydrophilic	Neutral	NH_3^+, amide	H-bonding, hydrophobic
Oil	No change	Hydrophobic	Hydrophilic	No change	None	None, hydrophobic

the charge on the pelt will be at any pH value; the higher the IEP, the more cationic or less anionic the charge on the pelt will be at any pH value.

3. The relative reactions at the sidechains and the backbone amide links of the protein will determine the type of reaction and hence the degree of stability of the tannage: the fastness of the reagent will influence the interaction between reagents and the substrate. Here, the important factors are the effects of free tanning agent and charge on the leather; reactions due to covalency, electrostatic interaction, hydrophobic interaction and hydrogen bonding will all have effects on leather properties.

The nature of the tanning agent will impact differently on the different aspects of post-tanning.

1. Retanning may involve interaction with the first tanning agent or it may be independent, because the chemistries are incompatible or the reaction sites may be different.
2. Dyeing outcome depends on the chemistry of the dye and the chemistry of the substrate. They may work together, as is the case with metal mordanting, or colouring may be adversely affected by similarity of chemistry, such as syntan tanning, or the colour may be dulled or 'saddened' by the presence of vegetable tannins. The HHB properties of the substrate will influence uptake of dyes, because the dyes can exhibit the full range of HHB properties.
3. Fatliquoring has two aspects that can be influenced by the substrate. First, there is the question of depositing the

neutral oil by damaging the emulsifying mechanism. This is clearly affected by charge, but also by the availability of charged sites on the leather. Second, there is the degree of compatibility of the neutral oil with the HHB properties of the leather, modified by the particular chemistries of the previous process steps.

15.3 Chrome Retanning

A common retannage for chrome-tanned leather is more chrome tanning. This raises the question: why? It is not obvious why the leather should be rechromed when the reaction could and should have been completed during primary tanning. Possible reasons are as follows.

1. To increase the shrinkage temperature.
   Since rechroming is typically conducted with an offer of about 1% Cr_2O_3 (4% chrome tan powder) or less, *i.e.* half of or less than the original offer, usually over a period of about 1 h, *i.e.* an order of magnitude shorter than the original tanning process, it can be understood that the effect of chrome retanning will be much less than that in the original reaction. Moreover, the time allotted is unlikely to result in an effect all through the cross-section. Furthermore, since retanning is conducted under conditions that are more astringent than in primary tanning, fixation is less likely to be uniform through the cross-section. Therefore, the retannage is more to do with an effect on the surfaces.
   It is a common experience that rechroming has little or no effect on the shrinkage temperature of the final leather. This is understandable, because of the conditions, and the resulting change in chrome content contributes little to the tanning effectiveness.
2. To increase the chromium content.
   For many applications of chrome-tanned leather, the specification typically includes the requirement for 4% Cr_2O_3 on dry weight, *e.g.* Defence Specifications. Note that this is a sort of quality assurance specification, because at that level of chrome, the tannage is *likely* to be complete and the shrinkage temperature is *likely* to be >100 °C. Note also that there is no guarantee that those properties will be met – it is possible to fix that amount of chrome and still obtain a shrinkage temperature considerably lower than 100 °C.

From the conditions of retannage set out above, the fixation of chrome is not going to be as effective as the primary reaction. Indeed, it is possible to discharge more chrome in the effluent stream than is offered in the retannage, by not fixing much chrome and by actually stripping chrome from the wet blue during retanning.

3. To even up the colour.

When bovine wet blue is split, it is common to observe variations in colour over the split surface. Although this may not adversely influence the overall properties and performance of the leather, cosmetically it looks better to make the colour more even or uniform.

Rechroming is often used when wet blue is purchased from different sources, in an attempt to make the colour more uniform between batches of leather.

4. To change the reactivity of the leather.

Incorporating fresh chrome into aged wet blue has the effect of creating new cationic sites, which might be useful in fixing anionic reagents later in the post-tanning process.

More fundamentally, the IEP will be moved to a higher value, so that at any pH value the charge on the leather is either less anionic (negative) or more cationic (positive) than that on the unretanned leather. This also will influence the reactivity towards the post-tanning reagents, which are often anionic.

Note that the effect of IEP is more important than the introduction of cationic charges, because they can be discharged chemically, whereas the influence of IEP with pH on the charge of the leather is a permanent effect.

5. To modify the properties of the leather.

Fixing chrome under conditions at the limits of basicity in solution means that polymeric chrome is bound to collagen (see Chapter 11). In this way, a degree of filling and softening is achieved, without using reagents that might adversely affect light fastness or water resistance, *etc.* However, considering the amount of chrome likely to be fixed, the effect will be small.

15.4   Sequence of Post-tanning Steps

Although there are many variations on the procedures within post-tanning processes, in general they tend to conform to the following sequence: retanning then dyeing then fatliquoring then special processes. The rationale can be expressed as follows.

- Retanning may not only modify the properties of the leather, but also will typically modify the reactivity of the pelt towards other reagents. Even if the processes do not include specific reactions to assist uniform colouring, it is possible to achieve such a side effect as a bonus. Alternatively, if the retannage includes mineral reagents, they may also have a mordanting effect on the dyeing step, to achieve modified colour or better fixation (see Chapter 16).
- Dyeing comes next, to take advantage of the changes to the reactivity of the pelt conferred by additional tanning. The relationship between dyeing and fatliquoring is controlled by the requirements of the dyeing compounds for reactivity towards the substrate. This reactivity is partly dependent on the HHB characteristics of the substrate, which can be positively or negatively modified by the presence of hydrophobic species such as fatliquors.
- Fatliquoring usually comes last, in order not to interfere with the colouring reaction. In particular, if the lubrication step includes reactions to confer water resistance, this can create a barrier to reaction of aqueous reagents with the collagen, so it must be done just prior to drying.
- If special processes are applied, they usually come last, so they are not affected or compromised by the reagents in the other steps.

15.5   Principles of Post-tanning

15.5.1   Mechanisms of Post-tanning

All the post-tanning process steps involve the fixation of a solute in solution onto a solid substrate. The reactions are made more complicated by the fact that the tanner is dealing with a substrate that has finite thickness. Therefore, in order to control the outcomes of these steps, it is necessary to understand the parameters that come into play. First, it is important to understand that there is a general mechanism that must be taken into account, which is the steps by which fixation occurs; this was introduced in Chapter 11. The steps involved in any general reaction in which a reagent in solution is fixed onto a solid substrate, *e.g.* in the heterogeneous system of post-tanning, may be defined as follows:[1,2]

1. transfer of the reagent from solution into the substrate;
2. hydrophobic bonding;

3. electrostatic interaction between the reagent and the substrate;
4. covalent reaction between the reagent and the substrate

The first controlling factor can be expressed as follows, where the components of the reaction interact directly with the solvent in a solvating manner:

$$\text{reagent} + \text{solvent} \rightleftharpoons [\text{reagent}]_{\text{solvated}} \qquad (15.1)$$

$$[\text{reagent}]_{\text{solvated}} + [\text{substrate}]_{\text{solvated}} \rightleftharpoons [\text{substrate} - \text{reagent}]_{\text{solvated}} \qquad (15.2)$$

The position of the equilibrium for eqn (15.1) depends on the affinity of the reagent for the solvent, *i.e.* the solubility or the solvating power of the solvent with respect to the properties of the solute. The position of the equilibrium for eqn (15.2) depends on the relative affinities for the solute for the solvent and the substrate. This is analogous to partitioning a solute between two immiscible solvents, where the equilibrium constant of transfer is analogous to the partition coefficient (see later). In the case of water as the solvent, the equilibrium for eqn (15.1) is defined by the degree of hydrophilicity or hydrophobicity of the solute/reagent. This has been defined as the hydrophilic–hydrophobic balance (HHB) or hydrophilic–lipophilic balance (HLB). This property is conventionally measured by chromatography in a variety of solvents, ranging from water to petroleum ether (light petroleum), *i.e.* highly polar to highly non-polar. Figure 15.1 shows the principle of chromatography, using the example of paper chromatography, in which the solvent travels upwards, up the paper, taking the reagent with it. The fraction of the way from the initial position of the reagent spotted on the paper to the solvent front is referred to as the R_f value and is a measure of the affinity of the reagent for the solvent. The closer the R_f value approaches 1.0, the higher is the affinity of the reagent for the solvent.

The equilibrium for eqn (15.2) depends on the relative affinities of the reagent/solute for the solvent and the environment within the substrate. That is, a hydrophilic reagent will tend to remain in solution, because it interacts favourably with water. On the other hand, a hydrophobic reagent will not be as soluble in water and will tend to move into a more hydrophobic environment, into the substrate. The extent to which this happens and the rate of transfer depend on the magnitude of the HHB/HLB value of the solute and the properties of the solvent.

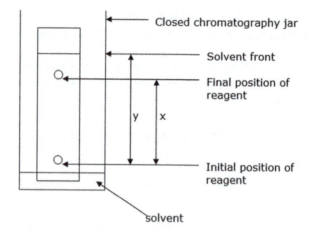

Rf = x ÷ y

Figure 15.1 Illustration of the paper chromatography method for measuring hydro-philic/lipophilic balance.

The concept of transferring a solute is familiar to chemists in the form of partitioning it between two solvents in the technique of solvent extraction. The notion can be extended to the free energy of transfer between two solvents, which determines the degree of preferential solvation when a solute is dissolved in a binary solvent.[3]

The step of hydrophobic bonding is essentially a special case for aqueous solution. If the driving force for transfer is dominated by the hydrophilic nature of the reagent, the first mechanism of reaction may be hydrophobic bonding. This has been suggested as the route of plant polyphenol fixation (see Chapter 13). Such bonding is likely to change to hydrogen bonding as its equivalent of the electrostatic interaction.

The concept of transfer from solvent to substrate should not be confused with the notion of penetration through the cross-section of the substrate. Tanners are constantly dealing with heterogeneous systems and balancing the rate of penetration with the rate of fixation. This is consistent with the model of stepwise reaction between a reagent and a substrate. Penetration is a consequence of favourable reagent–solvent interaction. The more favourable this is or the more soluble the reagent is in the solvent, the less favoured is transfer, so the reagent can penetrate through the substrate cross-section. Fixation is a consequence of favourable transfer into the environment of the substrate. Since the rate of transfer of a reagent is affected by the

HHB/HLB value, the reagent properties also affect the nature of the reaction. All reactions are dependent initially on some form of charge–charge interaction. Therefore, because electrostatic interactions are fast, reaction between the substrate and the reagent is initiated by the rate-determining step of transfer. Hence fast uptake means surface interaction, rather than penetration into the cross-section of the substrate. This has many implications in different processes in leather making, not only for the science of reaction efficiency, but also for the technological outcome.

Eqn (15.2), the transfer of the solute from the solvent to the substrate, will depend on whether or not the reagent is charged and whether or not the substrate is charged. This will affect the interaction with the solvent, particularly if the solvent is polar, such as water. Therefore, the relative charging of the reactants will influence the relative affinities of the solute for the solvent and for the substrate. This can be distinguished from the role of electrostatic attraction or repulsion, which determines the rate of fixation, following transfer from solution into the substrate.

It must follow that changes to the equilibrium for eqn (15.2) can be achieved by changes to each component of the equilibrium.

- Changing the HHB/HLB value of the reagent will alter its relative affinity for the solvent *versus* the substrate. This might be done using the chemistry of the reagent itself or by changing the reagent for one with a more appropriate HHB/HLB value.
- Changing the chemistry of the substrate will alter the relative affinity of the solute for the substrate *versus* the solvent. This might be achieved by manipulating the chemistry of the substrate: possibilities include chemical modification and charge change. Alternatively, the HHB/HLB value of the substrate may be changed by applying hydrophilic or hydrophobic reagents, *e.g.* water-resistant fatliquor.
- Changing the solvent will alter the position of the first equilibrium, with a consequent change to the relative affinity of the solute for the substrate. This change can be in either direction, depending on the nature of the change made to the solvent. Water can be made more or less polar by the presence of a neutral electrolyte or organic solvent. Similarly, any other solvent can be modified by the presence of other components, as solutes or other solvents. Ultimately, the solvent itself can be substituted.

The extent to which the creation of covalency occurs depends entirely on the chemistry of the reaction and for many tanning reactions may not apply at all. All chemical reactions result in bonding that lies somewhere on the scale between pure electrostatic and pure covalent. Few lie at the extremes, but it is clear that most reactions can readily be designated as one or the other.

In the course of reviewing the principles of tanning and advanced aspects of tanning and especially post-tanning, examples of all these possibilities will be encountered. An understanding of the variables in these reactions is a powerful tool for future developments in leather technology and predicting the outcome of change.

Leather technology has always relied on water as the delivery medium for reactions. Consequently, reagents have always had to have at least some affinity for water, in order to be delivered uniformly to the solid substrate. That affinity is responsible for the limits to reaction efficiency, since the solvating power can only be overcome in the limit by extreme conditions, which create their own problems for quality and performance. There are, however, new technologies available for delivering reagents in all process steps, discussed in full in Chapter 19.

15.5.2 Role of the Isoelectric Point

The concept of the isoelectric point (IEP) in proteins was introduced in Chapter 1, where it was shown that it could be defined as follows:

$$\text{IEP} = \frac{\sum_i f_i [\text{NH}_2]_i}{\sum_j f_j [\text{CO}_2\text{H}]_j}$$

There are two points relating to the IEP and its particular relevance to post-tanning that are very important and worth emphasising.

1. The IEP is a point on the pH scale, so it does not change with changes to the pH of the system. The IEP of collagen is the same whether it is in the limed state or in the pickled state: the fact that the charge changes does not change the IEP. The importance of this point is that the IEP can only be changed if there is a chemical change that alters the availability of active groups.

   In the context of charge on leather, care must be taken when ascribing the influence of charged reagents. For example, treating collagen with cationic chromium(III) species may be regarded as altering the IEP, because it may introduce positive charge. Although the introduction of charge is true, it should be noted that the charge can be lost by treatments other than pH change. From this point of view, the charge is immaterial as far as the IEP is concerned, although the binding of chromium(III) clearly impacts the IEP, by removing some carboxyl function from the IEP-determining ratio.

In the context of charge as a function of pH, the definition of IEP must be applied correctly. The introduction of charged species into the substrate clearly can influence the overall charge and hence modify the reactivity of the substrate towards specified reagents. An example is polycationic species as dye intensifying agents (see Chapter 16).

2. The charge on collagen is determined by the relative values of the IEP and the pH. If the pH is higher than the IEP, the collagen is negatively charged, and if the pH is lower than the IEP, the collagen is positively charged. Moreover, the further the pH is from the IEP, the greater is the charge: the change is not linear, but reflects the logarithmic pH scale. The clear consequence is the effect on the affinity of reagents that rely on charge for fixation to the leather. This can be modelled as shown in Figure 15.2, which also shows the effect of moving the IEP.

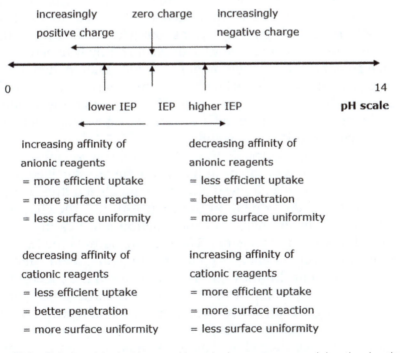

Figure 15.2 Relationship between pH and charge, governed by the isoelectric point.

From the model in Figure 15.2, it can be seen that the effects of moving the IEP are as follows.

- Change to higher IEP: at any pH value, the charge on the protein/leather is either less negative or more positive, depending on which side of the IEP is considered.
- Change to lower IEP: at any pH value, the charge on the protein/leather is either less positive or more negative, depending on which side of the IEP is considered.

The effects of charge on an electrostatic reaction can be understood from the following simple relationship:

$$\text{rate} \propto [\text{reagent charge}]^a[\text{substrate charge}]^b$$

where the rate refers to the rate of the fixation reaction, which will be controlled by the magnitudes of the charges, because they determine the degree of attraction. Clearly, if the charges are different, the affinity is favoured; this results in fast uptake, with a limitation on the extent to which the reagent can penetrate the substrate. On the other hand, if the charges are the same, the affinity is disfavoured and the reaction is slowed, because of the availability of reaction sites, so penetration is favoured. The measurement of the magnitude of the charge on leather is not easy. Attempts have been made to use charged dyes to determine charge interactions, but with little success. In the context of processing, the best we can do currently is to estimate the overall effect of the accumulation of process steps. In native collagen, the charge as a function of pH will reflect to some extent the swelling curve, since there is some cause and effect, although the swelling is also influenced, even controlled, by osmosis and lyotropy, depending on the specific conditions in solution.

The direction of change in IEP value is known for most leather-making process steps or can be judged from the reaction if conflicting reactions occur, such as in dyeing with premetallised dyes. However, in the absence of direct data on the change in IEP, the change may be estimated from the nature of the reaction and bearing in mind the typical offers of the reagents. The starting point is the IEP of raw collagen, the accepted change by liming and Gustavson's measurement of the effect of chrome tanning.[4] Estimates of the effects of post-tanning are given in Table 15.2. The numbers have to be estimates because, although it can be accepted that a covalent reaction will have a certain effect on the IEP value, it is not known how an electrostatic reaction influences the value. This is because it must be assumed that a pH change can have a reversing effect on the interaction and

Table 15.2 Known and estimated changes in the isoelectric point of type I collagen due to leather-making processing.

Process	Reaction	Estimated effect on IEP
None, native collagen		Physiological pH 7.4
Liming	Amide hydrolysis	5.0 to 6.0
Chrome tanning	Reaction with carboxylate	6.5 to 7.5
Retanning reactions:		
Aldehyde	Covalent bonding	Estimated ΔIEP = −1.0
Vegetable tannin	Electrostatic bonding	Estimated ΔIEP = −0.5
Syntan	Electrostatic bonding	Estimated ΔIEP = −0.5
Resin	No effect	Estimated ΔIEP = 0
Dye reactions:		
Acid dye	Electrostatic bonding with NH_3^+	Estimated ΔIEP = −0.5
1:1 Premetallised dye	Metal complexation at carboxyl, sulfonate reaction at amino	Estimated ΔIEP = +0.5
Reactive dye	Covalent reaction with NH_2 Electrostatic reaction with NH_3^+	Estimated ΔIEP = −1.0
Fatliquor, sulfo reaction at amino	Electrostatic bonding	Estimated ΔIEP = −0.5

Table 15.3 Estimated changes in isoelectric point, starting with chrome-tanned leather at 7.0, then retanning, dyeing (1:1 premetallised dye, acid dye or reactive dye) and fatliquoring.

Retan											
Aldehyde			Syntan			Vegetable			Resin		
6.0			6.5			6.5			7.0		
Dye											
1:1	Acid	Reactive	1:1	Acid	Reactive	1:1	Acid	Reactive	1:1	Acid	Reactive
6.5	5.5	5.0	7.0	6.0	6.5	7.0	6.0	5.5	7.5	6.5	6.0
Fatliquor											
6.0	5.0	4.5	6.5	5.5	6.0	6.5	5.5	5.0	7.0	6.0	5.5

hence the group on the protein may still be counted (to some extent) in the calculation of the IEP ratio. In addition, the influence of introducing charge is also not known. Using the model of Table 15.2, the changes in IEP might be estimated as shown in Table 15.3, starting with chrome-tanned leather with IEP 7.0, retanning with only one reagent, dyeing with only one type of dye, then fatliquoring. Furthermore, the actual changes to the IEP will be dependent on the amounts of reagents bound to the leather. Here the estimates are generalised, assuming typical processing and taking into account the chemistry of the interaction between the reagent and the collagen: in the absence of quantitative information, the estimated changes are based assumed small effects (0.5) from weaker interaction or larger effects (1.0) from stronger interaction. It is assumed that the effects of reagents on the IEP are accumulative.

The likely scenario in post-tanning is that the IEP will gradually move down, from pH 7 to pH 5–6. Therefore, at any given pH value, the leather will be less positive or more negatively charged than it was before. In other words, the leather has less and less affinity for anionic reagents, so penetration of these reagents is favoured. However, variations in this pattern must be recognised, especially when using premetallised dyes or if large offers are made in the process step.

15.5.3 Role of the Peptide Link

When considering the properties of collagen, reactions at the peptide link are important, since the link is the most common feature of protein structure. Hence its susceptibility to hydrolysis is a feature of beamhouse processing. In the case of post-tanning processing, its role depends on its ability to engage in fixation reactions. As described in Chapter 1, the bond can be drawn in two ways, showing that charge is somewhat separated between the amino and carbonyl groups: the interaction of the sulfonate group will induce charge separation to create a greater enol contribution to the structure, so it can be represented in two ways, as illustrated in Figure 15.3.

- *Hydrogen bonding:* The presence of negative charge on the oxygen allows hydrogen bonding with an active hydrogen-bearing group.
- *Electrostatic bonding:* The presence of positive charge on the nitrogen allows electrostatic bonding with anionic species.

The versatility of the reactivity of proteins originates from the partially charged nature of the peptide group. Despite the limitations of the partial charge, the peptide links have a powerful effect, because their concentration is higher than the other obvious reaction centres

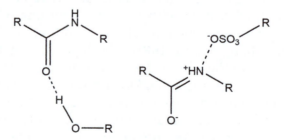

Figure 15.3 Illustrations of bonding at the peptide link.

on the sidechains. The contents of carboxyl groups and amino groups in dry collagen are 1.0 and 0.6 mol kg^{-1} dry weight, respectively, and may be compared with the amount of peptide links as follows. The molecular weight of the triple helix is 300 000, and it contains 3000 amino acids. Therefore, collagen contains 10 amino acids per kilogram, equivalent to nine peptide links. Hence there is an order of magnitude more peptide links than carboxyl groups and 15 times more than amine sidechains.

15.5.4 Role of the Sulfonate Group

A chemical theme that runs through the whole of the post-tanning reactions is the role of the sulfonate group. This theme is expanded in the individual sections below. Reagents that typically carry sulfonate groups include the following:

- syntans;
- modified vegetable tannins;
- dyes: acid, basic, direct, premetallised, reactive;
- fatliquors: sulfated fatliquors included.

In each case, the mechanism of fixation is the same. An electrostatic reaction can take place between the anionic sulfonate group and the protonated amino group. However, this can happen only when the collagen is acidified, to create the cationic group. Above the IEP, there are no cationic centres for reaction. At the IEP, the cationic groups are locked together with the anionic carboxyl groups: it is only when the salt links are broken by acidification that reaction with sulfonate can occur. The carboxyl groups are not in competition with the sulfonate groups, because the weak carboxylic acid groups become protonated, as the mechanism of breaking the salt link. This is the reason why acid dyes are so called, because they are fixed by acid. The general reaction mechanism is set out in Figure 15.4.

The mechanism of sulfonate fixation has sometimes been wrongly rationalised in terms of protonation of the sulfonate group, particularly as the mechanism for lowering the emulsifying power of the sulfo fraction of fatliquors (see Chapter 17). However, it is important to recognise that the mechanism actually does not include protonation of the sulfonate group: it is much too strong for that to occur under the aqueous conditions of leather making, as discussed in Chapter 9. A practical demonstration of the strength of the sulfonate group is to try

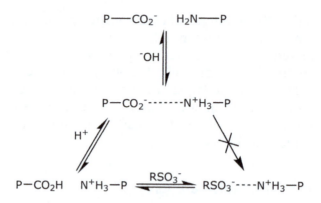

Figure 15.4 Mechanism of sulfonate fixation on collagen.

to create a non-swelling acid from an auxiliary syntan by acidification, *i.e.* the reverse of the reaction to create a syntan from a sulfonated compound. The reaction can be made to work only in concentrated sulfuric acid, a solvent that cannot be regarded as aqueous.

15.6 Coordinating Post-tanning Processes

Creating a process for post-tanning is not a case of merely putting together the steps for applying the required reagents in the right order. The following considerations must be addressed, and apply to every reagent that the tanner intends to fix onto the leather.

1. Surface reaction or penetration?
2. The required degree of penetration down the hierarchy of structure.
3. The requirement for uniformity of reaction: through the cross-section and over the surfaces.
4. The amount and suitability of the reagent necessary to achieve the requirement.
5. The rate of reaction, *i.e.* the process time yielding optimum efficiency and hence cost-effectiveness.
6. The affinity of the reagent for the substrate and consequent fastness properties.

The specific nature of each reagent and its function must be considered in the light of the following features of post-tanning processing.

1. Is this the best reagent for the purpose, using any criteria – cost, efficiency, safety, environmental impact, *etc.*?

2. Are there any options for 'compact processing', *i.e.* combining more than one process step into a single step? See later.
3. Do the process steps interact to create synergy, *i.e.* obtaining effects greater than the sum of the parts, *e.g.* one reagent influencing the fastness of another or together producing a new interacting species on the substrate that contributes to the leather stability in some way?

The following variables must be considered in order to achieve the required outcome.

1. The compatibility of the reagent for the substrate, *i.e.* the status of the substrate at any point in the process. It is not sufficient to consider the tanned starting material – it is important to consider how previous processes have modified the leather.
2. The compatibility of the reagent not only for the leather modified by previous processes, but also for subsequent reagents.
3. The role of the IEP. Although the relationship between processing and quantitative change to the IEP is not known with any precision, the direction of change due to reagent reaction can be judged qualitatively.
4. The role of pH. The relationship between pH and the IEP determines the charge on the leather. This controls transfer of the reagent onto the leather and the initial electrostatic interaction.
5. The role of the HHB/HLB value. The value for the leather will change as post-tanning steps progress. Although value changes may not be available, qualitative judgements about the directions of change can be made.

The conventional components of post-tanning are neutralisation, retanning, dyeing and fatliquoring (plus any special treatments); in each case, an industrial process will use one or more agents of that type. However, in post-tanning it is possible to combine process steps, to create 'compact' processing.[5] Here the principles of compact processing will focus on simple combinations of agents. The permutations and combinations are as follows, limited to some extent by the order in which they might have to be applied:

- neutralise + retan;
- retan + dye;

- retan + fatliquor;
- dye + fatliquor;
- retan + dye + fatliquor.

15.6.1   Neutralise and Retan

This is conventional technology, in which neutralising syntan can achieve both required outcomes at the same time. The mechanism relies on there being sufficient basifying power to raise the pH to the required level and sufficient tanning power to lower the IEP enough to ensure that the pelt is adjusted to be anionic.

15.6.2   Retan and Dye

The concept for this process is to colour the leather while conferring a retanning effect. Here, it might be assumed that the neutralising process will be separate, as the chemistries of the processes are likely to be different. In post-tanning, the retanning process is usually applied to modify the leather properties, especially the handle and the uniformity of handle over the hide or skin. However, the consequence for hydrothermal stability is typically relatively unimportant, because that is taken care of by the chrome tanning.

An available commercial option is the use of vegetable tannin, modified by fixing dye to the polyphenol. This has been done by the company Forestal Quebracho. The range of colours is limited and the presence of the vegetable tannins inevitably saddens the colour struck on the leather. An alternative approach is to use the chemistry of melanin formation:[6] using model reactant polyphenols and the aid of polyphenol oxidase, it is possible to achieve a tanning reaction and develop colour at the same time (see Chapter 16).

The chemistry of dyeing itself might be considered to be a retanning reaction, depending on the type of dye, which determines the nature of the binding interaction with leather, and the offer. Dyes are not normally considered to be a part of the stabilising system, simply because the offers of dyes are often less than the offers of conventional tanning compounds – molecular and equivalent weights notwithstanding. The use of reactive dyes should clearly constitute a form of retanning in our understanding, because they bond to basic sidechains *via* one or two covalent links. The effect will depend on the amount of reactive dye and the ability of the structure to interact with the aqueous supramolecular matrix.

15.6.3 Retan and Fatliquor

The concept in this combination is for the retanning reaction to lubricate the fibre structure, to prevent fibre resticking during drying and to soften the leather. To some extent this can be considered conventional technology, because polymeric tanning agents often confer softness, *e.g.* vegetable tannins, replacement syntans, resins and acrylate polymers. However, in some cases the tanning effect is weak: synthetic resins and polymers can bind to the collagen in a useful way, but may contribute nothing to the stability of the leather. Some specialist polymers, *e.g.* the water-resistant acrylic esters, certainly confer softness and their complexing reaction with bound chromium(III) might be considered to be a version of retanning.

15.6.4 Dye and Fatliquor

The concept of this combination is to colour the leather at the same time as introducing lubrication. To date, there is no industrial product available to do this. However, such a step is not impossible. It might be accomplished by using highly hydrophobic dyes (which do exist) and delivering them through the cross-section by an appropriate system, such as those set out in Chapter 19. The drawback is the use of an expensive dyestuff deep within the fibre structure, where its presence does not contribute to the perception of colour. However, such a combination might be useful in the contexts of grain layer or suede leather lubrication.

If this type of combination is desirable, using conventional dyes and fatliquors together, the compatibility of the components must be considered. One way of looking at the problem is to consider the impact of one reagent on the other:

current operation	compact processing	alternative processing
dye then fatliquor	dye with fatliquor	fatliquor then dye

Looking at the problem in this way highlights the effect of the HHB on the affinity of the dye for the substrate. Any change from current processing will tend to make the substrate more hydrophobic than before, hence changing the outcome of dyeing. Therefore, if compacting means merely combining the two steps of dyeing and fatliquoring, it is clear that the dyes will have to be changed to meet the new requirements of the substrate.

15.6.5   Retan, Dye and Fatliquor

The concept of this combination is to conduct the whole of post-tanning in a single step. It is possible that all of these require-ments could be met in a single chemical reagent. As a beginning, a compatible mixture for aqueous processing was described by Gaidau *et al.*[7]

However, the concept does raise the question of the role of the solvent. If the tanner could use a more exotic solvent, to solu-bilise simultaneously a wider range of reagents than is possible in water, it might be feasible to apply all of them at once. A candi-date solvent is liquid (supercritical) carbon dioxide. More recently, the introduction of ionic liquids and deep eutectic solvents into leather technology has widened the possibility for compatible and even incompatible mixtures of reagents to be applied to the leather simultaneously.

According to the choice of solvent, there may be no need for neu-tralisation. Alternatively, neutralisation may have to be retained as an aqueous process, to achieve the charge required by the new mixed process. See also Chapter 19.

15.7   Dynamic Mechanical Thermal Analysis (DMTA)

Although the basic properties of leather are created by the earlier pro-cesses in the beamhouse and in tannage, those properties are consid-erably modified by the latter processes of retanning and fatliquoring. Collagenous materials are viscoelastic,[8] which means that they exhibit viscous properties, they resist shear forces and they show elastic prop-erties, *i.e.* they return to the original state when a stress is released. These properties can be measured by dynamic mechanical thermal analysis (DMTA): under controlled conditions of humidity and tem-perature, a strip sample is anchored at each end and an oscillating force is applied. The technique measures the changing modulus of the material, λ, as the conditions change:

$$\lambda = \frac{\text{stress}}{\text{strain}} = \frac{\text{force per unit area}}{\text{ratio of change to original state}}$$

when the stress is sinusoidal, the lag phase, δ, will lie between $0°$ for a perfectly elastic material and $90°$ for a perfectly viscous material. The time lag between the displacement and response, the damping effect, is measured as $\tan \delta$. If the amplitude of the applied stress is σ_0 and the amplitude of the subsequent strain is ε_0, the storage modulus, E', is calculated as follows:

$$E' = \frac{\sigma_0}{\varepsilon_0}\cos\delta$$

the loss modulus, E'', a measure of the irrecoverable energy due to internal molecular motion, is calculated as

$$E'' = \frac{\sigma_0}{\varepsilon_0}\sin\delta$$

and

$$\tan\delta = \frac{E''}{E'}$$

Proteinaceous materials can exhibit glass transitions, when there is a change in physical properties at a particular temperature: above the glass transition temperature, T_g, the material has liquid/rubbery properties and consequently the molecules have high mobility; below the T_g, the material has amorphous/solid, glassy properties and consequently the molecules have low mobility. Glass transitions can be detected by a shift in the baseline in differential scanning calorimetry (DSC), when endothermic and exothermic changes in a sample can be measured as a function of temperature or heating rate, as illustrated in Figure 15.5.[8]

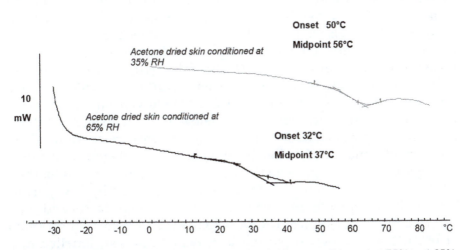

Figure 15.5   DSC thermograms of acetone-dried skins, conditioned at 35% and 65% RH, heated at 20 °C min^{-1}, showing the baseline shift indicating the glass transition temperatures. Courtesy of the *Journal of the Society of Leather Technologists and Chemists*.

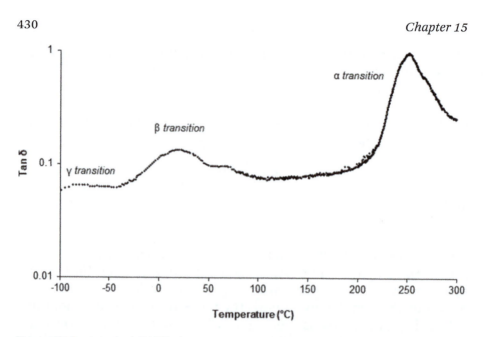

Figure 15.6 A typical DMTA thermogram of a leather tanned with vegetable tannin, conditioned at 0% RH. Courtesy of the *Journal of the Society of Leather Technologists and Chemists*.

Collagen has been regarded as a semicrystalline material owing to the presence of disordered regions, revealed by X-ray diffraction.[9,10] For this reason, leather may be expected to show thermal properties associated with both crystalline and amorphous materials. Some researchers consider the shrinkage temperature as melting of the crystalline regions, whereas the glass transition temperature is associated with the amorphous region of leather fibres.[11–13] Cot and co-workers reported the presence of a glass transition in leather tanned with chromium(III) salt: it was found to be at approximately 45 °C for leather conditioned at 65% relative humidity (RH) at room temperature.[14,15] Odlyha and co-workers reported two transitional events below the shrinkage temperature and they associated these transitions with relaxation of the polypeptide chains of collagen, considering collagen as a block copolymer.[16,17] The viscoelastic transition temperatures of leather, which may give practical information regarding physical performance, have typically not been measured and characterised. Figure 15.6 displays a typical DMTA response of a leather conditioned at 0% RH and shows three tan δ peaks of interest, labelled α, β and γ, associated with energy dissipation during stressing. Here, the β peak is assigned to the glass transition and the α peak to the shrinking-related transition. When α and β peak temperatures are plotted against wet shrinkage temperature obtained by DSC, shown in Figure 15.7, only the α peak temperatures show a correlation with shrinking transition.[8] Figure 15.8 presents the dynamic mechanical properties of chrome- and vegetable-tanned calfskin leathers, compared with acetone-dried untanned calfskin.

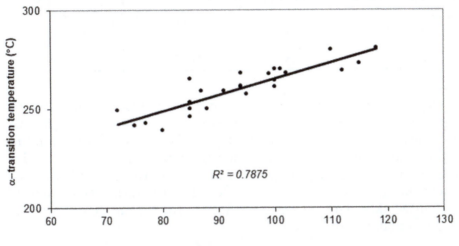

Figure 15.7 Relationship between wet shrinkage temperature measures by DSC and α-transition temperature measured by DMTA (samples conditioned at 65% RH) for various leathers with different shrinkage temperatures. Courtesy of the *Journal of the Society of Leather Technologists and Chemists*.

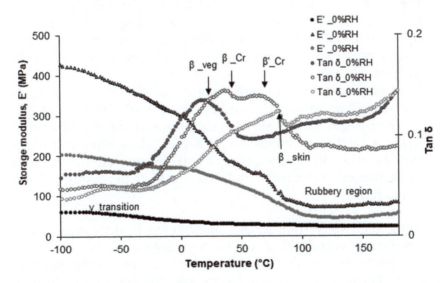

Figure 15.8 DMTA thermograms of chrome-tanned skin, vegetable-tanned skin and untanned skin, all preconditioned at 0% RH. Courtesy of the *Journal of the Society of Leather Technologists and Chemists*.

Table 15.4 Transition temperatures of tanned and untanned skin, conditioned at 0% RH.

Tannage	Collagen content (%)	γ Transition temperature (°C)	β Transition/glass transition temperature (°C)	β′ Transition temperature (°C)
None	93	−53 ± 7	81 ± 5	—
Chromium(III)	89	−66 ± 4	38 ± 8	65 ± 3
Vegetable	67	−78 ± 5	20 ± 3	—

The progressive decrease of storage modulus from −100 to +100 °C indicates a broad viscoelastic transitional region for all samples. However, the magnitude of the decrease in modulus is dependent on the tanning chemistry and corresponds to the regions where the β transition is observed. The tanned leathers show the β/glass transitions at a lower temperature than that of untanned skin. This demonstrates that tanning molecules are acting as a plasticiser.

1. From Table 15.4, hydroxyproline analysis of leathers conditioned at 0% RH shows that the chrome-tanned leather has 89% collagen content, whereas vegetable-tanned leather contains 67% collagen. A greater presence of vegetable tannins inside the collagen structure leads to greater depression of T_g.
2. Vegetable tannins are high molecular weight polyphenols and will give multipoint reactive sites for hydrogen bonding and hydrophobic interactions with the fibre. Hence they are effective in reducing the chain rigidity of collagen by intermolecular hydrogen bonds.
3. Chromium(III) binds to carboxyl groups, which occur along the surface of tropocollagen molecules: as a result, the temperature at which segment motion of collagen molecules becomes thermally activated is increased. However, chromium complexes that form an interpenetrating network between the collagen molecules act as 'spacers' and T_g is therefore depressed to lower temperature.

Tanning agents are traditionally understood to preserve collagen by introducing crosslinking between collagen molecules. Viscoelastic transitions would then be expected to move to higher temperatures. Because this was not observed for any of the leathers examined, it suggests that the mechanism of tanning is due to matrix or interpenetrating network formation around the collagen molecules. This is consistent with the latest thinking about the tanning mechanism, presented in Chapter 23.

The identification of the glass transition temperatures of leathers has several practical implications. This is especially so for post-tanning, where it may be suggested that processing above the glass transition temperature will be beneficial. This is because fibres are much more flexible and hence the uptake and fixation of dyes, fatliquor and retanning chemicals will be optimised.

References

1. A. D. Covington, *J. Am. Leather Chem. Assoc.*, 2001, **96**(12), 467.
2. A. D. Covington, *J. Am. Leather Chem. Assoc.*, 2008, **103**(1), 7.
3. A. K. Covington and K. E. Newman, *Pure Appl. Chem.*, 1979, **51**, 2041.
4. K. H. Gustavson, *J. Am. Leather Chem. Assoc.*, 1952, **47**(6), 425.
5. A. G. Puntner, *J. Am. Leather Chem. Assoc.*, 1999, **94**(3), 96.
6. A. D. Covington, C. S. Evans, T. H. Lilley and O. Suparno, *J. Am. Leather Chem. Assoc.*, 2005, **100**(9), 336.
7. C. Gaidau *et al.*, *Proceedings of IULTCS Conference*, Istanbul, Turkey, 2006.
8. S. Jeyapalina, G. E. Attenburrow and A. D. Covington, *J. Soc. Leather Technol. Chem.*, 2007, **91**(6), 236.
9. T. J. Wess, A. P. Hammersley, L. Wess and A. Miller, *J. Mol. Biol.*, 1998, **275**, 255.
10. D. J. S. Hulmes, T. J. Wess, D. J. Prockop and P. Fratzl, *Biophys. J.*, 1995, **68**, 1661.
11. C. A. Miles and T. V. Burjanadze, *Biophys. J.*, 2001, **80**, 1480.
12. A. D. Covington, G. S. Lampard, R. A. Hancock and I. A. Ioannidis, *J. Am. Leather Chem. Assoc.*, 1998, **93**(4), 107.
13. M. Komanowsky, *J. Am. Leather Chem. Assoc.*, 1991, **86**(7), 269.
14. T. Bosch, *et al.*, *J. Am. Leather Chem. Assoc.*, 2002, **97**(11), 441.
15. A. M. Mannich, *et al.*, *J. Am. Leather Chem. Assoc.*, 2003, **98**(7), 279.
16. M. Odlyha *et al.*, *ICOM Committee for Conservation*, 1999, vol. 2, p. 702.
17. M. Odlyha, *et al.*, *J. Therm. Anal. Calorim.*, 2003, **71**, 939.

16 Dyeing

16.1 Introduction

Dyeing is one of the more important steps in leather making: its importance lies in the fact that it is usually the first property of the leather to be assessed by the consumer or customer. He or she will make judgements at a glance: colour, depth of shade, uniformity. Therefore, it is critical that the science and hence the technology of coloration are well understood.

The origin of colour and its perception was reviewed by McLaren in the 1990 Procter Memorial Lecture for SLTC.[1] Later, Randall explored colour as it applied to leather[2] and reviewed the measurement of colour from surfaces, pointing out the difference between reflection from a relatively continuous surface, as is the case for grain leather, and the effects of reflection from the tips or the fibre of suede, resulting in a directional effect of colour perception.

The standard system for measuring colour defines the coordinates in the CIELAB (Commission Internationale de l'Éclairage $L^*a^*b^*$) colour space: a^* is the red–green component, b^* is the blue–yellow component and L^* is the black–white component, from which the other parameters can be calculated, such as the hue angle and the chroma. These measurements are useful in comparing colour and for colour matching.

Tysoe[3] reviewed the classification of dyes in the Colour Index and introduced the problem of colour matching for leather. Colour match

Tanning Chemistry: The Science of Leather 2nd edition
By Anthony D. Covington and William R. Wise
© Anthony D. Covington and William R. Wise 2020
Published by the Royal Society of Chemistry, www.rsc.org

prediction is relatively straightforward when dealing with pigments, but when the dyestuff confers transparent colour to leather, the outcome is dependent on the following parameters:

- dyestuff chemistry and mechanism of fixation;
- the relative affinities of the dyestuff and the substrate;
- the nature of the substrate, including its colour;
- illumination of the leather: the perceived colour depends on the light source, an effect called 'metamerism'.

The modern synthetic dyestuffs industry was initiated by the development of mauveine by Perkin in 1856: the primary markets were textiles, but the leather industry eventually took advantage half a century later, especially when it was realised that bright, deep shades could be achieved with the new tannage with chromium(III). Hitherto, all industries relied on vegetable colourants, which could only be made fast, with improvements in their weak coloration, by the use of metal mordants. Similarly, inorganic compounds were used as pigments in paints, but modern developments have introduced organic pigments that may spawn a new generation of dyes (Figure 16.1).

Causes of colour can be grouped under five main headings:[4]

1. excitations, *e.g.* from gas such as sodium or neon lights;
2. ligand field effects, *i.e.* complexation around transition metal ions;
3. band theory, from metals and semiconductors;
4. physical effects of dispersion, scattering, interference and diffraction;
5. molecular orbitals in organic molecules or charge transfer.

In the case of dyes, the mechanism is *via* molecular orbitals in organic molecules: the colour is generated by the creation of a system of delocalised electrons, made possible by chemically synthesising

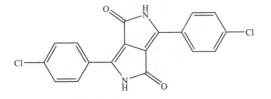

Figure 16.1   A diketopyrrolopyrrole.

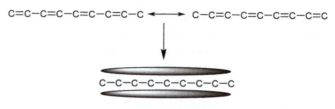

Figure 16.2   Formation of a molecular orbital for adjacent p orbitals.

molecules that contain a system of conjugated double bonds, *i.e.* alternate double and single bonds (Figure 16.2). What this actually means is that the π orbitals of the double bonds create a molecular orbital over the whole carbon chain. The system can include other delocalised groups, such as benzene rings or naphthalene rings, where molecular orbitals are already present above and below the plane of the ring. Note that often dye structures are presented in the form of the Kekulé structures of the aromatic rings, merely to emphasise the concept of conjugation over the molecule.

The groups with the delocalised electrons are called 'chromophores', because they are the primary sources of colour. Often the chromophores are linked by an azo group, N=N, which can contribute to the delocalisation of electrons over the chromophore system. The diazotisation reaction to create the azo group is conducted as follows.

An arylamine is reacted with nitrous acid under cold conditions, preferably close to 0 °C, to yield the diazonium salt:

$$Ar-NH_2 + NaNO_2 + 2HCl \rightleftharpoons Ar-N^+\equiv N\ Cl^- + 2H_2O + Na^+ + Cl^-$$

The diazonium salt then reacts with an active hydrogen on another aryl compound, to link the molecules *via* the diazo group:

$$Ar-N^+\equiv N + Ar-H \rightleftharpoons Ar-N=N-Ar + H^+$$

Many aromatic amines have been used to make the so-called azo dyes, but are now known to be toxic, particularly carcinogenic. Because there is a possibility of regenerating the amine from the azo derivative, they are now excluded from this application, notably benzidine (Figure 16.3) and its derivatives. Other banned reagents include 4-aminodiphenyl, toluidine, 2-naphthylamine, dianisidine and cresidine.

Another way of coupling chemical moieties is to use the reactivity of quinone derivatives, as discussed in Chapter 10; some options are given in Figure 16.4.

The delocalisation of electrons can be modified by the presence of substituents on the aromatic chromophores: these are called

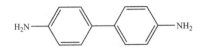

Figure 16.3 Benzidine.

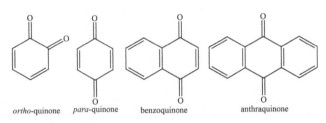

ortho-quinone   *para*-quinone   benzoquinone   anthraquinone

Figure 16.4 Quinone species for dyestuff synthesis.

'auxochromes'. They alter the distribution of electrons by electro-negativity effects or hyperconjugation. By changing the energy of the molecular orbital, the energy to excite an electron is changed and consequently the energy of the quantum released when the electron falls back to its ground state is changed. If the effect of the structural change is to cause the colour to move to longer wave-length, redder, it is called a 'bathochromic shift', and a change to shorter wavelength, bluer, is called a 'hypsochromic shift'. The principle is illustrated in Figure 16.5, where a quantum of energy excites an electron from the highest occupied molecular orbital (HOMO) to the lowest unoccupied molecular orbital (LUMO) in a π to π* transition; as described above, the larger the energy gap, the more blue is the colour.

Groups that donate electrons to the aromatic ring system (nucleophiles) include OH (phenolic), CH_3 (weak, hyperconjugation), OCH_3, NH_2, NHR and NR_2, groups that accept electrons (nucleophilic) include NO_2, COOH and SO_3H and groups that contribute to the solubility of the dye include SO_3^-, CO_2^- and NH_3^+ (also electron withdrawing)

In a leather context, the auxochromes react with the collagenic substrate in various ways, to confer colour fixation. In addition, the reaction can be considered a form of tanning: any chemical species that binds to the substrate alters the properties of the substrate, including contributing to the hydrothermal stability. This is not usually taken into account when considering the eventual outcome of wet processing.

Dyestuffs as supplied are typically not pure compounds. Owing to variations in synthesis, they may be formulated with other dyes

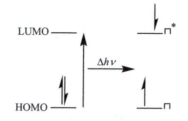

Figure 16.5 Generation of colour from molecular orbitals.

to remain on shade, which can lead to problems in colour matching (see later). They are also typically marketed with a diluent, which can include sodium chloride, sodium carbonate, dextrin, sulfite cellulose, naphthalene sulfonate and sodium sulfite. The reason is technologically sound: to allow the tannery operative to weigh quantities in kilograms, with little associated error. If the undiluted dyestuff were to be supplied, the quantities required for a pack would be grams; such weighings would have a large associated error, with consequent variations in dyeing consistency. It is the same argument that applied to formulations of enzymes, discussed in Chapter 8.

In the following sections, the structures of the different, commonly used dye types are illustrated. Each has its name, usually indicating the dye chemistry, the colour and a designated number: each has a Chemical Index number, where it is categorised by its structure. Not all of the dyes given as examples are designated particularly for leather applications: the structures are illustrative of the principles involved in creating dyes with different reaction characteristics and different degrees of applicability to leather.[5]

16.2 Acid Dyes

Acid dyes are so called because they are fixed under acidic conditions, as discussed in Chapter 15. This class of dyes is the most commonly used in the leather industry, particularly for chrome-tanned leather. The type of structure is illustrated in Figure 16.6.

The properties can be summarised as follows.

- Relatively small, typically hydrophilic molecules.
- Used for penetrating dyeing, producing level shades.
- Anionically charged, therefore with a high affinity for cationic leather.

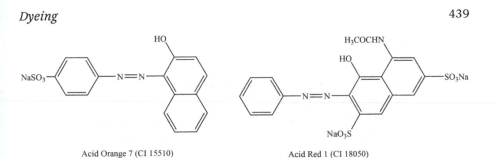

Acid Orange 7 (CI 15510)   Acid Red 1 (CI 18050)

Figure 16.6 Mono-azo acid dyes.

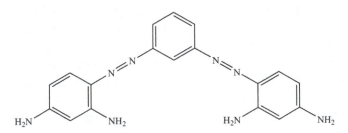

Figure 16.7 Disazo basic dye: Manchester Brown or Bismarck Brown.

- Fixed by acidification, owing to the presence of sulfonate groups, as discussed in Chapter 15.
- They react predominantly through electrostatic reaction between their sulfonate groups and the protonated amino groups of lysine.
- Secondary reaction is *via* hydrogen bonding through auxochrome groups.
- Some dyes may react with the bound chrome, using it as a mordant (see later).
- Good fastness properties: less complex molecules offer fewer opportunities for structural changes by free radical mechanisms, as discussed in the context of condensed vegetable tannins.
- Wide range of colours, offering bright, deep shades.

16.3 Basic Dyes

The structures of basic dyes are essentially the same as those of the acid dyes, except that they carry a net positive charge from the cationically charged amino substituents, even though they may also have anionic sites (Figure 16.7). Quaternary amino groups are less hydrophilic than sulfonate or carboxylate groups, so they are typically less water soluble than acid dyes.

The properties can be summarised as follows.

- They produce strong, brilliant colours (red, orange, yellow, green, blue, indigo, violet and black).
- They tend to be relatively hydrophobic, because they contain fewer solubilising groups than the acid dyes; consequently, they are often soluble in oils and non-aqueous solvents.
- They tend to bronze, *i.e.* produce a metallic sheen: this is due to surface reaction, when the dye molecules lie on top of one another, attracted by van der Waals dispersion forces, allowing light to be reflected from the layered structure. This is a consequence of hydrophobicity, where the dyes have greater affinity for themselves than for the aqueous solution or the substrate. The bronzing effect can be obtained with other dye types when the conditions favour very fast reaction on the leather surface; the layering mechanism of light reflection is always the same.
- Poor light fastness.
- Good perspiration fastness, because they are not displaced by elevated pH.
- High affinity for anionic leather, *e.g.* vegetable-tanned, anionic-retanned and acid-dyed leathers. The last property is exploited in 'sandwich dyeing': acid dye, then basic dye, possibly topped with more acid dye. The electrostatic attraction between the charged species creates deep shades, with good rub fastness.
- Precipitated by hard water and anionic reagents.
- Applied by mixing with acetic acid, then diluting with hot water.
- They react electrostatically through their protonated amino groups and ionised carboxyl groups on collagen.
- Secondary reaction is by hydrogen bonding.
- Since they are typically more hydrophobic than acid dyes, some reaction will be *via* hydrophobic bonding.

16.4   Direct Dyes

Direct dyes have the same sort of structural features as the acid and basic dyes, but with higher molecular weight. From Figure 16.8, the direct dyes have obvious similarities to acid and basic dyes. There is an additional category of direct dyes, the developed direct dyes, which

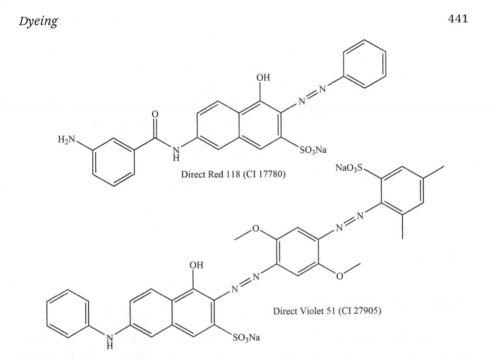

Figure 16.8 Direct dyes.

can be diazotised *in situ* with additional aromatic chromophores added to the dye as a route to black.

The properties of this type of dye can be summarised as follows.

- They are larger molecules than typical acid or basic dyes.
- Used for surface dyeing, with a consequent likelihood of uneven colouring.
- Acid is not needed for fixation, because they are more reactive, owing to the larger number of reactive sites on the molecule and the hydrophobicity of the overall structure.
- Fastness properties are average to good.
- Usually dark colours.
- Have the same sort of structures as acid and basic dyes, although typically with lower charge, resulting in a lesser importance of electrostatic bonding.
- High molecular weight means more direct reaction, not requiring fixation by pH adjustment.
- Relies more on hydrogen bonding from the larger number of auxochromes per molecule, similar to the relative astringencies of vegetable tanning agents, and more emphasis on hydrophobic bonding.

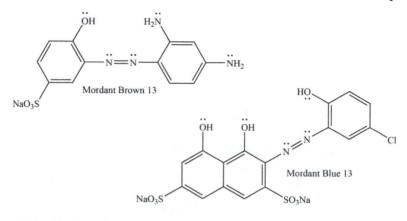

Figure 16.9 Mordant dye structures: lone-pair sites for possible complexation are indicated.

16.5 Mordant Dyes

The original mordant dyes were typically plant extracts, which produced relatively dull and pale shades and fixed poorly to textiles when used alone. In order to fix the colouring agent to a substrate, it was necessary to provide an additional fixing mechanism, by applying a metal salt either before or with the dye. Modern mordant dyes are similar to acid dyes, but usually with less anionic charge (Figure 16.9). They generally have a poor affinity for collagen, but rely on the presence of a metal ion with which they can complex, so the metal acts as the link between the leather and the mordant dye. In this reaction, the creation of five- or six-membered rings is the most energetically favoured.

Suitable metal salts include chromium(III), which of course may already be bound to collagen. Other metals that can be used include aluminium(III) and iron(III). The formation of metal complexes means that the final colour struck by the dye depends on the metal mordant, *i.e.* the colour developed is a combination of the molecular orbitals of the organic dye structure plus the colour of the metal complex created by the reaction with the dye and the reaction with the substrate. This is illustrated by alizarin, shown in Figure 16.10 and Table 16.1.

The technology is to apply the mordant to the leather, then treat with the dye. A common option used to be to treat the leather with chromium(VI) salt, usually as dichromate, and reduce it to chromium(III) *in situ*, in a process analogous to two-bath chrome tanning. This is no longer used in the leather industry, for two reasons.

1. The environmental impact of Cr(VI) is too high and discharges are strictly regulated.
2. The availability of premetallised dyes has simplified the process (see later).

The mechanisms of fixation are the same as for acid dyes, with the additional mechanism of covalent complexation.

- They have a low affinity for collagen, typically having fewer sulfonate groups, particularly if they are natural dyes, and few hydrogen bonding groups.
- Reliance on creating complexes with metal ions (mordants) previously fixed to collagen.
- Binding to mordant metals will vary in the degree of electrostatic and covalent character, depending on the metal, *e.g.* Al(III), Cr(III).

16.6 Premetallised Dyes

16.6.1. 1:1 Premetallised Dyes

The concept of premetallising dyes is to avoid the two-step process of mordanting then dyeing, by preparing the complex of dye and metal salt in advance. This is an example of a compact process, a concept discussed in Chapter 15. This principle is illustrated in Figure 16.11,

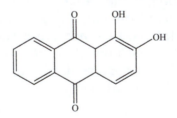

Figure 16.10 Structure of alizarin, a mordant dye.

Table 16.1 Effect of metal mordant on the colouring of alizarin dyeing.

Metal ion	Colour
Al(III)	Red
Sn(IV)	Pink
Fe(III)	Brown
Cr(III)	Puce–brown
Cu(II)	Yellow–brown

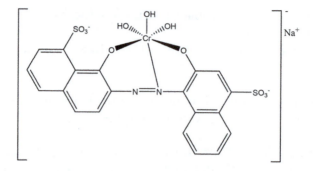

Figure 16.11 1:1 chromium(III) premetallised dye, Acid Blue 158 (CI 14880).

with a chromium(III)-complexed dye, showing how the availability of lone pairs of electrons can be used to create dative bonds in a complex. Other metal ions that are used in premetallised dyes include cobalt, iron and copper.

The mechanisms of fixation are the same as summarised for mordant dyes. The dyes exhibit the following properties.

- They have lower anionic charge than the corresponding anionic uncomplexed dyes.
- Penetration and levelness of colouring are good.
- They include pale, dull and pastel shades.
- Fastness properties are good to very good.
- They tend to be expensive, hence are used for premium leathers, *e.g.* gloving, clothing, suede, nubuck, aniline.
- They may be formed from either mordant dyes or conventional acid dyes precomplexed to a mordant metal ion, in the ratio of one dye molecule to one atom of metal.
- The primary fixation mechanism is through reaction between the metal ion and collagen carboxyl groups, which can vary in covalent character.
- Secondary reactions may be electrostatic and hydrogen bonding, depending on whether the dye is derived from a mordant or an acid dye precursor.

16.6.2. 1:2 Premetallised Dyes

From Figure 16.11, it is obvious that further complexation can occur in 1:1 complexes, because there are complexing sites remaining on the metal ion, illustrated in Figure 16.12. Note that the second complexing dye species does not have to be the same as the first. They are referred to as either 1:2 or 2:1 premetallised dyes.

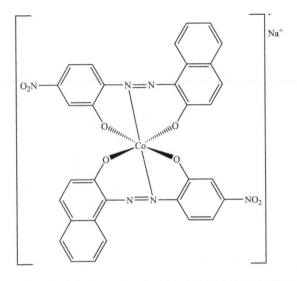

Figure 16.12 1:2 premetallised dye: Perlon Fast Violet BT (CI 12196).

The difference between 1:1 and 1:2 premetallised dyes is that the fully coordinated dyes are more anionic and can exhibit a wider range of reaction properties. For leather applications, there are three designated types. These designations and their consequences for reaction can be understood from the HHB/HLB mechanism of reactions, set out above.

1. *Type 1:* Water insoluble, soluble in alcohols, glycols, *etc.*
   Used for spray dyeing and finishing aniline and semi-aniline leathers.
   High covering power, with high tinctoral power.
   Good fastness properties to water spotting, wet rub, light.
2. *Type 2:* Low water solubility, more soluble in non aqueous solvents.
   Fastness properties inferior to those of Type 1.
3. *Type 3:* Higher water solubility.
   Used for dyeing woolskins; to colour the leather but not the wool.

In addition:

- The composition is similar to that of the 1:1 complexes, except for the ratio of dye to metal, hence reaction between the metal and collagen is not part of the fixation mechanism.
- Fixation is less dependent on electrostatic and hydrogen bonding than with acid dyes and 1:1 complexes.
- The fixation mechanism is similar to that for direct dyes, hence there are hydrophobic interactions.

- They may function more like pigments than dyes, with high covering power.

16.7 Aryl Carbonium Dyes

The structures of the aryl carbonium-based dyes give the appearance of being a modern development in dyestuff chemistry, but actually these dyes are the basis of the modern industry. Figure 16.13 shows the structures of mauveine, the first synthetic dye, and Malachite Green, as an example of a modern aryl carbonium dye. Dyes of this type give intense colour but have inferior technical performance.[4]

16.8 Reactive Dyes

Reactive dyes are typically acid dyes that have been covalently bound to a reactive group, capable of reacting covalently with collagen or leather. The principle behind this technology is that a covalently bound dye will be fast to chemical removal. This is especially useful in applications where the leather may be subjected to washing, dry cleaning or perspiration damage, *e.g.* clothing and especially gloving leathers. The original dyes in this class were called Procion, based on triazine chemistry, developed by ICI, and others followed, as illustrated in Figure 16.14, where the originating companies are indicated. In Figure 16.14 there are other examples of nitrogen heterocycle chemistries, but the principles are the same. Note that the nitrogen heterocycle may contain a second dye species, which is not identical with the first.

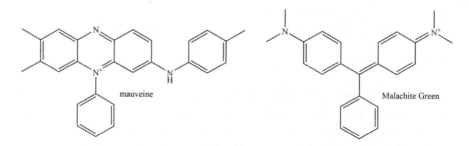

Figure 16.13 Structures of mauveine and Malachite Green (CI Basic Green).

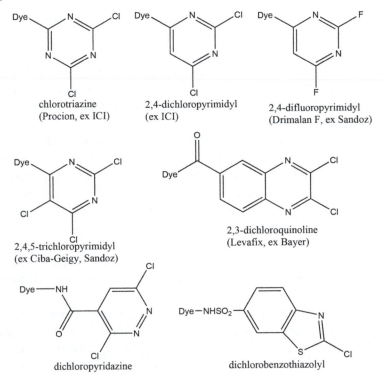

chlorotriazine
(Procion, ex ICI)

2,4-dichloropyrimidyl
(ex ICI)

2,4-difluoropyrimidyl
(Drimalan F, ex Sandoz)

2,4,5-trichloropyrimidyl
(ex Ciba-Geigy, Sandoz)

2,3-dichloroquinoline
(Levafix, ex Bayer)

dichloropyridazine

dichlorobenzothiazolyl

Figure 16.14 Reactive dye chemistry based on aromatic nitrogen heterocycles.

1. The heterocyclic ring system can react with available amino groups on collagen or leather.
2. The reaction is conducted under alkaline conditions, to ensure the availability of uncharged amino groups, by reacting with the acid produced by the reaction.
3. The reaction is driven by the addition of sodium chloride. This is an example of exploiting the stepwise mechanism of binding a solute to a substrate. Changing the affinity of the solute for the solvent, by making the solvent more hydrophilic due to the charged ionic content, decreases its affinity for the dye molecules and drives them into the more hydrophobic substrate.
4. There is competition between the reaction sites on the substrate and the hydrolysing effect of the solvent. The reactive dye is hydrolysed by water to the hydroxy derivative: the extent to which this happens depends on the specific chemistry of the reactive dye, but it always occurs to some extent.

In circumstances like this, it is important to remember that water usually has a competitive advantage over other reactants, because it is present at a concentration of 55.5 m. Because the resulting aromatic hydroxy group is much less labile than the chloride, the reactive part of the molecule essentially becomes deactivated. In effect, hydrolysis turns the reactive dye back to a conventional acid dye.

5. The tanner is then faced with a dilemma. Fixing the acid dye under typical conditions for that type of dye will maximise the depth of shade from the dyeing, but the fastness properties will be compromised. Removing the acid dye will reduce the depth of shade, but the fastness properties will be maximised. This is a commercial decision.

An alternative approach was developed later by Hoechst, *viz.* the Remazol series of dyes based on vinyl sulfone chemistry, illustrated in Figure 16.15.

The principle of the mechanism of fixation is the production of the vinyl sulfone group *in situ*, because sulfate is a good leaving group. The presence of the sulfone group also activates the vinyl group, allowing it to react with other groups that have an active hydrogen, *e.g.* hydroxyl or amino. Another approach is also shown in Figure 16.15, namely the formation of an ethylenimine derivative: the heterocyclic ring is readily opened to react with groups with an active hydrogen. In all cases, hydrolysis is still a competitive reaction.

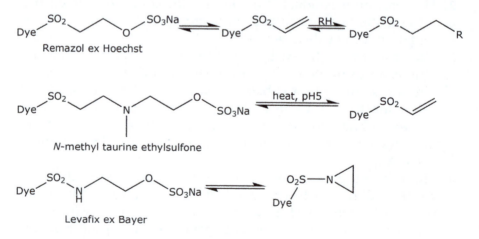

Figure 16.15 Reactive dye chemistry based on vinyl sulfones.

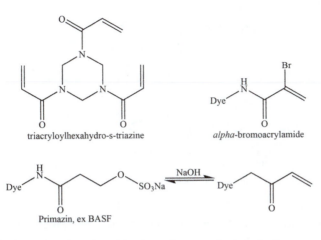

triacryloylhexahydro-s-triazine

alpha-bromoacrylamide

Primazin, ex BASF

Figure 16.16 Reactive dye chemistry based on acryloyl derivatives.

An alternative approach, using the acryloyl group, is illustrated in Figure 16.16. The presence of the carbonyl group activates the system in the same way as the sulfone group does in the same position in the molecule. The acryloyl group may be present from the start or it can be created from the sulfate derivative, as already seen in Figure 16.15.

Within the variations of chemistries of reactive dyes, some properties are common.

- Very good fastness to washing, dry cleaning and perspiration.
- Good light fastness.
- Limited range of colours: pale and medium shades.
- Expensive.
- Health hazard, owing to their reactivity towards organic substrates, therefore stricter COSHH (Control of Substances Hazardous to Health) regulations apply to their use.

It is not necessary to rely on a single chemistry in reactive dyes. Examples of mixed fixing systems are illustrated in Figure 16.17.

16.9 Sulfur Dyes

From the descriptions of the dye types already presented, it is clear that the dyes discussed so far are all chemically related. Sulfur dyes originate from 1893; they form a class of colourants that is different in chemistry and fixation mechanism. Like syntans, the

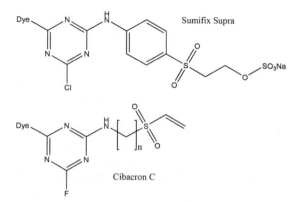

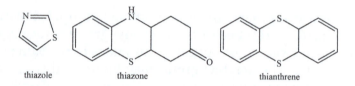

Figure 16.17 Reactive dye chemistry based on combinations of fixing mechanisms.

thiazole thiazone thianthrene

Figure 16.18 Chromophores in sulfur dyes.

structures of sulfur dyes are so complicated, because of the way in which they are prepared, that the structures are largely unknown. Consistency of colour therefore depends on consistency of production conditions.

Sulfur dyes are made by heating together aromatic compounds containing amino and hydroxy groups with a source of sulfur, in the following ways.

- Sulfur bake: the organic compound is heated with elemental sulfur to 160–320 °C.
- As for sulfur bake, but using polysulfide salt.
- Polysulfide melt: the reagents are heated with water, under reflux or in a pressure vessel.
- Solvent thionation or melt: as for polysulfide melt, but the solvent is butanol, dioxitol, *etc.*

Excess sodium bisulfite may be used to incorporate thiosulfate groups, *e.g.* $-S_2O_3^-$, to enhance water solubility. In this way, the preparation is cheap, although somewhat variable. The product often has low water insolubility, owing to its structure: Figure 16.18 illustrates

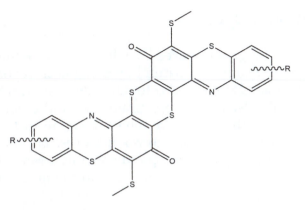

Figure 16.19 Phenothiazonethioanthrone chromophore.

a few of the chromophores identified, typically linked by disulfide bonds, as already encountered in cystine.

It has been suggested that some sulfur dyes contain the phenothiazonethioanthrone moiety as the chromophore, illustrated in Figure 16.19.[4]

In the technology of applying these dyes, they must be solubilised. This can be achieved using the chemistry already familiar in sulfide-based hair burning. The effect of (hydro)sulfide ion, a reducing nucleophile, on the insoluble dye is to break the disulfide links, creating the soluble, leuco (white) form. Subsequent oxidation re-forms the insoluble dye on and in the substrate leather.

$$Dye-S-S-Dye \underset{[O]}{\overset{S^{2-}}{\rightleftharpoons}} 2Dye-S^-$$

insoluble   soluble, leuco form

The properties of sulfur dyes can be summarised as follows.

- They are suitable only for leathers that can resist the high pH necessary for dye reaction, *e.g.* aldehyde or oil tanned. Note that chromium(III)- and vegetable-tanned leathers are severely damaged at pH 12–13; in each case the tanning is reversed.
- Good perspiration and wash fastness. Insoluble in dry cleaning solvents.
- Generally dull shades, with a limited range of colours – a true red shade is not available.
- Little or no affinity for wool.
- Little reactivity of the kinds set out for the other dyes, hence there is primary reliance on hydrophobic interactions.

16.10   Fungal Dyes

While the use of vegetable-based dyes has been superseded by synthetic dyes, a new source of natural dyes has been reported. Brazil-
ian workers[6] isolated and tested colourants derived from filamentous
fungi, specifically *Monascus* spp. Examples of the coloured compounds secreted by the fungus are shown in Figure 16.20.

These examples can function as leather dyes, but do not exhibit
a good enough performance for commercial use. Nevertheless, as a
source of new dyes, fungi in general offer considerable potential.

16.11   Dye Reactivity and Fixation

Dye fixation is subject to the principles of reaction from solution onto
a substrate, as set out in Chapter 15. The reactions are dependent on
the following conditions:

- the status of the substrate, including the impact of prior
  processes;
- the pH of the substrate and solution;
- the HHB/HLB values of the dyes.

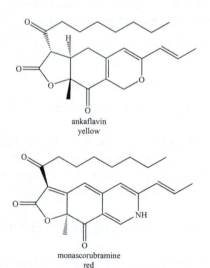

ankaflavin
yellow

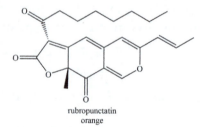

rubropunctatin
orange

monascorubramine
red

Figure 16.20   Coloured substances from *Monascus* spp.

Table 16.2 Dye fixation mechanisms: ✓✓, primary mechanism; ✓, secondary mechanism.

Dye type	Electrostatic	Covalent	H-bonded	Hydrophobic
Acid	✓✓		✓	
Basic	✓		✓	✓
Mordant	✓	✓		
Direct	✓		✓✓	
1:1 Premetallised	✓	✓✓		
1:2 Premetallised	✓		✓	✓
Reactive	✓	✓✓	✓	
Sulfur				✓✓

It is common for tanners to use a mixture of dyes; the trichromatic mixture for brown is common. However, if there are differences in the chemical properties of the components of the mixture, they will react independently. This often happens if the HHB/HLB values are not matched, so that the surface reactions and penetration will not be the same, hence the cross-section may be a different colour to the surface, leading to problems if the leather is scuffed or cut surfaces are visible in leather articles. This problem has been addressed in the Sellaset dyes, where the HHB values of the dyes are matched and the mixture behaves as though it were a single dye.

Dye fixation mechanisms are summarised for comparison in Table 16.2.

16.12 Role of the Substrate

16.12.1 Chrome-tanned Leather

One of the many advantages of chrome tanning is the minimal change to the collagen structure, particularly with regard to occupation of the sidechains and lack of inhibition of consequent reactions. There is a degree of cationic character, useful in fixing acid dyes. Also, the colour conferred to the leather is pale, hence there is little background colour to interfere with the colour struck by dyes.

The one area in which chrome has a major impact on dyeing is when there is too much surface reaction. It was shown in Chapter 11 that chrome fixes to a small extent at the peptide links. Ordinarily, the small amount and uniform distribution do not affect dyeing, but at high concentration it does interfere. This can be at two levels of effect.

- If the effect is shadowing, causing uneven colour, dyeing is likely to be uneven, owing to the variation in base colour of the substrate.
- If the effect is staining, reaction sites for dye fixation can all be taken up by the chrome, resulting in dye resist. Therefore, the fault of staining cannot be covered up by dyeing.

In each case, the fault cannot be remedied. The commonly used tannery option of 'put it into black' may work for chrome shadowing, but does not work for chrome staining. There the only solution is pigment colouring in finishing.

Chrome tanning is very lightfast, so any effect of light on the leather is due to the light fastness of the dyes themselves.

16.12.2   Vegetable-tanned Leather

The chemical basis of vegetable tanning is completely different to that of chrome tanning. The tanning agent is usually present at a much higher concentration within the leather and it interacts with both charged sidechains and peptide links and confers colour. In this way, all vegetable tannins block most of the dye fixation sites and produce colour themselves, which can be intense. The consequences are as follows.

- The dye will have to bind to the polyphenol molecules, at least in part, so the colour will be affected.
- The base colour will usually affect the overall colour of the dyed leather: this can vary from very pale cream, in the case of tara, to very dark brown, in the case of chestnut.
- The overall outcome is a marked dulling of the shade in comparison with the effect on chrome-tanned leather; this is referred to as 'saddening' the colour.
- Plant polyphenols may be susceptible to light damage, particularly the condensed tannins, which are likely to redden, although occasionally bleaching is observed with hydrolysable tannins such as tara. A change in base colour after dyeing will affect the perceived colour of the dyed leather.

16.12.3   Other Tannages

Syntan tannage will have a similar effect to vegetable tanning, especially if a replacement syntan is used. Aldehydic tannage deprives the substrate of sites for reaction by dyes relying on sulfonate fixation.

16.13 Dyeing Auxiliaries: Levelling and Penetrating Agents

16.13.1 Anionic Auxiliaries

Levelling agents are used when uniformity of colour on the grain surface of the leather is imperative, for example in aniline leathers. Anionic levelling agents are based on low molecular weight aromatic sulfonate condensates, as described in Chapter 14; some structures are illustrated in Figure 16.21. They function by competing with dyestuff for binding sites or occupying those sites by prior treatment. In either case, the kinetics might be expressed as follows:

$$\text{rate} = k[\text{dye}]^a[\text{collagen sites}]^b$$

Any reduction in availability of reaction sites reduces the rate of interaction with substrate, where the slower the reaction the greater is likelihood of the effect making it more uniform on the surfaces. Note that the presence of the auxiliary may also modify the HHB/HLB and the charge of the substrate, to contribute to the effect. Controlling the dye concentration is an option, as in vegetable tanning, but it is not used as a technological strategy.

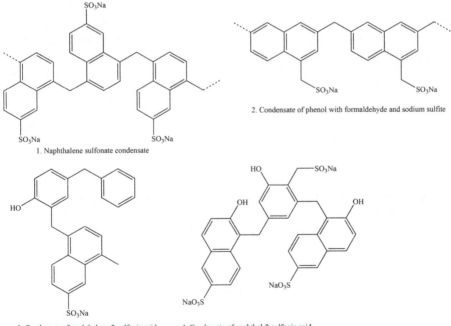

1. Naphthalene sulfonate condensate

2. Condensate of phenol with formaldehyde and sodium sulfite

3. Condensate of naphthalene-2-sulfonic acid and 4,4'-dihydroxy sulfone with formaldehyde

4. Condensate of naphthol-2-sulfonic acid and cresol with formaldehyde

Figure 16.21 Structures of some dye levelling agents.

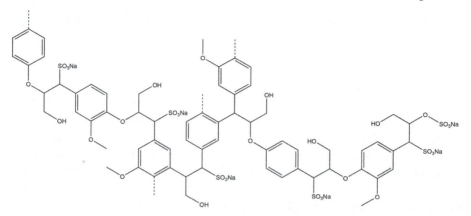

Figure 16.22   Illustrative structure of lignin sulfonic acid (sulfite cellulose).

In Figure 16.21, structure 1 is a naphthalene-2-sulfonate condensate: for levelling, the degree of polymerisation may be 2–10. Modifying the structure of the condensate by using an alternative crosslinker, such as a sulfone derivative, improves the light fastness of the auxiliary, as illustrated in structure 3.

An alternative approach is to use a by-product of the Kraft process for paper making, as introduced in Chapter 14. Since the structure of lignin, which is the basis of the material, is a non-uniform polymer, the structure can only be illustrated in the notional form presented in Figure 16.22. The similarity to the other synthetic reagents is clear. It was pointed out in Chapter 14 that such by-products have only a weak affinity for collagen or leather: in this case, there is no disadvantage, since what is required is for the reagent to interfere with dye fixation, to slow the reaction, not necessarily to prevent it.

In all cases, if the auxiliary is added before the dye, the colour intensity may be reduced by to up to one-third of what would be obtained on untreated leather. If it is added with the dye, there is less effect on the colour, but the levelling effect is slightly reduced. Conditions that favour penetration of the levelling agent, such as short float, increase the surface colour intensity and reduce levelness. The effect of the auxiliary is enhanced at higher temperature, *e.g.* 70 °C is better than 20 °C.

16.13.2   Auxiliaries that Complex with Dyestuff

These auxiliaries change the HHB/HLB properties and reduce the anionic charge. Examples include polyglycol ethers with weakly cationic amino groups. These will form strong complexes with high-affinity dyes, to improve penetration and levelness by making the dyes less reactive but retaining the colouring effect.

16.13.3   Auxiliaries that Have Affinity for Both Leather and Dye

Examples include poly(ethylene oxide) chains with strong cationic groups. These agents cannot be used with premetallised dyes containing anionic dispersing agents. They tend to produce very pale colours when offered at low levels. At higher offers, the colour intensity can be increased.

16.13.4   Intensifying Agents

The purpose of colour intensifying agents is to allow dyes to accumulate at higher concentration on the surface of the leather: the purpose is to achieve deeper shades, with greater covering power for defects in the grain surface. In this way, the fastness properties of the leather, particularly to rub or crock (see later), are not compromised as they would be by merely precipitating the dye on the leather surface by manipulation of the pH. These auxiliaries may also be referred to as dye fixing agents, since their function is to create additional sites at which dye molecules can be chemically rather than physically bound to the leather.

16.13.5   Cationic Tannages

The presence of cationic centres from the tanning reaction provides addition sites with which anionic dye can react. Such tannages include Al(III) chloride and sulfate, Zr(IV) sulfate, polyurethane and dicyandiamide retanning resins.

16.13.6   Cationic Auxiliaries

This class of auxiliaries includes cationic compounds with non-ionic ethylene oxide chains, illustrated in Figure 16.23. Short chains of this type increase the dyeing intensity, but levelness is reduced. The opposite applies to longer chains.

Other fixing agents are based on polyamines, such as dicyandiamide derivatives, illustrated in Figure 16.24.

Typically, this type of treatment is applied in a separate short float at low temperature before dyeing. It is followed by thorough

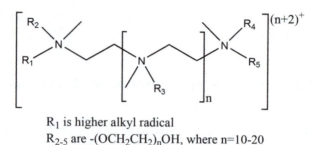

R₁ is higher alkyl radical
R₂₋₅ are -(OCH₂CH₂)ₙOH, where n=10-20

Figure 16.23 Cationic polyamines.

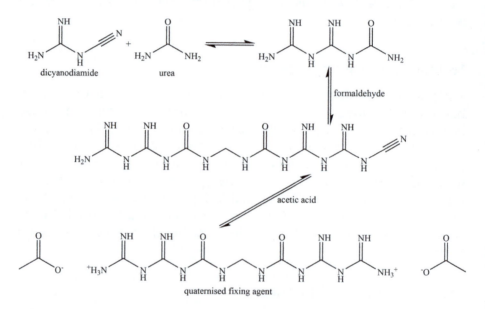

Figure 16.24 Dicyandiamide quaternary fixing agents.

washing, otherwise dyeing is likely to be unlevel. Leather treated in this way usually has poorer rub fastness properties than untreated leather.

16.14 Alternative Colouring Methods

In Chapter 2, the chemistry and biochemistry of the mammalian pigments, the melanins, were introduced. A potential route to colouring leather is to harness those reactions, to create colour *in situ* and possibly achieve some stabilisation of the collagen

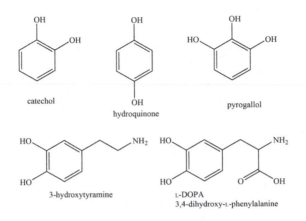

Figure 16.25 Phenols used with laccase in colouring studies.

Table 16.3 Shrinkage temperature (°C) of hide powder dyed with phenolic compounds and laccase.

Phenol	Phenol + laccase	Phenol alone
Catechol	71	61
Hydroquinone	74	61
Pyrogallol	59	60
L-DOPA	59	57
3-Hydroxytyramine	57	56

through the formation of quinoid species by polymerising phenols with enzymes. Laccase (benzenediol:oxygen oxidoreductase, E.C. 1.10.3.2) is a copper-containing polyphenol oxidase that catalyses the oxidation of electron-rich substrates such as phenols.[7–9] It is produced by some Basidiomycetes fungi that cause white rot of wood,[10] and certain reaction products of laccase-mediated oxidation and polymerisation of phenolic compounds can give coloured products, such as melanins.[11,12] In addition to producing colour, polymerisation products of phenolic compounds may also confer tanning properties, causing the coloured products to stabilise the collagen while at the same time binding covalently and thereby making the colour fast to washing.[13]

Some phenolic compounds, illustrated in Figure 16.25, give tanning effects with laccase, as shown in Table 16.3. The effects are relatively small, certainly in comparison with similar chemical reactions,[14,15] although the outcome is not unexpected.

Table 16.4 Kinetic parameters for laccase reaction with phenols at 30 °C and pH 6.0.

Substrate	K_M/mM	Relative V_{max}
Catechol	0.50	77
Hydroquinone	0.21	100
Pyrogallol	0.19	71
L-DOPA	0.89	92
3-Hydroxytyramine	0.82	74

The kinetic parameters K_M and V_{max} from the Michaelis–Menten equation are given in Table 16.4. Based on the K_M values, the order of affinity of laccase with phenolic compounds is pyrogallol > hydroquinone > catechol > 3-hydroxytyramine > L-DOPA.

The effects of laccase-mediated polymerisation in the presence of chrome-tanned pelt are as follows. L-DOPA is oxidised to L-dopaquinone by laccase in an oxygen environment and *in vivo* is polymerised to produce the final product melanin.[9,10] Hydroquinone is oxidised to *p*-benzoquinone by laccase under aerobic conditions[16] and the outcome is a weak purple colour conferred to the pelt. Catechol is oxidised by laccase to *o*-benzoquinone and this results in the deepest shade achieved in these tests, a dark brown–black colour. In this context, the colour struck on the leather is a reflection of the interaction between the phenol and the enzyme: the deeper shades resulted from higher affinity, but not the degree of blackness. The principle of colouring without dyes and achieving simultaneous covalent tanning appears to be feasible, although clearly it is not easy to transfer the biochemistry of melanin production into a biomimetic technology. The target in these trials was deep jet black, but the best colour achieved was a dark brown–black. However, although the principle was not proved completely, the approach to new compact processing is sufficiently interesting to warrant further study.

References

1. K. McLaren, *J. Soc. Leather Technol. Chem.*, 1990, **74**(3), 67.
2. D. L. Randall, *J. Am. Leather Chem. Assoc.*, 1994, **89**(10), 309.
3. C. S. Tysoe, *J. Soc. Leather Technol. Chem.*, 1995, **79**(3), 67.
4. R. M. Christie, *Colour Chemistry*, RSC Publishing, Cambridge, UK, 2015.
5. R. L. M. Allen, *Colour Chemistry*, Thomas Nelson and Sons Ltd, 1971.
6. W. F. Fuck, *et al.*, *J. Soc. Leather Technol. Chem.*, 2018, **102**(2), 69.
7. C. S. Evans, Enzymes of lignin degradation, in *Biodegradation: Natural and Synthetic Materials*, ed. W. B. Bett, Springer-Verlag, London, 1991.

8. N. Duràn and E. Esposito, *Appl. Catal., B*, 2000, **28**, 83.
9. E. Srebotnik and K. E. Hammel, *J. Biotech.*, 2000, **8**, 17.
10. A. Hattaka, *FEMS Microbiol. Rev.*, 1994, **13**, 125.
11. S. Chaskes and R. L. Tyndall, *J. Clin. Microbiol.*, 1975, **1**, 509.
12. H. Rorsman, G. Agrup, C. Hansson and E. Rosengren, Biochemical recorders of malignant melanoma, in *Malignant Melanoma, Advances of a Decade*, ed. R. M. McKie, Karger AG, Basel, 1983.
13. A. D. Covington, *Chem. Soc. Rev.*, 1997, **26**(2), 111.
14. O. Suparno, PhD thesis, University of Northampton, 2005.
15. A. D. Covington, C. S. Evans and O. Suparno, *Pubscie AEIF*, 2003, 3(1), 18.
16. A. Sànchez-Amat and F. Solano, *Biochem. Biophys. Res. Commun.*, 1997, **240**, 787.

17   Fatliquoring

17.1   Introduction

Contrary to popular belief among practitioners, the primary function of fatliquoring is not to soften the leather: this is only the secondary function. The primary function of fatliquoring is to prevent
the fibre structure from resticking during drying. As the leather
dries, the interfibrillary water is removed, allowing elements of the
fibre structure to come close together, which consequently allows
interactions to occur. In the limit, these interactions become
strong, because they are created by the Maillard reaction: this is
the reaction associated with 'browning' in cooking, presented in
Chapter 24. It is essential for leather quality to prevent this from
happening.

Figure 17.1 shows the structure of a wet blue fibre.[1] The photomicrograph is the result of cryo-scanning electron microscopy:
the sample is snap frozen in a slush of liquid and solid nitrogen
at about -200 °C, then the temperature is allowed to rise a few
degrees, so that the water can sublime away under vacuum to leave
intact structure. There are several aspects of the photomicrograph
to note.

1. Collagenic materials are at their softest when they are soaking
   wet.
2. The photomicrograph is a representation of wet leather, *i.e.* no
   artefacts of drying are present.

Tanning Chemistry: The Science of Leather 2nd edition
By Anthony D. Covington and William R. Wise
© Anthony D. Covington and William R. Wise 2020
Published by the Royal Society of Chemistry, www.rsc.org

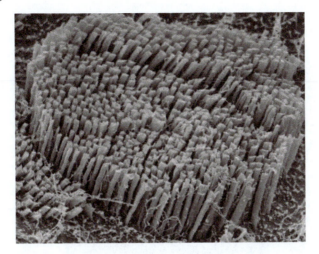

Figure 17.1 Cryo-scanning electron photomicrograph of wet blue leather.

3. The fine structure elements are fibril bundles: they indicate a high degree of opening up and the important level of the higher of collagen structure at which opening up occurs is at the level of the fibril bundles.
4. Taking points 1 and 3 together, the sites where lubrication is required are the fibril bundles: this implies that the conditions under which the lubricant penetrates the fibre structure are important.

Figure 17.2 shows a wet blue fibril bundle at higher resolution:[2] there are two aspects of this photomicrograph to note.

1. The structural elements are fibrils.
2. At the surface, where the sample is wettest, the fibrils are separated, but where evaporation of water can begin to take place it can be seen that the fibrils are much closer together.

The opening-up processes apply also to the fibril bundles themselves, although it is probable, because of the close association of the fibrils, that the physical properties are derived from the ability of the fibril bundles to distort and slip when stress is applied. Those properties are the elements of handle, including softness, and strength. Handle is a complex concept, because it relates to the way the leather feels when it is manipulated: it is an algorithm that combines density, softness, compressibility, stiffness, smoothness, springiness, stretchiness and probably other more subtle

Figure 17.2 Cryo-scanning electron photomicrograph of a wet blue fibril bundle.

parameters, easily solved in the brain of the experienced tanner, but much less easy to quantify objectively. Strength is the ability of a material to resist breaking or tearing stress: in the case of leather, it is the ability of the material to dissipate stress over an associated volume by movement of the fibre structure. In order to be able to do this, two criteria must be met.

1. The fibre structure must not be stuck together by the adhesions created during drying.
2. The fibre structure must be lubricated to allow the elements to slide over one another.

It is the purpose of the fatliquoring step to satisfy those criteria: it prevents fibre sticking during drying by providing an oil surface to the fibre structure, which then gives it the required lubrication. The effectiveness of the process step then depends on the extent to which the lubricant penetrates down the hierarchy of structure and its ability to allow contacting surfaces to slide. This can be illustrated by the photomicrograph in Figure 17.3, which shows the effect of drying without lubricating; the well opened-up fibre structure is stuck together at the level of the fibril bundles.

In Chapter 1, Figure 1.13, the feature of the banding pattern in collagen was discussed. It has been shown in modern studies of the

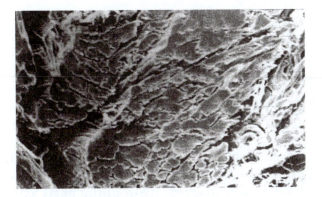

Figure 17.3 Chrome-tanned leather: well opened up in the beamhouse, but dried without lubricant after tanning.

banding pattern that the fundamental structure of collagen is affected by lubrication. In the absence of fatliquor, the *D*-period is 60.2 nm; following treatment with 10% lanolin, the *D*-period increased to 63.6 nm and the change was proportional to the lubricant offer.[3] This indicates a degree of relaxation in the fibrillar structure in the presence of lubricant. The relationship between fibrils is also affected by fatliquoring. Small-angle X-ray scattering (SAXS) studies of collagen under strain have shown that although lubricant may extend to the fibril level, it does not change the effect of strain. However, the organisation of the fibrillary structure is affected. Using the concept of orientation index (OI; scale of zero to one, where zero is isotropic, randomly organised, and one is parallel), Under 40–70% strain, the OI changed by up to 22% and the *D*-period changed by up to 1.8%, although inconsistent with fatliquor offer. The extensibility increased by 11% with 2% fatliquor offer: the elastic modulus decreased with fatliquor offer, but not in proportion.[4] Although the principles of fatliquoring are well known in the technology, the details of the relationship between collagen and modified collagen at the lower levels of the hierarchy of structure and the consequences for performance are less clear.

Typically, the leather industry employs partially sulfated or sulfited oils, which might be animal, vegetable, synthetic or, less commonly, mineral. Most commonly, the neutral oil is in the form of triglyceride (see later). Here, the so-called sulfo fraction is the emulsifying agent, keeping the neutral oil suspended in solution and thereby transporting it into the leather; it is the neutral oil that is the lubricant (see later).

Oil-in-water emulsions are created by the formation of particles, consisting of a small drop of oil surrounded by an emulsifier/

●⌇⌇⌇
hydrophobic tail
hydrophilic (charged) head

Figure 17.4   Model of an anionic oil-in-water emulsion particle.

detergent/surfactant/tenside (these terms are used interchange-ably). The emulsifying agent has a hydrophobic part, which is dis-solved in the oil, and a charged, hydrophilic part, which interacts with the solvent (water), to keep the particle suspended, as illus-trated in Figure 17.4.

The emulsion particles are prevented from coagulating or coalesc-ing because they are held apart by the repulsing effect of the high charges on the surface. Any chemical reaction that reduces the charge on the particle surface will allow the particles to come together, caus-ing the neutral oil particles to coalesce. High temperature can drive the particles together, breaking/cracking the emulsion. An important mechanism to make this happen and which is exploited in leather making is the ionic interaction between the sulfate/sulfonate group and the protein, in the way discussed in Chapter 15.

At the 1986 SLTC Conference, the theme was fatliquoring: Waite reviewed the state of understanding at that time and summarised the important contributors to the mechanism, as follows.[5]

1. The charge on the leather (neutralisation dependent) can affect fatliquoring effectiveness.
2. The penetrating power depends on the emulsion stability.
3. The particle size distribution of an emulsion is a variable that influences stability and penetration.
4. Sulfonation (to make partially sulfited oils) imparts more sta-bility than sulfation (to make partially sulfated oils). This is the basis for whether a fatliquor is used for grain lubrication or for the corium.
5. Softness depends on the ratio of emulsifier fraction to neutral oil fraction.
6. Softening is influenced by the viscosity and interfacial tension of oils.

Points 1 and 2 emphasise the importance of pH in this process step: if the emulsion is anionic (which it usually is), the charge on the leather should also be anionic to allow penetration. Neutralisation of the substrate means pH adjustment not only through the cross-section but also down the hierarchy of structure. The mechanism of fatliquoring can be expressed in simple terms as follows.

1. The neutral oil is transported into the pelt as an oil-in-water emulsion.
2. The emulsifying agent interacts with the leather, reducing or eliminating its emulsifying power.
3. The neutral oil is deposited over the fibre structure – the level of the hierarchy of structure depends on the degree of penetration.
4. The water is removed by drying, allowing the neutral oil to flow over the fibre structure. The distribution of neutral oil determines the degree to which fibre sticking is prevented, which in turn depends on the depth of penetration of the emulsion, which in turn depends on the pH of the system and the particle size of the emulsion. Additional factors are the nature of the sulfo fraction carrier, sulfated or sulfited, the mechanism of distributing the oil, where the oil is distributed within the hierarchy of the fibre structure and the nature of the oil. It is the last parameter that contributes to the softness of the leather: in general, the higher is the viscosity of the neutral oil, the better are the lubricating properties. However, there is a limiting effect, which is the ability of the oil to be emulsified and hence carried into the leather: at high viscosities, the oil tends to deposit on the surface, leaving the leather empty.[6]

It is clear from the analysis of the mechanism that the application of a fatliquor must incorporate the notion of creating as small an emulsion particle size as the formulation will allow. This is accomplished by applying the following conditions.

- High temperature of the diluting water: ideally, heating the fatliquor to the same temperature as the water allows the oil to break up into small drops due to the decreased viscosity. Note that there is a limit to the temperature, since heating is a powerful mechanism of coagulation; a practical limit is 60 °C. This would be a maximum value to avoid damaging vegetable-tanned leather, although chrome-tanned leather is typically capable of withstanding much higher temperatures.
- Mechanical action: for mixing the fatliquor into water, mechanical action must be maximised, because this is the way in which

the oil is dispersed into drops. Ideally, a motorised stirrer should be used, with the fatliquor added into the vortex. The common practice of adding cold fatliquor to hot water in a blue barrel, stirring with a broom handle, is not optimum and therefore best use will not be made of the fatliquor.

- Some tanners have favoured adding water to the fatliquor. In this way, a water-in-oil emulsion is formed. Continued additions of hot water reverse the emulsion to oil-in-water. There is no clear evidence to indicate that this a better way of optimising the eventual emulsion.

The particle size of the emulsion is determined by the degree to which the oil is converted to the sulfo derivative: the higher the sulfo-to-neutral oil ratio, the more solubilised the oil becomes and the particle size is reduced. Most fatliquor formulations contain about 50% neutral oil, 25% sulfo fraction and 25% water – this will produce emulsion particles about 30 nm in size. Higher sulfo fractions can create microemulsions, where the oil is on the boundary of suspension and solubilisation, when the particle size is about 5 nm. The influence of particle size can be seen in Figure 17.5, comparing the typical emulsion with a microemulsion: the oil particles can be seen as black dots, visualised in transmission electron microscopy by osmium tetroxide.[1,2]

The relative effects of the neutral oil and the sulfo fraction have been distinguished by separating the components of synthetic fatliquors and applying them individually.[1] For comparison of equal quantities of reagent, 20% fatliquor was compared with 5% sulfo fraction and 10% fatliquor was compared with 5% neutral oil. In order to allow the

(a)   (b)

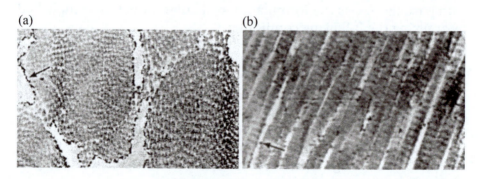

Figure 17.5   Penetration of oil emulsions, shown by transmission electron microscopy: (a) 25 nm reaches the fibril bundles; (b) 5 nm reaches the fibrils. Courtesy of the *Journal of the American Leather Chemists Association*.

Table 17.1 Relative effects of neutral oil and sulfo fraction. The softness of the leather is measured by the loop test, where lower weight/magnitude means softer leather.

Offers on wet weight	Sulfited oil	Sulfated oil
20% Fatliquor	44	56
10% Fatliquor	127	123
5% Sulfo fraction in water	450	590
5% Neutral oil in acetone	69	66
5% Neutral oil + 5% non-ionic surfactant	70	53
Controls:		
5% Non-ionic surfactant	164	
Acetone, then rehydrate	781	
Water	1220	

neutral oil to penetrate, it was formulated with either 5% non-ionic surfactant or acetone. The results are given in Table 17.1 and indicate the following.

- Softness is conferred by the neutral oil, with no contribution from the sulfo fraction.
- The alternative ways of delivering the neutral oil appear more effective than the original fatliquor.

It is clearly not feasible to use organic solvents in fatliquoring, but alternative emulsifying agents are of interest. An obvious feature of the emulsion of neutral oil and non-ionic surfactant is the instability of the emulsion: in attempting to measure the particle size by laser light scattering, it was apparent that the rate of coagulation was faster than the rate of measuring the particle size dimensions. This raised the issue of why the fatliquoring effect was so good.

From Figure 17.6, the wet leather fatliquored with sulfated oil has the appearance of droplets over the surface, which would be expected if the oil was deposited as emulsified drops. However, the leather fatliquored with neutral oil emulsified with non-ionic surfactant looks completely different: the surface is coated with what appears to be a water-in-oil emulsion. In both cases, the dry leathers do not show any difference in structure, so the oil is assumed to be uniformly distributed in the regions where it was deposited. Therefore, it can be inferred that the unstable emulsion exhibits a different way of delivering the oil. Once a stable emulsion particle hits the fibre structure, it is likely that the sulfo fraction will interact with it, depending on the charge conditions, causing the emulsion to lose its emulsifier and the neutral oil will be deposited;

(a)   (b)

(c)   (d)

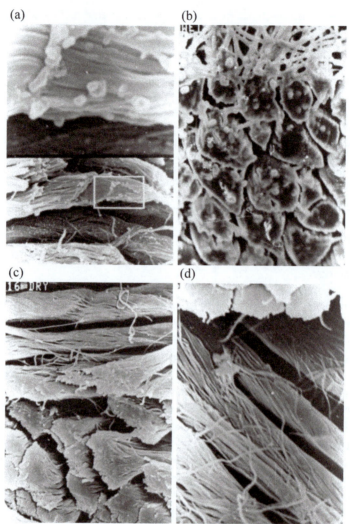

Figure 17.6   Leathers treated with sulfated synthetic oil fatliquor and with neutral
synthetic oil emulsified with a non-ionic surfactant, from Table 17.1.
(a) Leathers treated with sulfated synthetic oil, wet leather; (b) leathers
treated with non-ionic surfactant and neutral oil, wet leather; (c) leathers treated with sulfated oil, dry leather; (d) leathers treated with non-
ionic surfactant and neutral oil, dry leather. Courtesy of the *Journal of
the American Leather Chemists Association*.

there is no mechanism for the neutral oil to move within the aqueous medium, so it is static until the water is removed in drying
and it can flow over the leather surface. However, if there is residual emulsifier in solution, because of the weak emulsifying effect,

(a) (b)

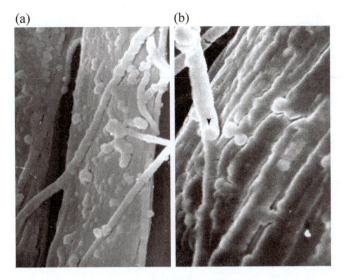

Figure 17.7 Effect of sulfation or sulfonation on the mode of deposition of neutral oil on leather. (a) Globular sulfated oil; (b) globular and smeared sulfited oil.

continuous interaction can occur between the deposited oil and the solvent water, creating a water-in-oil system that allows the oil to flow over the surface in the presence of water.

A comparison can be made between sulfated and sulfited oils. Sulfated oils are less stable and hence less penetrating because the sulfo fraction reacts more readily with the leather, so the emulsions break easily over the leather. On the other hand, sulfited oils are less reactive towards the substrate, so the emulsions are apparently more stable in the environment of the leather and hence the emulsion particles penetrate more deeply into the hierarchy of structure. It is for this reason that the technological approach to fatliquoring is to use a mixture of the two types, to achieve both surface and internal lubrication. However, there is another consequence of the lower affinity of sulfited oil for the fibre surface, namely that the sulfo fraction remains available to interact with oil deposited on the leather surface. This is illustrated in Figure 17.7, where it is apparent that the neutral oil is deposited in a globular manner by the sulfated emulsifier, but in both a globular and a smeared manner by the sulfited emulsifier.

The difference in mechanism exhibited by sulfated and sulfited oils explains the effect of the unstable emulsion of neutral oil with

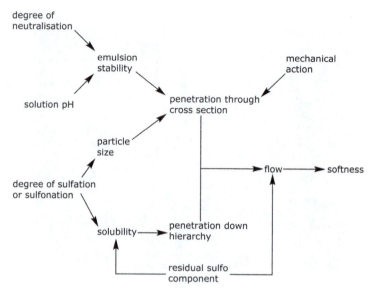

Figure 17.8   Model of the fatliquoring mechanism.

non-ionic surfactant. The presence of the weak detergent in solution provides a continuing mechanism for distributing the neutral oil, analogous to the sulfited oil.

The mechanism of fatliquoring can be summarised by the interaction of the contributing parameters, as shown in Figure 17.8.[1]

The development of softness in leather is intimately associated with mechanical action, because the process of drying causes adhesions to occur, weak adhesions in the case of well-fatliquored leather, which must be broken to soften the leather. This is achieved by the process of staking, in which the leather is mechanically stressed; the traditional process used to involve bending the leather by hand through an acute angle over a blunt blade, but now machines pummel the leather automatically and with precisely controlled force. A crucial feature of the staking process is the water content of the leather: more than an equilibrium moisture content is required, 15–20%, to provide additional lubrication of the fibre structure.

Figure 17.9 demonstrates the relationship between softness and strength.[7] In Figure 17.9a, there is a positive correlation between softness and strength, indicating that higher strength is obtained from more effective fatliquoring. There will be a maximum value of the strength for any leather, which is dependent on the chemical processing up to the point of the fatliquoring: it is the fatliquoring process that determines whether the optimum strength or less is displayed. In Figure 17.8b there is a negative correlation, demonstrating that as the stressing continues the fibre structure begins to

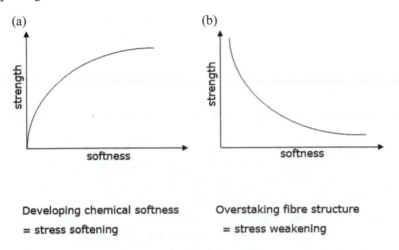

Figure 17.9 (a) Chemical softness and (b) stress softness.

break down. This can be achieved by too much mechanical action, either too much energy applied to the leather or less energy applied for too long, such as multiple passes through the staking machine. Alternatively, and additionally, if the moisture content is too low, the fibre structure is less flexible and the components slip less easily, so the brittle structure tends to loosen and break, seen as increasing weakness.

It is common practice to introduce the fatliquor(s) as the last wet process step, as discussed in Chapter 15, but that is not a hard and fast rule. Lubricant is routinely applied at all stages of processing, including prior to tanning: the only proviso is that the conditions, *e.g.* pickling before chrome tanning, should be compatible with the chemistry of the fatliquor formulation. The argument for early fatliquoring is to extend the period during which the oil might migrate down the hierarchy of structure and improve the impact on the leather properties.

17.2 Anionic Fatliquors

17.2.1 Sulfated Fatliquors

For sulfating, the oil must be unsaturated, with a minimum iodine value of 70, where the iodine value is defined as the number of grams of iodine absorbed by 100 g of oil or fat. Oils that have been used in this regard are castor, neatsfoot, soya, groundnut and cod. Note that after 1980, sperm whale oil was voluntarily abandoned by the global leather industry.

Figure 17.10　Lactonisation.

The chemistry of the preparation of a sulfated oil is in three steps.

1. *Preparation:* 10–20% concentrated sulfuric acid on the weight of the oil is added slowly to the oil, with constant stirring. The temperature of the exothermic reaction must be controlled to <28 °C, otherwise the oil can char, causing darkening, and the triglyceride oil may be hydrolysed to release free fatty acids. This latter effect can give rise to the problem of 'spue', when the longer chain carboxylic acids can migrate from the internal structure of the leather to the grain surface, visible as a white efflorescence: diagnosis is confirmed when the efflorescence melts with the heat of a match. Problems associated with overheating can be catalysed by the presence of iron salts. An alternative is to use a mixture of sulfuric and phosphoric acids (usually in the ratio 0.8 : 1.0). This removes the effect of iron (by reaction with the phosphoric acid) and the reaction is less exothermic, so the acid mixture can be added faster. The process is more expensive and results in a mixture of sulfated and phosphated oils.
2. *Brine wash:* Excess free acid is removed by washing the partially sulfated oil with brine, which also separates the oil fraction from the aqueous fraction. Brine is used to avoid creating an emulsion, which would occur if water alone were to be used. Alternatively, sodium sulfate, ammonium chloride or ammonium sulfate could be used. Some hydrolysis of bound sulfate may take place. Additionally, there may be lactonisation of adjacent hydroxyl groups with carboxyl groups, as shown in Figure 17.10.
3. *Neutralisation:* Bound and free acid groups are neutralised with alkali:

$$-OSO_3H \rightarrow -OSO_3^-Na^+ \text{ or } NH_4^+, \textit{etc.}$$
$$-COOH \rightarrow -COO^-Na^+ \text{ or } NH_4^+, \textit{etc.}$$
$$\text{free fatty acid} \quad \text{water-soluble soap}$$

Also, free sulfuric acid is converted to salts: $2Na^+ SO_4^{2-}$ or $2NH_4^+ SO_4^{2-}$, *etc.* Any lactones will be hydrolysed.

1. Combination of sulfuric acid with double bonds

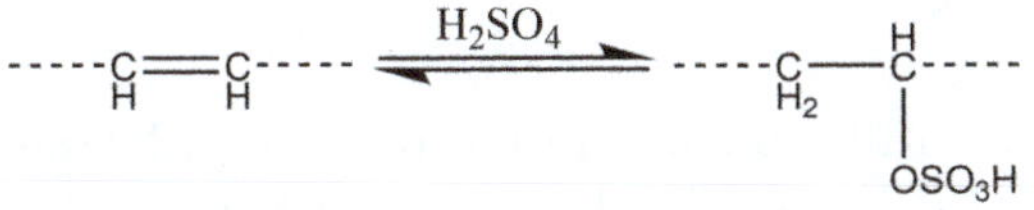

2. Reaction of sulfuric acid with hydroxyl groups e.g. ricinoleic acid in castor oil

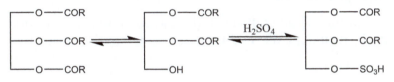

3. Hydrolysis of triglycerides and reaction of sulfuric acid with a hydroxyl of a glycerol derivative

O—COR	O—COR	O—COR
O—COR ⇌	O—COR $\xrightarrow{H_2SO_4}$	O—COR
O—COR	OH	O—SO$_3$H

Figure 17.11 Sulfating reactions.

Bound SO$_3$ may range from 2%, considered to be a low level, to 3–4%, considered to be a medium level, to 6–8%, considered to be a high level; fatliquors may therefore be designated low, medium or high sulfated. Figure 17.11 shows some sulfating reactions.

As the level of sulfation increases.

- The anionic charge increases, hence greater affinity for cationic leather.
- The lubricating effect decreases, owing to the lower concentration of neutral oil.
- The emulsion particle size decreases, ultimately to the point of forming a microemulsion (<5 nm) or even actually dissolving in water.
- The stability of the emulsion to coagulation by acid or metal salts increases.
- At high levels, the oil functions more like a wetting agent than a lubricant, hence the leather becomes more hydrophilic.
- The leather becomes looser, in terms of break, possibly owing to the damaging effect of the sulfate species on collagen.
- The likelihood of hydrolysing the oil to create free fatty acids increases, thereby creating the possibility of chrome soaps, fatty acid spue, poor wetting back, uneven dyeing and poor finish adhesion.

Sulfated oils are used as follows.

- *Low level of sulfation:* Low stability of the emulsion to coagulating (cracking) by acids or metal salts. Typically used to lubricate the outer surfaces, particularly the grain. May be used for drum oiling of vegetable-tanned leather.
- *Medium level of sulfation:* More stable to coagulation, therefore greater potential for penetration. Used for surface neutralised chrome leather.
- *High level of sulfation:* Used for complete penetration, *e.g.* through neutralised chrome leather for gloving, clothing, softee leathers.

17.2.2   Sulfited Fatliquors

The requirement for the oil is unsaturation, as for sulfating. Options include cod oil, neatsfoot oil, *etc.* Preparation is conducted in two steps.

1. *Sulfitation/oxidation:* Using the example of cod oil, air is blown through a mixture of 100 parts of oil and 50 parts of 40° Baumé sodium bisulfite solution with stirring at 60–80 °C. Alternatively, hydrogen peroxide may be used instead of air. The chemistry is illustrated in Figure 17.12.
2. *Brine wash:* Washing with brine removes excess sodium bisulfite.

1. Reaction between bisulfite and double bonds

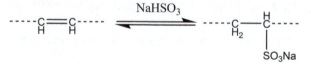

2. Oxidation of double bonds, then reaction with bisulfite

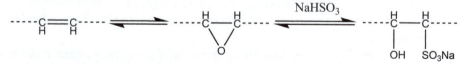

3. Reaction between carbonyl groups and bisulfite

Figure 17.12   Sulfitation and oxidation.

Note that no pH adjustment is required, owing to the use of bisulfite rather than sulfur dioxide or sulfurous acid.

In comparison with sulfated oils, sulfited oils exhibit the following properties.

- No charring or darkening.
- Higher emulsion stability to acids, hard water salts and metal ions, *e.g.* Al(III), Cr(III), owing to the presence of sulfonate and hydroxysulfonate groups and the low level of free fatty acids or soaps, because there is little hydrolysis of triglyceride during synthesis.

The fatliquor may be formulated with non-ionic detergent, to increase emulsion stability, to promote better penetration and to make the leather softer and fuller. There is a danger of giving the leather loose break.

Sulfited oils are used as follows:

- for softness and strength (see later) for all leathers by deep penetration;
- for woolskins and furskins in mineral tanning baths;
- in shrunken grain production, to minimise loss of tensile strength in the acidic tanning bath.

17.3   Soap Fatliquors

Raw oil, *e.g.* neatsfoot (100 parts), is emulsified with soft soap, *e.g.* potassium oleate (30 parts). Note that stearate is likely to cause fatty acid spue. The emulsions have a large particle size, owing to their tendency to have low stability towards water hardness and acid. The formulation typically has pH ~8. At pH < 6, the soap is increasingly converted to free fatty acid (Figure 17.13), which does not act as an emulsifier, so the emulsion coagulates.

The uses of soap fatliquors are limited, owing to the low emulsion stability. They have traditionally been used for surface fatliquoring calfskins for shoe uppers and formaldehyde-tanned sheepskin for gloving leather. There is an American practice of improving the emulsion stability by formulating a soap fatliquor with a small amount of low-sulfated oil (up to 0.5% SO_3 on a moisture-free basis), together with a stabilising colloid, *e.g.* starch or natural gum.

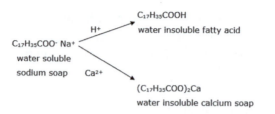

Figure 17.13 Soap reactions.

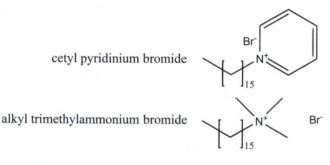

Figure 17.14 Examples of cationic emulsifying agents.

17.4 Cationic Fatliquors

Raw oil is emulsified with a cationic agent, exemplified by the structures in Figure 17.14, where the hydrophilic group is typically straight-chain aliphatic C_{10}–C_{18}. The cationic fatliquors have the following features.

- Low affinity for cationic charged leathers, *e.g.* chrome-tanned leathers.
- High affinity for anionic charged leathers, *e.g.* vegetable-tanned leathers, for lubricating the outer layers.
- Incompatible with anionic reagents, *e.g.* anionic dyes, fatliquors, retans.
- High stability to acid, but unstable to alkali.
- Good stability to metal ions and salts.
- May have poor shelf life.
- Emulsion stability can be improved by formulating with non-ionic detergents, *e.g.* alkyl ethylene oxide condensates.
- As a second fatliquor after anionic fatliquor (*cf.* 'sandwich dyeing').

It is well known in the detergent industry that a combination of anionic and cationic surfactants that creates a total carbon chain length of >24 will produce a water-insoluble precipitate. This can be easily observed by mixing any domestic anionic liquid washing-up detergent with any domestic cationic fabric softener – but be aware that it does make a sticky mess!

- On chrome-retanned leather to increase surface oil and aid paste drying.
- On vegetable-tanned leather for bags or cases, to lubricate the grain layer and to make it more pliable.
- For oiling vegetable-tanned leather.
- In combination with basic aluminium(III) chloride, for retanning chrome leather for suede. This will increase the cationic charge, which gives a higher affinity for anionic dyestuff and produces greasy nap.
- For woolskins in the mineral tanning bath.
- For white gloving leather, which might be made with a combination of aluminium(III) and sulfonyl chloride.

17.5 Non-ionic Fatliquors

17.5.1 Alkyl Ethylene Oxide Condensates

These fatliquors are emulsified with compounds made by condensing ethylene oxide in the presence of an aliphatic alcohol:

$$C_nH_{2n+1}-(OCH_2CH_2)_xOH$$

The properties of the emulsifying agent depend on the value of n, the aliphatic carbon chain length, and the value of x, the degree of polymerisation of ethylene oxide. These compounds are not very good emulsifying agents, because the hydrophilic end, the ethylene oxide chain, does not have a high affinity for water, relying only on hydrogen bonding *via* the ether oxygens.

The properties are as follows.

- High stability towards metal ions, salts and hard water and wide pH tolerance.
- Miscible with cationic and anionic reagents.
- Little or no affinity for anionic or cationic charged leathers.
- The non-ionic emulsifier increases the hydrophilicity of leather.
- For fatliquoring zirconium(IV)- or aluminium(III)-tanned leathers, which are usually highly positively charged.
- As a crusting fatliquor for suede splits, *i.e.* merely to prevent fibre resticking on drying, to aid rewetting.

- May be formulated with anionic or cationic fatliquors, to improve stability.

17.5.2 Protein Emulsifiers

Uncharged proteins, at their isoelectric point, can act as emulsifiers. Globulin proteins, *e.g.* albumin, produce a 'mayonnaise' type of fatliquor. These products are limited to a narrow pH range, close to the isoelectric point, otherwise they become significantly charged and function as anionic or cationic formulations.

17.6 Multi-charged Fatliquors

These fatliquors are formulations of non-ionic, anionic and cationic fatliquors, in which the presence of the non-ionic species prevents precipitation of the anionic and cationic species. They are more stable to a wider pH range than singly charged fatliquors and hence more stable to variations in leather charge. The proportions of the constituents can be varied, depending on the leather properties required, *i.e.* depth of penetration, surface lubrication and ease of removal from paste drying plates (where it would be an alternative to two-stage fatliquoring).

17.7 Amphoteric Fatliquors

Raw oil is emulsified with an amphoteric reagent, *i.e.* containing both acidic and basic groups, for example as shown in Figure 17.15.

The point of neutrality, the isoelectric point, depends on the numbers of acidic and basic groups, as discussed for proteins. A typical value of the isoelectric point for this type of structure is pH 5. At pH > 5 the emulsifier is negatively charged and at pH < 5 the emulsifier is positively charged. The choice of pH of the fatliquor depends on the charge of the leather and the requirement for surface reaction or penetration.

17.8 Solvent Fatliquors

These are typically anionic fatliquors, containing a high boiling point, polar petroleum solvent with a minimum flash point of 60 °C and preferably odourless. The function of the solvent is to replace the water as

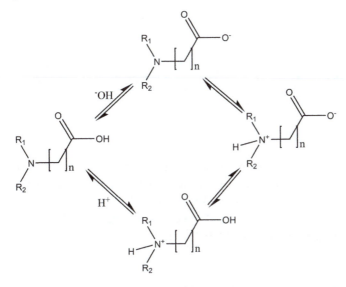

Figure 17.15 Amphoteric emulsifier.

it is removed during drying. With lesser affinity for the leather than the polar triglyceride oils, it can spread further and deeper. Therefore, these products are used for making soft leather. Note that the current attitude to the use of solvents in leather making means that these products are used less now than in the past.

17.9 Complexing Fatliquors/Water Resistance Treatments

A comprehensive review of water resistance is presented in the following, in which the chemistries of the modern approaches are discussed. Before these treatments were developed, a simpler approach to making leather hydrophobic was adopted, which was also part of the lubricating step in leather manufacture. There are two traditional forms of these reagents, designed to be applied to chrome-tanned leather.

1. An old option was to use a long-chain fatty acid dimer, *e.g.* oleic acid dimer, applied as a soap at pH 8.[8,9] This was used as a method of introducing water resistance, where the carboxylic groups are complexed with bound chromium(III), leaving the hydrophobic moieties exposed over the fibre structure. More often than not, the technology did not work in industrial practice.

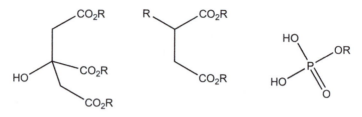

Figure 17.16 Complexing lubricants.

2. Another traditional option was to use dibasic or tribasic acid esters, such as succinic, phthalic, citric or phosphoric acid, bound to an aliphatic chain with 15–24 carbon atoms, as illustrated in Figure 17.16. These also were unreliable reagents for making water-resistant leather.

17.10   Silicone Fatliquors

Silicones have a wide range of applications in modern life, so it is surprising that the applications in leather technology have been limited to a role in water resistance, discussed later. There is little reference to the use of silicones in the lubrication of leather, limited to the work reported by Tang *et al.*[10] Aminosilicone was modified with acetic anhydride, to convert 50% of the amino groups into amide groups to reduce the cationic character. It is reported that applying a stable emulsion with conventional non-ionic surfactants with glycerol and dimethyldioctadecylammonium chloride also formulated with beeswax and cottonseed oil produced a 7% increase in tensile strength, a 3% increase in tear strength, 31% elongation at break and an increase of 10 °C in shrinkage temperature.

17.11   'Solid' Fatliquors

In general, tribology concerns the interaction of surfaces including lubrication to reduce friction: the two mechanisms of importance are lubricant liquid and rolling interface, *i.e.* the use of grease or the use of a roller bearing. The first is the effect used for many years in conventional fatliquoring technology. By analogy, Marriott[11] demonstrated that the second option was feasible, using inert polymers in the form of emulsions, *e.g.* polyethylene. This may be the mechanism by which 'sulfur tannage' has a lubrication effect. This is a very different approach to preventing fibre sticking and softening of leather – it is a concept worth pursuing, to research the impact of properties and performance.

17.12   Water Resistance

17.12.1   Introduction

The general function of post-tanning operations is to confer specific performance properties to the leather. One of the important features of that part of leather making is to make the leather water resistant. This is typically applied only to shoes and boots, with the possible inclusion of clothing leathers. Such processes are typically applied to chrome-tanned leather, since other methods of tanning make the leather too hydrophilic to allow high levels of water resistance to be conferred. With the increasing emphasis on wet white and organic leathers, the challenge is to make them water resistant.

It is common to refer to treatments for leather with hydrophobic reagents as waterproofing. This implies that water is completely excluded from the cross-section of the leather, so that no transmission of moisture occurs. In practice, the traditional methods of achieving this were to fill the structure with, for example, wool grease, so-called 'stuffing', or to apply a water-impermeable finish coating. In the first case, the treatment is not actually completely effective, since moisture can still find its way through the fibre structure.In the second case, the product is more like plastic than leather. The role of finishing in creating water resistance is recognised in *Official Methods*[14] of testing, where a requirement is to remove the finish by buffing prior to testing. It is important to recognise that leather is expected to 'breathe', *i.e.* its water vapour permeability must not be significantly reduced by the treatments. Therefore, some of the techniques used hitherto fail by this criterion. It is also useful to bear in mind that the success of the possible treatments is relative, ranging from small to large effects, hence the application of hydrophobing treatments is preferably referred to as conferring water resistance – the term waterproof is typically not used in a leather context.

In considering ways to confer water resistance, it is useful to consider how a leather can fail a water resistance test or fail in use. There are three mechanisms:

- Water may flow through the fibre structure, because of the voids in the material.
- Water may wet the surfaces of the fibre structure and flow across the solid surface.
- Water may travel through the fibre structure, in the way that wax flows through a wick in a candle: this mechanism is referred to as 'wicking'.

The first mechanism is resisted by the barrier approaches. The second mechanism is resisted by making the fibres less wettable. The third mechanism is a special example of the second, since wicking depends on the inner elements of the fibre structure being unaffected by the water resistance treatment. Therefore, the second mechanism is addressed by a relatively unsophisticated hydrophobing treatment, which is required to penetrate through the cross-section, but may not penetrate down the hierarchy of structure. The third mechanism can only be countered if the water resistance treatment penetrates down the hierarchy of structure; it is relatively easy to reach to the level of the fibril bundles, since that is achievable in well-conducted fatliquoring, so this may confer a good degree of water resistance. Consequently, it may be assumed that the highest degree of water resistance is conferred by treating the fibre structure at the level of the fibrils. From fatliquoring science, it is known that penetration to this level requires the reagent to be solubilised or, at the limit, to be in the form of a microemulsion.

The water resistance of a material such as leather depends on the way in which water (or any other solvent) interacts with the surface. In the case of leather, this means not only the actual surfaces, but also the surface of the fibres within the fibre structure and lower down the hierarchy. This is illustrated in Figure 17.17.

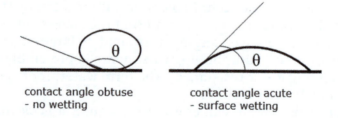

contact angle obtuse
- no wetting

contact angle acute
- surface wetting

Figure 17.17   Interaction between a liquid and a surface: the contact angle, θ.

Table 17.2 Surface tension, γ, of some liquids and the critical surface tension, γ_c, of some solid surfaces.

Surface/liquid	Chemistry	Surface tension, γ or γ_c/ mN m^{-1}
Water	H_2O	74
Ethanol	CH_3CH_2OH	23
Acetone	CH_3COCH_3	24
n-Octane	C_8H_{18}	22
Leather	Collagen	>70
Polyurethane		33
Silicone polymer	$[(CH_3)SiHO]_n$	24
Fluorinated polyacrylate	$[CH_2CH(CO_2C_nF_{2n+1})]_m$	11

The surface tension of the solvent and the surface determines the wetting of the surface by the solvent. Some values are presented in Table 17.2.

The critical surface tension of a solid, γ_c, is equivalent to the surface tension of a liquid, γ, if the contact angle is zero. Liquids that have $\gamma > \gamma_c$ will not wet the surface, but if $\gamma < \gamma_c$ the surface will be wetted. The purpose of modern water resistance treatments is to change the critical surface tension of the available fibre surface sufficiently to prevent water wetting it. Note that the presence of neutral electrolyte in aqueous solution does not change the surface tension markedly.

These concepts are also important in the chemistry of finishing, addressed in Chapter 21.

17.12.2   Principles of Conferring Water Resistance

The ability of a material such as leather to resist the absorption and transmission of water can be expressed in terms of weak and strong resistance. Weak resistance can be regarded in clothing terms as 'shower resistance': this means a garment does not immediately absorb rain water, but prolonged exposure will result in both rapid absorption of water and its transmission through the cross-section. This applies to both clothing and shoes: in the case of the latter, almost any wet circumstances will result in wet feet. Shower resistance may be sufficient to prevent water spotting of double-face or suede clothing leathers; this is the effect of wetting causing the surface fibres to stick, giving rise to a visible non-uniformity, typically in the form of spots from the raindrops. Strong resistance means that there is a significant delay between exposure to water and its transmission through the cross-section: the accompanying amount of water absorption reflects the movement of water within the material, although the relationship between wetting and transmission depends on the nature of the material, including methods of leather processing. Although the resistance of leather to wetting can be high, notably it cannot strictly be called 'waterproof', because that term must be reserved for materials that provide a complete barrier to water, *e.g.* a plastic sheet. The importance of water resistance was demonstrated when the American Leather Chemists Association (ALCA) held a symposium in 1995 in which it was the sole topic.[12,13]

The water resistance specifications for defence leathers vary from country to country all over the world, but in no case might the requirement be regarded as 'high performance'; *e.g.* in the UK, resistance to water penetration in the Bally penetrometer test (IUP 10)[14] is required to be only 3 h.[15] For the modern soldier in the field, in a war theatre for periods of 24 h or more, such performance is insufficient. However, it is possible to raise the water resistance to over 120 h on the Bally penetrometer or >5 00 000 flexes on the more stringent Maeser tester. It is likely that lack of failure after 1 week in these tests means that the leather is unlikely to fail the test at all, although the fibre structure may fail. See later for technologies to exceed that specification.

There are many types of treatment traditionally offered to the leather industry and also some more modern ones, considered later. However, it has been a feature of water resistance treatments that there are two shortcomings, which for many years resulted in the technologies being kept a closely guarded secret in each tannery making such leather.

- The effects were often moderate, conferring relatively poor water resistance.
- The effects were often inconsistent, sometimes working and sometimes not working.

In the modern leather industry, such inconsistency and poor performance are easily overcome, particularly if the tanner is aware of the factors that contribute to the outcome of modern treatments. Very high degrees of water resistance can be achieved by careful control of appropriate chemical treatments; even treatments not known for conferring high performance do have the potential to work better than might be expected. These modern chemistries are specifically designed to retain the ability of the leather to allow transmission of water vapour, the 'breathability' property. Although the treatments are technically feasible and relatively simple, the outcome of the treatments depends on many factors that can either prevent the development of water resistance or contribute to its creation.

Some workers have addressed the factors that influence water resistance;[16] for any chemical treatment, the factors that influence the developing of water resistance can be listed and rationalised as follows (not necessarily in order of importance).

17.12.2.1 Opening Up

Opening up is critical, in order to allow the reagent to penetrate deeply down the hierarchy of structure. Here, the important effect is to split the fibre structure finely, separating the fibres at the fibril bundle level, to allow the penetration of reagents. In the case of reagents that must be presented in the form of an emulsion, the particle size requirement may mean taking particular steps to solubilise materials that would ordinarily be applied as an emulsion, thereby contributing to the performance of the reagent. This may be accomplished simply, depending on the chemistry of the reagent, for example adjusting the pH to a higher value than specified or as supplied. Alternatively, if the hydrophobic agent can only be solubilised by means of detergents, this is clearly counterproductive (see later).

The degree of opening up is important, because unsplit fibres can act as a wick, to transport water through the leather cross-section. The more finely split the fibres are, the less easily they can be wetted and the wicking mechanism is reduced. Clearly, if the reagent penetrates to the level of the fibril surfaces, this is the best treatment that may be achieved.

17.12.2.2 Looseness

If opening up goes too far or too much mechanical action is applied, particularly in the alkaline swollen condition, the corium structure will become loosened. The open structure will facilitate water transmission, because it can occur without wetting the fibres. Similarly, it is more difficult to confer high water resistance to leathers with a naturally open structure, *e.g.* sheepskin.

17.12.2.3   Substrate

The primary tanning chemistry can influence the ability of the leather to achieve high water resistance; in essence, there are two practical possibilities in this regard.

- *Chrome tannage:* The hydrophilicity/hydrophobicity of chrome-tanned leather can be controlled by the way in which the tannage is conducted. In this case, the parameters can be set to obtain the maximum hydrophobic nature, *i.e.* slowest rate of change of pH and temperature, within the constraints of the processing time.
- *Hydrophilic tannage:* The use of vegetable tannins (plant polyphenols), resins or syntans, either for low-stability tannage, such as wet white (syntan plus glutaraldehyde), or in combination for high hydrothermal stability tannage, depends on the formation of hydrogen bonds between the tanning agent and the collagen. This reaction necessarily produces hydrophilic leather – in the case of vegetable tannins, hydrolysable tannins are worse than condensed tannins in this regard. These tannages make it more difficult to make water-resistant leather: this is mostly due to the hydrophilicity, but also relates to the lack of metal species in the leather, which is essential for some powerful water resistance treatments.

17.12.2.4   Reagent Chemistry

There are many strategies for incorporating hydrophobic groups into the leather structure (see later). These range from the so-called hydrophobic fatliquors through to specific agents. Modern high-performance agents are based on partial esters of acrylate polymers, offered as solo reagents or formulated with silicone derivatives: these acrylate species can react covalently with bound chromium(III) in chrome-tanned leather. If there is no metal 'mordant' for this function, other reagents may have to be used, *e.g.* metal complexes of fluorochemicals or siloxane derivatives.

17.12.2.5   Reagent Offer

The degree of water resistance conferred usually correlates with the amount of the reagent used, although there is an effective upper limit to the amount that is necessary to achieve high performance. This can be understood in terms of the extent of the treatment over the available fibre structure surface. If the whole surface is inadequately treated, there is a mechanism for water to pass through the wetted regions.

17.12.2.6 Neutralisation

Most modern high-performance water resistance reagents are anionic compounds and hence rely for their effect on penetration into the fibre structure, which in turn depends on the degree of neutralisation of the leather. There are four components to this step that must be considered.

1. The isoelectric point of the leather determines the nature and magnitude of the charge on the leather at any point on the pH scale. In the case of acrylate reagents with $pK_a \approx 4$, the leather must be neutral or anionic at pH 6. Any processing that raises the isoelectric point significantly above pH 6 would be counter-productive in this regard; this therefore rules out tannages that target the basic amino sidechains.
2. The pH must be high enough to ensure that there is minimal interaction between the anionic reagent and the leather, *i.e.* cationic centres: these include both protonated basic groups and residual cationic character in the bound metal ions. In the case of applying a cationic agent, such as a fluorocarbon complex of chromium(III), the pH must lie below the isoelectric point.
3. Ideally, there should be a balance between the charge created by neutralisation, dependent on the isoelectric point, and the charge introduced by the reagent: the more balanced the reaction is, the less charged the leather is, so the ionic mechanism of failure is minimised.
4. The process must be prolonged, to allow neutralisation to take place down the hierarchy of fibre structure. The conventional neutralisation period is relatively short, because for most post-tanning processes neutralisation is less critical in this regard.

17.12.2.7 Temperature

The temperature at which the reagent is prepared can be important, whether the condition refers to emulsification or solubilisation. When making emulsions, it is critical that the solvent and the product should be heated, to reduce the viscosity of the hydrophobic compound, so that it can break up into as small particles as the chemistry allows. It is not possible to define an optimum temperature for this operation, because high temperature is an effective way of destabilising emulsions, so there is a practical upper limit to the temperature used, usually 60–70 °C. Destabilisation occurs when the temperature is high enough to make the particles collide with sufficient energy to overcome the charge repulsion, hence they coalesce. Also, elevated temperature can help the reagent to penetrate and, depending on its chemistry, to fix.

17.12.2.8   Neutral Electrolyte

The wettability of a leather is the starting point for water resistance: the easier it is to wet the fibres, the easier it is for water to be transmitted through the leather. The presence of neutral electrolyte makes the fibres easier to wet, because ions attract water into their solvation shell, providing nuclei for additional water fixation: the more water soluble the salt is, the worse is the effect on water resistance. Therefore, it is important to minimise the amount of salts in the leather, first by ensuring that there is minimum neutral salt in all post-tanning processing and then by washing prior to applying the water resistance treatment and subsequently washing, before drying the leather.

17.12.2.9   Surfactants

It is common to include surfactants of various types in leather-making processes, particularly in the earlier steps of beamhouse processing. Since their function is to facilitate wetting, their presence in leather is incompatible with water resistance. Cationic detergents are unusual and typically not used in the leather industry. Anionic detergents are the most common industrial and domestic agents, because they are the most effective wetting and cleaning agents. It is particularly important to exclude these agents from processing if water-resistant leather is the required product. Sulfates and sulfonates have high affinity for leather and they can remain within the structure long after they have been applied. If a wetting agent is needed in the beamhouse, non-ionic surfactants should be used. Although they are less effective at wetting and cleaning, they have little affinity for collagen and therefore do not persist in the leather.

17.12.2.10   Post-tanning

It is usual to apply a water resistance treatment at the end of the post-tanning sequence of operations, because of the likely interference that it would have on other processes if the fibre structure becomes hydrophobic. Hence it is important to recognise the influence of the other post-tanning reagents on the final performance of the leather as a water-resistant material:

- *Retanning:* Just as the primary tanning chemistry can influence water resistance, so the retanning can control it. If a hydrophilic retanning agent is used, it will confer wettability to the leather. A less hydrophilic retannage, such as with chromium(III), is more compatible with water resistance and this may be further controlled by appropriate masking.
- *Dyeing:* The choice of dyes can influence the wettability of leather, because their chemistries can vary, even within a group of the same or similar colour. Indeed, the concept of hydrophilic–hydrophobic balance (HHB) as a quantifiable parameter is already established for dyestuffs and used to control the degree of surface reaction or penetration. The presence of hydrophilic dyes in the leather will act as a wetting agent, allowing water to be transmitted through the leather.
- *Fatliquoring:* The extent of reliance upon lubricants other than the water resistance agent depends on the nature of the latter agent, since some can operate effectively as the lubricant. For others, it is necessary to include some conventional fatliquoring agents. Fatliquor formulations typically contain partially sulfated or sulfited (sulfonated) oil as the emulsifying agent for the neutral oil. Therefore, they act like surfactants, possessing hydrophilic and hydrophobic functionality, and consequently they also have the ability to wet surfaces. Therefore, despite the oily nature of the fatliquor, it can have the same effect as an anionic detergent treatment.

    The choice of fatliquor should be confined to the so-called 'water-resistant fatliquors': they typically have a weak effect on water resistance, but will offer the least adverse effect on a more powerful water resistance treatment.

17.12.2.11   Capping

Acrylic-based water resistance treatments include a final step called 'capping', in which the residual anionic character of the leather is neutralised by applying a cationic agent. If this step is omitted, the water resistance treatment will merely confer a weak effect, *e.g.* resistance to water penetration for a few minutes or 1 h on the Bally penetrometer. Effective capping can extend the period of resistance to many hours. It is possible to use a variety of cationic agents. Those that merely interact electrostatically, such as quaternary ammonium compounds, are not very effective. Similarly, metals that react electrostatically with the carboxyl groups of the acrylic polymer, *e.g.* Al(III), Ti(IV) and Zr(IV), are not optimally effective. The preferred reagent is chromium(III), which forms covalent complexes; this has less charge separation in the bonding, and thereby does not exhibit the equivalent effect of the presence of neutral electrolyte.

17.12.2.12 Solvent

Water resistance agents may be formulated with non aqueous, water-miscible solvent, as part of the manufacturing process or as an aid to emulsification. Its presence in leather can also facilitate wetting and the consequence may be observed as time-dependent water resistance development following crusting. That is, if the leather is tested immediately after crusting and conditioning, it may exhibit poor water resistance because the solvent remains in the leather. However, if the leather is stored for several days, for example in the conditioning room, further testing would show an improvement in water resistance, because the storage period allows residual solvent to evaporate.

17.12.3 Chemistries of Water Resistance Treatments

17.12.3.1 Vegetable-tanned Leather

At the time when leather was made predominantly using vegetable tanning, it was difficult to confer a high degree of water resistance, and filling the fibre structure of the leather with hydrophobic, greasy material was the only traditional option in this regard. For example, Tim Severin's successful experiment to reproduce the transatlantic voyage of Saint Brendan from Ireland to America in a leather boat was achieved by constructing the boat from oak-tanned hides, 'stuffed' for water resistance with wool grease.[17] However, it is clear that this approach to creating water resistance from a hydrophilic substrate offers limited wider applications.

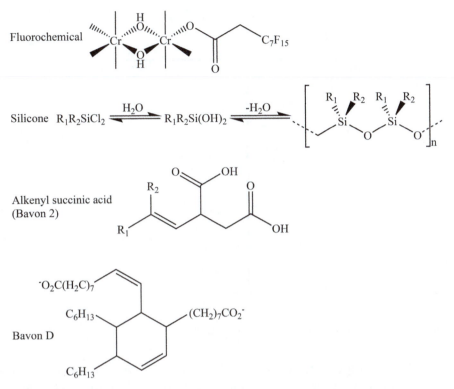

Figure 17.18 Chemistries of some water resistance treatments.

17.12.3.2 Chrome-tanned Leather

Chrome-tanned leather is naturally relatively hydrophobic, at least in comparison with untanned collagen. Hence it provides a good start for making water-resistant leather.

Compared with the simplistic, barrier approaches to water resistance of the past, what is needed in the modern industry is chemical treatments that do not adversely affect the handle and 'breathability' properties of the leather. Some of those are presented in brief in Figures 17.18 and 17.19, where it can be seen that they tend to rely on the presence of bound chromium(III) species, acting as a type of mordant for the hydrophobic agent.

The silicone reaction is readily understood: the chain of Si–O moieties provides many oxygen atoms that are capable of interacting electrostatically with collagen or leather. In this way, the fibre structure becomes coated with a surface of water-repellent alkyl groups. This

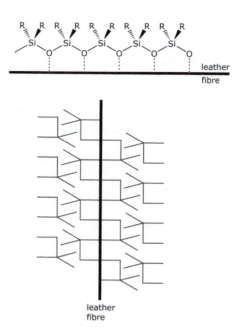

Figure 17.19   Modes of action of silicone and the Bavon process.

is an effective type of treatment, although it is more commonly used in conjunction with other chemistries, which facilitate the delivery of the silicone.

The Bavon process was used commercially for some time, but is the archetypal inconsistent process. The principle lies in a stepwise mechanism.

1. Reaction between the alkyl succinyl dicarboxylate and bound chromium(III).
2. Creation of a second layer of the reagent, when the alkyl groups interact in a hydrophobic bonding arrangement.
3. When the dry leather is wetted, a water-in-oil emulsion structure is formed at the carboxylate surface, preventing the passage of water through to the fibre surface.

At least, that is the theory! Sometimes it worked, but success could never be guaranteed. Hence for many years tanners closely guarded the details of processing for water-resistant leather.

The current industry standard for making water-resistant leather from chrome-tanned stock is partially esterified acrylic polymers,

Table 17.3 Glass transition temperatures (T_g) of some polymers.

Polymer	$T_g/°C$
Polydimethylsiloxane	−127
Polyethylene	−125
Polythiofluoromethylene	−118
Polytetrafluoroethylene	−113
Polyisobutylene	−73

exemplified by the Lubritan series of agents from Rohm and Haas. The process is simple, if the conditions set out above are adhered to. Here the mechanism involves complexing available carboxyl groups in the polymer with bound chromium(III), therefore presenting the surface of the fibre structure as covered with alkyl groups. These groups also act as spacers within the fibre structure, functioning like a lubricant. In this way, processing is made more compact (see later). The esterified acrylic polymer can be used alone or in combination with silicone chemistry, to provide two mechanisms of water resistance.

The role of silicones in this context is interesting, because there is added versatility due to the potential for introducing additional functionality into the molecule, such as amine, amide, carboxyl and epoxide groups, which can enhance the use of these reagents in retanning.[18] The reactivity of silicones is assisted by the unusually flexible nature of the molecule: the rotational energy for the Si–O bond is close to zero, compared with 14 and 20 kJ mol^{-1} for polyethylene and polytetrafluoroethylene, respectively. In addition, polydimethylsiloxane exhibits the lowest known glass transition temperature, making it the most flexible polymer known, illustrated in Table 17.3.

17.12.3.3 Non-chrome-tanned Leather

The alternative substrate for water resistance is leather tanned with reagents other than chromium(III), *i.e.* all the organic tannages, such as aldehyde and syntan, including vegetable. These tannages have the common feature of making the leather highly hydrophilic. This makes the task of introducing water resistance much more difficult that for chrome-tanned leather. The options available are as follows.

- Silicones, which can bind to the hydrophilic groups, but are not very effective as solo agents.

- Fluorocarboxylate complexes of chromium(III). The 3M company offers such a product. However, if the purpose of the tannage is to make mineral-free, specifically chrome-free, leather, this is counterproductive. Also, fluorochemicals are increasingly coming under pressure for their environmental impact.

The development of new methods of delivering reagents into leather (see Chapter 19) will open up the possibility of using highly hydrophobic reagents to confer high water resistance, even to leathers that start out as highly hydrophilic.

References

1. A. D. Covington, *et al.*, *Proceedings of IULTCS Congress*, Porto Alegre, Brazil, 1993.
2. A. D. Covington and K. T. W. Alexander, *J. Am. Leather Chem. Assoc.*, 1993, **88**(12), 241.
3. K. H. Sizeland, *et al.*, *J. Am. Leather Chem. Assoc.*, 2015, **110**(3), 66.
4. K. H. Sizeland, *et al.*, *J. Am. Leather Chem. Assoc.*, 2015, **110**(11), 355.
5. T. Waite, *Proceeding of Annual Conference*, Society of Leather Technologists and Chemists, 1986.
6. R. M. Koppenhoeffer, *J. Am. Leather Chem. Assoc.*, 1952, **47**(4), 280.
7. A. D. Covington and K. T. W. Alexander, *J. Am. Leather Chem. Assoc.*, 1993, **88**(12), 252.
8. P. S. Briggs, *J. Soc. Leather Trades Chem.*, 1968, **52**(8), 296.
9. P. S. Briggs, *J. Soc. Leather Trades Chem.*, 1969, **53**(8), 302.
10. K. Tang, *et al.*, *J. Soc. Leather Technol. Chem.*, 2009, **93**(6), 217.
11. A. G. Marriott, PhD thesis, University of Surrey, 1978.
12. *J. Am. Leather Chem. Assoc.*, 1995, **90**(4).
13. *J. Am. Leather Chem. Assoc.*, 1995, **90**(5).
14. SLP22 and IUP10, *Official Methods of Analysis*, Society of Leather Technologists and Chemists, UK, 1996.
15. Defence Standard, UK/SC5611, Issue 1, 1998.
16. R. Palop and A. Marsal, *J. Soc. Leather Technol. Chem.*, 2000, **84**(4), 62.
17. H. A. Birkin, *et al.*, *J. Soc. Leather Technol. Chem.*, 1978, **62**(3), 55.
18. D. Narula, *J. Am. Leather Chem. Assoc.*, 1995, **90**(3), 93.

18 Enzymology

18.1 Introduction

Despite enzymes having been known for over 100 years and their use in the leather industry extending much further back than that (albeit without the tanners' knowledge), it is only recently that their use has expanded outside simple bating processes. Bating (see Chapter 8) has a history of using digestive enzymes, but now that chemists, biochemists and biologists have improved their understanding of enzymes, and as biotechnology has improved, the range of available enzymes is becoming increasingly affordable and accessible on the industrial scale. They are efficient ways of reducing the energy use in a processing step while maintaining the specificity of the reaction and, in many cases, they reduce the level of pollution in waste water; for this reason, it is envisaged that the use of enzymes in the leather industry will continue to increase.

This chapter sets out to furnish a basic understanding of enzymology and deals with the comparatively new requirement that practitioners need a good understanding of the basics of enzymology, allowing them to understand common pitfalls and misconceptions and to pick their way through the terminology.

18.2 Kinetics

An enzyme is a biological catalyst; it increases the rate of a reaction by providing an alternative reaction pathway with a lower activation energy than that of the uncatalysed route, without itself undergoing a permanent change in the process.

Tanning Chemistry: The Science of Leather 2nd edition
By Anthony D. Covington and William R. Wise
© Anthony D. Covington and William R. Wise 2020
Published by the Royal Society of Chemistry, www.rsc.org

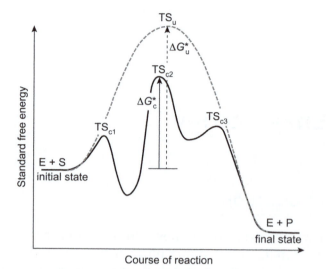

Figure 18.1 Energy profile diagram of a general exothermic reaction; the upper line is the uncatalysed reaction, the lower line is the catalysed route. Courtesy of Martin Chaplin.

All reactions (whether they are endo- or exothermic) that conform to classical mechanics must surmount an activation energy barrier when transitioning from reactants to products (shown in Figure 18.1). The difference in energy between the reactants and the highest point on the graph denotes the activation energy (also called the energy barrier – denoted ΔG) and represents the energy required to 'push' the reaction through an inherently unstable 'transition state' (TS). This is known as transition-state theory (TST).

It is the height of the activation barrier that determines the rate of reaction: this is due to TST treating the absolute rate of reaction in terms of a hypothetical thermodynamic equilibrium of the ground state with the transition state and the partitioning of this hypothetical transition-state species between the two ground-state structures. From this treatment, the absolute rate of a reaction is related to the hypothetical concentration of the transition state and can be readily calculated by applying equilibrium thermodynamics.

An analogy with the activation energy would be rolling a ball down a hill: it does not matter how long the downhill on the other side is, energy must be put in to get the ball to the very top of the hill first, and if insufficient energy is put in to get to the top of the hill, the ball will simply roll back to where it started. This activation energy

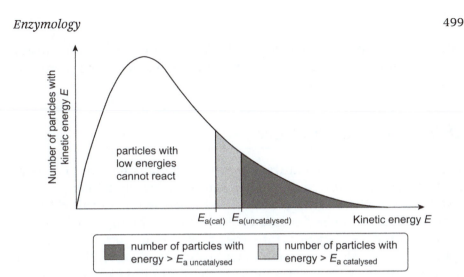

Figure 18.2 Maxwell–Boltzmann distribution curve, depicting both a catalysed reaction and an uncatalysed reaction.

also explains why flammables at room temperature, *e.g.* natural gas, do not spontaneously combust: they require a source of ignition first, and this source of ignition provides the energy required to get the reaction over the energy barrier.

In classical theory, an enzyme-catalysed reaction still has an activation energy and still has transition states; however, importantly, it provides a different reaction pathway, with a different transition state that is lower in energy. Enzymatic reactions normally have multiple transition states (as depicted by the lower line in Figure 18.1), but it is always the one highest in energy that dictates the activation energy. Relating this back to rolling a ball down a hill – in this case the ball will still start and finish in the same place as the uncatalysed route (upper line in Figure 18.1), but it is key that the hill that it has to go over is now lower and requires less energy. In order to explain why reducing the energy required to surmount an energy barrier increases the rate of reaction, a Maxwell–Boltzmann distribution of particles must be considered (Figure 18.2).

In Figure 18.2, the *y*-axis represents the number of molecules/particles with a specified kinetic energy (*x*-axis). It should be noted that in some versions of this graph the *x*-axis is also labelled as velocity. As the activation energy of a reaction is decreased, depicted by the labelled vertical line moving to the left, a greater number of reactants have the requisite energy to react. The increase in the number particles that are above the required energy provides an apparent increase in the rate of reaction. The Maxwell–Boltzmann distribution curves are also often used to depict why an increase in reaction temperature results in an increase in rate of reaction.

18.3   Enzyme Structure

There are thousands of natural enzymes and many remain undiscovered or have unresolved structures; however, nearly all enzymes fit a standard model. In this model, all active enzymes, so-called 'holo-enzymes', are composed of two or more parts. The most substantial component of the enzyme is (are) the proteinaceous part(s), called the apoenzyme, which in the quaternary structure can often be made of multiple subunits. These subunits are then combined with a cofactor that can fit into one of three subgroups.

1. A coenzyme: a non-protein organic substance that is loosely attached to the protein.
2. A prosthetic group: an organic substance that is firmly attached to the protein.
3. A metal ion: including (but not limited to) K^+, Fe^{2+}, Fe^{3+}, Cu^{2+}, Co^{2+}, Zn^{2+}, Mn^{2+}, Mg^{2+}, Ca^{2+} and Mo^{3+}.

Only when the apoenzyme and the cofactor are combined can the enzyme be deemed to be active (holoenzyme).

18.4   Mechanisms

It is also important for users of enzymes not only to understand that the enzymes lower the energy barrier and how that affects the rate of reaction, but also to understand the basic mechanism of how this might be achieved. The additional understanding will allow extrapolations of results obtained in practice and begin to allow users to formulate how reaction conditions have affected the enzymes and thus predict the outcome.

18.4.1   Lock and Key

Enzymes are highly specific molecules that catalyse only certain reactions;[1] such specificity might come in the form of absolute specificity (*i.e.* catalyse only one reaction), group specificity, linkage specificity

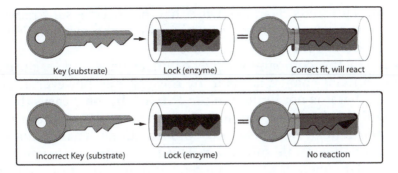

Figure 18.3 The lock and key analogy. © Shutterstock/Aldona Griskeviciene.

or even stereospecificity. This specificity comes from the shape of the active site and the groups that are in this active pocket; however, it gives rise to the first conceptual mechanism discussed, the lock and key. This concept uses an analogous situation to help to explain and understand the specificity of enzymes and later on helps in understanding enzyme inhibition (see Section 18.6.3). A lock (enzyme) can be opened only if the correct key (substrate) is inserted (see Figure 18.3). It also helps the reader to visualise the three main stages of an enzyme mechanism: separate lock and key, lock and key complex and finally the open lock.

Continuing with the lock and key analogy, it is also important to address at a molecular level what gives the lock its shape.

18.4.2 Induced Fit

In 1958, a modification to the lock and key model was suggested:[2] while the lock and key theory is good at explaining the specificity of the enzyme, it is less good at explaining how the enzyme provides the reduction in energy over an uncatalysed reaction (see Figures 18.1 and 18.2). The only information it provides are the groups, the amino acids that make up the active site, which might lead to a hypothesis on binding and reaction mechanisms.

The induced fit model extends the understanding gained from the lock and key theory. The globular structure of enzymes provides a degree of flexibility and allows the structure to change shape as it interacts with the substrate. As a result, the substrate is unlikely to bind to a fixed active site, but instead will experience the conformation continuously changing, until the substrate is completely bound and the enzyme–substrate complex is formed.[3] Although the extent to which a shape change is observed changes for each enzyme, the role of the mechanism is clear: to ensure that the enzyme–substrate complex is in the prime conformation to facilitate the reaction with the lowest activation energy.

Owing to the comparative size of an enzyme, the majority of conformational change is taken by the enzyme; however, there is evidence that it is not limited to the enzyme: in some cases the substrate will also flex to accommodate a lower energy transition state.[4] It is important to note that this conformational change does not happen for all possible substrates. This may help explain why seemingly similar substrates have markedly different reaction rates with the same enzyme; structurally, the substrate may limit conformational change and thus substantially increase the activation energy required for the reaction.

The possibility of conformational change within both the enzyme and the substrate may also help to explain why enzymes typically appear to have several transition states on the energy profile diagram. The uncatalysed reaction might involve the transfer of an electron in a single, high-energy step. However, for an enzyme, in addition to the electron transfer step, it will also have a transition state where the enzyme and substrate bind, followed by multiple transition states where the complex might undergo conformational change.

18.4.3 Quantum Mechanical Tunnelling

There is another reaction mechanism that is so interesting that it deserves a special mention. Originally considered to be a rare mechanism, it is now increasingly being associated with enzyme-catalysed reactions, particularly those that involve electron transfer. To understand it, the tanner must enter the world of quantum mechanics. This chapter will do this in a light-touch manner, as even the great scientist Niels Bohr once said:

> ... *those who are not shocked when they first come across quantum theory cannot possibly have understood it.*

The discussion of enzyme mechanism so far has centred on TST, where a reaction must surmount an energy barrier (a hill) as the reaction progresses through the high-energy transition state (the highest point on the hill) to reach the products (the other side of the hill). Although TST has been used to picture enzyme-catalysed reactions over the last 50 years,[5] recent developments imply that this 'textbook' illustration could be, at least in some circumstances, fundamentally flawed.[6] The problem is that TST considers only the particle-like properties of matter; however, matter (especially those particles with smaller mass, *e.g.* electrons and hydrogen atoms) can also be considered as having wave-like properties: this is known as the wave–particle duality of matter. Hence an alternative picture to TST has emerged for some enzyme-catalysed reactions involving electron or hydrogen transfer.

One important feature of the wave-like properties of matter is that it can pass through regions that would be inaccessible if it were treated as a particle, *i.e.* the wave-like properties mean that matter can pass through regions where there is zero probability of finding it (Figure 18.4). As the name of the theory suggests, the analogy is that instead of climbing a 'hill' one simply 'tunnels' through it.

Supporting this theory is the fact that, because TST is derived from the Arrhenius equation, it has long been known that it is a temperature-dependent process where, as temperature increases, so does the rate of reaction.[7] In contrast, with only a few exceptions, quantum tunnelling has been considered to be significantly less temperature dependent.

It is possible to calculate the tunnelling rate for a particle with a known mass, as there is a directly proportional relationship between the tunnelling rate (k), reduced Planck's constant ($\hbar$), the width of a

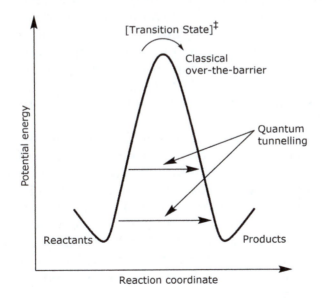

Figure 18.4   Reaction profile graph showing quantum tunnelling.

barrier (l), the height of the barrier (V) and the mass of a particle (m) with a kinetic energy (E), as shown in the following tunnelling rate expression (given a rectangular potential energy barrier):[7–9]

$$k \propto \exp\left[\frac{-2l\sqrt{2m(V-E)}}{\hbar}\right]$$

What this equation demonstrates is that there is an exponential decrease in the tunnelling rate for a particle of a constant mass, if either the width or height of the barrier is altered. This in turn helps to provide an insight into why each enzyme has a unique shape and such high specificity. As an example, consider that proteins are electrical insulators; nevertheless, it is documented that electrons can travel large distances (up to ~3×10^{-9} m or 30 Å) through proteins when being transferred from a donor to an acceptor redox centre.[10] In contrast, with electron transfer *via* quantum tunnelling, the quantum-mechanical transfer of other particles in enzymes has not been documented extensively. It was the ground-breaking work of Bahnson and Klinman in 1995 that provided the first experimental indication of hydrogen tunnelling in enzyme molecules.[10,11]

As the above rate equation shows, the probability of tunnelling decreases with increasing mass of the particle, which reduces significantly the probability of hydrogen tunnelling compared with electron tunnelling (the mass of the hydrogen nucleus is ~1840 times greater than that of the electron). Using the same expression for proton tunnelling gives a transfer distance of only 0.58 Å. This distance is similar to the length of a reaction coordinate and is thus suggestive of a high tunnelling probability. The larger masses of deuterium and tritium lead to corresponding transfer distances of 0.41 and 0.34 Å, respectively, thus making kinetic isotope effect (KIE) studies attractive for the detection of hydrogen tunnelling in enzymes.[6,10,12]

The KIE in proton transfer reactions is present in all cases when the rate of proton transfer is not the same as the rate of deuterium transfer, *i.e.* $k_H > k_D$. Considering only classical over-the-barrier or transition-state theory, the k_H/k_D ratio is dependent on both the temperature of the reaction and the size of the activation barrier that the reaction has to surmount.[13] The difference in activation energies is only slight and occurs because the C–D/C–T bonds are shorter than C–H, resulting in their extra strength and ultimately leading to a higher energy required to break the C–D/C–T bond compared with the C–H bond (Figure 18.5).[6]

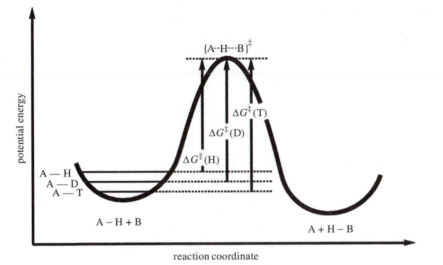

Figure 18.5 Schematic representation of the energetics for the enzyme-catalysed reaction A–H + B → A + H–B.

If the reaction proceeds *via* a quantum mechanism, then the increase in mass from hydrogen to deuterium substantially reduces the probability of a quantum mechanism. The subsequent increase in energy outstrips the expected increase in energy that would be observed by the increase in bond energies.

18.4.4   Michaelis–Menten

In enzyme catalysis, there are two equations that define the process:

$$E + S \underset{k_{-1}}{\overset{k_1}{\rightleftharpoons}} ES$$

$$ES \xrightarrow{k_2} E + P$$

where E is the enzyme, S is the substrate or reactant, ES is the intermediate enzyme–substrate complex (the Michaelis–Menten complex) and P is the product.[14] The resultant rate equations are as follows:

$$-\frac{d[S]}{dt} = \frac{d[P]}{dt}$$
$$= k_1[E][S] - k_{-1}[ES] = k_2[ES] \tag{18.1}$$

$$\frac{d[ES]}{dt} = k_1[E][S] - k_{-1}[ES] - k_2[ES] \tag{18.2}$$

The concentration of the complex reaches a steady state when

$$\frac{d[ES]}{dt} = 0$$

Applying this condition to eqn (18.2):

$$[ES] = \frac{k_1[E][S]}{k_1 + k_2} \tag{18.3}$$

The initial conditions are as follows:

$$[E]_0 = [E] + [ES] \tag{18.4}$$

Substituting [E] in eqn (18.4) into eqn (18.3):

$$[ES] = \frac{k_1[E]_0[S]}{k_1 + k_2 + k_1[S]} \tag{18.5}$$

From eqn (18.1), the rate of formation of the product becomes

$$\frac{d[P]}{dt} = \frac{k_2 k_1[E]_0[S]}{k_1 + k_2 + k_1[S]} \tag{18.6}$$

The combination of rate constants is called the Michaelis–Menten constant, K_m:

$$K_m = \frac{k_1 + k_2}{k_1} \tag{18.7}$$

$$-\frac{d[S]}{dt} = \frac{k_2[E]_0[S]}{K_m + [S]} \tag{18.8}$$

If the substrate concentration is high, the rate equation becomes

$$v = k_2[E]_0 \tag{18.9}$$

If the enzyme concentration is low, the rate equation becomes

$$v = \frac{k_2[E]_0[S]}{K_m} \tag{18.10}$$

or

$$v = \frac{k_1 k_2[E]_0[S]}{k_1 + k_2} \tag{18.11}$$

Hence the kinetics are first order in substrate concentration. Alternatively, if $[S] \gg K_m$, then the kinetics are zero order.

If k_2 is much larger than k_{-1}:

$$v = k_1[E]_0[S] \tag{18.12}$$

then the kinetics refer to the formation of the enzyme–substrate complex.

If $k-_1$ is much larger than k_2:

$$v = \frac{k_1 k_2 [E]_0 [S]}{k_{-1}} \tag{18.13}$$

then the activation energy calculated from a plot of log v against $1/T$ will refer to the activation energies for the two reactions set out at the beginning of this argument:

$$E = E_1 + E_2 - E_{-1}$$

From eqn (18.9), the rate of the enzyme-catalysed reaction is a function of the substrate concentration and the Michaelis–Menten constant: the lower the value of K_m, the greater is the affinity between the enzyme and the substrate. Hence measuring the kinetics of the reaction provides information regarding the way in which the reaction is conducted. An example of such data is presented in Chapter 13, concerning enzyme-catalysed polyphenol polymerisation.

The traditional view of the mechanism of enzyme catalysis is expressed as the 'lock and key' theory. The reaction site of the enzyme is designed to receive the specific chemical structure of the transition state of the intermediate between reactant/substrate and the product, as a key is shaped to fit into a lock. However, there are two ways of looking at the model. The mechanism of enzyme catalysis has been argued in terms of the way in which the reactant is converted *via* a transition state to the product. This has been expressed as stabilisation of the transition state by the electric field of the active site of the enzyme relative to the reactant, *i.e.* the enzyme is altered to accommodate the reactant or the enzyme has a greater affinity for the transition state than for the reactants. Alternatively, the reactant may be arranged by the enzyme in the active site, leading to the transition state. The mechanisms are summarised in Figure 18.6, relating the free energies of the interaction or binding between the enzyme and the Michaelis–Menten complex (MC) and the transition state (TS).[15]

Martí *et al.*[15] argued that these apparently contradictory views can be reconciled in a single mechanism. The enzyme favours progress in the reaction, because it has a reaction site that is preorganised to

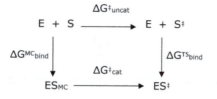

Figure 18.6 Relationship between the roles of the Michaelis–Menten complex and the transition state in enzyme catalysis.

accept the substrate. The enzyme is likely to have a predisposition to interact with those substrate conformations that resemble the transition state more closely than others, thereby incurring a lesser energy penalty because deformation of the substrate and/or the enzyme's reaction site is not required.

18.5 Active Site

The active site is the region of an enzyme where the substrate molecule will bind to the enzyme and undergo a specified chemical reaction. The active site consists of at least two parts: the catalytic site, where the amino acid residues facilitate the reaction, and the binding site, where the amino acid residues form temporary bonds with the substrate during the reaction. Both parts of the active site and the remaining quaternary structure of the enzyme help to provide the active site with a specific shape and enhance the reaction specificity. It is normal for the active site to be small in comparison with the number of residues that make up the whole enzyme.[16]

The interactions at the binding component of the active site are familiar to any chemist and can they involve electrostatic interactions, hydrogen bonding, van der Waals forces, hydrophobic interactions or any combination thereof. The interactions vary in energy, but can be considered generally weak, to ensure easy release of the product once the catalytic reaction has taken place. It is for this reason that 'strong' covalent bonding between the substrate and the enzyme active site does not occur.

The residues in the active site responsible for facilitating the catalytic reaction are normally close to the residues responsible for the binding process. This restricts the number of possible conformations that the substrate can have, which would move it too far away from the relevant area for the reaction to take place. The mechanism by which each enzyme catalyses a reaction varies between enzymes and is dependent on the reaction they are catalysing, the substrate and the amino acid residues in the active site.

18.6 Factors Affecting Enzyme Catalysis

As with any reaction, there are some common factors that affect the rate of reaction for enzyme-catalysed reactions, but not always in the way that might initially be expected. To be able to exploit the potential of enzymes, it is important to understand the most important factors that affect their productivity.

As has previously been discussed in this chapter, enzymes are complex proteins with a very specific globular shape, and as a result they are affected by changes in the environment in the same way that other proteins might be, *e.g.* collagen. In all the mechanisms detailed in Section 18.3, shape is crucial to the function of the enzyme, hence any factor that changes the shape will affect the efficacy of that enzyme.

18.6.1 Temperature

As with all reactions, temperature has a profound effect on the rate of an enzyme-catalysed reaction. It is for this reason that cooling skins is an effective short-term means of preservation: it reduces the activity of intra- and extracellular enzymes to a point where the hides can be transported and used in a reasonable timeframe.

As might be expected, increases in temperature generally increase the rate of an enzyme-catalysed reaction: a 10 °C rise in temperature might increase the rate of reaction by as much as 50–100%. There is, however, an optimal temperature for enzyme function: for the majority of enzymes derived from mammalian sources this is approximately 37 °C. Above this temperature, the enzyme can become denatured, changing the shape to a point where the enzyme becomes completely inactive. This is an irreversible process and as a result the observed reaction rate decreases significantly and will never recover unless additional 'fresh' enzyme is added.

An analogous situation arises with collagen, which undergoes something similar at (normally) higher temperature, which in industry is often called 'shrinkage'. The effect is the same in both proteins: a fundamental and irreversible change in the shape of the protein that prevents it from functioning.

18.6.2 pH

Changes in pH are utilised throughout the leather manufacturing processes (see Chapters 5–14). However, this has never been more crucial than when enzymes are employed. The pH range that an enzyme can tolerate is different for each enzyme and that range can be wide. However, it is important to recognise that each specific enzyme has a well-defined, normally narrow pH range in which it functions best. Outside the working pH range of the enzyme, two things are possible. The most catastrophic is irreversible denaturation of the enzyme: this can be due to the breakup of internal peptide bonds or disruption of the numerous intrachain links, *e.g.* hydrogen bonds that give the unique shape of the enzyme. Such alterations to the shape of the enzyme alter the active site and prevent the substrate from fitting and curtailing all catalytic activity. Tanners are familiar with the analogous situation in which altering the pH of a liquor to the extremes will cause collagen to swell and ultimately denature.

Alternatively, mild changes in pH alter the crucial interactions in the active site. This can prevent the substrate from binding or prevent the reaction, as the efficacy of processes such as proton or electron transfer are heavily reliant on protonation of residues within the enzyme structure. Importantly, should a reaction narrowly stray outside the optimal pH range then the situation can be easily corrected, as it does not cause irreversible damage to the enzyme.

The narrow pH range over which enzymes function is of particular importance to tanners when they consider the bating enzymes they might use. There is increasing pressure on tanneries to move away from ammonium-based deliming salts to improve their effluent discharge. There are other technologies that can be used, indeed any acidic compound could be employed to lower the pH, each with pros and cons. One of the higher profile technological candidates is carbon dioxide deliming. The premise for CO_2 deliming is that as the CO_2 dissolves into the aqueous liquor it reacts to form carbonic acid, which in turn deprotonates and reduces the pH. One of the reasons why this technology is increasing in popularity is that, as with ammonium-based deliming processes, the point at which the buffer forms is in a pH range that is not detrimental to the integrity of the collagen. However, although the pH of the buffer will not damage the pelt, the two technologies crucially buffer at different pHs; ammonium salts buffer at pH 8–9 and CO_2 buffers at pH 5–6. The result is that it is necessary for the following step, bating, to use different enzymes that reflect the difference in pH.

18.6.3 Inhibition

In addition to temperature and pH, there is another important factor that can affect enzyme activity, namely the presence of an inhibitor. In systems as complex as those inside a processing drum, the presence of an inhibitor is both difficult to identify and difficult to control. If the other conditions within the system (namely pH and temperature) are optimal, but the reaction is still not proceeding at the expected rate, there are two likely possibilities: either the enzyme has lost its activity before being added to the system or it is losing its activity once added. Owing to the complexity of inhibition, this section will address only two mechanisms by which inhibition can occur.

Inhibition, as the name suggests, is the process by which a molecule binds to an enzyme and reduces the observed catalysed rate of a reaction. The two most commonly observed mechanisms of inhibition are competitive and non-competitive, although it should be noted that there are alternative mechanisms that have been proposed in the literature. Although the two inhibition mechanisms have the same effect, the mechanisms for both processes are quite different.

Consider the lock and key mechanism for enzyme catalysis. Competitive inhibition (Figure 18.7) involves an inhibiting molecule that has a similar shape to the substrate and an affinity for the active site; however, unlike the substrate, it cannot undergo a reaction. As the inhibitor and the substrate cannot bind to the enzyme at the same time, they 'compete' for the active site and the apparent rate of reaction decreases. The decrease in reaction rate is dependent on the

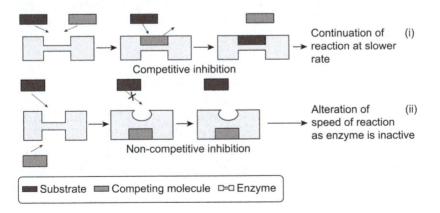

Figure 18.7 Mechanisms of enzyme inhibition: competitive and non-competitive. Adapted from ref. 17 with permission from Elsevier, Copyright 2009.

relative concentrations of the enzyme, inhibitor and substrate. Relating this to the analogous 'lock and key', the inhibitor is a similar key to the substrate; however, although it 'fits in the lock', it will not turn and open the lock, so until this inhibiting key is removed the real key cannot be put into the lock.

Non-competitive enzyme inhibition employs a different mechanism. In this case, the inhibitor does not have any affinity with the active site of the enzyme; instead, it binds to a different location of the enzyme and in doing so it changes the shape of the active site so that the substrate no longer reacts. Owing to the difference in mechanism, the decrease in reaction rate is not dependent on the concentration of the substrate and is affected only by the enzyme and inhibitor concentrations. Relating this to the 'lock and key' model, the inhibitor does the equivalent of 'changing the locks', so that the key no longer fits.

18.7 Enzyme Assays

There are several factors that affect the performance of an enzyme, so how does the user know what will produce the optimal reaction conditions? This is where enzyme assays play a role. The concept of an enzyme assay is that it quantifies the maximum velocity of an enzyme-catalysed reaction with respect to a change in a variable. The information allows the user to determine the most active enzyme for the reaction and under which conditions it performs best. Often enzyme assays are conducted during the research and development phase of the enzyme before it is released by the manufacturer; however, in some cases there are also easy ways of obtaining similar data in a tannery or laboratory that might be of help in exploring conditions outside the 'norm'.

18.7.1 Enzyme Assays for Bating Activity

The analysis is based on enzymes attacking a solid, insoluble protein substrate that is dyed with a reactive, covalently bound dye: the rate of reaction is measured by the rate at which the substrate is solubilised, the rate at which colour goes into solution. Some commonly used substrates are as follows.

- *Undyed casein:* The Lohlein–Volhard test, conducted at pH 8.2, for protease activity.

- *Azo-albumin:* Blue substrate made from bovine serum albumin, for general protease activity.
- *Undyed hide powder, for collagenase activity:* Since the preparation of standard hide powder includes liming, the collagen is not pristine and hence the assay is not specific.
- *Hide Powder Azure, denatured or undenatured:* Blue acid-dyed collagen, for general protease activity or collagenase activity. The substrate may give problems of too high background colour in solution. Denaturing is achieved if the substrate is washed hot, to remove excess or loosely bound dye. An alternative or additional washing treatment is with a hydrogen bond-breaking solvent, such as dimethyl sulfoxide, which also causes denaturation of the collagen. Even if the Hide Powder Azure is allegedly undenatured, the argument about the specificity of enzyme action on non-pristine collagen also applies.
- *Hide Powder Black, denatured or undenatured:* Collagen dyed black with reactive dye having two sites for binding, so gives less colour in controls. It is used for general protease activity or collagenase activity.
- *Azocoll:* Blue substrate for general protease activity.
- *Elastin Red:* Red substrate for elastase activity.
- *Keratin Azure:* Blue substrate for keratinase activity.

The principles are as follows.

1. Prepare a solution at the pH for testing (test-tube size), preferably using a buffer.
2. Add the substrate: coloured material makes the test more sensitive. The amount is theoretically unimportant, because it is not in solution and will not be completely used up. However, in practice, because dye may be released by the control, it is preferable to use consistent amounts of the substrate.
3. Agitate the solution at the temperature for testing for a specified length of time: this can be determined experimentally, but typically it might be 1 h.
4. At the end of the specified time, rapidly filter the solution, preferably through inert glass-wool, to remove residual substrate.
5. The solution contains the products of the degradation reaction. Measure the amount, either with a spectrophotometer or with some other device for measuring the colour or optical density.

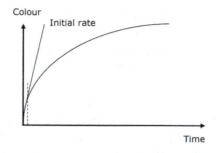

Figure 18.8   Typical trace of enzyme kinetics.

The rate of reaction is relatively constant at the initial rate, when it can be assumed that the kinetics are zero order, as shown in Figure 18.8, where the dashed line indicates a suitable time for the assay to run. In this way, the relative activities of enzymes can be quantified.

The method can also be used to detect the presence of enzymes. For example, using Elastin Red and typical conditions, such as 35 °C and pH 8.5 for 1 h, the presence of a red colour in the filtered solution demonstrates the presence of elastase in the bate, depending on how much colour is generated in the control. It should be noted that the status of the protein can lead to false-positive results in the assay for collagenase: denaturation or degradation of pristine collagen is difficult to avoid, so it is necessary to view the effects of proteases on (allegedly) intact collagen with some scepticism.

All these assays can be automated: the solution can be circulated through the cuvette of a spectrophotometer and the optical density can be recorded continuously. All that is required is to isolate the suspended substrate from the cuvette, which could be accomplished with a sintered-glass barrier. If required, the quantitative kinetics of the reaction can be calculated.

18.8   Enzymology of Leather Making: Current, Potential and Future

Although the main bulk of this chapter has been a 'light-touch' introduction to enzymology, thought must be given as to where enzymes are, could be and should be used in the leather industry. The use of enzymes in the leather industry will increase as advances in the technology continue, so thought should be given to the benefits that might be gained by wider usage. The matter of where the enzymes are currently, could be or should be used is often a matter of personal belief based on the perceived benefits in a given situation. Some concepts and ideas with a basis in the literature are presented here in a short 'list'-style format.

18.8.1   Enzymes in Soaking

Although the use of proteolytic enzymes in the soaking processes is now considered common, there is much scope for additional biotechnology.

- *General proteases:* For initiation of opening up, to attack the non-structural proteins. The use of proteolytic formulations (often the same as bates) during a long soak could eliminate the need for a subsequent bating step.
- *Lipase:* To attack triglyceride grease, thus aiding rehydration. Supplied for the purpose as a grease removal agent for epidermal surfaces: this is of doubtful value, since epidermal grease is sebaceous, although it does contain some triglyceride grease.
- *Chondroitinase A:* To attack hyaluronic acid, a necessary reaction because its participation in chrome tanning is likely to have a consequent effect on stiffening the fibre structure. In addition, the move away from salt-preserved rawstock necessitates an easy way of removing hyaluronic acid without adding salt to the soak or soaking for extended periods of time.
- *Cellulase, hemicellulase (xylanase), ligninase:* To degrade the main components of dung. A clean hair surface allows green fleshing (higher yield of better quality tallow), allows hair shaving (lower environmental impact and hair save effect from hair burning chemistry) and allows green splitting (higher value by-product split).
- *Amidase:* There is evidence that for chrome tanning, the asparagine should be converted to aspartic acid, using asparaginase. The alternative is to use a general amidase, which will include glutaminase.
- *Amylase:* Sold as a soaking enzyme, but not an obvious enzyme for the purpose, owing to a lack of starch in skin (starch is a plant polysaccharide). However, it is conceivable that the right amylase could generically attack other saccharide linkages.

18.8.2   Enzymes in Unhairing

This is a logical target and some technologies are commercially available, but these are still evolving.

- *Proteases:* In hair burning used to accelerate degraded keratin of the cortex and epidermis and in hair saving used to improve opening up, to facilitate the penetration of unhairing agents through the cross-section of hide or skin. They may also play a useful role in hair burning by attacking the medulla structure in hair.

  Under typical hair burning conditions, the alkalinity of the system means that only some enzymes have sufficient activity to be useful, so-called alkali-stable enzymes; they usually have the added property of elastolytic activity. This is a form of enzyme-assisted chemical unhairing process. Conversely, chemical-assisted enzyme unhairing uses chemicals to degrade the epidermis, to allow enzymes to penetrate to the follicle base from the hair side; this is speculative and has not been demonstrated on an industrial scale.

  Natural, endogenous proteases can effect hair release by a 'sweating' process, which is no longer widely practiced. They are applied only when grain damage is unimportant, *e.g.* for merino or merino cross pelts. The consequence of grain damage by enzyme unhairing in general is the reason why it is no longer a commercial option.

- *Dispase:* Used to target type IV collagen in the basement membrane of the grain surface and the follicle base for hair saving unhairing. It also has elastolytic activity.

- *Keratinase*: Not currently a commercial process, allegedly because of fears of attack on the epidermal layer in operatives. The targets are the epidermis and hair cortex, but probably not the more resistant cuticle.

  A formulation of keratinase, elastase and protease can be created by *Bacillus* fermentation and may be used to clean the grain by degrading residual hair and gain area.

Although not yet proven technologies, there are some other potential areas for exploration:

- medulla specificity;
- prekeratinised zone;
- inner layers of follicles, to ease hair removal.

18.8.3 Enzymes in Liming

For obvious reasons, this processing step is often closely allied to that of unhairing and it might be possible to combine the enzymes for both steps (depending on the processing method being used), but the role of the enzymes would be different.

- *Proteases:* Used to accelerate opening up, targeting non-structural protein, possibly allowing the user to reduce the concentration of unhairing chemicals. Typically, proteases are received as mixtures, so they tend to produce non-specific, mixed reaction(s) that need to be carefully observed, to avoid unwanted damage to the pelt.
- *Elastase:* Used to enhance area yield, especially in upholstery leather production. Problems of looseness may be encountered, particularly because elastase is rarely found in isolation and is traditionally mixed with protease.
- *Chondroitinase B:* The necessity for removing dermatan sulfate is not clear, since it is part of the structural makeup of the fibrils. Perversely, it may be desirable to leave it in the pelt as it functions *in vivo* as a hydrating mechanism.

Although not yet proven technologies, there are some other potential areas for exploration:

- epidermis specificity;
- collagen type VII;
- reticulin, linking fibril bundles;
- chondroitin C;
- decorin;
- collagen type III, the grain enamel, for nubuck.

18.8.4 Enzymes in Deliming

This might seem strange, as deliming is normally about simply adjusting the pH in a manner that is easy to control and not detrimental to the pelt. However, with the rise of CO_2 deliming there is an obvious target for enzyme catalysis.

- *Carbonic anhydrase:* To accelerate the hydration of dissolved carbon dioxide and thus accelerate the acidification reaction. The proof of concept has been demonstrated; remaining problems are the source of the enzyme and its robustness under the violent conditions of drum processing.

18.8.5   Enzymes in Bating

Enzymes were used for this particular leather processing step before technologists even knew that enzymes existed. However, for this processing step there are advances that could be made, which may facilitate developments elsewhere in the industry.

- *Proteases:* Traditionally used to attack non-structural protein, but typically not specific. In practice, conditions (pH, temperature and particularly time) allow reaction only within grain and flesh layers, not through the cross-section.
- *Lysyl oxidase:* To mimic the formation of natural covalent cross-links. Speculatively, this reaction may be used to enhance the resistance to hydrolytic damage.
- *Transglutaminase:* Already used in other industries to provide stability to protein, in terms of the physical properties of protein products, by reacting at the glutamine sidechain. The benefits to the leather industry have not yet been demonstrated, except for dyeing (see later). It has been shown that stability increases, but not T_s.

Although not yet proven technologies, there are some other potential areas for exploration.

- proteolytic specificity at different pH values, particularly in the acid range, to improve options for alternative deliming systems;
- reticulin: collagen type III.

18.8.6   Enzymes in Pickling

There are enzymes that can be used in the pickling process, but they are essentially bating enzymes that are stable under acidic conditions and consist primarily of protease: acid-acting proteases could be useful for opening up and making differently sourced material more consistent, and acid bates are currently commercially available. Acid bates have also been applied to wet blue with claimed success, although such claims are possibly exaggerated (see later).

18.8.7 Enzymes in Degreasing

Degreasing fatty skins, such as sheepskins, can be problematic and becomes an obvious target for enzymatic action.

- *Proteases:* Could be used for degreasing woolskins where the protease would attack the protein component of the cell walls of the lipocytes. Although trialled, to date this has not been successful owing to collateral damage to collagen in the form of looseness.
- *Phospholipase:* An alternative approach to degreasing, targeting the phospholipid component of lipocyte cell walls and thus does not attack the collagen (as above). This technology shows some promise and there are possibilities for obtaining industrial-size quantities by growing an appropriate microorganism on vegetation.

Although not yet proven technologies, there are some other potential areas for exploration.

- Is it possible to use acid-stable enzymes to mimic all alkaline processes, circumventing the need for alkaline processing and avoiding risks associated with swelling and waste streams?

18.8.8 Enzymes in Wet Blue Production

- *Proteases:* This approach has often been researched in an effort to achieve an additional bating effect, but all attempts failed. Unsurprisingly, this was likely due to the resistance of tanned collagen to enzyme attack, which is part of the definition of tanning.
- *Elastase:* This is now commonplace for certain types of leather and can be applied to wet blue in combination with protease. In this case, the protease may actually be useful as it will contribute to the breaking down of elastin once it has been partially degraded by elastase. Chromium-tanned collagen is unaffected by the protease component, but elastin is much less modified by chromium(III) chemistry, and so remains vulnerable to elastase degradation. The effect is to increase the area yield, but because the fibre structure of the corium is set by the tannage there is

no accompanying loosening effect. Other tannages do not allow the effect to occur, *e.g.* vegetable tannins interact hydrophobically and so tan both collagen and elastin.

- *Degreasing:* By analogy with the area gain strategy of elastolytic action on wet blue, grease may be removed from woolskins by applying a mixture of protease and lipase to chrome-tanned leather. The function of the protease is to break open the lipocytes; the function of the lipase is to hydrolyse and self-emulsify the triglyceride grease. Elevated temperature can be safely used to mobilise the grease above its melting point of 42 °C.

18.8.9   Enzymes in Other Tanning Processes

Although the majority of leather produced today is chromium tanned, other tannages are still prevalent; however, as the chemistries of the tanning processes are different, so must the enzymatic action also be different.

- *Vegetable tanning:* The use of polyphenol oxidase or laccase may provide enough crosslinking between vegetable tannin molecules to provide a degree of hydrothermal stability. The effect has been demonstrated with laccase and low molecular weight phenols.
- *Iridoid compounds:* Tanning can be conducted with iridoid compounds, iridoid glycosides, which are terpene/polyphenol derivatives, such as oleuropein, activated by a β-glucosidase, to remove the saccharide component of the natural product derived from privet or olives.

18.8.10   Enzymes in Post-tanning Processes

'Post-tanning' is normally considered to comprise three separate steps, retanning, colouring/dyeing and fatliquoring, where the potential for enzymatic activity to enhance the processes is great. However, much of this is still at a conceptual stage and is yet to be proven on either a laboratory or industrial scale.

- Retanning could be enhanced enzymatically by addressing the points above. This has even more relevance given the inherent stability of the tanned leather to enzymatic activity, meaning that much more active and aggressive enzymes could be used to enhance the retanning properties.

- It might be possible to colour without dyes which could utilise the melanin synthesis type of reaction. Here natural polyphenols can be used; smaller phenols and naphthols can also be used. The advantage is that the polymerised phenolic species are bound covalently to the collagen *via* a quinoid reaction.
- Transglutaminase for binding dyeing auxiliaries – additional fixation sites – and fixing dyes themselves, *via* amino groups, to give enhanced fastness as a result of covalent bonding. This has been patented by BLC.

18.8.11 Enzymes in the Treatment of Untanned Waste

The leather industry is reducing its waste footprint, but it is possible that enzymes could help this further, particularly for untanned waste.

- Recovery of protein from hair; selective degradation to allow breakdown as fertiliser.
- Recovery of tallow and protein (intact), by selective degradation of protein content.
- Use of anaerobic methanogenic bacteria to create biogas from organic materials.

18.8.12 Enzymes in the Treatment of Tanned Waste

Following on from Section 18.8.11, tanned waste must be treated in a different manner, owing to its increased resistance to enzymatic action.

- Recovery of chromium by breaking down tanned waste with a cocktail of proteolytic enzymes. The peptides can subsequently be recovered for gelatine, animal feed or fertiliser.
- Similarly, recovery of protein from vegetable- or organic-tanned waste, as the enzymes can target the protein, leaving the tannins undegraded and thus possibly recycling them.
- Use of methanogenic bacteria to create biogas: this applies to denatured chrome-tanned waste, because the chromium(III) is effectively deactivated by reaction with collagen, but vegetable-tanned waste remains resistant to breakdown, because the tannins remain intact after denaturation and can therefore deactivate the bacteria/enzymes.

18.8.13   Enzymes in Other Areas of Processing

There are some other areas within the industry that might benefit from the use of enzymes.

- Enzyme treatment for detergents in leather, which would provide enhanced water resistance of leather.
- Targets for enzyme degradation reactions for effluent treatment include the following:
  - sulfide oxidation;
  - sulfate reduction (to sulfur);
  - ammonia fixation;
  - residual organic tanning agents;
  - residual dyes;
  - residual lubricant.

References

1. R. Porter and S. Clark, *Enzymes in Organic Synthesis*, Wiley, 2009.
2. D. E. Koshland, *Proc. Natl. Acad. Sci. U. S. A.*, 1958, **44**(2), 98.
3. R. Boyer, *Concepts in Biochemistry*, Wiley, 2002.
4. A. Vasella, G. J. Davies and M. Bohm, *Curr. Opin. Chem. Biol.*, 2002, **6**(5), 619.
5. J. Kraut, *Science*, 1988, **242**, 533.
6. M. J. Sutcliffe and N. S. Scrutton, *Philos. Trans. R. Soc., A*, 2000, **358**, 367.
7. M. Buchowiecki and J. Vanicek, *J. Chem. Phys.*, 2010, **132**, 194106.
8. J. Basran, S. Patel, M. J. Sutcliffe and N. S. Scrutton, *J. Biol. Chem.*, 2001, **276**, 6234.
9. W. J. Bruno and W. Bialek, *Biophys. J.*, 1992, **63**, 689.
10. N. S. Scrutton, *Biochem. Soc. Trans.*, 1999, **27**, 767.
11. B. J. Bahnson and J. P. Klinman, *Methods Enzymol.*, 1995, **249**, 373.
12. N. S. Scrutton, J. Basran and M. J. Sutcliffe, *Eur. J. Biochem.*, 1999, **264**, 666.
13. C. G. Swain, *et al.*, *J. Am. Chem. Soc.*, 1958, **80**, 5885.
14. L. Michaelis and M. L. Menten, *Biochem. Z.*, 1913, **49**, 333.
15. S. Martí, *et al.*, *Chem. Soc. Rev.*, 2004, **33**, 98.
16. T. Bugg, *Introduction to Enzyme and Coenzyme Chemistry*, Blackwell, 2004.
17. E. M. Aldred, Pharmacodynamics: How drugs elicit a physiological effect, in *Pharmacology A Handbook for Complementary Healthcare Professionals*, ed. E. M. Aldred and C. Buck, Elsevier, 2009, ch. 19, pp. 137–143.

19 Reagent Delivery

19.1 Introduction

The very nature of leather manufacture necessitates the 'delivery' into the pelt of the chemicals required to convert a putrescible piece of skin into a non-putrescible piece of leather. Exactly how this is achieved has been experimented on for a century or more, with each system tested having both pros and cons. This chapter summarises the findings of this ongoing area of development and also elaborates on what the results reveal about the relationship between solvent and substrate and reagent as a system.

Around 5000 years ago, leather was made in the most rudimentary manner, where the solvent used for delivery of chemicals was water, and this continued until about a century ago. Although over the years many different means of reagent delivery have been addressed, the industrial method has remained an aqueous process, but nowadays there is a drive to move away from water for environmental and economic reasons. Leather is a very high consumer of water, sometimes using up to 100 tonnes of water per tonne of rawstock. Progress has been made in reducing its use, so a typical operation in the developed economies will use about 30 tonnes of water per tonne of rawstock and some modern industrial processes can use less. There are some tanneries that claim to function with zero discharge: this should not be confused with zero float, it does not eliminate water and it is still an aqueous process, in which the water used is purified, treated and recycled, and thus a smaller demand is placed on fresh water supplies.

Tanning Chemistry: The Science of Leather 2nd edition
By Anthony D. Covington and William R. Wise
© Anthony D. Covington and William R. Wise 2020
Published by the Royal Society of Chemistry, www.rsc.org

The ideal solution in leather manufacture would be to use pure reagents and not have to solubilise them, but given the requirement of water in the structure of collagen as discussed in Chapter 1, is this possible?

19.2   Low Float

The leather industry is always looking for new ways to reduce water consumption without adversely affecting the quality and performance of the leather. Many new technologies are marketed as having some degree of water saving over 'conventional' processing or promote the fact that they have a low requirement for water. Although it is also true to say that all 'low-float' systems are water saving, it is important to distinguish that not all water-saving technologies are 'low float', *e.g.* compact post-tanning. In addition, the term 'low float' is undefined and can cover a wide range of percentage (by mass) floats.

Realistically, the majority of 'low-float' technologies that are currently marketed focus on the pickling, tanning and post-tanning steps. Current technology has not managed to obviate the need for comparatively large floats in other processing steps, *e.g.* soaking and liming, where the technology has instead focused on recycling the solvent rather than reducing it, *e.g.* countercurrent soaking.

It is also important to consider that a change in float length can have a profound effect on aspects beyond the amount of water used during that processing step.

- Mechanical action on hides within the processing drum. Such an effect is often unquantifiable and unpredictable and relies on trial and error to determine it.
- Although technicians and technologists do their best to eliminate issues, it is not unusual to see an uneven distribution of chemicals throughout the hide. This is partly due to the inevitably higher concentration of chemicals, which increases the rate of penetration (note: not the rate of fixation), exploiting any subtle variations in pelt structure.
- Increased loading on drum motors due to the more uneven or 'lumpy' movement of the skins in the drum.

It is possible to reduce the effect of some of these drawbacks by addressing the drum design. In some cases, it might be that a compartmentalised drum offers an improvement, but this might not be suitable in all cases and requires a commitment to the technology.[1]

19.3 Zero-float Processing

Although the term 'zero float' is well known in the industry, tanners are aware that it is meaningless. In practice, it refers to processing after draining as much of the float as possible. Since the degree of draining is variable, across the range of process vessels and indeed within factories, all that can be said of this approach is that it is low-float processing and usually means 10–30%. The presence of residual water in the process vessel distinguishes this condition from zero water in non-aqueous float, discussed later. The residual free water exerts some control over the transfer step of the fixation reaction by solvation, so that transfer remains incomplete. Therefore, low float alone cannot produce completely efficient chrome uptake. That can only be achieved in aqueous solution by adopting extreme conditions such that all the available water is absorbed with the chrome or involving high temperature and pH. Tanners have recognised that reducing the float, *i.e.* increasing the concentration of the tanning agent, results in faster reaction, which is of course a consequence of exploiting the kinetics. Note that although the kinetics of tanning are difficult to model, the kinetics of complexation are clear.

19.4 Non-aqueous Floats

19.4.1 Introduction

The notion of tanning in a non-aqueous solvent is not new. The literature contains many examples of technologies based on the use of non-aqueous solvents for leather processing. They fall into two categories.

1. The pelt is solvent dehydrated, then processing continues in a chosen solvent (although not strictly non-aqueous, the recent work of Bacardit's group[2] employed a similar process).
2. The non-aqueous solvent is applied to the pelt, but the solvent system has to include at least one component that would allow the solvent to be miscible with the water in the pelt from the previous process.

The primary motivation in these processes was to save energy when drying the leather. All manner of conventional organic solvents have been proposed, from paraffins to alcohols to chlorofluorocarbons, but each suffered from the same fundamental drawback of volatility and/or flammability. Hence solvent tanning has never been used commercially.

19.4.2 Paraffin Processing

Wei[3] proposed a process based on a non-miscible organic solvent, in this case paraffin, developed further by Silvestre and co-workers,[4,5] in which an alternative solvent is used only as the tumbling and heat transfer medium, and the substrate is conventionally processed wet pelt. This approach has the advantage that the organic solvent has low flammability and the efficiency of reagent uptake is practically complete, but the savings in drying energy are relatively small.

Using an organic solvent tanning medium and water wet pelt is an extreme case, strictly an example of low-float tanning, but it is still covered by the proposed four-step mechanism of fixation of a reagent onto a substrate. Previously, it was assumed that the interaction between the solvent and the reagent/solute was finite. However, by choosing a highly hydrophobic solvent, the affinity for a hydrophilic chromium ion is negligible. Now, consider the situation when wet pickled pelt is tumbling in liquid paraffin and chrome tan powder is added into the system. It is reasonable to suppose that the chrome salt is insoluble in the solvent, but water is available in the form of wet pelt. Therefore, the solute is effectively partitioned between an inert solvent and the water within the pelt. Consequently, uptake of the chrome is very rapid, controlled by the rate of dissolution in the pickle liquor in the pelt: uptake is effectively total and is achieved in a matter of minutes. Uptake does not, of course, mean fixation: the kinetics of complexation are controlled by the conditions in the aqueous solution, not how the solution was created. Therefore, the remainder of the reaction proceeds at the typical tanning pace.

The way in which the chrome binds is also governed by its ability to diffuse through the substrate. In ordinary tanning, diffusion through the cross-section of the pelt dominates the process; ordinarily, diffusion is regarded only in terms of penetration, but lateral diffusion is relatively unimportant, because diffusion occurs evenly across the pelt. However, the non-aqueous solvent case is different. Chrome salt crystals will dissolve where they contact the wet pelt and penetration through the cross-section will occur locally. The process is fast and therefore is necessarily uneven over the surface. The rate of fixation will depend on the net pH conditions at the site of dissolution. The higher the pH, the faster is the fixation, the slower is the lateral diffusion and hence the more uneven the reaction becomes. Clearly, the solution lies in controlling the relative rates in the reaction, the rate of diffusion *versus* the rate of fixation, which is analogous to the circumstances in conventional tanning.

Controlling the reaction must be done by controlling the net pH. Usual basification cannot be employed, since the problem of reaction at the site of contact will be the same as for the chrome salt, resulting in greater unevenness in reaction and hence greater non-uniformity of colouring/T_s. This applies to any form of basification, so the preferred approach would be an analogue of the ThruBlu process, discussed in Section 11.7, when the alkalinity within the pelt provides the basifying function. This is illustrated in Table 19.1.[6]

The tanning reaction can be driven by heating the pelt, where the outcome is independent of the variations in chrome basicity and initial pelt pH, but is highly dependent on chrome offer, as discussed in Chapter 11. It is possible to basify the tannage, but it is necessary to employ an agent that will not react on contact; however, there is no advantage over self-basification by the pelt.

Any suggestion that a non-aqueous solvent might be employed in the leather industry would appear to run counter to the aim of ridding all tannery operations of reliance on organic solvents. However, many sectors of industry use organic solvents successfully in an environmentally sound manner, *e.g.* dry cleaning (although this too is

Table 19.1 Chromium tanning of wet pickled pelt in paraffin: conventional salts offered, tumbled for up to 6 h, then aged overnight at 50 °C.

Chrome offer/ % Cr_2O_3	Chrome basicity/%	Pelt pH	Shrinkage temperature of leather as a function of time in process (°C)						
			0.5 h	1.5 h	2.5 h	3.5 h	4.5 h	6.0 h	Over-night
1.0	33	4.0	79	87	95	93	95	97	98
2.0	33	4.0	94	103	113	115	117	125	123
2.0[a]	33	3.0	94	96	98	107	105		123
2.0	50	5.0	93	101	103	101	105		124

[a]+ 0.4% MgO.

moving away from organic solvents to supercritical fluids). Therefore, it is not unreasonable to imagine processing in an alternative solvent, although that would presuppose a fundamental change in operating vessels. The use of supercritical carbon dioxide as a processing medium has attracted some attention, but to date the potential of the technology has not been fully reviewed (see Section 19.4.4).

19.4.3   Organic Solvents

The use of an organic medium in the leather industry is simply an extension of the paraffin concept. By altering the hydrophilic/hydrophobic balance of the system it is possible to increase the rate of penetration, increase exhaustion values and gain access to technologies not accessible in an aqueous medium. In addition, many of the organic solvents that have been evaluated in the literature are more volatile than water and thus reduce the energy involved in removing the solvent *via* evaporation. However, as mentioned in Section 19.4.1, the volatility of the solvents also presents a significant drawback, so although they have been extensively researched and often offer benefits over conventional aqueous systems, they remain unused on an industrial scale.

The use of organic compounds is now reserved for modification of the properties of the aqueous solvent rather than functioning as solvents in their own right. The chemistry of the organic additive is simple: to adjust the hydrophilic/hydrophobic balance of the system, potentially adjusting solubility, stabilise emulsions and/or improve uptake efficiency. Many examples of this kind of solvent modification exist; one such example is the use of ethanolamine to increase the hydrophobicity of the solvent and improve chromium uptake during tanning.[7,8]

19.4.4   Supercritical Carbon Dioxide

Supercritical fluids are a concept that most within the leather industry will not have come across; as Figure 19.1 shows, it is a phase of a substance where its temperature and pressure take it above its critical point, the point at which it becomes supercritical. The exact conditions under which a specific substance becomes supercritical are not fixed, but vary depending on the substance's physical properties.

The properties of a supercritical fluid are complex, because defined liquid and gas states do not exist, and the substance can behave as

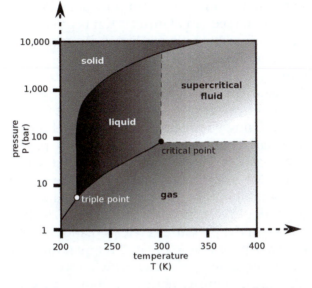

Figure 19.1 A pressure–temperature phase diagram, defining the critical point where conditions allow the formation of a supercritical fluid.

either. For example, in this supercritical state the substance can penetrate through solids as though it were a gas, but it can also dissolve materials as though it were a liquid. It was for this reason that supercritical carbon dioxide has been used in other industries, *e.g.* for dry cleaning,[9] and was explored as a potential solvent to replace water in the leather industry. It has a clear advantage in that it progresses technology away from reliance on water and it is also non-flammable.[10] It is possible to recover the gas and reuse it, without a significant problem of contamination, and in turn it provides access to 'closed-loop' processing. However, supercritical CO_2 cannot dissolve everything and difficulties with solubility are the reason why it is not suitable for every step in the leather-making process. There are also some serious issues with expense: the pressures required to obtain supercritical CO_2 are high and large chambers (required to process a number of hides) would have to have thick, solid metal walls to cope with the pressure, resulting in a large capital outlay. Finally, such a system would require a highly trained specialist technician to run it.

19.4.5 Compressed Gases

This technology can be considered as analogous to the supercritical fluid technology; the only difference is that the pressures are lower, so that the gas in question does not reach the critical point. Although this difference seems minor, it has major implications for the behaviour of the solvent – as can be seen in Figure 19.1, it is possible for a gas to be compressed to a point where it has a liquid phase (at the triple point), but without it becoming supercritical.

This technology was most recently explored by AkzoNobel, who developed the DeMythe LDD process, which utilises dimethyl ether (DME) under pressure to remove water and grease from pelts.[11] It was an example of utilising chemistry from outside the leather industry to explore alternatives to the accepted aqueous processing within the leather. The basis of the technology stemmed from the fact that under pressure, DME's amphiphilic properties are accentuated – that is, it possesses both lipophilic and hydrophilic characteristics – and it could dissolve both grease and water from a pelt. The benefits are further enhanced when the pressure is released and the DME becomes a gas at normal atmospheric pressure and temperature, leaving behind semi-solid fats and greases and water, which, owing to their immiscibility, are separable in a purity not easily obtained with conventional processing. The DME can then be recycled ready for the next batch processing of pelts.

Although this is not a supercritical fluid technology, the issues and drawbacks are similar, in that DME requires a markedly different reaction vessel to the normal processing drums seen in a modern tannery. This vessel still has to cope with high pressures and as such might involve a high capital outlay, in addition to the requirement for a specially trained technician. Furthermore, the technology seems compatible only with the early stages of the leather manufacturing process and no demonstration has been shown for other stages, *e.g.* tanning.

19.4.6   Bead Processing

This technology was originally developed for the laundry industry, where plastic beads replace some or all of the water used for cleaning. This technology has been proven and is now widely used in the large-scale laundry cleaning industry.

The transfer of this methodology to leather processing is an excellent example of cross-fertilisation of industrial methodology and is now in the early stages of also being used in the leather manufacturing industry. The concept[12] is the same in the leather industry as it is in the laundry industry – to use the beads (simple solid, spherical polymer balls) to reduce the need for water during the processing step to which they are applied. Although the balls themselves are solid, the large quantity of them in a moving drum gives a quasi-fluid-like motion.[13]

A quirk of this technology is that the inert beads do not have any dissolution properties, but instead become 'coated' in the active additive, *e.g.* dye, which then drives the compound into the pelts by means of mechanical action and physical processes, *e.g.* hydrophobic/hydrophilic balance.

As has been discussed on a number of occasions, the exact role (or roles) of water in many of the processing steps is not clear; however, it is increasingly clear that water is an important component of the transport mechanism and solvation in the system. The upshot is that, although the bead technology is excellent at saving water, it is unlikely to become a truly zero-float process and will almost certainly rely on the inherent water within the pelt structure or the addition of a small float along with the beads.

Developmental research is returning some interesting and surprising results that are supplying supporting evidence on the role of water within the structure of collagen during the tanning processes. These results have yet to be published, but are likely be published after this book has gone to print.

19.4.7 Ionic Liquids

Ionic liquids (ILs) are non-aqueous solvents that are, as the name suggests, highly ionic in nature. They are known more simply as liquid salts, a term that accurately depicts what they are, as they consist solely of ions, but are liquid at ambient temperatures, *i.e.* below 100 °C.[14]

The melting point of an ionic species is related to the strength of the bonds between the ions holding it in a solid lattice structure, called the 'lattice enthalpy': this can be measured in two ways, depending on whether the calculation is based on the lattice being broken up (lattice dissociation enthalpy) or formed (lattice formation enthalpy). It is usual to specify which is being used, but in practice both return the same value, just with opposite signs, *i.e.* plus or minus.

When an ionic compound is melting, the heat is used as a source of energy to break the ionic lattice; therefore, the higher the melting point of a compound, the higher is the lattice energy and *vice versa*. The lattice enthalpy is related to the charges on the ions within the system, the radius of the ions and the regularity of their shapes. Hence systems with low opposing charges (*e.g.* +1 cation and −1 anion), where the ions have large radii and irregular or non-symmetrical shapes, will have low lattice energies and thus low melting points. In ILs, the lattice energy is low enough that only ambient temperatures are required to melt the solid. This is demonstrated in Table 19.2, which details the specific cation and anion of some ionic compounds with their corresponding melting points.

In Table 19.2, it can be see that sodium chloride has a melting point of 801 °C, but by exchanging the sodium ion for a cation with a larger radius – potassium – the melting point is reduced. Exchanging the sodium or potassium ion for a much larger and more irregularly shaped cation (1-ethyl-3-methylimidazolium or 1-butyl-3-methylimidazolium) substantially lowers the melting point. Table 19.2 also demonstrates the same trend to be true for the anion, where the larger and more irregular the shape, the lower is the melting point.

Table 19.2 Melting points of some ionic compounds.

Cation	Anion	M.p. (°C)
Na^+	Cl^-	801
K^+	Cl^-	772
(1-ethyl-3-methylimidazolium structure)	Cl^-	87
(1-butyl-3-methylimidazolium structure)	Cl^-	65
(1-butyl-3-methylimidazolium structure)	NO_3^-	38
(1-butyl-3-methylimidazolium structure)	$AlCl_4^-$	7
(1-butyl-3-methylimidazolium structure)	$CF_3CO_2^-$	−14

ILs have been utilised in leather research recently with varying degrees of success.[15] Although some biocompatible ILs exist, the majority are frequently toxic, are often expensive and can sometimes react violently upon mixing with water. Whereas ILs are used in other industries, these drawbacks have so far precluded their use at full scale within the leather industry. The industries that currently use ILs have a much smaller scale requirement and a more obvious way of recycling the liquid.

Such drawbacks have not prevented ILs being used as part of an aqueous system to enhance the process. These are normally very weak solutions of ~0.1–2% of an IL in water and so should not strictly be treated as a different solvent system. Fathima and co-workers published data relating to the use of a range of ILs to enhance the effect of 'opening up' during processing.[16]

19.4.8 Deep Eutectic Solvents

Historically, the terms ionic liquid and deep eutectic solvent (DES) were used in an interchangeable manner; however, as understanding improves, there are increasingly large differences between the two systems and, although not officially defined, DESs are now treated as different to the ILs.

DESs are eutectic mixtures of Lewis or Brønsted acids/bases and contain a range of anionic and cationic species. The commercially applicable DESs typically consist of quaternary ammonium halides with either metal salts and/or hydrogen bond donors; mixing the two components helps to break down the lattice structures (similarly to ILs), which results in compounds that would otherwise be solid at ambient temperature turning to liquid. As can be seen in Figure 19.2, the ratio of the two components being combined affects the melting point of the final mixture; however, there is an optimum ratio that maximises the depression of the melting point of the system – this is called the eutectic point. When there is a deep depression, this is a 'deep eutectic point', which gives rise to the name of the solvents.[17]

Although DESs have a very high ionic component, unlike ILs they do not consist solely of discrete anionic and cationic components. However, in many cases DESs have properties that are comparable to those of ILs and they have been used in a wide variety of applications, including metal processing, synthesis and control of phase transfer.[17] In addition to their hygroscopic nature, many DESs often have a small percentage of water as part of their formulation and

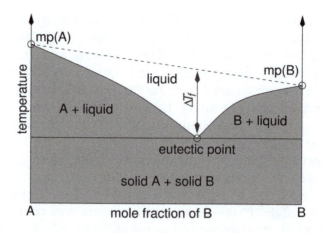

Figure 19.2 The formation of a deep eutectic.

so are rarely deemed anhydrous; however, the water fraction is sufficiently small that in most cases the liquids are still considered non-aqueous.

The advantage that DESs have over ILs is that they are, for REACH purposes, considered as mixtures rather than new compounds; this, alongside the lower costs of raw materials, results in the solvents being much cheaper than ILs. Many DESs are formulated from benign compounds, avoiding the toxicity issues associated with ILs, and in addition, as they have often been complexed with a hydrogen bond donor, they rarely react violently with water.

The differences between DESs and ILs are further highlighted in part by their use at a non-commercial level in leather science. It has been demonstrated at a laboratory scale that DESs can be used in a 'pure' form, *i.e.* without further addition of water, for a range of different stages within the leather manufacturing process. This is arguably the most significant outcome from this work[14] – a paradigm shift from convention, with a move away from solution chemistry to utilising pure or highly concentrated, active ingredients in a liquid form. One benefit from this new approach is that the user no longer has to consider concentration gradients between the solution and the pelt/leather – which is often the most important factor affecting rate of reaction in industrial-scale processing, where percentage uptake also has to be considered.

Recent research has focused on the tanning and post-tanning stages and in many cases DESs have demonstrated shorter processing times compared with conventional processing methodologies. It is theorised that the decrease in processing time could be due to a number of factors depending on the chemistry of the system being used. With vegetable tanning (the most significant processing time improvement), it is thought that the processing improvement is as a result of better solution chemistry. With chromium tannages, it is thought that the opening up of the fibre structure could be the largest contributing factor.[18,19]

Another advantage to using DESs is that they may provide access to chemistries not currently available to the leather industry. A good example is the incorporation of graphite material into the leather structure;[19] this had not been possible to date, as graphite is highly hydrophobic, so that when graphite powder is added to water it simply aggregates into large insoluble masses. Although the DESs do not dissolve the powder, they do increase the stability of the colloidal suspension and prevent the rapid aggregation of particulate matter. This extra stability allows the particles to be pushed into the open structure of leather by mechanical action.[18]

19.5   Direct Injection

The technologies addressed so far are all concerned with the medium for carrying the solute into the substrate, where the success depends on the relative chemistries of the reagent, the substrate and the solvent, which define affinities, penetration rate and efficiency of interaction/reaction. An alternative approach is to inject the reagent in solution directly into the substrate. One way in which this might be accomplished is to lay the pelt on a porous plate and reduce the air pressure below the plate while flooding the pelt with reagent solution: this would rely on the solution being sucked through the pelt. Unfortunately, this simple procedure does not work, because the pressure difference across the pelt can only be 1 atm.

In the 1990s, Pauckner at the Westdeutsche Gerberschule developed an injection process using jets, like an array of miniature shower heads: the pilot equipment could deliver an aqueous solution at 20 bar (atm) or more and was capable of blowing holes in bovine hide. The principle was demonstrated that chromium tanning salt could be injected through hide in seconds, although equilibration of concentration within the pelt and the fixation reaction were separate processes. The technology was scaled up to a pilot version by the Krause Company and designated the Penetrator, illustrated in Figure 19.3. The success was limited, because

Figure 19.3 Trials with the Krause Penetrator: Heinz Meyer on the left, Tony Covington on the right.

the machine was based on a perforated plate, capable of delivering solution at only 6 bar. However, despite problems with jet blockages, it was demonstrated that sheepskin could be dewooled in 20 min, limed hide could be chrome tanned directly and chrome-tanned leather could be post-tanned (retan, dye, fatliquor) in a single step. The conclusion was that '… the injection principle is fundamentally sound, with the future viability depending partly on mechanical design improvements'.[20]

19.6   Overview

The traditional mode of reagent delivery in leather making, using aqueous solution, is inherently inefficient, because the reagents have to be soluble in water and there is inevitable partitioning of the reagent between the substrate and the solvent. This means that, in order to achieve a high uptake of the reagent, the tanner must resort to conditions that accelerate the reaction: the consequence of high reactivity is surface reaction, which is inevitably non-uniform. This limits the options open to the tanner to make the best use of reagents and minimise environmental impact. The solution is to rethink the mode of delivery, whether that means changing the medium, modifying the profile of the kinetics over the period of reaction or altering the way in which the solution is driven into the substrate.

All those options for change are available, at least all have been tried, with greater or lesser success. Other solvents have been tried over many decades, but all have drawbacks, not least of which is flammability. It is more likely that a more radical approach is required

– ILs. This new technology offers the promise not only of new solvents but also of new systems in which the reagent is a component of the solution – proven principles, but requiring development for industrial application. There is room for more creativity in controlling and varying the chemistry of reactions, for example in chrome tanning by the use of complexing agents. Mechanical action has a role to play: it cannot change the thermodynamics, but it can change the apparent kinetics, by helping to get the reagent to the substrate reaction sites more effectively.

The modern leather industry is no longer constrained by convention in processing. Options may not find favour universally, but they do offer the promise of technologies that have niche roles for speciality production and new products.

References

1. V. J. Sundar, *et al.*, *J. Sci. Ind. Res.*, 2001, **60**(6), 443.
2. G. Baquero, *et al.*, *J. Am. Leather Chem. Assoc.*, 2017, **112**(3), 102.
3. Q.-Y. Wei, *J. Soc. Leather Technol. Chem.*, 1987, **71**(6), 195.
4. F. Silvestre and A. Gaset, *J. Soc. Leather Technol. Chem.*, 1994, **78**(1), 1.
5. F. Silvestre, *et al.*, *J. Soc. Leather Technol. Chem.*, 1994, **78**(2), 46.
6. A. D. Covington, unpublished results.
7. A. D. Covington, *Tanning Chemistry: The Science of Leather*, RSC Publishing, Cambridge, 2009.
8. A. D. Covington, *J. Soc. Leather Technol. Chem.*, 2011, **95**, 231.
9. J. M. DeSimone, and W. Tumas, *Green Chemistry Using Liquid and Supercritical Carbon Dioxide*, Oxford University Press, 2003.
10. A. Marsal, *et al.*, *J. Supercrit. Fluids*, 2000, **16**(3), 217.
11. J. H. Berkhout and J. R. Garcia del Rio, *World Pat.*, WO-2005059184–A2, 2005.
12. J. E. Steele, *World Pat.*, WO 2014/167358, 2014.
13. S. Rostami, J. Steele and A. D. Covington, *Proceedings Asian International Conference on Leather Science and Technology*, Okayama Japan, 2014.
14. W. R. Wise, A. P. Abbott and J. Guthrie-Strachan, *World Pat.*, WO 2015/159070, 2015.
15. A. Mehta, J. Raghava Rao and N. Nishad Fathima, *J. Phys. Chem. B.*, 2015, **119**, 12816.
16. G. C. Jayakumar, *et al.*, *RSC Adv.*, 2015, 5, 31998.
17. E. L. Smith, A. P. Abbott and K. S. Ryder, *Chem. Rev.*, 2014, **114**(21), 11060.
18. A. Abbott, A. P. M. Antunes, A. D. Covington, B. Mmapatsi and W. R. Wise, *Leather Int.*, 2016, **217**(4860), 26.
19. A. P. Abbott, O. Alaysuy, A. P. M. Antunes, A. C. Douglas, J. Guthrie-Strachan and W. R. Wise, *ACS Sustainable Chem. Eng.*, 2015, 3, 1241.
20. N. J. Cory, A. D. Covington and M. D. Mijno, *J. Am. Leather Chem. Assoc.*, 1994, **89**(9), 289.

20 Drying

20.1 Introduction

From Chapter 1, it should be clear that water plays a major part in leather making, whether it is considered from the point of view of structural features or from the part it plays in processing: as expressed by Bienkiewicz, leather and water are a 'system'.[1] In her studies of the effects on the viscoelastic properties of collagenic biomaterials of moisture content and heat, Jeyapalina[2] approached the system from a materials science point of view. In this chapter, the discussion of the effect of drying regime draws on two of Jeyapalina *et al.*'s subsequent ground-breaking studies,[3,4] in which important implications of the water–leather relationship for leather properties were defined, in particular the effect of drying on leather softness.

20.2 Viscoelastic Materials

The concept of viscoelastic material is not a new one – it refers to a material that exhibits both a viscous component and an elastic component. Although the ratio between the viscous component and the elastic component may alter, all materials exhibit viscoelasticity, *i.e.* there is no such thing as a perfectly elastic material or a perfectly viscous material (although under specific parameters some materials do come close).

Tanning Chemistry: The Science of Leather 2nd edition
By Anthony D. Covington and William R. Wise
© Anthony D. Covington and William R. Wise 2020
Published by the Royal Society of Chemistry, www.rsc.org

Elasticity is the ability of a material to resist permanent deformation when under stress. Although the strain will increase under stress, when the stress is removed the strain will return to its original point (normally considered to be zero). An elastic/rubber band is the most obvious commonly available material that exhibits such properties – although it should be noted that elastic itself is still not an ideal elastic material. Viscous materials resist shear flow and strain linearly with time when a stress is applied. Most liquids demonstrate viscosity: a classic example is black treacle (also known as molasses), which is a very thick liquid derived from the sugar refining industry.

The two components of a viscoelastic material have their own moduli, as illustrated in Figure 20.1:

- elastic component = storage modulus;
- viscous component = loss modulus.

Figure 20.1 relates to something that is familiar, namely a bouncing ball, but also aptly demonstrates the terms where they come from and what they mean in practice. If a ball is dropped, it will not return to the same height; its rebound height is always lower than that of the initial height from which it was dropped. The height to which it rebounds denotes the elastic component of the material; when the ball hits the ground it deforms, stores energy and then releases that energy as it returns to its original shape. Likewise, the difference in

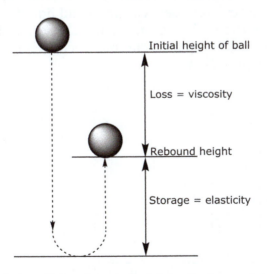

Figure 20.1 Depiction of 'dropping a ball', defining the loss and storage moduli.

height between the starting point and the rebound height denotes the loss modulus, which can be related to the viscous component of the material.

Although the analogy of dropping a ball is in practice imperfect, because other uncontrolled factors might affect rebound, *i.e.* air resistance, mass, *etc.*, it is still a useful thought experiment. Most people can relate to the fact that different balls are likely to have different properties and therefore are likely to rebound to different heights – compare what might happen with a golf ball *versus* a tennis ball *versus* a squash ball: each has a different rebound height because each ball has different viscoelastic properties.

20.3 Leather Drying Stages

Drying of leather has been intensively studied and reviewed.[5-17] The process has been explained in terms of different phases of drying, applying to the different states of water that are found within leather, bulk water and different degrees of bound water.[18] In its natural state, skin contains 150–190% of water on dry weight. Chemical processing changes the fibre structure at the molecular level and alters the relationship between structure and moisture content in the pelt, causing variations in the capillary water. The water in collagen can be divided into three main groups: structural water, bound water and bulk water.[19] Bulk water has a liquid-like character and is capable of forming ice crystals at 0 °C. Bound water exhibits a structure between solid and liquid,[20] so it does not freeze at 0 °C. Structural water molecules are part of the fibre structure and behave like a solid. The observed progress of drying is illustrated in Figure 20.2.

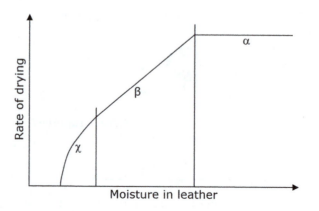

Figure 20.2   Rate of drying as a function of moisture content.

In the earliest stage of drying, there may be sufficient bulk water on the surface of a leather for it to act like a liquid water surface.[1] Evaporation occurs at a constant rate, which is directly proportional to the surface area of leather (A), mass transfer coefficient (K_g) and the difference between the vapour pressure of the water at the surface temperature (P_s) and the partial pressure of the water vapour in the air (P_a):

$$dW/dt = AK_g(P_s - P_a)$$

This indicates that the rate of drying is dependent on the temperature and the relative humidity of the air. In theory, fast drying conditions can be used during this phase of drying, provided that the losses of the bound and structural water molecules are prevented. After this constant rate of drying period is completed, the rate of drying is controlled by the diffusion of moisture/vapour from the higher concentration in the centre of the leather to the lower concentration at the surface. During this period, the rate of drying is entirely dependent on the ease of diffusion of water vapour through the structure. Consequently, the factors that affect the diffusion rate will control the drying rate.

Friedrich attributed the constant rate period, α, to the removal of water from between fibres, the first falling rate period, β, to the water which is situated between the fibrils and the second falling rate period, χ, to the water located within the fibrils.[19] He also reported that the constant drying rate and the first falling rate cease at the critical points of 50 and 30% moisture content, respectively, later confirmed by Elamir.[20] Leather is a viscoelastic material and, as a result, the mechanical response to loading involves both a viscous component associated with energy dissipation and an elastic component associated with energy storage. Water acts as a plasticiser in leather and hence introducing or removing this plasticiser will inevitably change the viscoelastic properties. It follows that, during drying, the viscoelastic properties of leather change continuously with the removal of water. Therefore, the measurement of the dynamic mechanical behaviour of leather during drying can give vital information on changes occurring to the internal fibre structure.[2]

20.4 Role of Temperature with Moisture

Figure 20.3 shows the dynamic mechanical thermal analysis (DMTA) drying curve for partially processed chrome-tanned leather at a constant air temperature of 70 °C, where modulus and tan δ are expressed as a function of time. Tan δ is the ratio of the loss modulus (E'') to the storage modulus (E'):

$$\tan \delta = E''/E'$$

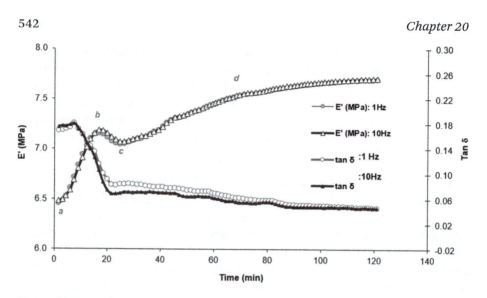

Figure 20.3 DMTA drying trace of partially processed wet blue at 70 °C: the initial and final moisture contents of the sample were 162 and 7% on a dry weight basis, respectively. Courtesy of the *Journal of the Society of Leather Technologists and Chemists*.

and as it measures the energy dissipation of a material it is often referred to as defining the damping effect, as also discussed in Chapter 15.

The storage modulus E', indicating the stiffness of the leather, shows an overall increase as the leather changes from a wet state to a dry state. The general shape of the curve is the same regardless of the drying temperature, where the only difference is the point at which inflections are observed.

If the profile of the DMTA drying curve is related to the molecular motions associated with leather fibres, then at the second inflection point, c, tan δ is expected to give rise to a peak. None of the results show such behaviour and hence suggest that the observed changes in modulus must be related to the moisture content of the leather.

Figures 20.4 and 20.5 show the DMTA drying curves for chrome-tanned leather at 20–70 °C. The moisture contents at the inflection points b and c of the modulus curve remain constant regardless of the drying temperature: the inflection point b corresponds to ~60% moisture content and inflection point c to ~30% moisture content.

Figure 20.6 shows the drying profile of vegetable (tara)-tanned leather. The general shape of the curve is same as that obtained for

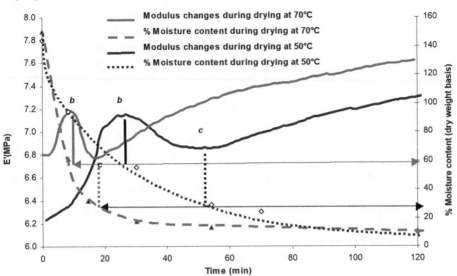

Figure 20.4   Changes in storage modulus (E') and moisture content as a function of time during drying of wet blue leather at 50 and 70 °C at 65% relative humidity. Courtesy of the *Journal of the Society of Leather Technologists and Chemists*.

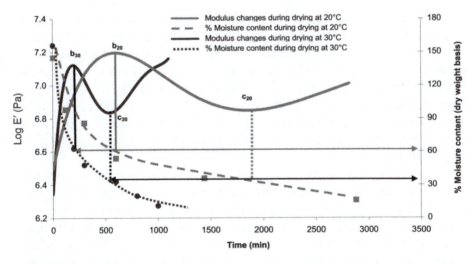

Figure 20.5   Changes in storage modulus (E') and moisture content during drying of wet blue at 20 and 30 °C at 65% relative humidity. Courtesy of the *Journal of the Society of Leather Technologists and Chemists*.

chrome-tanned leather, but the moisture contents at the inflection points differ: the critical moisture contents are 42% at inflection point b and 27% at inflection point c. However, when the moisture contents at the inflection points are calculated on dry collagen

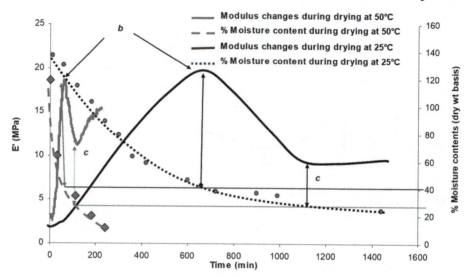

Figure 20.6 Changes in storage modulus (E') and moisture content during drying of partially processed vegetable (tara)-tanned leather at 25 and 50 °C at 65% relative humidity. Courtesy of the *Journal of the Society of Leather Technologists and Chemists*.

weight, the chrome- and vegetable-tanned leathers show similar values: ~60% at inflection point b and ~34% at inflection point c. The observations from Figures 20.4–20.6 suggest that the inflection points on the drying curve must relate to the collagen–water relationship, rather than be a reflection of the chemistry of the leather type. The changes in the modulus during the dehydration process of a leather must be related to the different states of water and, therefore, a generic master curve for drying of leather can be compiled, as shown in Figure 20.7.

From differential scanning calorimetric (DSC) analysis, it is possible to measure the amount of freezable water in the leather, and this is plotted as a function of total moisture content in Figure 20.8.

From Figure 20.8, the non-freezable portions of water in chrome- and vegetable-tanned leather are ~61 and ~42%, respectively. When these values are normalised to the weight of collagen in the leathers, both leathers contain 60 ± 3% moisture on dry weight. Therefore, the initial increase in bend modulus from a to b in Figure 20.6 is attributed to the removal of freezable water from the leather.

Heidemann and Keller[21] analysed the equatorial X-ray diffraction peaks of hide at various moisture contents and observed that the intensity of the diffraction goes through a maximum with the water content. The implication is that the crystallinity/orderly packing of

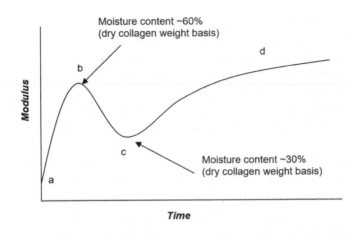

Figure 20.7 Generic curve of changes in bend modulus during drying of leather.

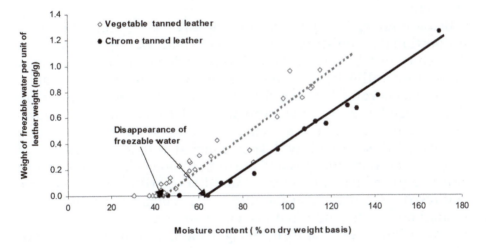

Figure 20.8 Weight of freezable water plotted against total water content of the leather sample.

collagen molecules has a high dependence on the moisture content of the material and the optimum order is when it contains ~33% moisture on wet weight or ~49% moisture on dry weight. Hence, as the drying proceeds, the storage modulus of the material is expected to increase with decrease in moisture content until the optimum molecular order is reached at 49% moisture content. The initial increase in storage modulus from a to b in Figure 20.7 is therefore attributed to an increase in the crystallinity of the collagen.

There are two possible explanations for the decrease in modulus from b to c in Figure 20.7. First, it comes from decreasing crystallinity of the collagen molecules with removal of water. Second, it might be due to changes in fibril dimensions. Komanowsky[5] demonstrated that, as water is removed from the space between fibrils, a rapid, longitudinal and lateral shrinking of fibrils takes place; this process occurs as the average moisture content decreases from ~50 to ~27% on dry weight. These dimensional changes would allow more freedom of movement to the fibre network and, hence, a decrease in bend modulus with decreasing fibre size would be expected.

A number of workers have suggested that, below 23% moisture on dry weight, water content is associated with the collagen structure. Pineri *et al.*[22] argued that a water concentration between 11 and 23% corresponds to the association of water molecules between triple helices: removal of such associated water would decrease the free volume of the collagen molecules and thus an increase in modulus would be expected.

Below point c, the equatorial spacing changes are almost directly proportional to moisture content.[23] This indicates that these water molecules may be directly bound to the sidechains of the collagen helix or microfibrils, so removal of this type of bound water molecules creates voids between the fibrils. To fill the voids, fibres are drawn together and the alignment of sidechains may form covalent crosslinks: a molecular dynamic simulation by Mogilner *et al.* illustrated this possibility.[24] They concluded that the absence of associated water produces a distortion of molecular conformation and also introduces an increased number of hydrogen bonds. This will restrict the molecular mobility of the collagen and will result in an increase in bend modulus. Once all of the water that can be removed from the structure of collagen has been removed, the modulus will plateau. Therefore, an increase in modulus between c and d is attributed to a decrease in free volume, and also an increase in crosslinking between collagen molecules. The DMTA drying profile of leathers under various isothermal conditions indicates that differences in storage modulus are found only between c and d. The implication is that the stiffness of leather is governed only by the final phase of drying.

It can be concluded that leather may be dried at high temperature to give a softer product, provided that the humidity is maintained to prevent the structural water molecules from being lost to cause stiffness. From a practical point of view, controlling the humidity of the drying environment at high temperatures presents problems. The implication is that there would be a benefit in having a two-stage drying process: during the first stage, high temperature may be used to remove moisture rapidly until ~30% moisture on dry weight is achieved, after which the leather may be dried in a conditioned environment to give a final residual moisture content of 14%.

20.5 Drying in Phases

Given the evidence that leather has three distinctive drying stages, it raises the question of whether drying in phases might yield benefit. To examine the concept, Jeyapalina *et al.* dried chrome-tanned leathers at various temperatures for 48 h, but in an atmosphere controlled at 85% relative humidity, when the moisture content fell to 35%, after which they were conditioned at 65% relative humidity and 20 °C, so the moisture was equilibrated at 14%. Control chrome-tanned leathers were dried at constant temperature for 48 h in an atmosphere without humidity control, after which they were conditioned in the same way. The effect of drying regime on the bend modulus of the leathers is shown in Figure 20.9.[2,4]

Under normal drying conditions, when leather is dried at a high temperature the stiffness is generally found to be high. Therefore, it is commonly believed that stiffness has a high dependence on the

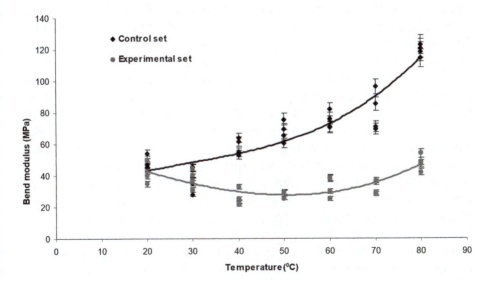

Figure 20.9 Effect of drying temperature on bend modulus: higher value means stiffer leather. Courtesy of the *Journal of the Society of Leather Technologists and Chemists*.

drying temperature. However, on the contrary, Figure 20.9 shows that the two-stage drying process can produce softer leather. This confirms that the water–leather relationship during drying plays an important part in determining the softness properties of leather.

Drying of leather at elevated temperature is a diffusion-controlled process.[6,7] The rate of drying is entirely dependent on the concentration gradient: when the evaporation of water is faster than the diffusion of water from interfibrillary regions to the surface, because the temperature of drying is too high, then the outer surfaces may become irreversibly crosslinked as the structural water of leather fibres is lost. Once fibre sticking is introduced into the collagenous matrix, rehumidifying will not replace the lost structural water to the same extent as before, resulting in the high bend modulus observed for leather samples dried at high temperature. When leather is dried with controlled humidity, the rate of drying will still be a diffusion-controlled process at high temperature and the gradient of water vapour pressure between the leather and the external atmosphere decreases with drying time. Therefore, when the equilibrium moisture content is approached, the rate of drying slows. If the relative humidity is high enough to stop any structural water from being replaced, then drying the leather at a high temperature will be faster, with no undesirable effect on stiffness.

Komanowsky[5] explained why a high degree of shrinkage of fibres occurs when the moisture content drops from about 50 to 27% (on a dry weight basis). During the initial stage of drying, the water lost from smaller capillaries is quickly replaced by capillary 'suction' from larger capillaries: larger capillary regions vary in size and hydrostatic pressure is reported to be lower than the 3.5 atm osmotic swelling pressure. The hydrostatic pressure generated during the evaporation of water molecules will therefore not exceed the osmotic swelling pressure, hence no shrinkage of fibre dimensions is expected. Therefore, removal of water at a higher rate at high temperature will be without undesirable consequences. After the larger capillaries have lost all the water, during the second stage of drying, water from the inter fibril region is removed. The hydrostatic pressures of these regions are reported to be −18.2 and −78 atm, respectively.[5] These values are greater than the osmotic swelling pressure, hence a noticeable shrinking of fibres is expected. However, employing high temperature to remove moisture from this region will not cause adverse effects, because closely bound, structural water is not lost. In practice, it is not easy to predict the precise end of this drying period and prolonging these drying conditions into the next stage

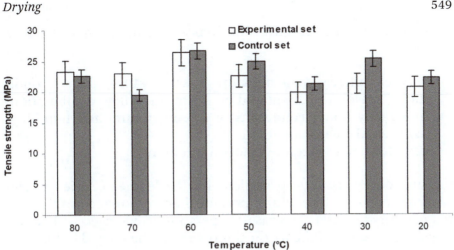

Figure 20.10 Effect of drying temperature on the tensile strength of wet blue dried by a two-stage procedure. Experimental samples were dried to 30% moisture content at various temperatures and then conditioned to 14% moisture at 65% relative humidity and 20 °C. Control samples were dried at various temperatures, after which they were conditioned to 65% relative humidity at 20 °C. Courtesy of the *Journal of the Society of Leather Technologists and Chemists*.

of drying may cause damage at the surface. During the final stage of drying, water molecules that are removed must come from the intra-fibril region, the bound structural water.[18] If the leather is to be dried at elevated temperature during this phase of drying, while fibre shrinking is also proceeding due to higher hydrostatic pressure, fibrils have a greater possibility of coming close enough to form intermolecular crosslinks and hence fibre sticking. High temperature also facilitates permanent crosslinking between collagen molecules by supplying the activation energy. Therefore, in order to produce softer leather, more moderate drying conditions should be employed during this final stage of drying.

Figure 20.10 shows the effect of drying regime on the strength of chrome-tanned leather. Drying by the two-stage process was shown to produce softer leather than the conventional drying approach, but not at the expense of adversely altering its tensile strength.

In the context of drying, Corning[25] demonstrated that leather area is controlled only by the moisture content and not by the rate of drying, which is controlled by the temperature of drying. Therefore, if the drying regime involves introducing structural changes that cannot be reversed by rehydration, area is lost.

The commercial alternatives for drying have traditionally been as follows.

1. Dry completely to a moisture content >10%, then recondition ('season') with sprayed-on water and allow equilibration, but rarely under conditions of a temperature- and humidity-controlled environment.
2. Dry down to the required moisture content.

The first option is easy to implement: the leather is dried at elevated temperature to equilibrium and then rehydrated with a limited quantity of water. The ease of operation is balanced with a loss of quality in terms of the physical properties. The second option has always been viewed as impractical, because it would involve sophisticated process control. However, it has now been shown that this option is actually practical, because the only change to conventional hard drying technology is the addition of humidity control in the drier. By adopting the use of a high air temperature, the initial stages of drying to achieve rapid drying down to 30%, followed by a slower rate of drying for the removal of the remainder of the moisture from the leather, the quality of the leather is improved, but without major alteration to the operational aspects of processing. In addition, the two-stage drying process does not result in irrecoverable area loss, which is reason enough to consider adopting this development in technology.

20.6   Leather as a Material

The viscoelastic properties of leather, as with other materials, can be measured using a variety of methods; most commonly dynamic mechanical (thermal) analysis [DM(T)A], which addresses the viscoelastic properties of the leather as a function of another variable, most commonly temperature or humidity.[26] Understanding this relationship in terms of the material properties is crucial to area gain during manufacture, addressing 'in-use' performance and, perhaps increasingly importantly, tackling prediction.

As might be expected, the viscoelastic properties of leather are complex: the reason for this complexity is that leather is a multi-part system, each 'part' affecting the overall observed properties of the material – species of animal, moisture content, bating, tannage, retannage, fatliquoring, elastin content, amount of protein degradation and drying parameters, to name a just few parameters, and there are many more. However, although quantification of the effects that these parameters have is limited, general trends have been known for decades: for example, most tanners know that vegetable- and chrome-tanned leathers have fundamentally different mechanical behaviours. Tanners have used this knowledge of the properties of each when considering the end use of the leather.

Leather is a material that is extremely responsive to its environment. Changes in humidity have been shown to have a rapid effect on the viscoelastic properties of the leather. This was addressed while investigating drying parameters, but subsequent work has proved that the same response can be observed when 'dry leather' (moisture content ~14%) is subjected to similar fluctuations in humidity, as shown in Figure 20.11.[27]

Figure 20.11 demonstrates possible progress beyond two-phase drying into a multi-phase drying programme, although different leathers will respond differently to conditions. Undoubtedly, more work is needed to understand the mechanism of these changes:

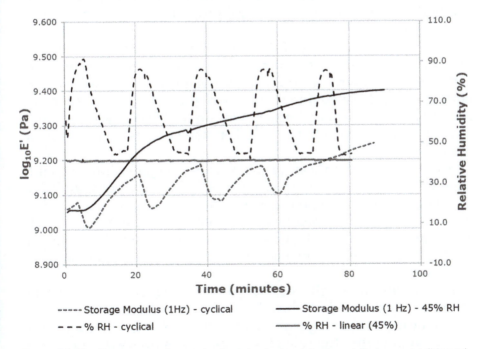

Figure 20.11 Response of wet leather to 'pulses' in humidity: preconditioned, chromium-free crust (115% moisture) dried at 55 °C, using either a relative humidity cycle (cycled between 80 and 45%) or a linear relative humidity set at 45%.

how leather parameters (*e.g.* tanning agent) affect the observations and whether they alter the overall mechanism of viscoelastic change.

20.7 Overview

This chapter has focused on the complex relationship between water content and viscoelastic properties of leather with a particular focus on the drying phase of leather manufacture.

Although there have been numerous publications in this area, because of the way such information is presented, it can be difficult to digest and can subsequently be overlooked. However, most can relate to what they observe: a wet leather is more flexible and elastic than a dry leather. This distinction is supported by the research detailed in this chapter – importantly, though, the properties do not 'suddenly switch' or even follow a linear progression from one to another. As a result, the parameters by which a leather is dried – heat *versus* humidity *versus* time *versus* tension – can have a significant impact on the physical properties of the leather and on area yield.

Owing to the complexity of the multicomponent system of leather, there are still some outstanding questions that need to be answered and a greater comprehension must be obtained of the whole system during drying. How do different tannery processes and methodologies affect the drying of a leather? Why do they have this effect? And what adjustments should be made between systems to maximise or minimise an effect?

Research to answer these questions is ongoing; machinery manufacturers have recognised the advantages of this improved knowledge and the potential impact on tannery economics. They have begun to implement changes to their machinery that reflect the latest research and exploit the materials' inherent natural vulnerabilities during drying, to gain area or provide the tanner with beneficial physical properties.[26,28]

In addition, although not immediately obvious, the impact of the research detailed in this chapter extends beyond the practical application of drying parameters – it helps in gaining a greater understanding into the role of water, how it moves within the system and what effects that might have. This in turn may allow the development of new water-saving technologies that are unrelated to drying.

References

1. K. Bienkiewicz, *J. Am. Leather Chem. Assoc.*, 1990, **85**(9), 305.
2. S. Jeyapalina, PhD thesis, The University of Northampton, 2004.
3. S. Jeyapalina, G. E. Attenburrow and A. D. Covington, *J. Soc. Leather Technol. Chem.*, 2007, **91**(3), 102.
4. S. Jeyapalina, G. E. Attenburrow and A. D. Covington, unpublished results.
5. M. Komanowsky, *J. Am. Leather Chem. Assoc.*, 1990, **85**(1), 6.
6. J. Lamb, *J. Soc. Leather Technol. Chem.*, 1982, **66**(1), 8.
7. K. Bienkiewicz, *Physical Chemistry of Leather Making*, Krieger Publishing Co., Malabar, Florida, 1983.
8. L. Buck, in *The Chemistry and Technology of Leather*, ed. F. O'Flaherty, W. T. Roddy and R. M. Lollar, Reinhold Publ. Corp., New York, 1962.
9. K. T. W. Alexander, D. R. Corning and A. D. Covington, *J. Am. Leather Chem. Assoc.*, 1993, **88**(12), 252.
10. E. Heidemann, *Fundamentals of Leather Manufacturing*, Eduard Roether KG, Darmstadt, 1993.
11. A. W. Landmann, *J. Soc. Leather Technol. Chem.*, 1994, **78**(2), 44.
12. M. Komanowsky, *J. Am. Leather Chem. Assoc.*, 1990, **85**(5), 131.
13. S. A. J. Shivas and A. Choquette, *J. Am. Leather Chem. Assoc.*, 1985, **80**(5), 129.
14. B. Abu el Hassan, A. G. Ward and S. Wolstenholme, *J. Soc. Leather Technol. Chem.*, 1984, **68**(5), 159.
15. J. Monzo-Cabrera, *et al.*, *J. Soc. Leather Technol. Chem.*, 2000, **84**(1), 38.
16. E. P. Lhuede, *J. Am. Leather Chem. Assoc.*, 1969, **64**(8), 375.
17. K. Liu, N. P. Latona and J. Lee, *J. Am. Leather Chem. Assoc.*, 2004, **99**(5), 205.
18. A. R. Haly and J. W. Snaith, *Biopolymers*, 1971, **10**, 1681.
19. E. Friedrich, *Handbuch der Gerbereichemie und Lederfabrikation*, ed. W. Grassmann, Springer-Verlag, 1955.
20. F. M. Elamir, PhD thesis, Leeds University, 1975.
21. E. Heidemann and H. Keller, *J. Am. Leather Chem. Assoc.*, 1970, **65**(11), 512.
22. M. H. Pineri, M. Escoubes and G. Roche, *Biopolymers*, 1978, **17**, 2799.
23. M. A. Rougvie and R. Bear, *J. Am. Leather Chem. Assoc.*, 1953, **48**(12), 735.
24. G. I. Mogilner, G. Ruderman and R. Grigera, *J. Mol. Graphics Modell.*, 2002, **21**(3), 209.
25. D. R. Corning, BLC The Leather Technology Centre, unpublished results.
26. W. R. Wise, A. D. Covington, K. B. Flowers and A. Peruzzi, *J. Am. Leather Chem. Assoc.*, 2016, **111**(1), 24.
27. W. R. Wise and K. B. Flowers, unpublished results.
28. K. B. Flowers, A. Peruzzi, W. R. Wise and A. D. Covington, *J. Am. Leather Chem. Assoc.*, 2015, **110**(10), 317.

21   Finishing

21.1   Introduction

The finishing of leather is the final stage of leather manufacture and refers to the application of polymer film(s), often containing a colourant and other additives, being applied to the surface of leather, in a manner analogous to conventional painting. It is considered by some to be the most important step in leather manufacture as it can have the greatest influence on the 'handle' of the leather (the tactile feel) and visual appearance.

Treating finishing as a 'straightforward coating' oversimplifies the process: there are many different polymers that can be used, a plethora of additives and numerous finishing techniques, where each variation changes the appearance and feel of the leather. The recipe, techniques and even the purpose of the finish itself depend on the requirements for the end use of the leather.

The subject of finishing is vast, so the aim of this chapter is not to provide a complete review of leather finishing but rather, by focusing on the chemistries of the polymers and additives, to act as a starting point for those wanting to familiarise themselves with the basics.

The analogous situation to finishing leather is 'painting' – preparation of the surface (often by roughing it to maximise the contact surface area), application of a primer (in leather terms, this is the first or base coat) and then application of subsequent undercoats and topcoats. If any of these stages are performed incorrectly, then the painting often fails before it might be expected to; for example, it might not have the aesthetics, feel or surface performance that are

Tanning Chemistry: The Science of Leather 2nd edition
By Anthony D. Covington and William R. Wise
© Anthony D. Covington and William R. Wise 2020
Published by the Royal Society of Chemistry, www.rsc.org

desired (*e.g.* scratch resistance) or the paint could peel as a result of poor 'keying' into the surface, *i.e.* penetration and mechanical interlocking.

Fundamentally, the range of chemistries is the same for top coats and base coats of the same polymer system, *i.e.* types of bonding, crosslinking, *etc.* The properties differ only to reflect the specific requirements of the coating. In this context, all finish resins are designed for particular applications: the formulations and film properties depend on the choice of chemistry used in the manufacture of the product, together with other contributors to performance, including the degree of crosslinking, the molecular weight of the polymer and sundry additional components.

21.2 Polymers for Finishing

Four main classes of polymers are used in finishing: urethanes, acrylics, nitrocellulose and proteins; however, by no means should these be considered the only polymers that can be used. These polymers are often sold as preformed dispersions (usually water based) and are more commonly known by finishers as 'resins' or 'binders'.

21.2.1 Urethane

Polyurethanes (PUs) include polymers with a broad range of properties that comprise organic units joined by the urethane–carbamate linkage. This linkage is achieved by step-growth polymerisation of two different monomers, each containing either two (or more) isocyanate groups or two (or more) hydroxyl (alcohol) groups (see Figure 21.1).

The exact properties of the polymer formed are largely dependent on the types of isocyanates and polyols used to make it. If the monomers incorporate long aliphatic sections with minimal crosslinking sites, then this tends to yield a softer polymer; conversely,

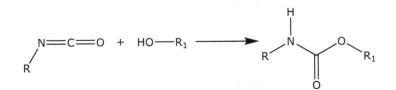

Figure 21.1 Combining an isocyanate with an alcohol to form a carbamate (urethane) bond.

if they include only short aliphatic regions or fixed aromatic areas or have a high crosslinking density, then the polymer will tend to be hard.

In general, the polyol is responsible for incorporating the long aliphatic components and the isocyanates tend to be smaller, less flexible molecules. As a result, once combined with a chain extender (normally a diol or diamine), there are hard and soft segments within the polymer chain (see Figure 21.2). This leads to a quasi-two-phase morphology within the polymer, where each section will aggregate into a microdomain: highly ordered hard segments and generally amorphous soft segments. It is the relative ratio and compatibility of these two domains that dictate the overall properties of the PU.

The crosslinking within PU leads to an interlinked three-dimensional network, often with a high molecular weight. As a result, PUs used in industries outside leather finishing are mostly

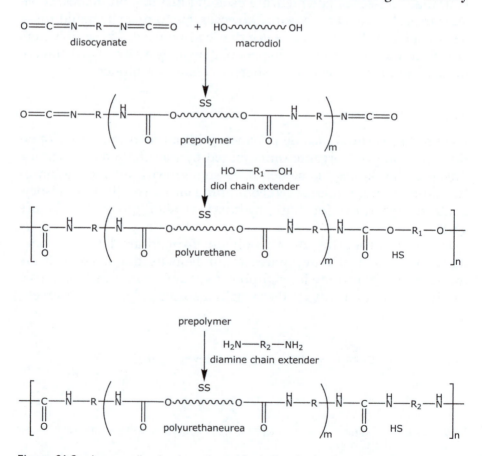

Figure 21.2 A generalised polyurethane depicting the hard and soft segments.

thermosetting polymers – that is, they undergo thermal decomposition before becoming a viscous liquid and, as a result, do not have a defined melting point. Thermoplastic varieties exist and, although both types are used in the leather industry, these are the types more commonly found in finishing. Most crosslinking within the polymer is controlled at the point of manufacture, but it is possible for users to apply additional crosslinking agents to the finishing recipe and alter the final properties of the polymer. In practice, only small additions of crosslinker are used to make minor modifications to the physical properties; if major changes are required, then a different starting product is normally used.

It is important to note that the user has little control over the properties of the PU – this is fixed by the chemical company that manufactures the product. However, it is still important to have a basic understanding of the science that leads to the varied properties observed between different PUs, as this will lead to rationalised product selection when constructing a finish that needs to meet precise specifications.

21.2.2 Acrylic

Acrylics are routinely used in finishing and, like PUs, are a family of polymers that can exhibit wide and varied properties, depending on their construction and monomeric starting point. The monomers are all derived from acrylic acid but could otherwise be significantly different in their chemical construct, as Figure 21.3 demonstrates.[1,2]

Depending on the properties of the monomer and on the desired properties of the final polymer, acrylics can be formed by either chain growth or step growth at the point of manufacture.

- Chain-growth polymerisation involves the opening up of the unsaturated carbon–carbon double bond to form a homopolymer with a saturated backbone containing only carbon.
- Step-growth polymerisation occurs when the unsaturated acrylic ester is reacted with monomers containing an amine by nucleophilic conjugate addition (see Figure 21.4) of the amine groups to the acrylic C=C bonds. This method of polymerisation yields backbones containing nitrogen, oxygen and carbon, but the exact configuration of the polymer is dependent on the monomer.

As with PUs, the acrylics are often crosslinked at the point of manufacture, creating thermosetting polymers instead of the thermoplastic polymers that might be expected from linear polymers. Various

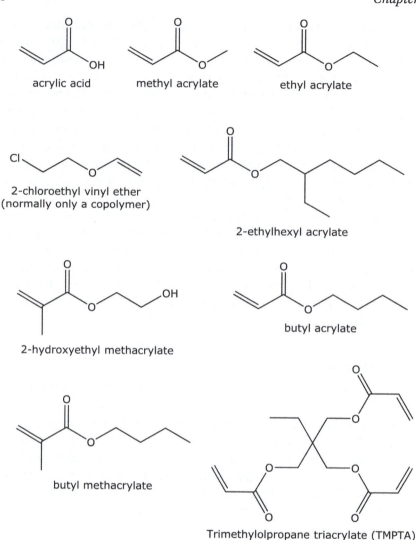

acrylic acid

methyl acrylate

ethyl acrylate

2-chloroethyl vinyl ether
(normally only a copolymer)

2-ethylhexyl acrylate

2-hydroxyethyl methacrylate

butyl acrylate

butyl methacrylate

Trimethylolpropane triacrylate (TMPTA)

Figure 21.3 A selection of possible derivatives of acrylic acid.

crosslinking agents can be used, with a range of functionality, each imparting their own properties on the final polymer; an example is provided in Figure 21.5.[3]

In addition to the crosslinking inherently designed into the polymer at the point of manufacture, the degree to which acrylics are crosslinked can be adjusted by the user *via* the addition to the finishing recipe of slow-acting crosslinking agents. However, just as with other polymers, the crosslinking agents are often toxic, difficult to handle and expensive, hence their use by the finisher is typically kept to a minimum.

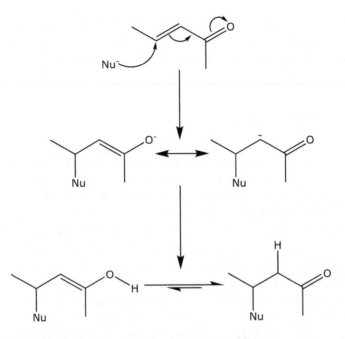

Figure 21.4 Mechanism of nucleophilic conjugate addition.

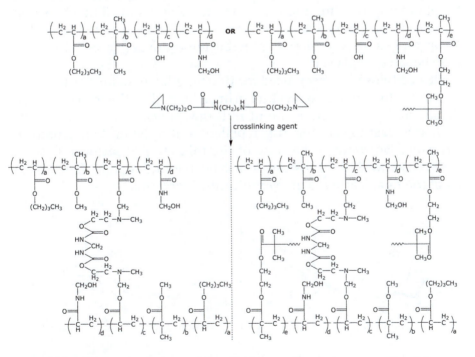

Figure 21.5 Schematic diagram of crosslinking reaction between acrylic chains.

The range of possible monomers, copolymers and crosslinking agents, combined with a 'choice' in the method of polymerisation, goes some way to explaining the wide range of properties that are exhibited by acrylics.

21.2.3 Nitrocellulose

Nitrocellulose has been known for many years: since its discovery in 1832, it formed the basis for early explosives, often referred to as flash paper, flash cotton or gun cotton. It is prepared by mixing a source of cellulose (*e.g.* cotton, paper or wood) with a mixture of concentrated sulfuric and nitric acids so that one or more hydroxyl groups in the cellulose structure become nitrated (see Figure 21.6). The recipe for its preparation continued to be developed from the point of discovery, addressing safer ways of producing the inherently unstable compound. In 1862, it became one of the first synthetic plastics, when it was synthesised, dissolved in alcohol and then cast as a film.[4] Although the process for manufacture continued to be developed, the basis of the polymer remained unchanged. In 1880, its use became widespread, as it was used by companies such as Kodak as the basis for photographic film, motion picture films and X-ray films; however, numerous accidents – some of which were fatal – meant that by the 1930s more stable and less flammable alternatives (*e.g.* cellulose acetate) began to replace it.

Nitrocellulose has been used for finishing leathers for decades and continues to be used today. It is primarily used in the 'top coat' of a finish, as it yields a transparent film that is water resistant, readily cleanable, fast to rubbing, dirt repellent and resistant to mechanical stressing.[5] Fortunately, because of the thin layer of nitrocellulose in comparison with the substrate, it has been found that the flammability of the final leather is not significantly increased.[6]

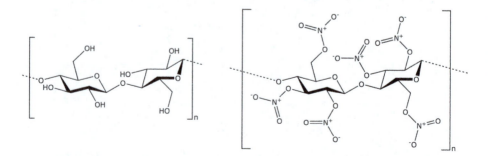

Figure 21.6   The repeating unit for cellulose (left) and nitrocellulose (right).

In a similar way to the original development of nitrocellulose, although the amount of nitration might alter between products for leather finishing, the basic polymer has largely remained unchanged. The key to changing the properties of the nitrocellulose polymer is in the additives that are a part of the final film – especially plasticisers. Details of exactly what plasticisers are present in each product are, unsurprisingly, difficult to find, owing to commercial confidentiality, but is likely that the technology has moved away from the older, natural additives, such as camphor, dammar, kauri and shellac.[7] Instead, the synthetic plasticisers aid consistency and reduce the quantity of volatile components inherent in any natural system.[7]

21.2.4 Protein

Proteins in finishing tend to be referred to as binders and, as the name suggests, are derived from naturally available proteins, which are made from many amino acid monomers. The proteins have to be cheap, readily available and compatible with being used as a finishing medium. These criteria tend to limit the selection to casein (from mammalian milk), albumins (from egg white) and blood serum; of these, the most commonly used protein in commercial finishing products is casein.[8] This is a traditional finish, used for more than a century for making high-end ladies' shoe uppers, glacé kid. Although there are many reasons why a finishing technician might use proteins in their finish, for example for aiding plating/embossing, it is the fact that they provide an excellent 'polished' look when subjected to a glazing machine/jack that clearly differentiates them from other finish reagents. Greater durability of the protein finish can be achieved by tanning the film *in situ*, typically with an organic agent such as an aldehyde.

Some efforts to utilise proteinaceous tannery wastes as protein finishing components, most notably from wet blue shaving wastes, have been detailed in the literature.[9,10] This is normally achieved by hydrolysing or partially hydrolysing the proteinaceous waste before incorporating it into a finishing process. For the technology to become commercial, the recovery of the protein must be cheaper than with the use of casein, it has to be miscible with other components of the finish, it should be easily delivered and it must not have a detrimental effect on the desired leather properties. The results published so far show some promise, but the technology requires further development.

21.2.5   Other Components Including Dyes and Pigments

As has already been alluded to in this chapter, the finish 'mixes' are composed of numerous other components. There are ratios of film-forming (resins) to non-film-forming components (everything else) that are part of the technology, but exactly what makes up the non-film-forming component varies greatly depending on the intended final specification of the leather. This section addresses possible additives, but the options are too wide and diverse to be covered in a comprehensive manner. However, the importance of one non-film-forming component does merit a more detailed mention – the colourant.

Colourants are subdivided into two categories, dyes and pigments, which are differentiated by their molecular size and solubility in water. Dyes are soluble in water and provide a coloured but transparent solution, whereas pigments are comparatively large molecules, are poorly soluble in water and therefore produce a coloured translucent or opaque mixture. Owing to these properties, pigments exhibit what is termed 'covering power' – the ability to mask/hide minor defects, *e.g.* changes in underlying leather colour or minor changes in surface topography. In contrast, dyes offer no covering power.

The use of dyes and pigments in finishing has even defined the terminology used in the industry.

- Aniline leathers are finished with a light coating (the natural texture of the skin, including the grain pattern, must be visible) that contains no pigment – the addition of the dye is merely to make small colour adjustments and provide a more even colouring.
- Semi-aniline leathers also have a light finish, but may incorporate a small quantity of pigment; however, there should not be so much that the grain pattern is not visible.
- Pigmented leathers: as the name suggests, this heavier finish incorporates a much larger quantity of pigment, which, because of its covering power, will obscure all the natural surface features of the leather.

The chemical structures of the active colourants in either dyes or pigments containing products sold by chemical manufacturers are, unsurprisingly, difficult to find. In many cases, a colour may be made up of more than one component, adding further complexity to the overall finish composition.

Table 21.1 summarises as a guide the functions of typical materials that might be included in a finish, but it is not possible here to list all

Table 21.1 Primary functions of finish components.

Finish component	Covering power[a]	Colour	Filling	Handle modifier	Releasing agent	Dry rub resistance	Wet rub resistance	Water repellence	Oil repellence	Finish flow out	Penetration agent	Stain resistance	Aesthetics	Physical properties
Filler/duller	✓		✓		✓				✓				✓	
Oils/waxes	✓		✓	✓	✓			✓	✓				✓	✓
Butadiene			✓	✓			✓							
Surfactant										✓	✓			
Dye		✓											✓	
Pigments	✓	✓	✓			✓	✓	✓					✓	
Rheology modifier						✓	✓	✓		✓				
Crosslinker									✓			✓		
Silicone				✓	✓									

[a]This could include covering of minor defects.

of the possible components and variations. There are also specialist components that impart many different properties to the finish and resultant leather. A good example is the system developed solely to reduce the temperature increases on the surface of the leather from radiative heat. As it is mostly infrared light that is responsible for increases in temperature from radiation, the systems available rely on a pigment and dye embedded in the finish and/or leather that reflect light in the infrared region of the electromagnetic spectrum. Importantly, the human eye cannot see infrared light, so the inclusion of this colourant does not affect the colour struck on the leather.

21.3   Adhesion

Although the finish has many roles, the primary function is to offer some degree of protection to the leather. This is possible only because the polymer finish adheres to the surface and, even under extreme flexing, such as the bending of a shoe, does not detach from the leather. Adhesion of the finish is therefore key to the overall performance of the leather. Interestingly, as a finish usually comprises two to five different layers, it is only the first layer, the basecoat, that must adhere to the leather itself; other layers of finish need to adhere to the underlying polymer (intercoat adhesion).

There are four main factors that affect adhesion of the finish: wetting of the surface, *i.e.* the so-called wettability, types of adhesive bonds, drying parameters and diffusion. There are also other influences, but these are outside the scope of this chapter.

21.3.1   Drying of the Finish

It is the drying of the finish that forms a film through a process called fusion (Figure 21.7) and it is crucial to the integrity of the finish film.

- Phase one involves the evaporation of the solvent, although not in all cases, since generally the polymers are dispersed in water. Owing to the excess of solvent in the system, the rate of evaporation remains constant in this phase.
- The rate of evaporation slows significantly in phase two, as it is now dependent on the mass transport/diffusion of the solvent (no longer in excess). This phase is also important, as the overall volume of the finish now begins to decrease significantly, bringing the polymer droplets closer together; when they meet, they begin to diffuse and coalesce.

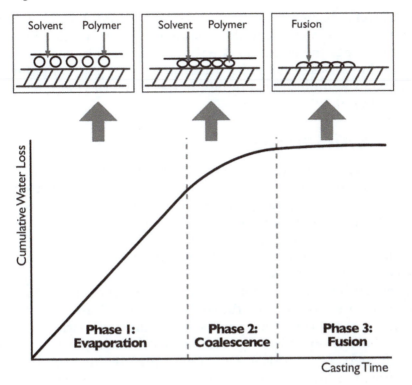

Figure 21.7   The three phases of water loss on polymer drying. Courtesy of World Trades Publishing Ltd.

- By phase three, most of the solvent has now evaporated and what remains must diffuse through the polymer itself or utilise microscopic interparticle channels to reach the surface. The coalescing particles become homogeneous and the film forms.

Both drying temperatures that are too high and incorrect finish viscosities can lead to problems with coalescence and consequently film formation (collectively known as autohesion). These issues may not be immediately apparent in the tannery, but could quickly become an issue once the final article is in use.

Understanding the process of film formation can also help the user appreciate why adding too much non-film-forming components can cause premature failure of the finish coat. The film-forming component of the finish is the only component that can exhibit coalescence of particles/droplets; more 'solid' components such as pigments cannot coalesce. This will then cause a 'break' in the continuous polymer film; by increasing the amount of non-film-forming component, there is an increase in the amount of 'breaks' in the film itself, which disrupts the continuous nature of the finish. Each break in the film imparts a 'weak point' that can act as a point of failure for the finish; the more points of failure there are, the more likely the finish is to fail.

21.3.2   Surface Wetting

An important factor in adhesion is how the wet finish interacts with and flows on the surface of the leather. It is these two properties that will define how easily the particles of polymer coalesce, how easily they will penetrate the surface and how easily they will occupy macroscopic/microscopic voids within the leather surface.

It is important to understand what wetting or wettability is and how it is measured. The wettability of a surface relates to how well a liquid (often water) interacts with a surface. It is not uncommon for a finisher to put water or saliva on the surface of a crust leather; using their experience to interpret the way in which it interacts with the leather as a measure of how a hydrophobic the leather is. From a controlled, scientific perspective, this interaction is measured in terms of a contact angle, that is, the internal angle formed when a liquid meets a solid surface, as demonstrated in Figure 21.8 and also addressed in Chapter 17.

If the interaction between the liquid and solid is favourable, then the droplet will quickly 'flatten' and the contact angle will be small. If, on the other hand, the interaction is not favourable, then the opposite will be true and droplet will be pronounced, with a large contact angle. This is demonstrated in Figure 21.9, which also shows that it is possible, with poor wetting or a highly unfavourable interaction, for the contact angle to exceed 90°.

Whereas those who work with leather can use the simple contact angle test as a guide to practical outcome, the scientific test takes place under controlled conditions and uses glass as the solid. The use of glass instead of leather has an advantage in that it eliminates the

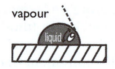

Figure 21.8   A liquid interacting with a solid, providing a contact angle θ. Courtesy of World Trades Publishing Ltd.

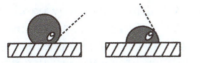

Figure 21.9 Relationship between contact angle and the favourability of the solid–liquid interaction. Left: unfavourable interaction, large contact angle, no wetting and no spreading. Centre: contact angle below 90°, wetting occurs but no spreading. Right: favourable interaction, very low contact angle with wetting and spontaneous spreading. Courtesy of World Trades Publishing Ltd.

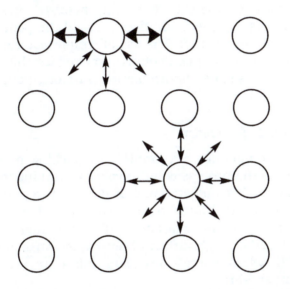

Figure 21.10 Depiction of the differences in intermolecular forces that give rise to surface tension.

effect that surface topography might have on the results. The properties of the glass, in terms of its hydrophobicity, can also be adjusted so that it can replicate a range of possible substrates.

Alongside the relationship of the interaction between the solid and the liquid, the surface tension of the liquid can also affect the apparent wettability of the substrate, in addition to the flow of the wet finish. Surface tension arises because the molecules of a liquid are attracted to each other, to a greater or lesser extent. In the bulk of a liquid the molecules are equally attracted to one another, but at the surface of the liquid the molecules cannot be pulled equally in all directions and this magnifies the forces between them (Figure 21.10). This attraction at the surface pulls the molecules more closely together and yields a more densely packed layer of molecules that form a quasi 'skin' on the liquid.

Surface tension is utilised by Nature – it is why pond skaters (*e.g. Aquarius remigis*) can apparently sit on and skate across still water and, because surface tension will try to pull a liquid into a sphere that has the smallest volume-to-surface area ratio, it also explains why rain falls as almost spherical drops. The tendency for a liquid to form spheres also affects the apparent wettability: the lower a liquid's surface tension, the more likely it is to flow across the surface and fill the voids. As a general rule, surface tension decreases with increase in temperature, but the addition of solvents and/or surfactants to the system has a far greater effect. This is the chemistry upon which so-called 'penetrators' is based, additives that aid the penetration of a finishing mix into the leather. The use of simple water-miscible surfactants/solvents (*e.g.* simple alcohols) that reduce the surface tension of the water-based finish formulation promotes penetration into the substrate.

21.3.3   Types of Adhesion

Although adhesion might seem simple – the 'sticking' of two surfaces to one another – the process could involve a combination of mechanisms. In the leather industry, one of the most important mechanisms of adhesion of the finish to the leather is mechanical adhesion. This is crucial for the first coat and involves the interlocking of the polymer finish to the rough surface of the leather. An analogous situation is that with Velcro, where loops on one fabric are interlocked with hooks on another fabric.

Key to the effective utilisation of this mechanism is surface wetting (see Section 21.3.2). As depicted in Figure 21.11, the more easily a finish wets the surface, the more easily it can penetrate into the substrate.

Good penetration is key to large gains in mechanical adhesion; as Figure 21.12 depicts, this allows the first finish coat to fill voids within the structure of the leather, including areas where there might be an 'overhang'. Not only does this process maximise the mechanical interlocking, it also substantially increases the surface area from which other methods of adhesion can act.

Figure 21.11   How surface wetting is related to penetration. Left: poor wettability provides poor penetration. Right: good wettability provides excellent penetration. Courtesy of World Trades Publishing Ltd.

Figure 21.12   How good wetting can lead to excellent penetration, even filling 'overhangs' (indicated with an arrow). Courtesy of World Trades Publishing Ltd.

The relative strength of chemical bonding as a mechanism of adhesion is also highly dependent on the area of the leather–finish interface; even so, chemical bonding is not considered to be a major part of the adhesion of finishes to leathers. In this case, the term 'chemical bonding' refers to all types of possible bonding, including van der Waals forces, dipole interactions, hydrogen bonding, covalent bonding and ionic bonding. Although van der Waals forces, dipole interactions and hydrogen bonding are generally considered to be weaker forces, the large number of bonds over a large surface area provides a significant effect.

There is an experiment that can conducted at home to demonstrate this effect. Take two books of roughly equal size (old telephone books are ideal) and overlay one page from one book on top of a page from the other. If this is done just a few times, the books are easily pulled apart; however, if it is conducted repeatedly for the entirety of each book then it is impossible for a human to exert sufficient force to pull the books apart. This elementary experiment demonstrates how a seemingly weak force can become significant when it is multiplied many times.

Covalent bonds are regarded as strong bonds, but rarely play a significant role in adhesion of the finish to the leather. It is possible that they play a secondary role in diffusion adhesions, where crosslinking between the two layers of finish is possible, but this is not the main use for crosslinking agents.

Ionic bonding is used in leather finishing with the application of cationic finishes, which bind to anionic leather. Surprisingly, although this type of finish provides good coverage and reduces chemical usage, it often suffers from lower adhesive properties than other finishes. The poor adhesion arises from the fact that the oppositely charged leather and finish quickly interact, to bond electrostatically and thus prevent effective penetration. This is analogous to attempting to chrome tan a hide at too high a pH, whereby the chromium salt reacts on the surface of the hide and so fails to penetrate through the cross-section.

The remaining mechanism that is important in leather finishing is diffusion adhesion. This has a very similar mechanism to autohesion, although strictly autohesion can occur only between identical

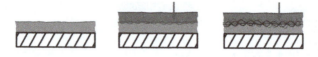

Figure 21.13   Intercoat adhesion by diffusion. Courtesy of World Trades Publishing Ltd.

polymers,[11] which may not be the case in leather finishing. As the name implies, it involves the diffusion of a newly applied 'wet finish' diffusing into the previous coat (Figure 21.13). This involves initial swelling of the first coat at the interface between the two layers, then diffusion of one into the other. During drying, this process effectively partially combines the two layers and allows them to behave as if they were only one. As Figure 21.13 indicates, this process of diffusion can occur only between two polymer films and not between a film and the leather and, as a result, this is often termed intercoat adhesion.

All forms of adhesion are normally enhanced by mechanical compression and heating, *e.g.* embossing, provided that the parameters of this mechanical action are compatible with the finish being applied.

21.4   Overview

A discussion of the methods by which the finish is applied is outside of the scope of this book, other than to mention that they can also affect the way in which the film is created and formed. There is an intimate relationship between the chemistries of the system, mechanisms of adhesion, the desired properties of the finish (in both wet and dry states) and method of application. These methods, well known in the art, include padding, spraying, roller coating and curtain coating – each of which has its own technology of variations, all delivering the finish formulation in a different way and each providing a different starting point for the mechanisms of film formation. It is therefore paramount for a finisher to consider the application method in addition to the specification of the final article when designing a finishing recipe.

References

1. E. M. Pearce, *Kirk-Othmer Encyclopedia of Chemical Technology*, 1978–84, John Wiley and Sons, New York.
2. Y. Yu, B. Liao, G. Li, S. Jiang and F. Sun, *Ind. Eng. Chem. Res.*, 2014, **53**(2), 564.
3. Y. Yoo, G. Hong, S. Hur, Y. Lee, H. Kim and J. Lee, *J. Appl. Polym. Sci.*, 2009, **112**, 1587.

4. S. Fenichell, *Plastic: The Making of a Century*, 1996, HarperBusiness, New York.
5. S. M. Gumel and B. B. Dambatta, *Int. J. Chem. Eng. Appl.*, 2013, **4**(4), 249.
6. Y. Gong, W. Chen, J. Chen and H. Gu, *J. Soc. Leather Technol. Chem.*, 2007, **91**(5), 208.
7. B. K. Brown, *Ind. Eng. Chem.*, 1924, **6**, 568.
8. A. Bacardit, *et al.*, *J. Soc. Leather Technol. Chem.*, 2009, **93**(4), 130.
9. J. Zarlok, K. Smeichowski and M. Kowalska, *J. Soc. Leather Technol. Chem.*, 2015, **99**(6), 297.
10. Z. Chunhui and Q. Lu, *J. Soc. Leather Technol. Chem.*, 2014, **98**(6), 269.
11. F. Awaja, Autohesion of Polymers, *Polymer*, 2016, **97**, 387.

22   Environmental Impact

22.1   Introduction

Leather is an industry that has faced criticism over the years and this continues to be the case today, as exemplified in 2011 by the Blacksmith report,[1] claiming that leather tanning is the fourth most polluting industry. This was based on misinterpretation and a lack of understanding of the fundamental chemistries involved in leather manufacture, including the role of chromium(VI): the immediate response from the leather industry was a robust rebuttal by Catherine Money.[2] While many who work within the industry feel that some of the criticism is unjustified or unfair, over the last half century the global industry has spent much time and money creating improved practices that have had a positive effect on environmental impact.

The recent increase in veganism has put the leather industry's source of raw material under scrutiny, with mischievous suggestions that the making of or using leather means being responsible for the slaughter of millions of animals per year. In practice, abattoirs exist only because of the meat industry and the skin from the animal is seen as a by-product. In using that by-product, the leather industry is in fact improving the environmental impact of the meat industry and potentially removing a major environmental problem. It is likely that, for as long as there is a meat industry, there will be a leather industry utilising its waste: the simplicity of this relationship means that it is difficult to see where improvements can be made.

Tanning Chemistry: The Science of Leather 2nd edition
By Anthony D. Covington and William R. Wise
© Anthony D. Covington and William R. Wise 2020
Published by the Royal Society of Chemistry, www.rsc.org

Regardless of opinions on the current environmental impact, it is certainly fair to say that historically the leather industry was one of the most polluting industries. As a consequence, the consumption of water, the use of chemicals and prevention/upcycling/downcycling of waste – including what happens to the material at end of life – tend to be considered the largest areas where gains in the environmental impact of the leather industry can be made. The reason for research efforts being applied in these three fields is that they are areas that the leather industry can control and therefore in which improvements can and have been made.

The environmental impact of the leather industry is now deeply entrenched in everything it does. Much of the research relating to chemical use and water consumption is implicit in the sector and covered throughout this book. For this reason, this chapter has a greater focus on the waste that is produced by the industry. It is important to recognise that not every solution suits every scenario – one size does not fit all – but an understanding of the pros and cons of each technology, in addition to reviewing with a critical eye, can certainly assist in making the best decisions.

22.2   Waste or Co-products

Waste products generated by the leather industry can take many forms, each having its own source in the production process. There are many ways of categorising, assessing and treating the waste streams, but detailed discussion is outside the scope of this book. A simple scheme is shown in the following list, where the waste is categorised according to its state and then sub-categorised according to its content.

- Solid waste.
  - Tanned waste: for example, end-of-life leather, offcuts/cutting waste, shaving and buffing waste.
  - Untanned waste: for example, fleshing waste and hair.
  - Chemical waste: for example, salt.
  - Effluent waste: for example, filtered solids, also referred to as 'filter cake'.
- Liquid waste.
  - Aqueous effluent: normally with dissolved impurities.

- Gaseous waste.
  - Unwanted gases produced by processing, rarely at toxic levels: for example, hydrogen sulfide, ammonia and substituted volatile amines.

It should be noted that this way of categorising has many flaws owing to its simplicity, not least because there are some waste streams that could cross over between categories. Nevertheless, it is from this list that the sections below were selected.

22.2.1 Solid Waste

22.2.1.1 Tanned Waste

In the Glossary at the beginning of this book, the definition of 'Leather' states that it should be 'imputrescible' – however, this property makes it difficult to deal with at the end of life. The issue of the disposal of leather waste or leather products is an issue that must be addressed, since both are contributors to the general problem of landfilling.

In the case of chrome-tanned leather, one perceived problem is the chromium content, with regard to leaching and possible oxidation *in situ*. Although chrome-tanned leather still dominates the marketplace, the current trend is to favour non-chrome tanning methods, on the basis that it is better to use renewable tanning agents, particularly vegetable tannins, assumed to be more recyclable, *i.e.* readily returned to the earth. However, controlled degradation of waste using microorganisms is not straightforward for leather; an approach would become more feasible if a simple method could be found to reverse the effects of tanning or if leathers can be produced that have an inbuilt weakness towards biodegradation (see Chapter 25). Model laboratory studies have been conducted on the use of anaerobic microorganisms that can degrade organic matter to produce biogas; the amount produced is used as a measure of susceptibility to degradation.[3]

Biogas is a mixture of methane and carbon dioxide, resulting from the anaerobic decomposition of organic waste.[4] The conversion of complex organic matter to methane requires a consortium of microorganisms, comprising several interacting metabolic groups of anaerobic microorganisms.[5] Biogas formation occurs at four overlapping stages of reaction: hydrolysis, acidification, acetification and methanogenesis. Methanogenesis is the last step in the complete anaerobic decomposition of organic compounds and therefore plays an essential role in the mineralisation of organic material.[6] There are two distinct routes by which the methane is produced, which are

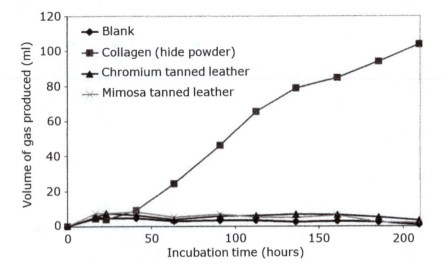

Figure 22.1 Rate of biogas produced from untanned collagen, chrome-tanned collagen and mimosa-tanned collagen.[3] Courtesy of S. Yagoub.

determined by the types of methanogenic bacteria present.[7] The ace-toclastic methanogens convert acetic acid to methane and carbon dioxide [eqn (22.1)]: this route accounts for approximately 70% of the methane formed during anaerobic degradation. The second route involves hydrogenotrophic methanogens that utilise carbon dioxide and hydrogen to form methane and water [eqn (22.2)]. The action of these methanogenic bacteria is critical to achieving complete degradation, as their substrates have an inhibitory effect on earlier phases of degradation.

$$CH_3CO_2H \rightarrow CH_4 + CO_2 \qquad (22.1)$$

$$CO_2 + 4H_2 \rightarrow CH_4 + 2H_2O \qquad (22.2)$$

Chrome-tanned collagen is as degradable as the raw collagen, but mimosa-tanned collagen remains resistant to degradation (Figure 22.1). It is clear why these results were observed. The reactivity of chrome is spent by the tanning reaction, so the denatured collagen behaves just like any other protein but containing some form of inert chromium(III). On the other hand, the vegetable tannins react more weakly, retaining their reactivity towards protein and the denaturing reaction does not alter their chemistry, so the polyphenols remain reactive to protein (Figure 22.2). The outcome is that the degraded protein is protected against biodegradation and the polyphenols

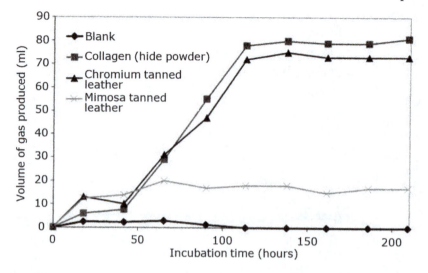

Figure 22.2　Effect of heat denaturing the tanned collagen prior to treating it with the anaerobic bacteria.[3] Courtesy of S. Yagoub.

Table 22.1　Effect of condensed tannins on inhibition of anaerobic bacteria.

Vegetable tannin	Minimum inhibitory concentration $(g\ L^{-1})$	Hydroxylation pattern	
		A-ring	B-ring
Myrica	3	5,7	3′,4′,5′
Gambier	3	5,7	3′,4′
Mimosa	1	7	3′,4′,5′
Quebracho	1.5	7	3′,4′

retain reactivity towards the bacterial proteins, causing them to malfunction. This is in agreement with previous studies that have shown condensed tannins to be toxic to variety of microorganisms[8] and it is the vegetable tannins that form the least biodegradable constituents of tannery effluent.[9] This toxicity is thought be due to enzyme inhibition, substrate deprivation, action on membranes and metal ion deprivation.[10] This runs counter to the perceived reversibility of tannages and the biodegradability of differently tanned collagenic materials. Therefore, if recycling leather to the earth is a preferred option, chrome tanning is better.[11]

Using starch as a reference biogas producer, the condensed vegetable tannins were added at concentrations ranging from 500 to 300 mg L^{-1}: Table 22.1 sets out the lowest concentrations causing almost complete inhibition to biogas generation.

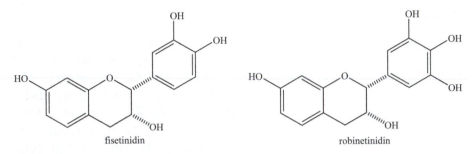

Figure 22.3 Monomeric structures of fisetinidin and robinetinidin.

The inhibitory effect of the condensed vegetable tannins may be explained in part by the pattern of hydroxylation found in the tannin monomers. Quebracho and mimosa are constructed from (epi)fisetinidin and (epi)robinetinidin monomers (see Figure 22.3) respectively, both of which are hydroxylated only on the A-ring at position 7. The monomers of the procyanidin polyphenol gambier and the prodelphinidin polyphenol myrica, (epi)catechin and (epi)gallocatechin, respectively, are hydroxylated on the A-ring at positions 5 and 7. A trihydroxy B-ring makes those species more reactive in terms of crosslinking and quinoid tanning reactions, by reaction at both the A- and B-rings.[11] That implied reactivity towards protein is not reflected in an ability to inhibit microbial attack. However, it is clear that any inhibitory effect is dependent upon the structure–reactivity relationship leading to the hydrogen bonding capability of the polyphenols.

Recent progress in solid waste disposal has also been made by Lanxess AG, who developed a process for degrading tanned waste *in situ*, that is, on the tannery site. The products generated by these processes can be recycled back into the leather-making process as retanning agents.

Breaking the leather down, either chemically or biologically, is not the only method of reducing tanned waste – it is possible to convert the solid waste into another material. Although it is difficult to find published details of products that use this method of reducing waste, it is not a new concept or technology. Some of these products compress solid leather wastes and impregnate with a binder, which produces a board-like material, whereas others use alternative processing to produce a material that looks like leather, but has the properties of a textile.[12]

22.2.1.2   Untanned Waste

Untanned wastes are much easier to process and utilise elsewhere. There is often no need for chemical treatment as the material is not tanned, so it is susceptible to biological attack, *i.e.* it is putrescible. Currently, the most common source of untanned waste is the beamhouse processes of leather manufacture, notably fleshing and unhairing. However, as the technology of removing tanning agents from leather becomes more commercially viable, the biological/proteinaceous products from these processes may act as another source of untanned wastes.

22.2.1.3   Chemical Waste

This area of waste is open to interpretation and is a 'grey area'. It can cover a number of different sources, *e.g.* undissolved waste in the floats – although this is largely treated by an effluent plant and commonly forms part of the cake (see the following section). It could also refer to the unwanted/unused chemicals that periodically must be disposed of or perhaps the solid matter removed during 'deep cleaning' processes.

The finishing of leather often falls into this category, as many finishing chemicals commonly have a short shelf life in comparison with wet end chemicals. In addition, the very nature of mixing and then applying a finish to the leather surface results in waste generation.

Irrespective of the source of the waste, good practice within a tannery, *e.g.* proficient stock management, will help minimise it. However, the real change needs to stem from the chemical companies that manufacture the chemical components, building recycling mechanisms as part of the services they offer, and focusing on research that will eventually lead to more economically viable benign chemicals replacing those more harmful ones currently used.

22.2.1.4   Solid Effluent Waste

It is usual for the treatment of tannery wastes to produce a 'sludge'. It is common for this sludge to be 'dewatered', so that only a minimum amount of moisture is left in the final waste product (often called a 'cake' at this point), before it is sent on to landfill. Depending on the effluent treatment process, this sludge can be rich in metals (normally metal salts) and organic materials, but current purification methodologies struggle to isolate these compounds in a commercially and economically viable manner. If this could be achieved, then it would provide a way of significantly reducing the solid waste generated by the tannery industry.

22.2.2 Liquid Waste

This is one of the areas of 'waste' that could also be considered one of the aforementioned 'grey areas', neither black nor white in definition, but in the context of this book it refers to liquid that is contaminated by (non-solid) wastes that cannot be removed by filtration, *i.e.* wastes that are dissolved. It is also intentional that this section does not cover the topic of 'effluent treatment' – another wide area that can cover chemical, biological and physical treatment methodologies.

There used to be a saying: 'the solution to pollution is dilution'. The suggestion is that, rather than dealing with a problem pollutant at source, one simply dilutes it so that it ceases to be an issue, *e.g.* reducing its concentration to below a toxic level. Assuming that the pollutant is not gaseous, the issue is that the most common diluent is water – a precious resource the use of which industry is trying to reduce – and of course the source of pollution is never actually addressed.

There is now a concerted effort to move in the opposite direction to the above saying – by making the liquid wastes ever more concentrated, they become easier to contain and it is often easier to remove the contaminants (by chemical or biological means), and this supports the industry efforts to reduce water consumption. However, this approach extends beyond simply trying to use less water; nearly all processing involves solution-phase chemistry, so any changes made to the system will have an effect on the liquid waste stream. For example, efforts at improving uptake efficiencies will have a direct impact on the concentration of pollutants.

One of the more significant areas for improvement is the reduction or removal of simple inorganic salts (most commonly sodium chloride) from processing. Currently, sodium chloride is the cheapest form of preservative and the most common anti-swelling additive to a pickle solution. Methods to remove salts from the effluent do exist, *e.g.* reverse osmosis, but these are reactive rather than proactive measures. The only real way forward is to replace salt altogether. New and old technologies to replace salt in preservation and pickling were

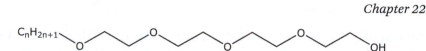

Figure 22.4   Alkyl alcohol condensed with four molecules of ethylene oxide.

discussed in Chapters 3 and 9, but are not currently in mainstream use owing to the increased expense.

Understanding what is in liquid waste will not only help to improve the efficiency of processes, but may also help to avoid issues and/or utilise certain phenomena (*e.g.* precipitation) to reduce the pollution within the effluent waste. A good example of this is understanding and/or utilising the incompatibility of vegetable tannins with non-ionic detergents/surfactants, because most detergents of this type are based on ethylene oxide condensates. They have the structure shown in Figure 22.4. It can be seen that the oxygens are lined up, albeit in an apparently staggered pattern, although there is free rotation about all bonds. This allows plant polyphenols to engage in multiple hydrogen bonding, so the polyphenols become coagulated, resulting in insolubility.

22.2.3   Gaseous Waste

This might be a moot point, but most workers in the industry or those who live near a tannery will testify that gaseous wastes are perceptible and constitute a nuisance. Those in the industry are all familiar with the smell of a tannery – this is the combination of gaseous wastes that are most commonly discussed and are much harder to deal with. Gases are inherently difficult to contain, can be difficult to treat and all, with one exception, have very little processing or commercial value. The key is in appreciating how and why these gases are formed and utilising this information to prevent their formation and subsequent release.

The one waste gas that could have some commercial value is carbon dioxide. Although inroads are being made at reducing the amount of energy that making leather uses, it is nevertheless an energy-intensive process. The production of this energy (some of which may be produced on-site) usually results in the formation of carbon dioxide – consider calculations on the carbon footprint of leather.[13] With the increase in the use of carbon dioxide deliming technology (see Chapter 7), it might be possible to utilise the production of carbon dioxide in this process.

22.3 Overview

The environmental impact of leather making exercises the minds of tanners, not least because so many of the aspects of production are subject to legislated consents for discharges. The minimum treatment that they can apply to liquid streams is to mix and balance, to adjust the pH and allow settling of the precipitate, to form sludge. Thereafter, the options for treatments can vary widely, from membrane filtration or molecular species separation to high-rate biological filters, through upflow anaerobic sludge blanket (UASB) reactors to multiple-approach systems. All these methods take the tanner way from chemistry into technology – hence the absence of their comprehensive inclusion in this book.

It should be clear from reading this book that environmental impact is embedded in the science of leather making. Every step in processing has to be viewed in the light of the developmental steps as follows:

current → cleaner → clean → cleanest

where the definitions are somewhat vague and the steps are relative in an undefined way and cannot be defined in absolute terms. This is often 'baggage', where methodology can be a matter of personal preference, unrelated to the local requirements or the underlying science.

It is not possible to address the principles of tanning and associated processes without including considerations of the nature of the chemical reactions involved, the efficiencies of those reactions and the implications of consequent discharges to air, water or land. In this way, difficulties in defining 'cleanness' notwithstanding, leather science and leather chemistry in particular contribute to the sustainability of the global leather industry.

References

1. Blacksmith Institute, *Pure Earth*, 2011, http://www.blacksmithinstitute.org.
2. C. A. Money, Open Letter - Chrome Tanning Maligned, *Leather International*, 2011.
3. S. Yagoub, PhD thesis, The University of Leicester, 2006.
4. P. N. Levett, *Anaerobic Bacteria*, Open University Press, 1990.
5. J. G. Ferry, *Crit. Rev. Biochem. Mol. Biol.*, 1992, **27**(6), 473.
6. M. Blaut, Metabolism of Methanogens, *Antonie van Leeuwenhoek*, 1994, **66**, 187.

7. M. Aloy, *et al.*, *J. Am. Leather Chem. Assoc.*, 1989, **84**(4), 97.
8. T. K. Bhat, *et al.*, *Biodegradation*, 1998, **9**, 343.
9. A. Folachier, *et al.*, *Das Leder*, 1978, **29**(4), 49.
10. J. D. Reed, *J. Anim. Sci.*, 1995, **73**, 1516.
11. A. D. Covington, *Chem. Soc. Rev.*, 1997, 111.
12. R. Kanigel, *Faux Real*, Joseph Henry Press, Washington DC, 2007.
13. F. Brugnoli, *Euroleather*, Cotance, Brussels, 2018.

23 Theory of Tanning: the Concept of Link–Lock

23.1 Introduction

Among practitioners in the art, there is little understanding of the fundamental nature of the tanning process in terms of the principles that underpin the reactions and even less understanding of the mechanisms upon which that they rely. It has been acceptable to express the reactions in terms of a 'crosslinking' effect, which somehow contributes to the stability of the collagenic species by holding it together, to resist the effects of temperature, which somehow breaks down the structure. In order to improve upon current technology, it is fundamental to understand the tanning reaction at the molecular level: the traditional view does not allow that to happen and is therefore not good enough. What is required is an overview of tanning, a comprehensive theory that explains every tannage – without exception.

A review of the literature indicates that the theoretical basis of leather making, particularly the impact of tanning technologies, had received little attention prior to the work of Gustavson.[1] The suggestion that the mechanism of tanning depends on a crosslinking effect had been made and the breaking of the artificial bridges was claimed to be the basis of the shrinking transition.[2] This view received some criticism because some stabilising reactions actually weaken collagen, but later it was demonstrated that tanning reactions may either weaken or strengthen collagen.[3] In fact, there is no direct correlation between hydrothermal stability and physical stability. The concept

Tanning Chemistry: The Science of Leather 2nd edition
By Anthony D. Covington and William R. Wise
© Anthony D. Covington and William R. Wise 2020
Published by the Royal Society of Chemistry, www.rsc.org

of crosslinking seems to have been generally accepted by the scientific and technological community, although as Gustavson himself remarked:[1]

> *It must be admitted that the concept of crosslinking has been applied in many instances without sufficient experimental support, even purely speculatively. Critical appraisal of this concept is therefore appropriate.*

In his studies on chrome tanning, Gustavson thought that he had provided the proof required, that the proportion of crosslinks in bound chrome could be measured by wet chemistry analysis. It was shown in Chapter 11 that the assumptions and conclusions of that analysis were flawed, so consequently the only evidence that remained to support the concept is indirect.

Chemically unmodified collagen has attracted much attention regarding the origin of its high hydrothermal stability relative to other proteins. The stability of intact, native collagen can be attributed in the first instance to its hierarchy of structure, within which hydrogen bonding[4] and inductive effects[5] are likely to operate. In addition, it has been postulated that the packing of the triple helices into fibrils constitutes a 'polymer-in-a-box' model,[6] providing a further level of stabilisation. Studies on unmodified and chemically modified collagen confirmed that the observed hydrothermal stability is dependent on the moisture content: reducing the water content causes the fibres to approach more closely, preventing them from collapsing into the interstices, and this effect is correlated with elevated denaturation temperature.[7] Therefore, a reduced ability to shrink is the same phenomenon as increased hydrothermal stability. Consequently, chemical modification may reduce the ability of collagen to shrink, which results in a higher observed denaturation temperature.

It has been demonstrated that the triple helix of collagen is surrounded by a supramolecular water sheath nucleated at the hydroxyproline sidechains,[8] in which the water molecules binding the sheath to the triple helix are labile, but constitute a stable structure.[9] In studies of the effects of basic chromium(III) sulfate and tara extract (hydrolysable plant polyphenol) on skin collagen, dynamic mechanical thermal analysis of leather during drying indicated that the amount of structural water is independent of the nature of the stabilising chemistry.[10] When collagen is wet, the matrix can be degraded at elevated temperature, and at the same time hydrogen bonds in the triple helix are broken, allowing the α-helices to become detached from the triple helix structure, observed as shrinking, leading to gelatinisation.

If the supramolecular structure is made up only of water, the process is facilitated to the greatest extent, causing the lowest denaturation temperature. However, if the supramolecular water sheath is chemically modified, resistance to shrinking is introduced, observed as an elevation of the denaturation temperature, that is, higher hydrothermal stability – this can be observed in the effect of sodium sulfate solution on raw collagen, where the sulfate ions can insert into the water structure because they are tetrahedral in shape, resulting in an increase in the shrinkage temperature of a few degrees.

The hydrothermal stability of collagen can be altered by many different chemical reactions, well known in the fields of histology, leather tanning and other industrial applications of collagen.[11] The effects of some of these chemical modifications can be summarised as shown in Table 23.1, where the denaturation temperature is typically measured by the perceptible onset of shrinking.

From these data, well known in the art of leather making, it appears that the stabilising effects fall into two groups: moderate increase or large increase in hydrothermal stability. This has been rationalised in terms of the entropic and enthalpic contributions to the modified collagen structure,[12] where the shrinking/denaturation kinetics are controlled by the enthalpy of activation, moderated by the entropy of activation.[11] In this way, most chemical modifications confer only the moderate shrinkage temperature rise that is observed in the majority of cases, because they contribute only a minor enhancement to the structure. Here, the assumed basis of hydrothermal stability was the assumption that the unravelling of the triple helix was controlled by the supramolecular effect of crosslinking the triple helices, primarily

Table 23.1 Typically observed effects of some chemical modifications on the denaturation temperature ranges of collagen.

Chemical modification	Denaturation temperature (°C)
None	65
Metal salts: *e.g.* Al(III), Ti(IV), Zr(IV)	70–85
Plant polyphenol: gallotannin or ellagitannin	75–80
Plant polyphenol: flavonoid	80–85
Synthetic tanning agent: polymerised phenols	75–85
Aldehyde: formaldehyde or glutaraldehyde	80–85
Aldehydic: phosphonium salt or oxazolidine	80–85
Basic chromium(III) sulfate	105–115
Semi-metal: hydrolysable polyphenol + *e.g.* Al(III)	110–120
Synthetic polymer + aldehydic reagent, *e.g.* melamine–formaldehyde + oxazolidine	105–115

by interacting with the amino acid sidechains. Using that conventional view of the tanning mechanism, it was possible to construct a theoretical model of the tanning mechanism in which hydrothermal stability was correlated with the extent to which the collagen could be held together by stiff crosslinks, which were bound to the triple helices by high-energy interactions.[11,13,14] This view seemed to explain the observed effects of known tanning systems, although it did not provide an explanation for the relationship between the chemistry of tanning and the mechanism of hydrothermal shrinking.

In his John Arthur Wilson Memorial Lecture of 2001, Ramasami reviewed the history of tanning theories over the last century in an attempt to create a unified theory.[15] He too accepted the notion of cross-linking, but focused his thinking on the effect of tanning compounds to alter the structure of collagen at different levels of the hierarchy of collagen structure. Since the definition of tanning is based on resistance to biochemical attack, he considered the effects of tanning reagents on the reaction between collagenase and collagen. He concluded that there is a correlation between hydrothermal stability and conformational changes at the fibrillar–triple helix level of collagen structure and the consequent effect on resistance to enzymatic degradation.

There is a clear discrimination between those tanning reactions which confer moderate hydrothermal stability and those which confer high hydrothermal stability to collagen. This begs the question: if most chemical modifications can increase the denaturation temperature only by up to 20 °C, what is the mechanism by which collagen can achieve high hydrothermal stability when the denaturation temperature is increased by over 60 °C? To put this into context, assuming an activation energy for the shrinking/denaturation reaction of 518 kJ mol^{-1},[16] elevating the shrinkage temperature from 65 to 85 °C enhances the hydrothermal stability by four orders of magnitude, but the rise from 85 to 110 °C represents a further five orders of magnitude.

In this chapter, an answer to that question is proposed, by synthesising an explanation, a mechanism of tanning, based on modern thinking in leather-making science, the observation of known tanning reactions and experimental investigations of novel reactions for stabilising collagen. The requirements of any useful theory are threefold:

1. it must explain all known stabilising reactions, without exception;
2. it must be capable of predicting the requirements for new reactions;
3. it must allow accurate predictions of the outcomes of new reactions.

23.2 Discussion of the Evidence

The combination reactions in Table 23.1 have some features in common.

- The primary reaction between hydrolysable plant polyphenols, the gallotannins and ellagitannins, and collagen is through multiple hydrogen bonding. The subsequent reaction is to crosslink the tannin molecules together, *via* the pyrogallol moieties locked together by the complexation reaction with metal ions.[17,18]
- There is a similar primary reaction between collagen and the flavonoid tannins, which is weaker in terms of the availability of phenolic hydroxyl reaction sites on the tannin, but which also includes some covalent bonding between the lysine sidechains of the collagen and the aromatic A-ring of the flavonoid, *via* quinoid reactions. The next reaction involves crosslinking the tannin molecules by oxazolidine, an aldehydic reactant; this occurs at the 6- and 8-positions on the A-rings of procyanidin or prorobinetinidin polyphenols, and additionally at the 2'- and 6'-positions of the B-rings of prodelphinidin or profisetinidin polyphenols,[19–21] where the latter polyphenolic reactions yield higher denaturation temperatures.
- Similar high-stability combination reactions have been observed with melamine–formaldehyde polymer as primary reactant, with tetrakis(hydroxymethyl)phosphonium sulfate as the aldehydic crosslinker[22] and with low molecular weight phenolic compounds crosslinked with aldehydic compounds or laccase (polyphenol oxidase).[23]

In each case, the synergistic combination reactions create matrices of crosslinked polyphenolic or synthetic polymeric species, which means that they act in concert, effectively working as a single chemical moiety. In the case of the flavonoid combination stabilisation reaction, it has been shown that an important feature of the reaction is the linking of the polyphenol species to collagen *via* the aldehydic crosslinking reaction.[15] It is also a feature of these high hydrothermal stability reactions that they exhibit a strong reaction with collagen: in some cases multiple hydrogen bonding occurs and in others there is some covalent reaction. In this way, the matrix is firmly bound to the collagen structure. It has been proposed that this aspect of the combination reaction is an important element in the overall requirements for matrix stabilisation of collagen that lead to a high shrinkage temperature.[23] The notion does not invoke the requirement for resistance to displacement of the tanning reagent in chemical terms, but does incorporate the idea that firmer bonding can influence the distortion of the structure, particularly during the shrinking transition. Supporting evidence of the role played by fixation of the matrix to collagen comes from the work of Holmes,[24] who showed that modifying collagen by binding chelating pairs of carboxyl groups to the amino groups of lysine can enhance the weak stabilising effect of aluminium(III) salts, to the extent of increasing the denaturation temperature from 75 to 95 °C.

In the group of high-stability reactions, the process involving chromium(III) salt might appear to be a chemical exception, having little in common with the organic combination reactions. This is deceptive. It is known that the chromium(III) species are covalently bound at carboxyl sidechains. Extended X-ray absorption fine structure (EXAFS) studies of chromium(III) bound to collagen showed that the dominant bound species are linear tetrachromium compounds, but the counterion, in this case sulfate, is not directly bound to the chromium as a ligand.[25] Furthermore, if the counterion is different, *e.g.* chloride or perchlorate, the effect of the stabilisation is only moderate, producing denaturation temperatures of about 85 °C.[26] Hence the chromium(III) species and the counterion must create a matrix with the supramolecular water structure, in a system analogous to the combination reactions. In the case of the inorganic reaction, the effect is a combination of the metal ions and the counterions and the water. The overall synergistic nature of the combined reactions depends on the functioning of the counterion as a structure maker in water; sulfate is known to act in this way, but chloride is a water structure breaker,[27] shown by circular dichroism to be capable of destabilising the collagen structure.[28]

Confirmation of the role of the counterion in the link–lock mechanism comes from the work of DasGupta:[29] treating wet blue with polycarboxylates for 1 h increased the shrinkage temperature commensurate with the ability to lock the bound chromium species together, as shown in Table 23.2.

The following points arising from Table 23.2 are worth noting.

1. The reaction cannot be covalent complexation, because the time of treatment is too short, hence the salts must be bound by electrostatics, as is sulfate.
2. Phthalate has a negligible effect because it can react only in a chelate manner because the carboxyl groups are too close together to crosslink.

Table 23.2 Effect of 5% polycarboxylate on the shrinkage temperature of wet blue leather.

Ammonium salt	Structure	T_s (°C)
None		114
Phthalate	1,2-Dicarboxybenzene	115
Isophthalate	1,3-Dicarboxybenzene	121
Terephthalate	1,4-Dicarboxybenzene	121
Pyromellitate	1,2,4,5-Tetracarboxybenzene	129

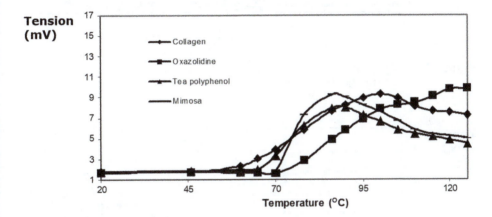

Figure 23.1 Hydrothermal isometric tension tests on skin treated with single organic stabilising agents. Courtesy of the International Union of Leather Technologists and Chemists Societies.

3. Isophthalate and terephthalate have similar effects because they can crosslink in more or less the same way.
4. Pyromellitate has the most powerful effect because it can crosslink through all four carboxyl groups, each carrying full charges. Hence the interaction is analogous to that with sulfate, but making a stronger supramolecular matrix.

A useful method for gaining insight into the structure of modified collagen is hydrothermal isometric tension (HIT), where samples are constrained against shrinking and the forces generated during transitions are recorded. The HIT curves of unmodified and chemically modified collagen are shown in Figures 23.1–21.3.[30]

The traces can be divided into three parts: the first is the tension-increasing process from zero to maximum tension, followed by a relatively constant-tension process, if present, and finally the tension will be constant or a relaxation process will occur, owing to the gradual destruction of collagen structure or rupture of some crosslinking

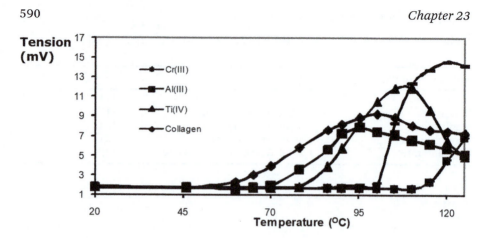

Figure 23.2 Hydrothermal isometric tension tests on skin treated with mineral stabilising agents. Courtesy of the International Union of Leather Technologists and Chemists Societies.

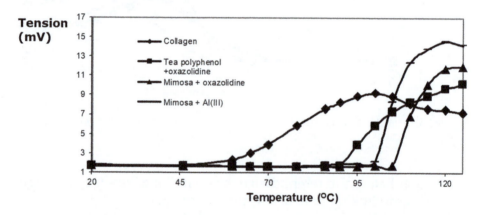

Figure 23.3 Hydrothermal isometric tension tests on skin treated with combinations of stabilising agents. Courtesy of the International Union of
Leather Technologists and Chemists Societies.

bonds. The slope of the curve in the tension-increasing process
accounts for the rigidity of the collagen fibre, caused by crosslinks: the
steeper the slope of the contraction curve, the more crosslinks should
be present in the collagen materials. Relaxation represents the stability of these connecting elements (crosslinking bonds): the steeper
the rate of the relaxation curve, the more unstable is the crosslinking.
In the case of collagen, the crosslinking does not refer to direct links
between triple helices, but to the structure of the intervening matrix.
Although we cannot quantify crosslinking, we still can obtain some
useful information about the relative crosslink density and stability
from the shapes of the HIT curves. The calculated results are given in
Table 23.3.[30]

Table 23.3 Relative rates of increase in tension after shrinking transition is initiated and relative rates of decrease in tension after the shrinking transition.

Stabilising agent	Slope of contraction	Slope of relaxation
None (raw collagen)	0.25	−0.13
Aluminium(III)	0.27	−0.20
Oxazolidine	0.28	0.00
Chromium(III)	0.29	Assumed zero
Green tea polyphenol	0.37	−0.13
Titanium(IV)	0.50	−0.52
Mimosa	0.67	−0.20
Mimosa + oxazolidine	0.91	Assumed zero
Mimosa + aluminium(III)	1.23	−0.05

Stabilisation by polyphenol reactions yields results that can be understood in terms of the availability of hydrogen bonding and subsequent localised crosslinking. However, the rate of tension increase during the shrinking transition of chromium(III)-stabilised collagen exhibits a result similar to that with raw collagen. The natural conclusion is that chromium does not crosslink collagen by joining adjacent sidechains. This runs counter to the traditional view of collagen stabilisation, in which the sidechains of the collagen structure are supposed to be linked by a single species, such as chromium(III) molecular ions; however, the only quantitative analysis in the literature leading to such a conclusion, the Gustavson model,[1] has been shown to be flawed.[31] Therefore, the traditional view of crosslinking does not have to be invoked in explaining chemical stabilisation reactions. The explanation lies in the creation of a chemical matrix, securely bound to the collagen.

Bone is an example of a situation where the stability of collagen clearly comes from the presence of a matrix. Here, the matrix is polymeric hydroxyapatite, not thought of as a typical crosslinker, but which can raise the shrinkage temperature of the constrained collagen to upwards of 155 °C.[6,32] In terms of the 'polymer-in-a-box',[6] this is a very strong 'box', an extreme case and probably the upper limit to collagen stabilisation, without dehydrating the protein, although the effect may be more like dehydration than stabilisation in an aqueous medium. Indeed, it is not yet clear that the mechanism of denaturation of collagen in bone is the same as that for free collagen, because at this elevated temperature the melting of the triple helix will be competing with hydrolysis of the peptide backbone, which has a lower activation energy, 173 kJ mol[−1].[33] This would explain the surprisingly low activation energies (~350 kJ mol[−1]) observed by Trębacz and Wójtowicz for the denaturation of mineralised collagen.[34]

The stability of unmodified or chemically modified collagen must depend on its ability to collapse or shrink by unravelling its chains into the available space between the chains. The ease with which this can happen depends on the constraints applied to the chains or the ease with which the intervening molecules can be displaced as the collagen shrinks. If those molecules are water, this matrix can be broken down and displaced relatively easily, since the reaction is the breaking of the hydrogen bonds in bulk water. If the matrix is stabilised by the inclusion of species bound to collagen, also by substituting some of the supramolecular water and interacting with the remaining water, the matrix is less easily displaced, observed as an increase in denaturation temperature. If the notion of direct crosslinking between triple helices is dismissed as the source of stabilisation, the matrix displacement model explains the similarity of outcome that is already familiar for the stabilising reactions. Some of the more compelling evidence comes from DasGupta,[35] in his studies of oxazolidine tanning. In comparing a monofunctional with a bifunctional derivative, it was shown that the outcome of solo tanning of sheepskin was the same; the shrinkage temperature was about 80 °C, *i.e.* the potential ability to crosslink did not affect the linking reaction. However, when used to retan vegetable-tanned leather, the monofunctional oxazolidine conferred only a small increase in T_s, but the bifunctional derivative produced a large increase in T_s. These observations indicate that the bifunctional derivative is capable of crosslinking, but does not necessarily do it in the context of solo tanning.

Therefore, the maximum effect of a single reagent, even with the potential to crosslink, is due to interference in the shrinking process, *i.e.* the bound species merely hinder the unravelling process by getting in the way. This view is confirmed by the HIT data and the specific case of chromium(III): if the crosslinking reaction is paramount, the stabilising effect of chromium(III) salts would be independent of the counterion – but it is demonstrably true that shrinkage/denaturation temperature is indeed highly dependent on the nature of the counterion.[31,36] Therefore, crosslinking is not paramount – at least not in the traditional way of thinking.

Even the relationship between the linking agent and the supramolecular water does not control the outcome of the stabilisation reaction. Table 23.1 indicates the range of stabilising reactions that yield only moderate hydrothermal stability: all have different relationships with water, in terms of their hydrophilicity or hydrophobicity, but the results are the same. The small variations in hydrothermal stability achievable by linking reagents appear to be related to the nature of their primary interaction with collagen: the weaker the interaction, the lower is the denaturation temperature. The order of increasing hydrothermal stability depends on the chemistry of the bonding:

electrostatic < hydrogen bonded < covalent

exemplified by the following selection of reactions:

Al(III) < hydrolysable polyphenol < aldehyde

It has been assumed that the shrinking reaction is independent of the stabilising reaction, because the enthalpy of denaturation is independent of the stabilising chemistry:[37] the assumption was that the process is controlled only by the breaking of hydrogen bonds within the triple helix. If collagen stabilisation is modelled in these terms,[38] the number of crosslinks required to elevate the melting temperature is unacceptably high. However, it is more satisfactory to regard the shrinking process of collagen, unmodified or modified, as being controlled by the breaking of hydrogen bonds of the water in the matrix. This is clear, since the relationship between denaturation temperature and moisture content is broadly similar (although different in detail) for unmodified collagen and highly stabilised, chromium(III)-tanned collagen,[39] where the difference in the denaturation temperatures of the wet materials is about 40 °C, but the denaturation/decomposition temperature for both of the dry materials is about 200 °C.

It has been shown that the water associated with the collagen structure is practically independent of the tanning chemistry: the amount is the same, about 23%, whether the collagen is stabilised with chromium(III), formaldehyde or high levels of vegetable tannin, hydrolysable or condensed.[10,39] It can be further postulated that some of this structural water must remain in the matrix modified by the tanning chemistry. Therefore, shrinking involves the unravelling of the triple helix, aided by the breakdown of the residual structural water associated with the triple helix. That unravelling is possible only if the intervening matrix can be rearranged around the new protein conformation, which presupposes that the hydrothermal event can also affect the matrix. In the case of a water matrix, the least hydrothermally stable, labile water can hydrate the unravelled chains and therefore there is no need to postulate a displacement of water or any other part of the supramolecular matrix out of the material.

As an approach to stabilising collagen, merely loading the structure with molecular species is not sufficient to confer high hydrothermal stability. For example, plant polyphenols confer only moderate stability, limited to about 85 °C, even when present at 30% on dry weight of collagen. Like most other stabilising reactions, the chemical reactions are limited to linking elements of the collagen structure to a relatively unstable matrix, in this case a packed array of unlinked molecules, but with some water still in the interstices. In the case of high offers of condensed polyphenols, the excess molecules can interact with bound molecules, forming 'reds' *in situ*. However, the interactions are weak and easily broken under mild hydrothermal conditions; therefore, the interactions do not constitute a stabilisation of the matrix, so the hydrothermal stability remains unchanged. The difference between the hydrolysable tannins and the condensed tannins can now be understood in terms of their abilities to function as linking agents: an element of covalency in the reactions of condensed tannins with collagen contributes to the stability of the matrix, resulting in shrinkage temperatures that are higher by 5–10 °C than the effects of a similar reaction based on hydrogen bonding alone.

In those chemical processes that result in high hydrothermal stability, the linking step must be combined with an additional step that locks the components of the matrix together. In this way, the matrix acts more like a single chemical compound, which is much less easily displaced. The higher energy required to achieve breakdown of the structure is observed as a higher temperature transition. It is an important aspect of the matrix-stabilising mechanism that the matrix should be bound to the collagen in a stable way, so that displacement of the interaction, which might lead to allowing shrinking, is prevented. This element of the high hydrothermal stability requirements may be conferred during the linking reaction or it may be conferred as a side effect, an additional aspect of the locking reaction. It is noteworthy that it was reported to be impossible to create boilfast leather if the pelt is in the swollen condition,[40] where it was assumed that the reason was the enhanced separation between the reaction sites. However, the observation can now be rationalised in terms of the separation between the chromium(III) linking agents and the consequently reduced effectiveness of the locking reaction.

Returning to the conventional concept of crosslinking between side-chains, it should not be concluded that such crosslinking does not occur. On paper, if the chemistry of a tanning system allows bonds to be created between two sites on adjacent triple helices, then crosslinks may be formed. Indeed, it can be shown that this happens, for example with 3,3'-difluoro-4,4'-dinitrodiphenyl sulfone reaction at lysine sidechains.[41] However, it is also worth recalling that glutaraldehyde, a reagent apparently capable of crosslinking to a high degree, does not exhibit any such reaction.[42] The conclusion from the link–lock view of tanning is that the hydrothermal stability does not depend on creating crosslinks of this type, because by themselves they are incapable of conferring high shrinkage temperatures.

23.3 Conclusion

All reactions that stabilise collagen conform to the link–lock model.

1. Single-component tannages are restricted by the thermodynamics of the shrinking reaction to confer only a moderate shrinkage temperature.
2. A high shrinkage temperature can be achieved only by tanning with at least two conventional components.
3. A high shrinkage temperature is achievable if there is strong primary chemical bonding between the tanning agent and the collagen and if the primary tanning agent is locked in place by a secondary tanning agent.
4. The chemistry of the reactions is less important than conforming to the requirements of the link–lock mechanism. Therefore, a high shrinkage temperature is achievable by many more routes than are currently known.

The link–lock theory turns traditional thinking on its head: for many years, it was thought that crosslinking was all-important, but the new theory says that, although it may occur, crosslinking itself cannot explain the outcome of tanning reactions. The old view required highly specific reactions between collagen and the tanning species to occur, an entropically disfavoured reaction; the new view says that the linking step is effectively non-specific. Only in the case of high hydrothermal stability tannages is crosslinking invoked, but this has nothing to do with collagen chemistry.

This new explanation of the stabilisation of collagen is simpler and more powerful than the currently accepted view. There is no evidence in the literature to support the old assumed view, but much evidence to indicate that the new, matrix-based view is right. The link–lock theory has the power to predict the outcome of processes yet to be developed, by the application of chemical understanding of reagent affinities. Moreover, it can be the basis for predicting the chemistries of new reagents to achieve high hydrothermal stability outcomes.

The link–lock theory was proposed in 2006 and definitively argued for in 2009 in the first edition of this book. Since then, the theory has received some support in the literature, but notably it has not been criticised, nor has an alternative or even a modified model been proposed.

References

1. K. H. Gustavson, *J. Am. Leather Chem. Assoc.*, 1953, **48**(9), 559.
2. K. H. Gustavson, *J. Am. Leather Chem. Assoc.*, 1946, **41**(2), 47.
3. K. H. Jacobsen and R. M. Lollar, *J. Am. Leather Chem. Assoc.*, 1951, **46**(1), 7.
4. L. Pauling, *J. Am. Chem. Soc.*, 1940, **62**(10), 2643.
5. S. K. Holmgren, *et al.*, *Nature*, 1998, **392**, 666.
6. C. A. Miles and M. Ghelashvili, *Biophys. J.*, 1999, **76**, 3243.
7. C. A. Miles, C. A. Avery, V. V. Rodin and A. J. Bailey, *J. Mol. Biol.*, 2005, **346**, 551.
8. H. M. Berman, J. Bella and B. Brodsky, *Structure*, 1995, 3(9), 893.
9. M. Melacini, A. J. J. Bonvin, M. Goodman, R. Boelens and R. Kaptein, *J. Mol. Biol.*, 2000, **300**, 1041.
10. S. Jeyapalina, G. E. Attenburrow and A. D. Covington, *J. Soc. Leather Technol. Chem.*, 2007, **91**(3), 102.
11. A. D. Covington, *J. Soc. Leather Technol. Chem.*, 2001, **85**(1), 24.
12. C. E. Weir, *J. Am. Leather Chem. Assoc.*, 1949, **44**(3), 108.
13. A. D. Covington, *et al.*, *J. Am. Leather Chem. Assoc.*, 1998, **93**(4), 107.
14. A. D. Covington, *J. Am. Leather Chem. Assoc.*, 1998, **93**(6), 168.
15. T. Ramasami, *J. Am. Leather Chem. Assoc.*, 2001, **96**(8), 290.
16. C. A. Miles, T. V. Burjanadze and A. J. Bailey, *J. Mol. Biol.*, 1995, **245**, 437.
17. J. F. Hernandez and W. Kallenberger, *J. Am. Leather Chem. Assoc.*, 1984, **79**(5), 182.
18. R. A. Hancock, S. T. Orszulik and R. L. Sykes, *J. Soc. Leather Technol. Chem.*, 1980, **64**(2), 32.
19. A. D. Covington and B. Shi, *J. Soc. Leather Technol. Chem.*, 1998, **82**(2), 64.
20. A. D. Covington, B. Shi, Y. He, H. Fan, S. Zeng and G. E. Attenburrow, *J. Soc. Leather Technol. Chem.*, 1999, **83**(1), 8.
21. A. D. Covington, C. S. Evans, T. H. Lilley and L. Song, *J. Am. Leather Chem. Assoc.*, 2005, **100**(9), 325.
22. A. D. Covington and M. Song, *Proceedings of IULTCS Congress*, London, 1997.
23. A. D. Covington, C. S. Evans, T. H. Lilley and O. Suparno, *J. Am. Leather Chem. Assoc.*, 2005, **100**(9), 336.
24. J. M. Holmes, *J. Soc. Leather Technol. Chem.*, 1996, **80**(5), 133.
25. A. D. Covington, G. S. Lampard, O. Menderes, A. V. Chadwick, G. Rafeletos and P. O'Brien, *Polyhedron*, 2001, **20**, 461.
26. K. Shirai, K. Takahashi and K. Wada, *Hikaku Kagaku*, 1975, **21**(3), 128.
27. A. Cooper, *Biochem. J.*, 1970, **118**, 355.
28. E. M. Brown, H. M. Farrell and R. J. Wildermuth, *J. Protein Chem.*, 2000, **19**(2), 85.
29. S. DasGupta, *J. Soc. Leather Technol. Chem.*, 1979, **63**(4), 69.
30. A. D. Covington and L. Song, *Proceedings of IULTCS Congress*, Cancun Mexico, 2003.
31. A. D. Covington, *J. Am. Leather Chem. Assoc.*, 2001, **96**(12), 461.
32. P. L. Kronick and P. Cooke, *Connect. Tissue Res.*, 1996, **33**(4), 275.
33. K. M. Holmes, K. A. Robson-Brown, W. P. Oates and M. J. Collins, *J. Archaeol. Sci.*, 2005, **32**, 157.

34. H. Trębacz and K. Wójtowicz, *Int. J. Biol. Macromol.*, 2005, **37**, 257.
35. S. DasGupta, *J. Soc. Leather Technol. Chem.*, 1977, **61**(5), 97.
36. K. Shirai, K. Takahashi and K. Wada, *Hikaku Kagaku*, 1984, **30**(2), 91.
37. A. D. Covington, R. A. Hancock and I. A. Ioannidis, *J. Soc. Leather Technol. Chem.*, 1989, **73**(1), 1.
38. M. J. Collins, M. S. Riley, A. M. Child and G. Turner-Walker, *J. Archaeol. Sci.*, 1995, **22**(2), 175.
39. M. Komanowsky, *J. Am. Leather Chem. Assoc.*, 1991, **86**(5), 269.
40. K. H. Gustavson, *Stiasny Festschrift*, 1937, p. 117.
41. H. Zahn and D. Wegerle, *Das Leder*, 1954, **5**(6), 121.
42. L. H. H. O. Damink, *et al.*, *J. Mater. Sci.: Mater. Med.*, 1995, **6**, 460.

24 The Future of Tanning Chemistry

24.1 Introduction: the Future of Chrome Tanning

The environmental impact of chromium(III) is low: as a reagent, basic chromium salts are safe to use industrially and can be managed efficiently, to the extent that discharges from tanneries can routinely be as low as a few parts per million in the liquid effluent. Therefore, it is clear that the industry should be able to meet all the future requirements of environmental impact. Consequently, we can reasonably assume that the future of tanning will include a major role for chrome. Nevertheless, the technology can be changed: efficiency of use can be improved, as can the outcome of the reaction, in terms of the performance of the leather (shrinkage temperature) and the effectiveness of the reaction (shrinkage temperature rise per unit bound chrome).[1]

The role of masking in chrome tanning is an important feature that can be exploited technologically. The use of specific masking agents, which gradually increase the astringency of the chrome species, by increasing their tendency to transfer from solution to substrate, is an aspect of the reaction that has not attracted scientific attention. Here, the requirement is to match the rate of diminishing concentration of chrome in solution with the rate of masking complexation, to maintain or increase the rate of chrome uptake. This type of reaction is already exploited technologically with the use of disodium phthalate,

Tanning Chemistry: The Science of Leather 2nd edition
By Anthony D. Covington and William R. Wise
© Anthony D. Covington and William R. Wise 2020
Published by the Royal Society of Chemistry, www.rsc.org

although the degree of hydrophobicity conferred by even a very low masking ratio can cause undesirably fast surface reactions. Other hydrophobic masking agents could be developed to give a more controllable increase in astringency.

It is already recognised that the chrome tanning reaction is controlled by the effect of pH on the reactivity of the collagen substrate: a second-order effect is the increase in the hydrophobicity of the chrome species by polymerisation. In addition, the rate of reaction is controlled by temperature. However, the efficiency of the process is limited by the role of the solvent in retaining the reactant in solution by solvation: this typically is countered only by applying extreme conditions of pH, which are likely to create problems of surface fixation, causing staining and resistance to dyeing.[1]

The role of the counterion in chrome tanning offers potential for change. It has been shown that the chrome tanning reaction is controlled by the particular counterion present.[2-4] It was fortunate for the leather industry that chrome alum [potassium chromium(III) sulfate hydrate] was the most readily available salt for the original trials of tanning ability: sulfate ion is highly effective in creating a stable supramolecular matrix, because it is a structure maker in water,[5,6] which is critical in the stabilising mechanism. The effect is very different if other salts are used, *e.g.* chloride or perchlorate, where the outcome is only moderate hydrothermal stability. However, even if these salts are used, the high hydrothermal stability can be acquired by treating the leather with another counterion. Since the effect is independent of a complexing reaction, the process of modifying the moderate tanning effect is fast. This opens up the chrome tanning reaction to modifications that exploit the separation of the link–lock reactions.

- The environmentally damaging sulfate ion might be replaced with other less damaging counterions, such as nitrate.
- The reactive counterions can then be applied; options include using the stoichiometric quantity of sulfate or organic anions.
- The counterion might be replaced with polymeric agents, including polyacrylates with the proper steric properties.

24.2 Other Mineral Tanning Options

It has been suggested that the environmental impact of chromium can be alleviated by replacing all or part of the offer with other metal tanning salts, where the likeliest candidates[7] are Al(III), Ti(III)/(IV),

Fe(II)/(III), Zr(IV) and lanthanide(III). The list of available options is limited to these few salts because of considerations of cost, availability, toxicity and reactivity towards carboxyl groups. All metals salts are mixable in all proportions in this context. However, the following general truism should be noted.

> *The damaging effect of any (alleged) pollutant is not eliminated and is barely significantly mitigated by reducing the degree to which it is used.*

The effects of substituting other metals into the chrome tanning process or even complete replacement have been discussed, and are briefly reviewed in the following. There are circumstances when other mineral tanning systems might be appropriate, but it should be clear that there is no viable alternative for general applications and particularly for those applications that require high hydrothermal stability.

24.3   Non-chrome Tanning for 'Chrome-free' Leather

In the current climate of emphasis on ecologically friendly processing and products, attention has been directed to the environmental impact of chromium(III) in tanning and leather. Therefore, this situation has fuelled the movement to produce so-called 'natural' leathers, marketed as and perceived to be 'chemical free', or using other appellations that are designed to suggest their ecological credentials. The perception of the abuse of resources in leather making has been exploited in the marketing of leather that is not mineral tanned. Hence it is typically assumed that *chrome free = organic*. Consequently, leathers and leather articles are offered to the market as non-chrome.

The use of the term 'chrome-free' leather implies that there is a problem underlying the inclusion of chromium(III) salt in the production of leather. This concept has its origins in the (alleged) environmental impact of chromium(III) waste streams, liquid and solid, and the mobilisation and viability of chromium(III) ions from tanned leather waste. However, latterly the accusation that some chrome leathers are contaminated with chromium(VI) or that chromium(VI) can be generated during the use of the leather article has heightened the notion that chrome-tanned leather may be undesirable.[8]

The decision to produce leather by means other than chrome tanning depends on elements of technology and elements of economics, and some of these aspects will clearly overlap. In each case, there are points for change and points against. If production methods are changed at the tanning stage, it must be accepted that every subsequent process step will be altered – some will benefit, others will not, and additional technological problems will have to be overcome:

Technological aspects:

- environmental impact of waste streams;
- 'just-in-time' quick response, wet white technologies;
- colour; may be counterproductive if vegetable tannins are used;
- changed properties, *e.g.* hydrophilic–hydrophobic balance (HHB), ionisation potential (IEP);
- process timings may or may not benefit;
- resource management will change, *e.g.* the continued programme of conventional retanning, dyeing then fatliquoring.

Economic aspects:

- waste treatment and cost of treatments, adding value and use of by-products;
- 'recyclability' of leather, which traditionally means compostability, *i.e.* returning the leather into the environment without adverse impact – although this must involve at least partial denaturation of the leather, by either pH or heat treatment or both;
- thinking beyond 'cradle to grave', extending to 'cradle to cradle';
- marketing of 'natural' leather;
- public perception of leather making – the use of 'chemicals', renewable reagents;
- eco-labelling, as a marketing tool for developed economies or just access for developing economies to developed markets.

It is important to retain a clear idea of what is required in the alternative leather. Typically, what is required is 'the same, but different'. Here the obvious difference is the absence of chromium(III) salt, but which elements or aspects of chrome leather character/performance/properties should or need to be retained is less simple. Here, the tanner needs to distinguish between the specific properties required in the leather and the more general property of 'mineral character': the former may be relatively easy to reproduce, at least with regard to individual properties/performance, but the latter is more difficult to reproduce, since mineral character usually means chrome-tanned character.

It is clear that no other tanning system will be able to reproduce all the features of chrome leather. Therefore, it is unlikely that a generic leather-making process with the versatility of chromium(III) will be developed; any new leather-making process must be tailored so that the resulting leather can meet the needs of the required application, to be 'fit for purpose'.

24.4 Single Tanning Options

- Other metals.
  For example, Al(III), Ti(III) or (IV), Zr(IV), possibly Fe(II) or (III), lanthanides(III). In each case, the leather is more cationic, creating problems with anionic reagents; the leather can be collapsed or over-filled depending on the metal. The HHB of leather depends on the bonding between the metal salt and the collagen. In all cases T_s is lower than for Cr(III).
- Other inorganic reagents.
  This category includes complex reagents, *e.g.* sodium aluminosilicate, silicates, colloidal silica, polyphosphates, sulfur, complex anions such as tungstates, phosphotungstates, molybdates, phosphomolybdates, uranyl salts, *i.e.* those that have application in histology.
- Aldehydic agents.
  This group includes glutaraldehyde and derivatives, polysaccharide (starch, dextrin, alginate, *etc.*) derivatives, analogous derivatives, for example of hyaluronic acid, oxazolidine, phosphonium salts. These reagents are not equivalent: they react in different polymerised states and their affinities are different, even though the chemistry of their reactivities is the same, similar or analogous. In each case, the leather is plumped up and made hydrophilic. The IEP is raised. The colour of the leather depends on the tanning agent – glutaraldehyde and derivatives confer colour, others do not. Oxazolidine and phosphonium salts may produce formaldehyde in the leather or there is a perception based on odour (which may be the original reagent itself).
- Plant polyphenols, vegetable tannins – hydrolysable, condensed. Reactivity is a problem for a drum process, so the reaction may require syntan assistance. Filling effects and high hydrophilicity

characterise these leathers. Colour saddening and light fast-ness may be a problem, particularly with the condensed tan-nins. Migration of tannin can occur when the leather is wet. The shrinkage temperature is only moderate, not more than 80 °C for hydrolysable and not more than 85 °C for condensed tannins.

- Syntans, *i.e.* retans or replacement syntans.
  As for vegetable tannins, although reactivity is more controllable by choice of structure. Light fastness is similarly controllable.
- Polymers and resins.
  This category includes melamine–formaldehyde, acrylates, sty-rene maleic anhydride, urethanes.
- Miscellaneous processes, *e.g.* oil tanning, *in situ* polymerisation.

24.5   Tanning Combinations

It is clear from the brief account of the individual tanning options set out above that the number of available combinations is very large. Hence it is important to understand the principles of combination tanning. Analysis can commence with the effects of the components on the shrinkage temperature, as an indicator of the combined chemistries.

1. The components may react individually, *i.e.* they do not interact, primarily because they react with collagen *via* different mecha-nisms and there is no chemical affinity between the reagents.

   Here the contributions to the overall shrinkage temperature can be treated additively. Also, the properties of the leather will be a combination of the two tannages, although the first tannage applied will tend to dominate the leather character.

   In Table 24.1, this is indicated by I = Individual, i = lesser effect.

2. The components may interact antagonistically, *i.e.* they inter-fere with each other. This may be due to competition for reac-tion sites or one component simply blocks the availability of reaction sites for the other component. It can be assumed either that the two components do not interact or that one component will react with the other in preference to reacting with collagen. Under these circumstances, the interaction does not constitute the creation of a useful new tanning species, *i.e.* the reaction of the second component does not involve crosslinking the first component. The outcome of the combined reaction is likely to be domination of the properties by the reagent which is in excess or has greater affinity for collagen.

Table 24.1 Indicative/possible interactions in combination tanning.[a]

First reagent	Second reagent						
	Metal salts	Inor-ganic	Alde-hydic	Vegetable tanning: hydro-lysable	Vegetable tanning: condensed	Syn-tan	Resin
Metal salts	I	I	I	A	I	*a*	I
Inorganic	I	*i*	I	*a*	I	*a*	I
Aldehydic	I	I	A	I	*s*	*s*	*s*
Vegetable tanning: hydrolysable	S	*s*	I	A	A	A	A
Vegetable tanning: condensed	I	I	S	A	A	A	A
Syntan	*s*	*s*	*s*	A	A	A	A
Resin	I	I	*s*	A	A	A	A

[a]For explanation of the letters in the table, see the text.

Here, the hydrothermal stability of the combination will be less than the anticipated sum of the effects of the individual tanning reactions. The effect of destabilising the leather is likely to be undesirable.

In Table 24.1, this is indicated by A = Antagonism, a = lesser effect.

3. The components may react synergistically, *i.e.* they interact to create a new species, which adds more than expected to the overall hydrothermal stability. Since a new tanning chemical species is created, the properties of the leather are likely to reflect the individual tanning effects, but to differ significantly from a simple additive outcome.

The additional elevation of T_s may constitute a high-stability process, as rationalised in the link–lock theory,[5] discussed in Chapter 23.

In Table 24.1, this is indicated by S = Synergistic, s = lesser effect.

In making choices of combinations, it is important to consider how the reagents combine to create the leather properties other than hydrothermal stability: these could include handle (softness, stiffness, fullness, *etc.*), HHB and strength (tensile, tear, extensibility, *etc.*). If the components react individually, it might be expected that the properties of each would be conserved in the leather, depending on the relative amounts fixed. If the components react antagonistically, the overall properties will be determined by the dominant reactant, although they may be modified by the presence of the second component, unless the antagonism means that the second reagent is not fixed. Synergistic reaction means that the tanning reaction is changed, so it can be assumed that the outcome will not be the same as the sum of the individual properties, although it can also be assumed that the leather will still reflect the individual properties to some extent.

In each case that might be considered, it is important to recognise that the order of addition of the reagents can make a significant difference to the outcome. The nature of the interaction in two-component combination tannages can be analysed as shown in Table 24.1, where upper-case letters indicate a greater degree of understanding of the effect and lower-case italics indicate a lack of clarity in the effects or that there may be exceptions.

Some general points can be made, as follows.

- The concept of the link–lock mechanism is useful in this context, because it offers a model of the tanning reaction that can be visualised and hence used in the analysis of reaction and outcome. This matrix model is easier to use than the previous models of specific interaction with the collagen sidechains.
- All metals are mixable in all proportions for tanning and all metals react primarily at carboxyl groups, typically electrostatically. It is unlikely that there will be any positive impact on the tanning reaction from mixed speciation. The filling/collapsing effects can be controlled by mixing the metal offers. There is no benefit with respect to T_s or cationic character.
- Any combination in which the components rely on hydrogen bonding for fixation is likely to be antagonistic.
- Hydrolysable vegetable tannins and metals: this is established technology. The product is full, hydrophilic leather. Covalent fixation of metal by polyphenol complexation modifies the properties of the metal ions, but there is still cationic character.
- Condensed tannins with aldehydic reagent: prodelphinidin and profisetinidin tannins or non-tans are preferred. Not all aldehydic reagents work; oxazolidine is preferred. The result can be a high T_s.
- Syntan with aldehydic reagent: the industry standard is to use unspecified syntan with glutaraldehyde. There is apparently no attempt to define the syntan in terms of reactivity to create synergy; in general, reliance is probably placed on simple

additive effects, so the rationale is not easy to discern. The technology could be improved by analogy with the condensed tannins, by applying the structure–reactivity criteria known in that context.

- Polymer with aldehydic reagent: melamine–formaldehyde polymer with a phosphonium salt can be a synergistic combination, depending on the structure of the resin, especially the particle size. There will be other polymeric reagents capable of creating synergistic tannages.
- Reliance on syntans and/or resins may carry the additional problem of formaldehyde in the leather, which is subject to limits in specifications. It is possible to avoid this problem by scavenging the formaldehyde by the presence of condensed tannins, preferably prodelphinidin or profisetinidin.[9]

It is a useful shift in thinking to consider all processing to be a combination of fixation reactions, to analyse the outcome of consecutive applications of different reagent chemistries, including all steps.

24.6　Leather Properties

In deciding what alternative tanning system to adopt, the following features of the new leather must be considered.

- The appearance of the leather, for example the presence of draw from surface reaction, and the fineness of the grain pattern (largely dependent on the closing of the follicle mouths by relaxing the fibre structure).
- The colour of the leather is usually predictable, but occasionally it is not, *e.g.* the combination of glyoxal and tara produces a plum-coloured leather.[10]
- Area yield: the more the leather is filled or the higher the astringency of the tannage, the higher will be the angle of weave in the corium and hence the lower the area yield.
- Stability fastness of the tanning agents: the response to conditions of use.
- Hydrothermal stability: the requirements of the conditions of end use by the customer, *e.g.* footwear manufacture.
- Post-tanning reactions: affinity of retans, dyes and fatliquors for the new substrate. In particular, the colour struck by dyeing systems may be affected.

- The effect of the new substrate on properties and performance after post-tanning: strength and handle.
- The affinity of the leather for the chemistries of finish coats, defined by the combination of fixation reactions.
- The response of the substrate to finishing: the effects on hardness, adhesion, rub fastness, *etc.*

Much of this information is predictable from leather science and from experience in leather technology.

24.7 Organic Tanning Options

24.7.1 Polyphenol Chemistry

The best known example of the exploitation of plant polyphenols for high hydrothermal stability tanning is the semi-metal reaction. Here, the requirement is for pyrogallol chemistry to create a covalent complex between the linking polyphenol and the locking metal ion.[11] In practice, this means using the hydrolysable tannins, but alternatively some condensed tannins can be used: the prodelphinidins (*e.g. Myrica esculenta*, pecan and green tea)[12] and prorobinetinidins (*e.g.* mimosa) each have the required pyrogallol structure in the B-ring. Many metal salts are capable of reacting in this way, so there may be useful applications for the future.

The condensed tannins can confer high hydrothermal stability by acting as the linking agent, with an aldehydic crosslinker acting as the locking agent; other covalent crosslinkers may also have application in this context. The reaction applies to all flavonoid polyphenols, where reaction always occurs at the A-ring. In the case of the prodelphinidins and profisetinidins, additional reaction can take place at the B-ring.[13] The effect is to increase the ease of attaining high hydrothermal stability. It is useful to recall that the combination tannage does not have to rely on conventional vegetable tannins, because it can work with the low molecular weight non-tans; here, there is advantage in terms of reducing or eliminating problems of achieving penetration by the primary tanning component.

The link–lock mechanism can be exploited in other ways, even using reagents that at first do not appear to be tanning agents; an example is naphthalenediols, as shown in Figure 24.1.[14]

The behaviour of naphthalenediols in the tanning process is highly dependent on the structure of the isomer, as indicated in Table 24.2. It is clear that the presence of a hydroxyl at the 2-position activates

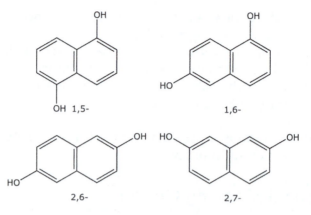

Figure 24.1 Structures of some naphthalenediols.

Table 24.2 Shrinkage temperatures of hide powder treated with dihydroxynaphthols (DHNs) and oxazolidine.

DHN	Shrinkage temperature (°C)	
	DHN alone	DHN + oxazolidine
None	57	75
1,5-	56	85
1,6- (2,5-)	64	90
2,6-	62	110
2,7-	62	79

the naphthalene nucleus: the 1-position does not work, as shown by comparing the 1,5- with the 1,6 (2,5)-diol. When there are two groups in the 2,6-positions, they act together. When the hydroxyls are in the 2,7-positions, they act against each other. The basis is the inductive effect of the hydroxyl on the aromatic ring, activating the *ortho* positions to electrophilic attack or allowing those positions to engage in nucleophilic attack at the methylene group of the *N*-methylol group of the oxazolidine, as demonstrated in Figure 24.2. The results in Table 24.2 also illustrate the principle that the linking agent may exhibit only a very weak effect in tanning terms, but successful locking of the linking species, combined with the ability to link the matrix to collagen in the locking reaction, can result in high hydrothermal stability. In the case in point, the polyphenol matrix is created from relatively unreactive (in tanning terms), low molecular weight phenolic species.

Non-chemical polymerisation is less effective. Applying laccase (polyphenol oxidase enzyme) to hide powder treated with

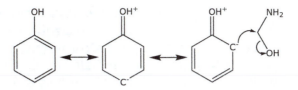

Figure 24.2 Reaction between the phenol and *N*-methylol moieties.

2,6-dihydroxynaphthalene produced the greatest increase in shrinkage temperature for the range of this type of linking agents tested, elevating the shrinkage temperature by 26 °C to 85 °C.[15] It is clear that the locking reaction is more easily and effectively accomplished by applying a second reagent, rather than relying on direct reactions between linking molecules.

24.7.2 Polymer and Crosslinker

Perhaps surprisingly, it is less easy to create high-stability tannage with polymers than it is by using oligomers or monomers, depending on the polymeric compound. It has been shown that high hydrothermal stability can be achieved using melamine resin crosslinked with tetrakis(hydroxymethyl)phosphonium salt. Several conclusions were drawn from these studies.[16]

- Not all melamine linking resins work. Therefore, the requirements for matrix formation are likely to be more important than the possession of specific chemical reactivity.
- Not all aldehydic locking agents work. The locking function depends on creating stable bonding between the linking molecules and forming a rigid species capable of resisting the collapsing triple helices.
- The linking reaction is dependent on physical parameters; particle size may be critical. It is not sufficient to provide space filling.
- The ability to form the basis of a supramolecular matrix must depend on the stereochemistry of the linking agent. This requirement is more easily satisfied with lower molecular weight species.

This reaction provided an interesting aspect of the matrix theory of tanning, when an attempt was made to accumulate hydrothermal stability by adding a matrix to a matrix. Here, the sequence

of reagent additions was as follows: melamine resin, phosphonium salt, condensed tannin (mimosa), oxazolidine. The observation was the achievement of high hydrothermal stability from
the melamine resin and phosphonium salt, added to by the condensed tannin, reaching a shrinkage temperature of 129 °C,
thereby matching the maximum shrinkage temperature achieved
by chromium(III) in the presence of pyromellitate (1,2,4,5-tetrac
arboxybenzene).[17] The melamine and phosphonium salt create a
matrix in which the melamine polymer reacts with the collagen *via*
hydrogen bonds and the phosphonium salt crosslinks the polymer,
while probably linking the matrix to the collagen. The introduction
of condensed polyphenol increases the shrinkage temperature by
a small additive effect; since it is applied after the matrix formation, it has no affinity for the resin, but has limited affinity for
the phosphonium salt. The subsequent addition of oxazolidine is
capable of forming a synergistic matrix with the polyphenol, but
the reaction causes the shrinkage temperature to decrease significantly.[10] The clear inference is that the new reaction is antagonistic
to the established matrix: by analogy with other antagonistic combinations,[1] there is competition with the mechanism that binds
the first matrix to the collagen, effectively loosening the binding
between the matrix and the collagen, allowing shrinking to occur,
with the consequence that the hydrothermal stability is lowered. It
is difficult to rationalise the observation by any mechanism other
than the formation of matrices.

24.8 Natural Tanning Agents

New tanning chemistries occasionally come to light and recent examples have the characteristic of being biomimetic, using natural reactions in a new context. Such organic tanning reactions are of interest
from three points of view.

1. They offer new methods of making leather, to yield new products,
   which may contribute to lessening the environmental impact of
   tanning.
2. They offer new opportunities for high hydrothermal stability
   tanning, by acting as new linking agents, then allowing manipulation of the chemistry of the locking step.
3. They may involve the novel use of enzymes in tanning, operating
   as catalysing activating agents, so the rate of reaction is highly
   controllable.

24.8.1 Carbohydrates

Derivatives of carbohydrates are well known in the form of dialdehydes, obtained by the oxidative effects of periodic acid, as discussed in Chapter 14. However, there can be direct reactions between carbohydrate and proteins, glycosylation-type reactions. Komanowsky effectively generalised these reactions in his studies of the reactions within collagen at low moisture content.[19] The basis of the formation of permanent bonds is the Maillard reaction, responsible for many of the transformations during the cooking of proteinaceous food. The reaction depends on reducing the moisture content, in order for the groups to approach close enough together to react; it works better at higher temperature and lower pH, in contrast to the high pH requirements of most aldehydes, although not all aldehydic agents conform to that criterion.[14,20]

The stabilising reaction is illustrated in Figure 24.3; it is assumed that the ketoamine product can react further with collagen, to form a crosslink (of unknown structure), a reaction that becomes more feasible when the interacting chains are close, unlike the situation in wet collagen and leather.

The natural products that are responsible for the cooking reaction include sugars and the glycosaminoglycans, hyaluronic acid, dermatan sulfate and chondroitin. These compounds or their derivatives might be applied as reagents as part of a two-step organic tannage.

Hyaluronic acid and dermatan sulfate offer the potential for useful new derivatives; they are by-products of the leather-making process. As indicated by their structures given in Chapter 2, analogous to the carbohydrate structure discussed in Chapter 14, it appears likely that aldehydic derivatives could be made. Although the reactivity of the

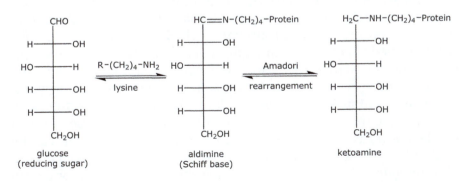

Figure 24.3 Maillard reaction.

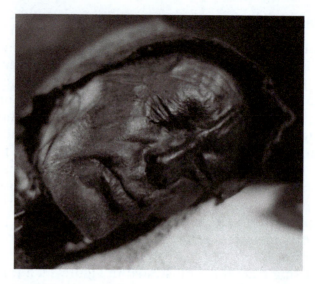

Figure 24.4 Tollund Man.

aldehyde function would not show any advantage, the overall reactive functionality is enhanced by the greater variety of chemical moieties in the molecule, such as carboxyl groups. Since the molecular basis of the tanning reaction would be a natural product, a component of skin, it might be expected that the leather would be hypoallergenic, useful for applications where metal content is to be avoided because of sensitisation, where close and prolonged contact with the body is involved.

24.8.2 'Bog Body' Chemistry

A related stabilising process for protein that has not previously been reviewed in a leather context is the preservation of so-called 'bog bodies',[21] illustrated by the 2400 year-old Tollund Man in Figure 24.4. Perhaps surprisingly, the preservation mechanism is not an example of vegetable tanning. The environment in which the tissues of the body are preserved is typically a sphagnum peat bog; the effect is thought to be due to a Maillard reaction between free amino groups in the proteins and reactive carbonyl groups in a soluble glucuronoglycan, sphagnan, containing residues of D-*lyxo*-5-hexosulopyranuronic acid, shown in Figure 24.5. It can form the furan derivative, which in turn can be converted into a species such as ascorbic acid, which can be oxidised to form a species such as dehydroascorbic acid, which resembles ninhydrin, the well-known colour-generating reactant for amino acid analysis.

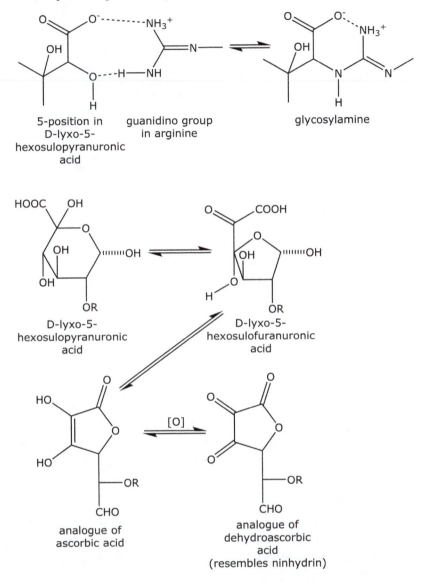

Figure 24.5 Reactions of decomposition products from sphagnum moss.

Sphagnan is a pectin-like polymer, bound covalently to cellulosic and amyloid chains in sphagnum moss; it is liberated by hydrolysis, as the moss turns into peat. As in periodic acid-oxidised carbohydrate derivatives, the chain length of the sphagnan products may influence the tanning potential, but this has not been investigated. In Nature the reaction is slow, owing to the low concentration of the active species in solution, but this is not a limiting factor for the leather scientist.

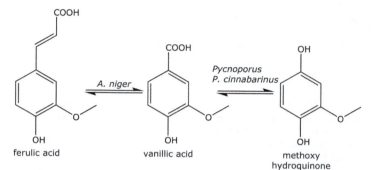

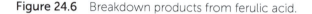

Figure 24.6   Breakdown products from ferulic acid.

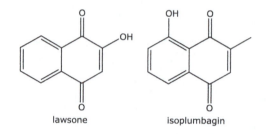

Figure 24.7   Secondary metabolites from *Lawsonia inermis*.

The chemistry of plant cell walls may offer a new source of chemicals from which tanning agents might be synthesised. The main breakdown product, ferulic acid, is polymerisable by laccase and can be converted by fungal species into other compounds that may be of interest to the leather scientist, as shown in Figure 24.6.[22]

24.8.3   Henna

Other sources of potential tanning agents might be exploited. Henna is a dye extracted from the leaves of *Lawsonia inermis*; lawsone and isoplumbagin are extracted from the leaves and bark respectively, as shown in Figure 24.7.[23]

The reactivity of quinone itself and derivatives is well known in tanning,[24] so the dual functionality of quinone and phenol reactivity is of interest in the context of combination processes.

Henna has been compared with wattle as an agent for retanning chrome leather, where the physical properties were judged to be the same, although not unexpectedly the dyeing intensity from henna was better.[25] Combinations of henna with aluminium salt show some interaction, but much less powerful than in a semi-alum reaction, as set out in Table 24.3.[26] Similarly, there is a weak interaction between 20% henna and 4% oxazolidine, included in Table 24.3.[27]

Table 24.3 Combination tannages with 20% henna and 2% Al_2O_3, aluminium(III).

Combination	T_s (°C)	Henna exhaustion (%)
Henna control	81	75
Al then henna	95	82
Henna then Al	97	86
Henna then oxazolidine	98	—

24.8.4 Nordihydroguaiaretic Acid (NDGA)

NDGA is a naturally occurring polyphenol, isolated from the creosote bush;[28] it has been proposed as a stabilising reagent for collagen, where it exhibits the unusual phenomenon of actually increasing the strength of the fibre, rather than the more common effect of lessening the weakening effect of processing. The mechanism of polymerisation of NDGA is presented in Figure 24.8: the polymer is thought to align with the protein chains, constituting a supporting and strengthening structure. Clearly, the polymeric species offers opportunities for additional reactions.

To date, the physical and thermodynamic properties of leathers have never been correlated, since they are regarded as properties based on different aspects of collagen structure; the former is dependent on the macro structure at the microscopic and visible levels and the latter is dependent on the molecular level. Since these properties have now been shown to be dependent on collagen modification, perhaps a useful relationship does exist. The strengthening aspect of the reaction is an interesting outcome of the introduction of this particular matrix into the collagen structure. This offers the opportunity of tailoring the tanning matrix for improvements in the physical properties.[29]

24.8.5 Genepin

Genepin (Figure 24.9) is an iridoid derivative, isolated from the fruits of *Gardenia jasminoides*. Its reaction with protein is characterised by the generation of deep-blue coloration under the alkaline conditions required for fixing. The reaction mechanism by which protein is stabilised is not clearly understood, but is thought to derive from opening of the oxy heterocyclic ring at the 3-position by a primary amino group on the protein to form a heterocyclic adduct, causing polymerisation by free radical reaction with a secondary amino group.[30,31] It has been shown that the shrinkage temperature of hide powder can be raised to 85 °C,[32] characteristic of a single tanning reagent. That in itself is not important, but the incorporation of new reactive groups into collagen by a linking reaction using genepin may provide new sites for

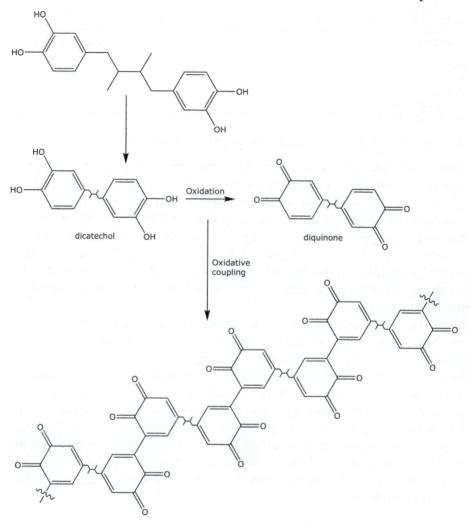

Figure 24.8   Reactions of NDGA, leading to the stabilisation of collagen.

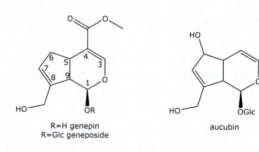

R=H genepin
R=Glc geneposide

aucubin

Figure 24.9   Structure of genepin.

reaction in a locking step. Alternatively, genepin may provide another approach to the locking step in a combination tannage. Notably, using genepin to retan aluminium-tanned collagen, 8% Al_2O_3 + 8–10% genepin, yields shrinkage temperatures of 101–104 °C.[33]

24.8.6   Oleuropein

Oleuropein is a natural product, a secoiridoid glycoside, found in privet and olive vegetation, where its function is part of the self-defence mechanism against infections and herbivores: it contains glucose, but not the cyclopentane ring in its structure. The mechanism of reaction with protein is presented in Figure 24.10; it is thought to involve the activation of oleuropein to the aglycone form, which reacts with lysine residues on proteins.[34]

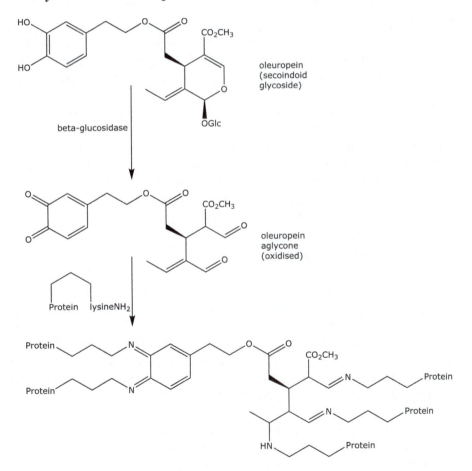

Figure 24.10   Schematic representation of the reaction of oleuropein with protein.

Table 24.4   Stabilisation of collagen film.

Reactant	Shrinkage temperature (°C)
Control	50
Oleuropein	67
Glutaraldehyde	68

From the work of Antunes *et al.*,[35] the stabilising effect can be compared with that of glutaraldehyde, as shown in Table 24.4, in which the collagen was recast as a film from solubilised collagen.

As with genepin, the chemistry of oleuropein offers the potential for acting as either a linking agent or a locking agent, using an appropriate complementary combination reagent.

24.8.7   Other Iridoids

There are many other iridoids, nearly 600 are known,[34] characterised by the presence or absence of glucose moieties and the presence or absence of a cyclopentane ring. Some of these are illustrated in Figure 24.11, together with the results of tanning studies presented in Table 24.5.[36]

These biomimetic approaches to tanning constitute potentially powerful options for biotechnology in leather making. However, it is equally clear that they do not offer the opportunity for a single-step, high hydrothermal stability tannage. The advantages lie in the new, covalent binding reactions, which create the linking part of the supramolecular matrix, which is firmly bound to the triple helix, and then is capable of undergoing a variety of locking reactions.

24.9   Other Reagents

Some sugars are known to have a stabilising effect on proteins, including a preserving effect on the activity of enzymes, whether in solution or in the freeze-dried state. This is an aspect of chemistry that has not been exploited in the leather industry. The effect is thought to be due to the interaction between the carbohydrate and water associated with the protein structure. Of the sugars studied, the most effective in this regard is trehalose.[37] Figure 24.12 shows the structures of sucrose and trehalose.

Clearly, there is a possibility of using low molecular weight carbohydrates as curing agents, as introduced in Chapter 3: trehalose can be obtained from starch, so this technology might be cost-effective. However, the technically more interesting application might be as part of a tanning combination.

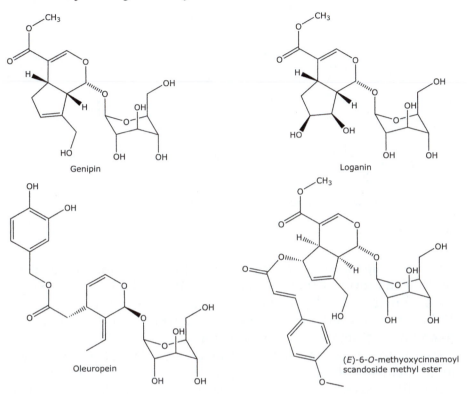

Figure 24.11 Structures of some iridoids.

Table 24.5 Tanning studies with iridoids. Data from the work of Ding and co-workers.[36]

Iridoid	Species source	T_s (°C)	Leather colour
Genepin	*Gardenia jasminoides* Ellis	80	Dark blue
Loganin aglycone	*Loncera japonica* Thunb.	82	Yellow–brown
Oleuropein aglycone	*Olea europaea* Linn.	85	Yellow
(*E*)-6-*O*-Methoxycinnamoyl scandoside methyl ester aglycone	*Hedyotis diffusa* (Willd.) Roxb.	83	Mauve

There is a remaining class of reagents that might be classified as either linkers or lockers; the requirement for new components for tanning is that they should have the ability to react covalently under the typical conditions of tanning.[6] There are options already known in the leather industry in the form of reactive dye chemistries, already discussed in Chapter 16; it is merely a matter of widening the exploitation. Those chemistries could be exploited in the form of multiple

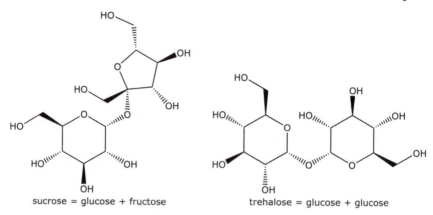

sucrose = glucose + fructose   trehalose = glucose + glucose

Figure 24.12 Structures of sucrose and trehalose.

reactive groups in the same molecule. This notion was initiated by BASF in the form of their product (dye) Fixing Agent P, triacryloyl-triazine; more recently, Clariant released their syntan product Easy-White, based on nitrogen heterocycle chemistry, but there is clearly potential for developing the approach further.

Around the same time, Lanxess AG launched X-Tan, based on isocyanate chemistry, discussed in Chapter 14. Similarly, the epoxide and dicyandiamide chemistries might prove to be useful, but only if they are applied according to the requirements of link–lock.

24.10 Compact Tanning

The concept of 'compact processing' is the condensing or shortening of processing by combining two or more reactions into a single process step. In this way, time is saved and consistency is created in the interaction between the process steps involved, because only one reaction takes place, instead of two or more. This is a powerful contribution to innovation in the tannery. The idea of compact processing can be applied at any stage in processing, although it is more commonly applied in the later stages of wet processing. Therefore, thinking about this concept should not be limited in scope. If it is accepted that high stability can be achieved only by a two-stage tanning action, then there are opportunities to exploit such an approach to tanning in compact processing.

The use of a non-swelling acid incorporates the notion of pre-tanning. It may be possible to substitute pretanning for retanning. However, it should be recognised that the first tanning agent controls the character of the leather. In the case of prime tanning being conducted with vegetable tannins, the requirement for further tanning is uncommon. An exception would be organic tanning, which might be based on polyphenols, where there would be a requirement for applying the second component of a combination process for high hydrothermal stability. Such an approach is feasible for the combination of gallocatechin-type condensed tannins with oxazolidine, where the reaction is activated by elevated temperature.[12,18,29,38]

The case of chromium(III) tanning is less clear. No technologies have been offered to the industry in which a retanning agent has been combined with a conventional chrome tanning salt. The closest to that situation are products referred to as 'chrome syntans', well known to the industry. However, these are primarily offered as filling versions of chrome retanning, rather than a new approach to prime tanning. Because chromium(III) has little affinity for phenolic hydroxide as a ligand,[39] chrome syntans either are mixtures of syntan and chrome salt or the syntan may be capable of complexation with chrome, by having some carboxyl functionality in its structure. In the latter case, the function of retanning might be accomplished with tanning. Similarly, an extended application of masking, to confer additional features, may be useful. The application of compact processing to post-tanning was discussed in Chapter 15. It is clear that there is potential for further development, using creative chemistry.

24.11 Alternative Technologies

It is conventionally assumed that tanning has to be conducted in aqueous solution. However, there are options available for the future, as discussed in Chapter 19. The technology of using different liquids for reagent delivery is feasible, but brings problems of its own, notably the requirement for diffusion across the pelt, unlike the conventional requirement for diffusion through the cross-section. However, such difficulties are solvable, if the willingness is there. The alternative is an essentially non-aqueous process, as indicated in the context of compact processing: this approach to processing, including the use of liquid carbon dioxide, has been discussed for some steps,[40,41] but industrial development has not yet followed. The advantage of such an approach to one-step post-tanning is that the substrate is dry (at least to the touch), so the moisture in the leather would have a limited influence on the process. It is less clear that such an approach would be feasible for primary tanning, in whatever way that might be done chemically. If chrome tanning were to be developed in that direction, it is not difficult to imagine the HHB properties of the chrome complex being tailor-made for the solvent by appropriate masking.

It is useful to speculate on the roles that enzymes might play in tanning. It is clear that biotechnology has an increasingly important part to play in the beamhouse, but it is less clear if it can contribute to collagen stabilisation. From the matrix theory of tanning and collagen stabilisation, the inability of transglutaminase to increase the hydrothermal stability is predictable and understandable, even though it is clearly capable of introducing crosslinking in the conventional sense.[42] Similar reactions, which can introduce crosslinks into collagen, are unlikely to function as useful tanning agents, beyond altering the texture of the protein.

At the other end of the processing procedures, the role of drying remains to be exploited to advantage by industry, as discussed in Chapter 20. Some studies have linked the properties of leather, including area yield, to the programme of drying conditions:[43] softness does not depend on the rate of drying, only the moisture content. Using conventional plant, it is possible to modify the drying programme into two stages, to obtain advantage from the relationship between the viscoelastic properties of leather and its water content, so that the area gain of intact leather does not have to be at the expense of softness.

In considering alternative technologies, it is useful to consider whether there is an alternative mechanism to link–lock, *i.e.* whether or not there could be exceptions. If the mechanism of shrinking, as set out here, is correct, then the impact of the tanning reaction on hydrothermal shrinking is correct. Consequently, the use of conventional penetrating reactants, in the way also outlined above, is also correct. Therefore, the only exception to the link–lock mechanism would be a single-step reaction that combines both features. The requirement would be the formation of the stabilising matrix from a single-species polymerisation reaction, which would have to include reaction with the collagen. In this regard, polymerisation of cod oil is the closest reaction known in the art. However, in this case, despite the complexity of the chemistry of polymerisation by oxidation, there appears to be little direct interaction between the polymer and the collagen, despite the possibility of creating aldehyde groups.[44] Hence the leathering reaction does not elevate the shrinkage temperature significantly. Examples of chemistries apparently capable of multiple bonding with collagen do not react in that manner, because there is no evidence of an outcome that might be expected from such tanning reactions, as discussed in Chapter 14. Nevertheless, it is not inconceivable that such a polymerising tannage might be developed, particularly if an extended range of practically useful solvents is made available to the industry.

24.12   Overview

The range of chemistries available to the tanner is widening. By considering the molecular basis of the tanning mechanism, especially the requirements to confer high hydrothermal stability, the options open to the tanner are widened. If high-stability organic tannages are desired, they can be created just by observing the following rules.

1. Apply the first reagent, linking to the collagen with high-stability bonding, preferably covalently. The reagent must offer the potential for a second reaction, by possessing usefully reactive groups, but need not have a high molecular weight.
2. Apply the second reagent, to react with the first reagent by locking the molecules together; therefore, the second reagent must be multifunctional. Contributing to linking the matrix to the collagen is useful.

The properties of the resulting leather can be controlled by the choices of reagents: this applies to both mineral and organic tanning options. In the absence of a polymerising tannage capable of meeting the stabilising matrix criteria, we are limited to the two-step process. However, the two-step process does not have to extend processing times – this approach can contribute to compact processing: it is not difficult to envisage a two-component reagent mixture in which the affinities and kinetics of reaction are tailored to a stepwise process *in situ*.

The two-component approach to tanning can offer specifically required properties and therefore offers the basis for the production of bio-vulnerable or so-called recyclable leathers, so far not adequately explored. The definition of tanning refers to the resistance to biodegradation of a previously putrescible protein material; here the resistance refers to proteolytic attack. Therefore, it is feasible to consider tanning processes that incorporate a degree of vulnerability. In this way, high hydrothermally stable leather could be chemically or biochemically destabilised, to allow denaturation of the collagen at moderate temperatures, so that proteolytic degradation can be achieved. Targeting specific groups in the matrix may be sufficient to degrade its effectiveness: options are hydrolases, oxidases, reductases, *etc.*, depending on the chemistry of the tannage. Alternatively, tannages can be devised in which the stability is temporary, such as semi-metal chemistry, which is vulnerable to redox effects, as introduced in Chapter 13. Other approaches are possible and should not be beyond the wit of the leather scientist.

24.13 Conclusions

The expression of the link–lock mechanism has made it possible to take quantum steps forward in developments in tanning technology. This theory is a simpler and more powerful view of collagen stabilisation than the older model of direct crosslinking between adjacent sidechains; it is a more elegant view. It is no longer worth pursuing the single-reagent alternative to chrome tanning, because the alternative does not exist. Indeed, why should we seek an alternative to chrome tanning? It works well, it can be made to work even better and, anyway, by all reasonable judgements it causes little environmental impact. Nevertheless, organic options provide potential for new products from the leather industry. Now we can develop those options using sound logic, based on sound theory. The literature is littered with examples of multicomponent tanning systems;[45–49] the outcome of each is entirely predictable if we apply our new understanding of the theoretical basis of tanning.

The continuation of developing tanning and leather technology depends on constant reappraisal of all aspects of the subject. This is the role of leather science. Conventional, received wisdom should not be relied upon without critically reviewing exactly what it means, what it contributes to processing and products and what the wider implications are for the practical tanner. It is important to recognise that the scrutiny of current technology will often identify inconsistencies and misunderstanding of principles: the technology may work, but the science may not, as we have seen in the case of sulfide. However, this is not always a bad thing, because it can lead to new thinking, new developments and more profitability in an environmentally sound, sustainable industry.

References

1. A. D. Covington, *J. Am. Leather Chem. Assoc.*, 2001, **96**(12), 461.
2. A. D. Covington, G. S. Lampard, O. Menderes, A. V. Chadwick, G. Rafeletos and P. O'Brien, *Polyhedron*, 2001, **20**, 461.
3. K. Shirai, K. Takahashi and K. Wada, *Hikaku Kagaku*, 1975, **21**(3), 128.
4. K. Shirai, K. Takahashi and K. Wada, *Hikaku Kagaku*, 1984, **30**(2), 91.
5. A. D. Covington, M. J. Collins, H. E. C. Koon, L. Song and O. Suparno, *J. Soc. Leather Technol. Chem.*, 2008, **92**(1), 1.
6. A. D. Covington, *Proceedings of UNIDO Workshop*, Casablanca, Morocco, 2000.
7. A. D. Covington, *J. Soc. Leather Technol. Chem.*, 1986, **70**(2), 33.
8. R. M. Cuadros, *et al.*, *J. Soc. Leather Technol. Chem.*, 1999, **83**(6), 300.
9. A. J. Escobedo, MSc thesis, University College Northampton, 2002.
10. A. D. Covington, unpublished results.
11. R. A. Hancock, S. T. Orszulik and R. L. Sykes, *J. Soc. Leather Technol. Chem.*, 1980, **64**(2), 32.
12. A. D. Covington, C. S. Evans, T. H. Lilley and L. Song, *J. Am. Leather Chem. Assoc.*, 2005, **100**(9), 325.
13. L. Song, PhD thesis, University of Northampton, 2003.
14. A. D. Covington, C. S. Evans and O. Suparno, *J. Soc. Leather Technol. Chem.*, 2007, **91**(5), 188.
15. O. Suparno, PhD thesis, University of Northampton, 2005.
16. A. D. Covington and M. Song, *Proceedings of IULTCS Congress*, London, 1997.
17. S. DasGupta, *J. Soc. Leather Technol. Chem.*, 1979, **63**(4), 69.
18. A. D. Covington and B. Shi, *J. Soc. Leather Technol. Chem.*, 1998, **82**(2), 64.
19. M. Komanowsky, *J. Am. Leather Chem. Assoc.*, 1989, **84**(12), 369.
20. S. DasGupta, *J. Soc. Leather Technol. Chem.*, 1977, **61**(5), 97.
21. T. J. Painter, *Carbohydr. Polym.*, 1991, **15**, 123.
22. C. Faulds, B. Clark and G. Williamson, *Chem. Br.*, 2000, **36**(5), 48.
23. C. K. Liu, *et al.*, *J. Am. Leather Chem. Assoc.*, 2008, **103**(6), 167.
24. E. Heidemann, *Fundamentals of Leather Manufacturing*, Eduard Roether, Darmstadt, 1993.
25. A. E. Musa, *et al.*, *J. Am. Leather Chem. Assoc.*, 2008, **103**(6), 188.
26. A. E. Musa, *et al.*, *J. Am. Leather Chem. Assoc.*, 2011, **106**(6), 190.
27. A. Musa, *et al.*, *J. Am. Leather Chem. Assoc.*, 2009, **104**(9), 335.
28. T. J. Koob and J. Hernandez, *Biomaterials*, 2003, **24**, 1285.
29. A. D. Covington, C. S. Evans, T. H. Lilley and O. Suparno, *J. Am. Leather Chem. Assoc.*, 2005, **100**(9), 336.
30. H.-C Liang, *et al.*, *J. Appl. Polym. Sci.*, 2004, **91**(6), 4017.
31. M. F. Butler, *et al.*, *J. Polym. Sci., Part A: Polym. Chem.*, 2003, **41**, 3941.
32. K. Ding, M. M. Taylor and E. M. Brown, *J. Am. Leather Chem. Assoc.*, 2006, **101**(10), 30.
33. K. Ding, *et al.*, *J. Am. Leather Chem. Assoc.*, 2006, **101**(10), 362.
34. K. Konno, C. Hirayama, H. Yasui and M. Nakamura, *Proc. Natl. Acad. Sci. U. S. A.*, 1999, **96**, 9159.
35. A. P. M. Antunes, G. E. Attenburrow, A. D. Covington and J. Ding, *Proceedings of IULTCS Congress*, Istanbul, Turkey, 2006.
36. B. Zhang, *et al.*, *J. Am. Leather Chem. Assoc.*, 2011, **106**(4), 121.
37. J. K. Kaushik and R. Bhat, *J. Biol. Chem.*, 2003, **278**(29), 26458.
38. A. D. Covington and L. Song, *Proceedings of IULTCS Congress*, Cancun, Mexico, 2003.
39. S. Jeyapalina, A. D. Covington and G. E. Attenburrow, *Proceedings of IULTCS Congress*, Cape Town, South Africa, 2001.

40. Q.-Y. Wei, *J. Soc. Leather Technol. Chem.*, 1987, **71**(6), 195.
41. G. Gavend, *Industrie du Cuir*, 1995.
42. R. J. Collighan, S. Clara, S. Li, J. Parry and M. Griffin, *J. Am. Leather Chem. Assoc.*, 2004, **99**(7), 293.
43. S. Jeyapalina, G. E. Attenburrow and A. D. Covington, *J. Soc. Leather Technol. Chem.*, 2007, **91**(3), 102.
44. J. H. Sharphouse, *J. Soc. Leather Technol. Chem.*, 1985, **69**(2), 29.
45. N. N. Fathima, *et al.*, *J. Am. Leather Chem. Assoc.*, 2004, **99**(2), 73.
46. N. N. Fathima, *et al.*, *J. Am. Leather Chem. Assoc.*, 2006, **101**(2), 58.
47. N. N. Fathima, *et al.*, *J. Am. Leather Chem. Assoc.*, 2006, **101**(5), 161.
48. K. Ding, M. M. Taylor and E. M. Brown, *J. Am. Leather Chem. Assoc.*, 2007, **102**(5), 164.
49. K. Ding, M. M. Taylor and E. M. Brown, *J. Am. Leather Chem. Assoc.*, 2007, **102**(6), 198.

25 The Future for Leather

25.1 Introduction

Since leather is a by-product of the meat industry, an unwanted consequence of abattoir activity, there will be a future for leather – assuming
the cyclical fashion for vegetarianism and veganism does not become
the norm. Therefore, there will also be a requirement for change; for
improvements in process efficiency, chemical savings, environmental
impact, product safety and every other aspect of tannery operations
that make tanning sustainable.

Sustainability is a subject in its own right, comprising many features and topics, including corporate responsibility, but in the context
of leather science there are two main elements that are appropriate
for discussion here. The first is the way in which processing might
change, based on what is known and what is anticipated. The second
is putting the making of leather on a firm theoretical basis, when
understanding of the principles of each step allows the prediction of
the outcome of change or the prediction of the change necessary to
achieve a desired outcome in terms of properties and performance of
the product.

Tanning Chemistry: The Science of Leather 2nd edition
By Anthony D. Covington and William R. Wise
© Anthony D. Covington and William R. Wise 2020
Published by the Royal Society of Chemistry, www.rsc.org

25.2   Progressive Change

The drivers for addressing sustainability are as follows.

1. Green issues, which may arise from real or perceived environmental impact of production, which includes arguments concerning the carbon and water footprints of leather.[1]
2. Competition from other materials, which includes the economics of production and the availability of rawstock of adequate quality.

The technologies that tanners ought at least to consider can be separated into three groups.

1. *Development*
   These are technologies that arise from current and accepted practice that might be adopted in tanneries without major changes to plant and procedures. Such modifications to typical operations could also be characterised as natural or logical progressions from received and current understanding of leather science and leather technology. Developmental improvements tend to have a quick route to market, require a low investment of money and resources and normally have a quick return on that investment.
2. *Innovation*
   These are technologies that constitute a step change in thinking and ways of doing things: they would be part of a longer term strategy of development. Such changes may include plant and equipment, but not radical change.
3. *Revolution*
   There are technologies in the pipeline, already at the point of proof of concept, that would radically change the face of the leather-making industry: they would constitute a paradigm shift, 'thinking outside the box'. Such developmental transformations are necessarily for the long-term sustainability of the sector. Of the three groups, revolution requires the largest investment, often has the longest route to market and can be risky with regard to return on the investment.

25.2.1   Development

25.2.1.1   Tanned Waste

The disposal of tanned waste is an environmental problem faced by all tanners, because there is no economic way to deal with it, short of landfill, which is unacceptable and has an associated and ever-increasing cost. The alternative is to add some value. Tanned waste contains two main components, *viz.* collagen and (usually) chromium(III), both of which can be recovered for reuse.

Chromium can be recovered using three strategies: the first is to break down the protein at high pH, with or without the assistance of proteolytic enzymes acting on the collagen denatured at high pH,[2] and the second is to reverse the tanning reaction by oxidising the chromium salt to the hexavalent state, later reducing it back to chromium(III).[3] Both processes depend on reactions conducted in aqueous medium: they were developed by a collaboration between workers in Spain and the USA and could be applied commercially. The first process, alkali hydrolysis of the chromium–collagen bonds, creates a by-product of low molecular weight peptides that might be reused, *e.g.* for fertiliser. The second process, although it might be perceived to have a higher environmental risk, is also a route to recovering gelatine of sufficient gel strength to be of value. The third solution is to break the chromium–collagen bonds by direct action of the solvent system, for example the basis of the artificial perspiration test or even by the solvent itself;[4] the latter approach is more speculative and maybe should be included in the 'Innovation' part of the programme.

The use of waste leather as a fuel for biogas has been demonstrated.[5] In the case of chrome-tanned leather, denaturing the material at high pH creates a suitable feedstock. However, this does not work with vegetable-tanned leather, because the polyphenols remain intact and reactive towards protein, so they can still 'tan' the methanogenic bacteria and stop them from functioning. The same problem arises with wet white made with syntan. By this criterion, chromium is 'greener' than organic tanning!

25.2.1.2 Chromium Chemistry

A better understanding of the chemistry of chromium(III) and the mechanism of the tanning reaction allows the system to be exploited to the tanner's benefit, in terms of the environmental impact and the qualities of the leather produced. Optimising parameter control, *i.e.* pH and temperature, is well known and enables the tanner to improve tanning efficiency, particularly if those parameters can be controlled within the processing vessel (see Chapter 11).

An equally powerful aspect of process modification is exploiting the masking mechanism and its effects on chrome reactivity. The received wisdom that masking reduces reactivity, thereby enhancing penetration, does not apply under typical tannery conditions:

in fact, the masking ratios achieved by the usual offers of formate in the pickle increase rather than decrease the reactivity of the complexes and the rate of the masking reaction mirrors the rate of chrome fixation onto collagen.[6] Fortunately, this is exactly what the tanner needs, but it is not what he thinks he wants! The technology can be extended to make additional contributions to the reactivity of the chrome, using masking agents that increase molecular weight or hydrophobicity or both in the tanning complexes, for example using crosslinking or chelating ligands such as glutarate or phthalate, respectively. This is simple to apply.

A constant modern concern with chrome-tanned leather is the (alleged) conversion of chromium(III) to chromium(VI) within the leather and its consequent perceived impact on the environment.[7,8] While there is debate as to whether the observations are false positives, because the analytical methods using wet chemistry might be unreliable, until the matter is resolved unequivocally tanners would be well advised to take account of process conditions that might contribute to the problem, *e.g.* avoiding the use of unsaturated fatliquors,[9] as discussed in Chapter 11. Leathers that do not conform to current test requirements, whatever the accuracy might be, are causing a disservice to the leather industry, which relies on the continuing acceptability of chrome tanning. In addition, there is debate concerning the disposal of chrome leather to the environment at the end of life. One option is incineration, as discussed later.

25.2.1.3 Wet White

It has become increasingly common for industry to turn to wet white as an alternative to wet blue chrome-tanned leather, as discussed in Chapter 14. In this modern context, the term usually refers to a combination of glutaraldehyde and syntan: the assumptions are that the production of such leather is more environmentally friendly and the product is more recyclable. It is doubtful whether either assumption is justified, but it is clear that wet white is here to stay.

It is useful to consider the practicalities of this technology as the basis for future leather production. In the first instance, the combination of these reagents does not constitute a new, interacting reaction, as defined by the analysis of combinations and rules of conforming to synergistic chemistry, as discussed later, particularly because glutaraldehyde is relatively unreactive towards polyphenolic compounds, such as those used in the manufacture of syntans. Consequently, the outcome is a mixture of the individual tannages, yielding only a moderate shrinkage temperature and limiting the applications. This in itself is not a problem if the required/desired characteristics are conferred to the leather and if shrinkage temperature is not important. However, the continued practical use of glutaraldehyde as a tanning agent cannot be guaranteed.

Nevertheless, organic processing is a feature of current leather production, but is likely to develop and change. For many applications, achieving a high shrinkage temperature (>100 °C) in the leather is a priority. This can be addressed by a careful choice of reagents, requiring familiarity with the chemistry of the reagents. There are many combinations that might be used, but it is important to make the right choices and those choices must be informed by the required outcome and the theoretical principles of tanning science and technology.

A consequence of moving towards organic processing without adopting a scientific approach is to accept a reduced commitment to high T_s in the leather and the associated impact on the application of the leathers. This requires a shift in attitude and standards.

25.2.1.4 Materials Science

The physical properties of leather, particularly the softness and area, are controlled by the viscoelastic properties of the fibrous collagen, whether native or chemically modified, which in turn are controlled by the conditions of moisture and temperature. This is an area of research that has not received the attention it deserves, especially considering the importance of those physical properties.

As discussed in Chapter 20, Jeyapalina et al.[10] showed that significant changes to physical properties occur at critical moisture contents of 60 and 30% on dry collagen weight. Taking this into account, they demonstrated that benefits could be achieved by adopting a two-stage drying regime: first, fast drying to 35% moisture at high temperature and humidity, then conditioning to the required final moisture content at lower relative humidity. It is important to be able to define the conditions needed to optimise the properties of the leather at the end of drying under controlled regimes of humidity and temperature. This is currently practical, since industrial equipment for drying under controlled conditions of humidity and temperature is available.

25.2.1.5 Rethink Processing

Current practice in hide processing is to flesh in the limed state for most leathers, to split the hide in the limed state for upholstery leather and to split in the wet blue for shoe upper leather. Fleshing after liming is a waste of chemicals and reduces the value of recoverable tallow because of hydrolysis. Splitting in the limed state means wasting chemicals, particularly if the flesh split is disposed of, *e.g.* for food casing. Splitting in the blue means that the processing costs for the lower value flesh split are the same as the grain split and the shavings have negative value. The uneconomic situation can change if green splitting were to be a viable option. Typically this is not the case, since dung even at low degrees of contamination renders splitting impractical owing to pelt damage in the action of splitting. However, dung can be efficiently removed, which then allows green fleshing and green splitting. The former produces high-quality flesh for tallow recovery and more effective and efficient liming; the latter allows grain tanning to be separated completely from flesh split processing or disposal.

The physical nature of the green hide is such that splitting the soft material is difficult. A solution might be to adopt the old technology for wet white using silica to stiffen the pelt to allow splitting after pickling. It should work for green hide also, applying the reagent in the final soak float.

Green splitting would make enzymatic beamhouse operations practical, because penetration of the macromolecules becomes relatively fast through the split surface: this approach could allow biochemical unhairing, to avoid chemical unhairing and opening up and make carbon dioxide deliming quicker. Green splitting would make new technologies possible, *e.g.* unhairing cattle hide by painting as applied to sheepskins, which is an efficient use of reagents and a hair-saving process.

25.2.1.6 Closed-loop Processing

It is possible to envisage a closed-loop series of process steps, in which the inputs are rawstock, water and reagents and the outputs are leather and by-products to which value can be added. Some assumptions would make the approach viable, *e.g.* the use of fresh rather than salted hides, hair-save unhairing and enzymatic rather than chemical processing – the technologies are already known and in some cases have been demonstrated.

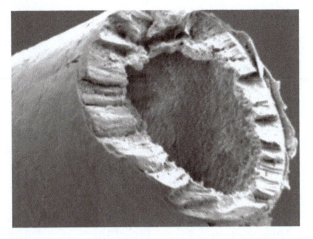

Figure 25.1 Partially composted hair, showing the degradation of the non-keratinous medulla, leaving the more intractable keratinous cortex and cuticle intact. Courtesy of A. Onyuka.

Water and electrolyte can be recycled into the beamhouse, including pickling for chrome tanning, chrome can be easily recycled[11] and post-tanning liquor can be recycled back into the beamhouse or tanning steps. All this has been demonstrated in practice, on the industrial scale.[12] By-products would be dung, hair and protein for fertiliser. In the case of hair, with appropriate enzyme action it can be converted into compost: Figure 25.1 shows the effect of proteolytic action, but complete breakdown requires the additional action of keratinase; suitable mixtures of enzymes can be isolated from the environment for this purpose.[13]

Of the skin components, usually discarded in the effluent, hyaluronic acid could be recovered and sold on for cosmetic or food purposes or for the creation of tanning agents analogous to dialdehyde carbohydrate products, as illustrated in Figure 25.2.

25.2.2 Innovation

25.2.2.1 Theory of Processing

The link–lock theory of collagen stabilisation has been proposed to explain how all tanning reactions work with regard to the hydrothermal stability conferred to the leather, as presented in Chapter 23. The value of this view of tanning chemistry is compounded by an understanding of the principles of fixation of reagents onto

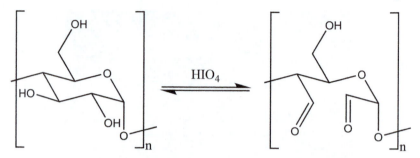

Figure 25.2   Making a dialdehyde tanning agent from starch.

a solid substrate in heterogeneous reactions, discussed in Chapter 15. This approach to leather technology is expanded upon in Section 25.4.

25.2.2.2   Rawstock Preservation

The industry standard of salt preservation for rawstock may not be sustainable: although it is very effective, the salt effectively can neither be disposed of nor reused without substantial cleaning up. Although much of the salt can be recovered, the contamination makes reuse uneconomic; moreover, there is no acceptable way of returning it to the environment.

Many other methods have been suggested, mostly for shorter term preservation, *e.g.* biocide spray or sterilisation by radiation or particle beam, but these are not yet safe or practical commercial options, as addressed in Chapter 3. Two other possibilities might be considered.

- The first is to use an alternative osmolyte, to confer the dehydrating, high osmotic pressure and bacteriostatic effect of neutral electrolyte: more useful salts such as potassium chloride or, for example, sodium or potassium nitrate or phosphate could be used, although they would not be as cheap as common salt.[14] Organic osmolytes are possible, such as carbohydrate/sugar: in an analogous scenario, honey is a traditional reagent for wound treatment used by the ancient Egyptians.
- The second is to revisit drying: this is a traditional preservation technique, used for many years, particularly in Africa, but has fallen out of favour in recent years. An alternative approach is use partial dehydration, but using chemistry rather than air flow, as

relied upon in traditional warm air drying. One way of doing this is apply superabsorbent polymer (SAP), acrylic polymer used for babies' disposable nappies/diapers.[15]

The newest approach is to use plant products, to exploit the biocidal action of the constituent terpenes, and, if appropriate, the accumulated salt from brackish water where the plants grew. This is clearly a practical option for preservation, but depends on the availability of suitable vegetable matter in sufficient quantity – it also assumes the ready biodegradability of the preserving material.

25.2.2.3 New (Organic) Tanning Chemistries

It is clear that the creation of new tanning agents can be better targeted by considering the stereochemistry of substituted aromatic species and the inductive effects that the substituents, particularly hydroxyl groups, have on the reactivity of the compound. An associated problem with natural plant polyphenols in the context of new organic tannages is the astringency of the condensed tannins and their high affinity towards collagen: one option is to reduce the tanning reactivity by decreasing the molecular weight, breaking them down into their monomers or dimers.[16] In synergistic combination tannages, this does not radically lower the effectiveness of the tannage.

Other sources of stabilising reactions of the biomimetic type include 'bog body' chemistry, involving the breakdown products of plant cell walls, rather than polyphenols, and carbohydrate chemistry and its derivatives.

25.2.2.4 Enzyme Targeting

Reference has already been made to the value of biochemistry in a sustainable leather industry, but in recent years there has been little development of note, with the exceptions of NovoCor AX for increased area yield and the Lanxess approach to coordinated multiple enzyme action: these processes can contribute significantly to environmental impact and the economics of leather production.

25.2.2.5 Reagent Delivery

Since the middle of the last century, there have been a few attempts in the literature to persuade tanners that there is an alternative to water for leather making. The latest solvent offering is dimethyl ether from AkzoNobel: this reagent is partially miscible with water but, with a boiling point of -25 °C and high flammability, it is not an ideal option.

Nevertheless, the efficiency of tannery processes does suffer from the presence of water, which competes in the equilibrium between any reagent in solution and the substrate onto which the tanner endeavours to fix the reagent. This subject was reviewed in Chapter 19 to address *inter alia* the new options of bead technology to alter the nature of the mechanical action and the potential of ionic liquids as process media (see later). It seems likely that any changes in these aspects of processing will come under the heading of 'Revolution', but that is not to say that the options described are not practical.

25.2.2.6 'Reinventing the Wheel'

Tanning technology has a long history, although in practice the literature extends back less than 150 years. However, within that recorded period, a lot of chemistry has been tested and applied at one time or another. Therefore, it is always a useful exercise to revisit old technologies for application today, using the current better understanding of leather science. There are many old processes that can be brought up-to-date, to contribute to the options available to the tanners and their customers, as suggested in Chapter 10.

25.2.2.7 New Products from Collagen

The use of collagen in sectors other than leather making is long established: the most familiar is its conversion into films for food purposes, such as sausage casings. An option for sustainability is the maximisation of the use of rawstock, which would include using the collagen not wanted for premium leather production for high value added products.

The versatility of collagen lies in its being able to form several types of solid, three-dimensional matrices, films and foams; in addition, its chemical reactivity allows a range of stabilities to be created, which means that many new products can be made. One of the more important is the formation of stable solids, foam scaffolds, which can support the growth of cells for tissue engineering. The difficult requirement of uniform pore size in collagen foams can be achieved by gas foaming, resulting in the correct pore size, but importantly the porous structure is interconnected. Stabilisation studies using glutaraldehyde showed that the scaffolds were not cytotoxic. The products have potential application in dressings for wound management.[17]

25.2.2.8 New Products from Leather

The end of life for leather articles has always been a recycling problem, particularly for shoes, because of the complex nature of the mixture of materials. There have been claims that it is possible to separate leather from the other components, but it remains to be seen whether the concept can be turned into a practical operation.

Haverkamp and co-workers[18] published a study of the conversion of chrome leather to char by pyrolysis at or above 600 °C in the absence of oxygen. From X-ray absorption near-edge structure (XANES) analysis, the product does not contain chromium(VI), nor is the chromium(III) extractable by strong acid solution: the main product is chromium(III) carbide, Cr_4C_3. This environmentally safe material can therefore be recycled back to the land as a soil improver.

25.2.3 Revolution

25.2.3.1 Bead Processing

Xeros Technology (the leather division is branded Qualus) has created a revolutionary new laundry technology in which the mechanical action and mechanism of transfer of soil from substrate to the cleaning fluid are enhanced by the presence of beads in a washing apparatus that are recycled during the laundering cycle. Surprisingly, it has been found by Xeros that these beads can also be used to advantage in leather-making processes such as tanning and dyeing.[19] The utility of beads for fixation reactions, tanning and dyeing of leather has been experimentally verified and many notable environmental and leather quality benefits have been observed. In addition, the practical problem of separating the beads from the leather has been overcome. The technology is addressed in Chapter 19.

25.2.3.2 Ionic and Deep Eutectic Liquids

Ionic liquids constitute a new class of chemical species that can be described in many ways,[20,21] as discussed in Chapter 19. This new approach to leather making allows leather scientists to rethink everything! It constitutes a new paradigm, a new approach to collagen stabilisation and the properties and performance of leather. Since it is estimated that there are 10^{18} possible ionic solvents, the applications are practically endless and the ability to tailor the solvent system to the delivery requirements can be precisely controlled.

25.2.3.3 Self-destructive Leather

An aspect of the perceived disadvantage of leather as a material for bulk consumer goods such as shoes is the question of waste management, *i.e.* controlled end-of-life disposal. Part of the problem is the dismantling of the articles, which are a complex mixture of materials – these might include chrome-tanned leather, vegetable-tanned leather, natural fabric, synthetic fabric, metal and rubber. The preferred alternative would be to recycle the leather as a biodegradable material, but this would appear to conflict with the definition of leather. After all, the tanner takes great care to build into the leather the highest degree of biochemical resistance. The alternative is to exploit chemistry that can in effect detan leather and hasten the breakdown of the protein structure. A well-known reaction that can operate in this environment is the Fenton reaction, the mechanism of which can be summarised as follows:[22]

$$Fe^{2+} + H_2O_2 \rightarrow Fe^{3+} + HO^{\cdot} + HO^-$$
$$Fe^{3+} + H_2O_2 \rightarrow Fe^{2+} + HOO^{\cdot} + H^+$$
$$HO^{\cdot} + H_2O_2 \rightarrow H_2O + HOO^-$$
$$HOO^{\cdot} + H_2O_2 \rightarrow O_2 + H_2O + HO^{\cdot}$$
$$Fe^{2+} + HO^{\cdot} \rightarrow Fe^{3+} + HO^-$$

Haslam[23] proposed the mechanism of autoxidation of vegetable-tanned leathers in terms of metal catalysis, as shown in Figure 25.3.

Studies at the University of Northampton[24] led to the concept of metal ion-catalysed redox degradation of vegetable-tanned leather, according to the general mechanism given in Figure 25.4. The mechanism has been applied to differently tanned leathers as a way of inducing damage, to allow biochemical degradation of the type likely to be encountered in landfill or modified for creating compost. In this way, end-of-life of leather articles may not be a long-term problem.

25.2.3.4 Alternative Materials

There has long been interest in industry in leather-like materials, such as reconstructed fabrics made from leather shavings bound into

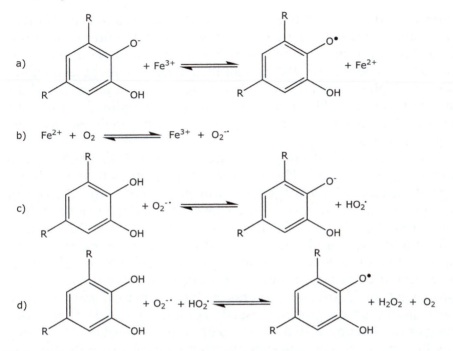

Figure 25.3 Iron-catalysed degradation of plant polyphenols.

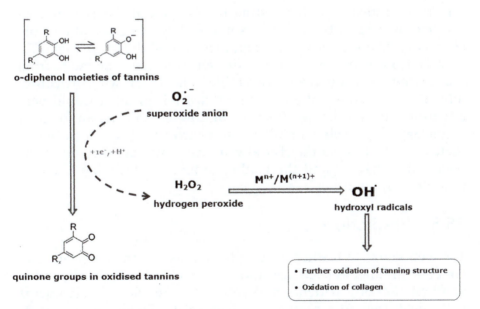

Figure 25.4 Redox detanning of vegetable-tanned leather. Courtesy of A. Duki.

a resin matrix: leather board and e-leather are examples. However, they cannot be called leather, which by the ISO definition requires it to be based on intact animal skin. A more recent area of interest is the creation of material that has all the characteristics of leather, but does not come from animals. This context is not a revisiting of the synthetic versions of leather, the poromerics, exemplified by DuPont's Corfam, but the growing of a collagen fibre matrix, analogous to growing skin *in vivo*.

The venture capital company Modern Meadow is known for growing collagen *de novo*.[25] This is a process analogous to the technology attempted in the Diagenesis project funded by the European Union and then extended by the University of Naples, although in that project the immortal cell lines were different and the extracellular matrix was produced on a scaffold of gelatine beads.[26] Both approaches succeeded in producing fibrous solid materials – they also cannot be called leather, but they may exhibit complementary properties to leather and therefore have practical applications. Such materials are not necessarily competitors to leather, since they would occupy very different places in the market. The whole field of creating new materials by tissue engineering in this way opens up the potential for making composites with other biomaterials, natural and synthetic, *e.g.* collagen with silk or manufactured polymers.

In this context, it is interesting to consider alternative materials that 'overlap' with leather, insofar as they make use of tanning chemistry. There are several companies around the world that offer organic fibrous materials from different sources: they range from identifying new sources of vegetable fibre to growing mycoprotein. The companies often assume that their product would benefit from the stabilising effect of tanning chemicals, although it is unclear exactly what stabilisation means in this context. Nevertheless, there is a market for alternative materials that might be regarded as 'green' and the leather industry could play a role in their development.

25.3 Prediction

The technology of leather making is founded on the application of craft and art and, as such, has always been based in the main on the accumulation of experience and observation to allow developments in the field. Therefore, progress has been relatively slow, because technologists have traditionally relied on testing small changes or they have engaged in the trial-and-error strategy of investigation. The main reason behind this history of change was a lack of a coherent, comprehensive view of the scientific principles underpinning the processes that constitute leather making, including the relationships between process steps. Consequently, it has been necessary for tanners to adopt a more conservative approach to process development than they might otherwise have wished.

At the heart of the requirements for the scientific understanding of leather technology lies a firm theoretical view of the mechanisms by which reagents are bound to collagen and the consequences of the reactions, in terms of the chemical and physical modifications that they confer on the leather, leading to specific properties and performance. In order to be able to predict the outcomes of process steps, it is important to address the following features of processing.

1. How might opening up be quantified?
2. What parameters control the fixation of reagents to collagen?
3. What are the principles of the mechanism?
4. In the case of multiple fixation reactions (which apply to every leather), what controls the overall cumulative effect of the multiple reagents?
5. Are there relationships between successive reagents?
6. How do multiple fixation reactions control the chemical and physical properties of the leather?
7. Is there a relationship between those accumulated properties?
8. How predictable is the outcome of processing with regard to properties and performance of the product?
9. Are we limited to qualitative prediction rather than quantitative prediction?

25.3.1 Processing Conditions

25.3.1.1 Float Length

The amount of float does not normally affect the chemistry of processing, but it will affect the rates of both penetration and reaction: the shorter the float, the more concentrated is the reagent and the faster that reagent will penetrate and react. These can be competitive processes, where a fast surface reaction may be comparable to the rate of penetration: knowledge of the conditions and chemistry should allow a judgement of the likely winner in the competition. Usually, the effect on penetration rate will dominate, compared with an increase in the kinetics of fixation. Technologically, there will be an effect on the uniformity of reaction through the cross-section, since pelt has a finite thickness. Some reagents are significantly changed by concentration, *e.g.* dyes and vegetable tannins can form aggregates, which penetrate more slowly and are more surface reactive than the separated molecules. This can result in non-uniformity of reaction over the surfaces. This is a general principle: the faster the rate of fixation, the less uniform the fixation will be, through the cross-section and visibly across the surfaces. This effect is most commonly met in chrome tanning/staining/shadowing and dyeing or any other colouring process step.

25.3.1.2 Mechanical Action

The amount of float influences the mechanical action experienced by the pelt: this contributes to the speed of penetration, because the squeezing and relaxing cycles experienced by the pelt increase in both frequency and intensity. The greater the degree of mechanical action, the higher the temperature increase due to friction is likely to be, and this may increase the kinetics of fixation: covalent reactions are typically temperature dependent, but electrostatic interactions, including hydrogen bonding, are not.

Mechanical action cannot affect the chemistry of processes. However, there is an observed correlation between mechanical action and efficiency of reaction: this is merely the result of getting the reactants together, maximising the availability to the reagent of reaction sites within the substrate and the ability of the reagent to reach them. In fact, strictly, the apparent mechanical action effect means *lower inefficiency* of uptake rather than greater efficiency of uptake.

25.3.2 Beamhouse

In contrast to the additive processes involving fixation of reagents to collagen, the beamhouse steps are primarily hydrolytic and therefore predominantly feature removal of species from the substrate. Quantification of these process outcomes is relatively imprecise: the mechanisms are broadly established and the results are typically more qualitative than quantitative, but they should be at least semiquantitative.

The beamhouse processes are typically conducted under alkaline conditions, which is the case for conventional technology. In the liming step, the targets are the non-structural proteins of skin: they do not possess a hierarchy of structure to the same extent as the molecular and macro structure of collagen, so they are more vulnerable to damage, to solubilisation when their amide linking groups are broken. The extent to which this occurs depends on the pH (usually 12.5 in the presence of lime), the temperature (usually <30 °C) and the period of the reaction (usually 18 h). The effects of variations in these parameters can be indicated using simple assumptions.

25.3.2.1 pH

If the hydrolysis reaction is first order in hydroxyl ion concentration:[2]

$$\text{rate} \propto [OH^-]$$

then the effect of varying the concentration, for example with temperature or the use of sugar to sequester calcium ions, can be calculated simply:

$$pH = -\log_{10}[H^+]$$

i.e. minus the logarithm to base 10 of the hydrogen ion concentration. The ionic product of water is

$$K_w = [H^+][OH^-] = 10^{-14}$$

and

$$pK_w = pH + pOH = 14$$
$$pOH = 14 - pH$$
$$[OH^-] = \text{antilog}(pH - 14)$$

For example, if the pH is 12, $[H^+] = 10^{-12}$ M, pOH = 2 and hence $[OH^-] = 10^{-2}$ M. At pH 11, $[OH^-] = 10^{-3}$ M. Therefore, comparing reactions at pH 11 and 12, the ratio of hydroxyl ion concentrations and hence the rate of catalysed hydrolysis is 10.

Similarly, at pH 12.5 and 13.0, $[OH^-] = 10^{-1.5}$ and 10^{-1} M, respectively: the ratio is $10^{0.5} = 3.2$. This might be the difference between conventional lime-buffered hydrolysis and caustic soda unhairing/liming or enhanced liming by sequestering calcium ions with sugar (see Chapter 6).

25.3.2.2 *Temperature*

The effect of temperature can be estimated in relative terms from the Arrhenius equation, as introduced in Chapter 3:

$$\ln k = \ln A - (E/RT)$$

or

$$k = A\exp(-E/RT)$$

Therefore, the relative temperature effect is given by

$$\frac{k_2}{k_1} = \exp\left[\frac{E}{R}\left(\frac{1}{T_1} - \frac{1}{T_2}\right)\right]$$

The well-known 'rule of thumb' is: *reaction rates change by a factor of two with a change in temperature of 10 °C*; that means: 10 °C higher, double the rate; 10 °C lower, halve the rate. This assumes that the exponential term is close to unity; however, the assumption can be tested by using some real figures, as follows.

If T_1 is 15 °C and T_2 is 25 °C, the term in parentheses is 1.1×10^{-4}, then substituting the gas constant and typical values of the activation energy at ambient temperature, assuming 10–100 kJ mol^{-1}, the temperature coefficient is 1–4, but is more typically 2–3.[27,28] In addition, since under the conventional conditions of processing the operating temperature ranges are relatively small, the temperature effect on the activation energy in the typical temperature range can be assumed to be small.

25.3.2.3 *Time*

If it is assumed that the reaction is not close to completion, whatever the conditions may be, the rate can be considered to be approximately linear, that is, the reaction can be considered to be zero order in the reactants other than the catalysing hydroxyl ion (or at the beginning of a first-order reaction), so the effect of time is easily calculated arithmetically, *i.e.* halve the time, halve the effect (Figure 25.5).

Prediction is made more difficult by the fact that pelt has finite thickness, so the chemistry cannot be assumed to be uniform through the cross-section. The thinner and more open the pelt is, the more it can be assumed that the pelt functions predictably, like a dissolved reactant.

Some useful comparisons can be made with the extent of the reaction under standard or typical conditions, when the outcome is already known. For example, in a given factory, the consistent effect of

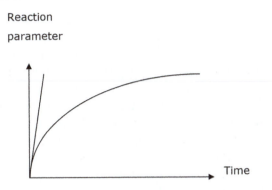

Figure 25.5   Notional effect of time using initial rate.

specified conditions should be known, so deviation from those conditions should be predictable, at least as a primary effect. What is less easy to predict is the outcome of a change in rate on the effect of the opening up, where the properties of interest are the strength and handle of the eventual leather and the impact of change is influenced by both the conditions and the nature of the raw material.

The handle of the final leather is affected by the beamhouse processes, where the following generalisations apply.

1. Longer liming means softer, weaker leather,[29] although the outcome is highly influenced by the nature and thickness of the pelt. This trend also applies to warmer liming, where the rate of hydrolysis is accelerated and hence the extent of hydrolysis is increased.
2. Higher temperature increases the possibility of actual or incipient denaturation: the effect is to weaken the leather.
3. Higher pH, whether from alternative 'liming' processes such as caustic soda unhairing or from calcium sequestration by polysaccharide, effectively extends liming in much the same way as higher temperature. It can also result in a greater degree of swelling; this may not be reversible to the same extent as with conventionally limed pelt, which will mean that the angle of fibre weave in the corium will not be as low as it might otherwise be, causing the grain to be less flat and the leather to be weaker.

25.3.3   Liming

Extending the opening-up effects in liming increases the relative effects of hydrolysis on collagen and the non-structural proteins: collagen is relatively resistant to damage, but increasing the removal of the non-structural proteins increases the looseness of the final leather. This cannot be disguised later in the process: if the fibre structure is allowed to restick after post-tanning, this is merely temporary and becomes apparent when the leather is put to practical use.[30,31]

The hydrolysis reaction can be modelled by the kinetics of hydroxide-catalysed hydrolysis of the sidechain amide groups: the rate is relatively linear over the first 20 h or so; thereafter, the indicative half-life is calculated to be 22 h.[29] This means that, in the first 22 h, potentially half of the groups may be hydrolysed, but it does not mean that half of all the amide links of the collagen in pelt are broken, because the reaction depends on the availability of the reaction sites to be broken. In practice, this means that thin sheepskin can be limed for only a few hours, but thick bull hides can withstand liming for several days.

One of the more important effects of liming is the change in the isoelectric point (IEP), which controls the charge as a function of pH (see later). The hydrolysis of the amide sidechains proceeds at roughly the same rate as the hydrolysis of the amide links in the protein chains: in a conventional liming process, about half of the asparagine and glutamine sidechains are hydrolysed, assumed to occur at the same rate. The rate of collagen damage can be measured from the rate at which aspartic acid is racemised:[29] the results indicate that the effects are initially slow, but accelerate markedly after about 20 h of liming. Therefore, after a day of liming, the hydrolysis reaction is slowing, but the damaging consequences are accelerating. In contrast, the technological effect of liming appears to be dominated by the first 6 h of processing, as indicated in Chapter 6.

Modelling changes resulting from variations in liming by comparison must be viewed with caution, because the result is so dependent on the nature of the substrate. Nevertheless, such comparisons might reasonably be made within consistent production in a tannery.

25.3.4 Bating

The outcome of bating depends on the following factors, assuming that the enzymes are of the same general type. For example, the effects of proteolytic pancreatic and bacterial bates differ insofar as only the latter exhibits elastase activity, which changes the composition and hence modifies the properties of the substrate.

25.3.4.1 Nature of the Enzyme and Its Specific Mode of Action

It is difficult to predict the relative effects of different enzymes if they attack protein at different sites along the chain, because this will cause different patterns of fragmentation of the structural proteins (and the collagen), which may alter the effects of the process of removal. This has not been systematically investigated.

The relative gross effects of enzyme action can be measured by comparing the results of assays conducted under the same conditions of concentration, temperature and pH, using dyed hide powder as the substrate, where the rate of reaction with the protein is determined by the colour from solubilisation of the solid (see Chapter 18). However, the assay can be influenced by the chemistry of dyeing and the extent to which the collagen substrate is denatured during its manufacture, so the actions of collagenase and general proteases can be distinguished. The assay itself does not discriminate between mechanisms of degradation and solubilisation of the collagen, which may influence the outcome of bating. This is the unusual situation in which the assay yields a result that is quantitative, but not qualitative.

25.3.4.2 Concentration

It might be assumed that the rate of hydrolysis is first order in enzyme concentration, *e.g.* double the concentration, double the rate of reaction, double the effect in the same time. It is important to recognise that this does not take into account the finite thickness of the pelt with regard to the effects on the physical properties.

25.3.4.3 Temperature

Enzymes function optimally at the temperature of the environment from where they (or their organisms) originate, which in the case of bates is typically around 40 °C. Although the Arrhenius equation can be used as an approximation of the effect of temperature below the optimum, the observed effect of changing the temperature (raising or lowering) may be greater or less than expected. Note that in the limit, raising the temperature results in denaturation of the enzyme, thereby deactivating it completely.

25.3.4.4 pH

The pH dependence of enzyme catalysis is likely to be more symmetrical about the optimum value than the temperature effect, *i.e.* reactivity can extend along the pH scale far beyond the optimum point: the shapes of the plots from different enzymes can be very different.[32] The precise relationship between pH and activity is not predictable: the pH–reactivity profile must be determined practically and then the relative rates applied.

25.3.4.5 Penetration Rate

As a 'rule of thumb', enzymes will penetrate through intact hide at an initial rate of about 1 mm h^{-1}, due to the high molecular weight and the charges on the fibre structure and the enzyme. This means that thin skins with a more open structure are likely to be bated throughout the cross-section, but bovine pelt is usually bated only on the surfaces. This is typically the case in industry, where the bating reaction is used primarily to remove hair debris and residual epidermis (scud) after hair burning.

25.3.5 Pickling

Pickling can continue the hydrolytic processes. However, the rate of acid-catalysed hydrolysis is about two orders of magnitude slower than hydroxyl-catalysed hydrolysis, so the effect of any change can be estimated. In order to make the reactions proceed at the same rate as they do in liming, the pickle pH would have to be lowered by at least two units.

This argument does not take into account the effect of finite pelt thickness, which will introduce a variable effect through the cross-section if the reaction is not allowed to reach equilibrium – which is rarely the case. This is more likely if the pickling process includes the use of formic acid, which penetrates faster than the hydrated hydronium ion of sulfuric acid, $[(H_3O^+)(H_2O)_6]$, because it is the electrically neutral form HCO_2H of a weak acid. Formic acid can also contribute a lyotropic contribution to opening up; this is a general feature of processing that has not been investigated.

25.4 Tanning/Post-tanning

In an analysis of the consequences of reactions involving fixation to collagen or leather, it is not necessary to distinguish between tanning and post-tanning or indeed pretanning (which are terms merely indicating the relative timing of the process step), because the outcome is the accumulation of effects from start to finish of wet processing. The overall effect will be determined by the nature of the reagents, the offers, their reactivity and reactions with the substrate and with each other.

Before considering those parameters, it is necessary to consider the affinity between the substrate and the reagent, because this will determine the rate of reaction and consequently influence the ability of the reagent to penetrate down the hierarchy of the substrate structure. At the same time, it is important to bear in mind that the substrate has finite thickness, so the possibility of variable fixation through the cross-section is a factor that must always be taken into account. Although there are technologies in which fixation at the surfaces is deliberately controlled, for the majority of situations it might be assumed that uniform reaction through the cross-section is desired and can be achieved by appropriate process conditions.

It is important to recognise the types of chemical bonding that typical tanning agents exhibit, indicated in the generalisations set out in Table 25.1. Here hydrogen bonding may occur at collagen sidechains or at the amide links of the protein backbone, hydrophobic bonding occurs at sites of non-polar sidechains, ionic bonding may occur at either anionic or cationic sidechains and covalent bonding is created at appropriately reactive sidechains.

The cumulative effect of fixation reactions depends on the three possible interactions between the reagents as follows, which should all be predictable from knowledge of the chemistry and reactivity of the reagents, as discussed in Chapter 24. Also, it is useful to recall the tanner's 'rule of thumb', which states that *the properties of a leather are controlled by the first tanning reaction*. This essentially means that the first tanning reagent has the advantage of interacting with reaction sites first or interfering with subsequent reactions; either way, the outcome is to influence the leather properties.

Table 25.1   Fixation mechanisms of common tanning agents (more ticks indicates a greater importance of contribution to reaction).

Tanning agent	Hydrogen bonding	Hydrophobic	Ionic	Covalent
Metal salts, oxyanions	✓		✓✓✓	
Chromium(III)	✓✓			✓✓✓
Aldehydes, aldehydic				✓✓✓
Hydrolysable polyphenols	✓✓✓	✓✓		
Condensed polyphenols	✓✓✓	✓✓		✓✓
Syntans	✓✓✓	✓✓	✓✓	
Resins/polymers	✓	✓	✓	
Oil		✓		
Epoxide, isocyanate				✓✓✓

Table 25.2   Generalised properties conferred to leather by common tanning agents.

Tanning agent	Stable tannage	T_s^a	Filling[b]	Soften-ing[c,d]	Hydro-philic[d,e]	Weaker leather[d,f]
Metal salts, oxyanions	No	Low	No	No	Nk	Yes
Chromium(III)	Yes	High	No	No	Yes	Yes
Aldehydes, aldehydic	Yes	Low	Yes	Yes	Yes	Yes
Hydrolysable polyphenols	No	Med	Yes	Yes	Yes	Yes
Condensed polyphenols	Yes	Med	Yes	Yes	Yes	Nk
Retanning syntans	No	Med	No	Nk	Yes	Nk
Replacement syntans	No	Med	Yes	Yes	Yes	Yes
Resins/polymers	No	Low	Yes	Yes	Nk	No
Oil, sulfonyl chloride	Yes	Low	Yes	Yes	Yes	No
Epoxide, isocyanate	Yes	Low	No	Nk	Nk	Yes

[a]Low means <75 °C, med means 75–85 °C, high means >100 °C.
[b]No does not necessarily mean that the structure becomes collapsed, which applies to only some metal tannages, *e.g.* Al(III). The degree of the filling effect depends on chemical structure, molecular weight and offer.
[c]The degree of softening effect depends primarily on the nature of the chemical reaction and molecular weight.
[d]Nk indicates not known, an uncertain outcome, depending on the chemical structure.
[e]Hydrophilicity depends on chemical structure and the chemical nature of the reaction.
[f]With a few exceptions, all reactions cause the leather to be weaker, depending on the chemistry and the offer.

Although it is possible to predict whether or not reactions are independent, antagonistic or synergistic and hence to predict the effect on the hydrothermal stability, as discussed in Chapter 15, other properties are more uncertain. However, predictions can be made from the generalised properties that they confer on the leather, as set out in Table 25.2. Tanners can construct their own matrix, based on their knowledge of specific reagents that they use.

25.4.1   Multiple Reactions

25.4.1.1   Combinations

In attempting to predict the outcome of a process step, it is useful to recall that, at any point in the series of steps, the tanner is dealing with a three-component system: the substrate, the reagent and the solvent/medium, and the outcome is controlled by the relationships between them – the affinities, based on charge, hydrophilic–hydrophobic balance (HHB) properties, chemistries. All of the components can vary in their impact on a particular step, depending on the history of the process, the previous steps and the current physical and chemical conditions. Everything that happened before the point of consideration will have an impact on the *status quo* and will have an impact on all subsequent steps. This must be taken into account. The tanner's task is to pick apart the factors that affect the step.

- How have all the previous steps affected the reactivity of the substrate?
- How will the reagent interact with the substrate in its current state of reactivity?
- How will the reagent interact with reagents already fixed on the substrate?
- What influence will the solvent have on the equilibrium between the substrate and the reagent?
- How will the resulting reaction affect the subsequent steps, especially the next one?

The answers to these questions depend on having a familiarity with leather technology and a fundamental understanding of leather science.

25.4.1.2   Tannage Stability

The stability or reversibility of the tanning effect depends on the chemistry of the interactions between the reagents and the collagen. This is subject to the effects of antagonism and competition between reagents that rely on electrostatic bonding (charge–charge, hydrogen or hydrophobic bonding). Covalent reaction can be considered permanent in conventional leather processing.

25.4.1.3   Shrinkage Temperature

Hydrothermal stability is defined either by the individual chemistries of the reagents or by their synergistic interaction. This is a property that is additive to only a limited extent; hence a low shrinkage temperature can be elevated to a moderate shrinkage temperature only by multiple reactions of individual tanning agents, but a moderate shrinkage temperature can be elevated to a high shrinkage temperature only by a synergistic interaction between two reagents.

25.4.1.4   Filling

Filling is a cumulative property: the expansion of the fibre structure by reagents can only increase, it is unlikely to be reversed by subsequent flattening reactions, even powerful reactions such as tanning with the strongly cationic aluminium(III).

25.4.1.5   Softening

The softening effect is analogous to filling, insofar as it depends on the effect of the chemistry of the reagent in keeping the fibre structure open and capable of easy distortion. However, the effect can be reversed by reagents that have a marked collapsing or filling effect. This property is dominated by fatliquoring, so tanning has only a second-order effect on softness.[31]

25.4.1.6   Hydrophilicity

The hydrophilic/hydrophobic characteristic of the substrate depends on the nature of the reagents, which can make the leather increasingly hydrophilic or increasingly hydrophobic by the cumulative effect of appropriate reagents. Similarly, reagents can have an antagonistic effect on either property.

25.4.1.7   Weakness

The more the natural structure of collagen is modified, the weaker it becomes, which is a cumulative effect, dependent on the extent to which the natural structure is disrupted by the bound reagents. Since this is not always an effect of high magnitude, useful leather can still be made, even though it may contain a significant fraction of bound chemicals. Nevertheless, it is useful to note the consequences of tannages on leather strength: in general they have a weakening effect, but some reactions are known that contribute to leather strength, notably synergistic or polymerisation tannages. These effects are modified by offers of reagents, since the magnitude of any effect is controlled by the relative offers and the ability of the reagents to confer effects and their relative efficacy, the intrinsic ability to produce change to the substrate.

25.4.2 Post-tanning

25.4.2.1 Dyeing

It is frequently ignored or forgotten that dyes function in the same way as conventional tanning agents, because of the ways in which they interact and bind chemically to collagen/leather. Although the tanning effects of dyes are rarely if ever measured or assessed and their effects are typically not apparent, particularly in chrome-tanned leather, the impact of these species on the properties of leather should be borne in mind when predicting the outcome of processing on subsequent process steps and on final leather properties and performance. Dyes can affect the IEP (see later) and hence change the charge at any given pH; they can affect the HHB, depending on their own HHB characteristics, and they can influence the reactions of subsequent reagents by the nature of their chemistries and reactivity towards the substrate. This is a consequence of not only the type of dye, but also the individual structural components, which can vary widely within each group of dyes.

Dyes affect the properties of the leather through their fixation reactions, but they can also influence properties by their residual reactivity *via* their auxochromes: they modify the colour as the structural components bound to the central chromophores, which define the base colour. This latter effect is often overlooked, although it is the basis of sandwich dyeing and colour-enhancing processes: it is the basis of mordanting reactions, which can be used as additional fixation, in the same way as 'capping' with metal salts such as chromium(III) is used to enhance the water resistance effects of fatliquors and other specific water resistance treatments, such as with acrylate polymers.

In general, the more penetrating the dye is (from aqueous solution), the more hydrophilic it is and the more it will confer hydrophilicity to the leather and the more surface active the dye is, the more likely it is to confer some degree of hydrophobicity.

Such considerations are important, because dyeing may be conducted in more than one step and it is typically followed by fatliquoring. In any case, there is no hard and fast rule that processing must follow the conventional pattern of tan, then retan the dye, then fatliquor. Indeed, it is always useful to consider all options of process step sequences.

25.4.2.2 Fatliquoring

The lubricating step is frequently the last wet step in processing, so its effect is rarely considered to extend further beyond the properties that it is designed to confer. However, it is possible to continue processing, either to introduce water resistance treatments or top dye for intense shades, or retan to prepare the dried leather for finishing.

The role of the sulfo fraction, the carrier for the lubricating neutral oil, should be recognised. The structure is invariably simple, whether the oil is sulfated, sulfited or cationic, so it functions like an auxiliary syntan. In this way, the tanning action is effectively zero, but the impact on the substrate may be marked because of the interaction with charged sites on the substrate. Nor can the effect of the neutral oil be completely ignored: it will alter the HHB characteristic of the substrate, depending on its own properties, because it will coat the fibre structure at the level corresponding to the mode of delivery[30] and may be regarded as a hydrophobic reagent, although this is not always the case.

The outcome of fatliquoring with regard to the chemical properties (as opposed to the physical effect of preventing fibre resticking in drying) of the leather depends on the chemistry of the fatliquor. Most conventional fatliquors make the leather hydrophilic, because of the effect of the sulfo fraction, where the magnitude of the effect depends on the degree of sulfation or sulfonation. Some fatliquors designated water resistant can confer a degree of hydrophobicity: although the effect is typically referred to as 'shower resistance', they can be much more effective if all other processing is conducted to minimise hydrophilicity, especially if the water resistance treatment includes 'capping' with a chromium(III) salt, where the residual anionic nature of the reagent is neutralised by cations.

25.4.3 Influence of Charge

25.4.3.1 Isoelectric Point

There is no mathematical model to allow calculation of the IEP as a function of process steps. This is unfortunate, because it is critical to the success of any reaction to know the IEP, the point on the pH scale at which the charge is zero. The sign and magnitude of the charge will determine the ability of a reagent to penetrate or react at the substrate surfaces, as discussed in Chapter 15. It is recognised that the changes depend on both the nature of the individual chemical species, the extent to which the acid–base function of the protein is affected by the fixation reaction and the offer. At least we can predict the direction of change, although the starting point may not be easily defined.

25.4.3.2 Applied Charge

Some reagents can add charge to the substrate, so that the effect of the IEP can be overridden. This was implied above, in terms of the impact of anionic reagents, particularly in conventional post-tanning, where the reagents are dominated by anionic charge, usually from sulfonate groups.

A more powerful effect can be achieved through cationic species, often metal ion complexes, but also, for example, by quaternary ammonium compounds. It is known that chromium(III) usually carries a positive charge, but its influence is weakened by covalent reaction with collagen, basification with hydroxyl ion and distribution of the charge over the large complex of metal nuclei and ligands. This is not the case with other metal ions, notably aluminium(III). Such metal ions react electrostatically and consequently there is considerable residual positive charge, and this can dominate the reactivity of the substrate, to the extent that all anionic species tend to react on the surface and fatliquor emulsions can crack. The charge can be effectively dissipated only by complexation, *e.g.* with polyphosphate or polyphenol.[33]

25.4.4 Finishing

In reviewing the wet processing as a continuously changing system, there is an implicit assumption that all considerations cease at the end of wet processing. However, that is not the end of leather making – the leather must be finished, as addressed in Chapter 21. Just as there has to be a match between the components of the system at any wet stage, there must be a match between the properties of the dried leather and the finish components, in order to maximise the properties of the finish, particularly with regard to adhesion, with consequences for rub fastness.

25.5 Reverse Analysis

If it is understood how reagents interact in a tanning context and it is known what properties they are capable of conferring, it is possible to reverse the analysis of outcome in processing. What this means is that a leather with a suite of properties and performance can be imagined and then a process to obtain those features in the leather can be devised. The logical steps can be summarised as follows in a guideline question sequence.

1. What features are desired, *e.g.* flatness, fullness, strength, softness, stability of tannage, physical and chemical or hydrothermal stability, and in what measure for each property?
   What is specifically not required?
2. Is the desired outcome the creation of a conventional leather or is extreme performance required?
   Is the process likely to be constructed from conventional reagents applied in conventional technology or are new approaches indicated?
3. Which reagents are potentially capable of conferring those properties, and are the relative importance and prominence desired and defined?
   How much of each reagent will confer the required features?
   What are the relative offers?
   Here there is need for leather technology input, based on known effects from reagents.
4. What interactions are required between the reagents, to refine the options?
   How do the properties of the reagents influence the effectiveness of other reagents?
   Are the choices consistent with what is both wanted and not wanted?
5. How do the reagents fit into a conventional, practical process, *e.g.* are they compatible with a logical programme of pH change and IEP/charge variation?
   Any new process must make technological sense, so that damaging swinging between pH extremes is avoided.
   Is there a need to review the relationships between process steps?
6. Can the desired product be made with conventional, available chemicals/biochemicals?
   What reactions might be facilitated by the use of enzymes, particularly those new to the leather industry?
   Will it be necessary to source or even design new reagents?

7. Can the process be designed to be compact, *i.e.* can steps be combined by rethinking the chemistry of the process steps?

8. In imagining new leathers with new properties, what are the limits to performance of collagen-based biomaterials and how might those limits be achieved if necessary?

   Leathers can be made to be effectively waterproof, the shrinkage temperature is probably limited to about 155 °C. But what are the upper limits for strength, fire resistance, *etc.*?

   Are the requirements consistent with known or realistic property limits?

25.6 Overview

The leather literature contains many examples of combination tannages, where the properties of leathers made by usually unrelated tanning agents are explored. The impression is given of testing combinations almost randomly, often with the avowed rationale of finding a commercially viable alternative to chrome tanning. The results are valid, but have limited value in the scheme of leather development. Here, it is proposed that the link between outcome and process (and in that order) can be predicted.

The thinking should not be confined to leather development, but can be extended to include all collagenic biomaterials, such as the requirements for 'smart' materials that react to conditions to change those conditions in a predicted way. The latter have clear application in sectors such as wound management: it has been demonstrated that active agents, such as antibiotics, can be chemically bound to reconstituted collagen and remain active. Therefore, the biomaterial as a dressing can offer an additional level of wound protection. It is not inconceivable that, with a suitable choice of substrate and linking chemistry, practically any chemically or biologically active agent may be bound to collagen in its natural or reconstituted forms.

25.7 Conclusions

Predicting the outcome of any technological change in the complex systems familiar in leather processing may be scientific, but it is not an exact science. Nevertheless, by adopting the kind of approach set out here, drawing on both scientific and technological principles, it is possible to go a long way towards accurate and useful prediction of the effects of concurrent and consecutive process steps. This can make developmental programmes much more efficient. It is a significant contribution to the profitability and sustainability of the global leather industry.

Perhaps more importantly, the ideas proposed here will allow the prediction of not only the processes designed to make leathers with desired properties, but also the constitution of reagents to achieve that end. This constitutes a *paradigm shift* in thinking and hence in creating strategies for advancing leather development and for defining and making new collagenic biomaterials. The sustainability of the global leather industry depends on its ability to change and move with the times: flexibility to address the fundamental issues as perceived by Society in general and to make the most of developments offered by leather science increasingly define sustainability. For many years, progress in leather technology has been slow, due in part to the changing face of science, the lack of financial support and the reduction in the numbers of leading scientists engaged in the subject. This has been characterised by studies and publications exhibiting narrow-scope, short-term programmes and a lack of fundamental research and scientific thinking. The time is ripe for that paradigm shift.[34]

With current developments that are under way, the potential to meet the challenges of the future is coming. All that is needed is a willingness support new thinking (which probably means investment) and to embrace the 'shock of the new'. We are rapidly approaching the point where the leather scientists can say to the tanners: 'What is on your wish list? We can show you how to get your wish!'

References

1. F. Brugnoli, *UNIDO, Leather and Leather Products Industry Panel*, Shanghai, PR China, 2012.
2. L. F. Cabeza, *et al.*, *J. Am. Leather Chem. Assoc.*, 1999, **94**(7), 268.
3. J. Cot, *et al.*, *J. Am. Leather Chem. Assoc.*, 2008, **103**(3), 103.
4. ISO 11641, IULTCS/IUF 426, *Official Methods of Analysis*, Society of Leather Technologists and Chemists, 2012.
5. S. Yagoub, PhD thesis, University of Leicester, 2006.
6. H. C. Holland, *J. Int. Leather Trades' Chem.*, 1940, **24**(5), 152.
7. Blacksmith Institute, *Pure Earth*, 2011. www.blacksmithinstitute.org.
8. C. A. Money, Chrome Tanning Maligned, *Leather International*, 2012.
9. R. Pallop, *et al.*, *J. Soc. Leather Technol. Chem.*, 2010, **94**(2), 70.
10. S. Jeyapalina, G. E. Attenburrow and A. D. Covington, *J. Soc. Leather Technol. Chem.*, 2007, **91**(30), 102.

11. A. D. Covington, *J. Indian Leather Technol. Assoc.*, 2000, **L**(8), 1.
12. R. P. Daniels, *et al.*, *J. Soc. Leather Technol. Chem.*, 2017, **101**(3), 105.
13. A. S. Onyuka, *et al.*, *J. Am. Leather Chem. Assoc.*, 2012, **107**(5), 159.
14. A. D. Covington, *et al.*, *J. Soc. Leather Technol. Chem.*, 2008, **92**(1), 1.
15. A. D. Covington, *J. Soc. Leather Technol. Chem.*, 2011, **95**(6), 231.
16. L. M. dos Santos, PhD thesis, Univ. of Northampton, 2017.
17. A. P. M. Antunes, *et al.*, *Proceedings of IULTCS European Congress*, Istanbul Turkey, 2006.
18. B. Johannessen, *et al.*, *ACS Sustainable Chem. Eng.*, 2014, **2**(7), 1864.
19. S. Rostami, J. Steele and A. D. Covington, *Proceedings of AICLST Conference*, Okayama, Japan, 2014.
20. A. Abbott, *et al.*, *Proceedings of SLTC Conference*, 2014, Northampton, UK.
21. A. Abbott, *et al.*, *UK Pat. Appl.*, PCT/GB2014/051148, 2014.
22. F. Haber and J. Weiss, *Proc. R. Soc. London, Ser. A*, 1934, **147**(861), 332.
23. E. Haslam, *Practical Polyphenolics*, Cambridge University Press, UK, 1998.
24. A. Duki, PhD thesis, University of Northampton, 2014.
25. A. Forgacs, The Art of Biofabricated Materials by Modern Meadow, YouTube, 2013.
26. P. Netti, *Lineapelle Innovation Square*, Italy, Milan, 2018.
27. W. P. Jencks, *Catalysis in Chemistry and Enzymology*, McGraw-Hill, 1969.
28. S. Glasstone, *Textbook of Physical Chemistry*, Macmillan, London, 1966.
29. A. D. Covington, O. Menderes, E. R. Waite and M. J. Collins, *J. Soc. Leather Technol. Chem.*, 1999, **83**(2), 107.
30. A. D. Covington and K. T. W. Alexander, *J. Am. Leather Chem. Assoc.*, 1993, **88**(12), 241.
31. A. D. Covington and K. T. W. Alexander, *J. Am. Leather Chem. Assoc.*, 1993, **88**(12), 252.
32. K. T. W. Alexander, *J. Am. Leather Chem. Assoc.*, 1988, **83**(9), 287.
33. A. D. Covington and L. Song, *Proceedings of IULTCS Congress*, Cancun, Mexico, 2003.
34. A. D. Covington and J. A. Wilson, *J. Am. Leather Chem. Assoc.*, 2012, **107**(8), 258.

Subject Index

Figures, Schemes and Tables are in **bold**.